Antibiotic Development and Resistance

Antibiotic Development and Resistance

Edited by

Diarmaid Hughes
Department of Cell and Molecular Biology
University of Uppsala
Sweden

and

Dan I. Andersson
Swedish Institute for Infectious Disease Control
Solna
Sweden

London and New York

First published 2001
by Taylor and Francis
11 New Fetter Lane, London EC4P 4EE

Simultaneously published in the USA and Canada
by Taylor and Francis Inc,
29 West 35th Street, New York, NY 10001

Taylor and Francis is an imprint of the Taylor & Francis Group

© 2001 Taylor and Francis

Typeset by Expo Holdings, Malaysia
Printed and bound in Great Britain by
St Edmundsbury Press, Bury St Edmunds, Suffolk

Every effort has been made to ensure that the advice and information in this book is true
and accurate at the time of going to press. However, neither the publisher nor the authors
can accept any legal responsibility or liability for any errors or omissions that may be
made. In the case of drug administration, any medical procedure or the use of technical
equipment mentioned within this book, you are strongly advised to consult the
manufacturer's guidelines.

British Library Cataloguing in Publication Data
A catalogue record for this book is available from the British Library

Library of Congress Cataloging in Publication Data

ISBN: 0–415–27217–3

1006026060

Contents

Introduction

It is now over fifty years since antimicrobial agents were introduced into clinical practice. From the beginning of this *antibiotic era*, as could have been anticipated, we have witnessed a dual evolution: on the one hand an increase in the number of different drug classes and in their activity, and on the other hand the emergence and dissemination of bacterial resistance. However, no new classes of antibiotics (by this I mean molecules not subject to cross-resistance with drugs already in use) have been found over the last twenty five years. In contrast, the evolution of bacteria towards resistance (a particular aspect of bacterial evolution) is both constant and inevitable. The net result of these contrasting evolutions is that multidrug resistance has become a major health problem. Although the reasons differ, antibiotic resistance is a problem in both developed and developing countries, and for both hospital and community acquired infections.

Under these circumstances, any book dealing with resistance to and development of new antibiotics is timely and welcome. There are numerous books and reviews dealing with antibiotic resistance or the discovery of new drugs. However, very few address both aspects of this question, which leads to current problems in developing suitable therapies for human infectious diseases. What makes *Antibiotic Development and Resistance* particularly valuable is that it does not consist of a long list of resistance mechanisms or recipes to find new drugs. It is rather composed of chapters addressing new concepts in the field. For example, the importance of resistance by active efflux of the antibiotic drugs has been recognized only recently. It can contribute to acquired resistance by hyperexpression of the structural genes for the pump. But efflux also contributes to resistance by delaying the death of the host under stress conditions (presence of the antibiotic in the environment) which induce a hypermutator state, which in turn allows time for the emergence of mutations associated with higher levels of resistance. Efflux also accounts for a large number of the *intrinsic* (natural) resistances of certain bacterial species or genera. Thus, efflux pumps have become a major target for new drugs. It is therefore all the more interesting that the various aspects of efflux are found here!

Although the current situation for therapy of human bacterial diseases does not encourage optimism, various approaches for finding new drugs have emerged in the past decade (combinatorial biochemistry, computer assisted drug design, virtual drug screening, genomics, post-genomics, proteomics, etc.). It is most appropriate that these various techniques, which are in contrast to previous stochastic screening of soil microorganisms, compose the last part of this book.

For those, like me, who believe that *comprehension should, ideally, precede action* in particular in the field of antibiotic therapy, *Antibiotic Development and Resistance* is a must.

Patrice Courvalin

Contributors

Dan I. ANDERSSON
Swedish Institute for Infectious Disease Control
Department of Bacteriology
S-171 82 Solna
Sweden

Richard H. BALTZ
CognoGen Biotechnology Consulting and Dow
AgroSciences
9330 Zionsville Road
Indianapolis
IN 46268
USA

Fernando BAQUERO
Servicio de Microbiologia
Hospital Ramón y Cajal
National Institute of Health (INSALUD)
28034 Madrid
Spain

Arnold S. BAYER
Division of Infectious Diseases
St. John's Cardiovascular Research Center
Harbor-UCLA Medical Center
1000 West Carson Street
Torrence, CA 90509
USA

Johanna BJÖRKMAN
Department of Cell and Molecular Biology
The Biomedical Center, Box 596
Uppsala University
S-751 24 Uppsala
Sweden

Lars G. BURMAN
Swedish Institute for Infectious Disease Control
SE-171 82 Solna
Sweden

Thomas A. CEBULA
Division of Molecular Biological Research and
  Evaluation (HFS-235)
Center for Food Safety and Applied Nutrition
Food and Drug Administration
200 C Street, S.W.
Washington, D.C., 20204
USA

Patrice COURVALIN
Institut des Agents Antibactériens
Institut Pasteur
28, rue du Dr Roux
75724 Paris Cedex 15
France

Otto DIDEBERG
Institut de Biologie Structurale *Jean-Pierre Ebel*
  (CEA-CNRS-UJF)
Laboratoire de Cristallographie Macromoléculaire
41 rue Jules Horowitz
F-38027 Grenoble Cedex 1
France

Santiago F. ELENA
Institut Cavanilles de Biodiversitat i Biologia
  Evolutiva and Department de Genètica
Universitat de València
46071 València
Spain

Elspeth GORDON
Institut de Biologie Structurale *Jean-Pierre Ebel*
  (CEA-CNRS-UJF)
Laboratoire de Cristallographie Macromoléculaire
41 rue Jules Horowitz
F-38027 Grenoble Cedex 1
France

Ruth M. HALL
CSIRO Molecular Science
PO Box 184
North Ryde, NSW 1670
Australia

Robert E.W. HANCOCK
Department of Microbiology and Immunology
University of British Colombia
#300-6174, University Boulevard
Vancouver
B.C. V6T 1Z3
Canada

Diarmaid HUGHES
Department of Cell and Molecular Biology
The Biomedical Center, Box 596
Uppsala University
S-751 24 Uppsala
Sweden

J. Eugene LECLERC
Division of Molecular Biological Research and
Evaluation (HFS-235)
Center for Food Safety and Applied Nutrition
Food and Drug Administration
200 C Street, S.W.
Washington, D.C., 20204
USA

Ving LEE
Microcide Pharmaceuticals Inc.
850 Maude Ave
Mountain View, CA 94043
USA

Dan D. LEVY
Division of Molecular Biological Research and
Evaluation (HFS-235)
Center for Food Safety and Applied Nutrition
Food and Drug Administration
200 C Street, S.W.
Washington, D.C., 20204
USA

Olga LOMOVSKAYA
Microcide Pharmaceuticals Inc.
850 Maude Ave
Mountain View, CA 94043
USA

Didier MAZEL
UPMTG, Départment des Biotechnologies
Institut Pasteur
28 rue du Dr. Roux
75724 Paris Cedex 15
France

Patty McMANUS
Department of Plant Pathology
University of Wisconsin
Madison, WI
USA

Nicolas MOUZ
Institut de Biologie Structurale *Jean-Pierre Ebel*
  (CEA-CNRS-UJF)
Laboratoire de Cristallographie Macromoléculaire
41 rue Jules Horowitz
F-38027 Grenoble Cedex 1
France

Barbro OLSSON-LILJEQUIST
Swedish Institute for Infectious Disease Control
SE-171 82 Solna
Sweden

Sally R. PARTRIDGE
School of Biological Sciences
Macquarie University Sydney
NSW 2109
Australia

Gilles POTEL
Laboratoire d'Antibiologie Clinique et Expérimentale
Faculté de Médecine
Centre Hospitalier Universitaire
44035 Nantes
France

Abigail A. SALYERS
Department of Microbiology
University of Illinois
Urbana, IL 61801
USA

Molly B. SCHMID
Microcide Pharmaceuticals Inc.
850 Maude Avenue
Mountain View, CA 94043
USA

Thierry VERNET
Institut de Biologie Structurale *Jean-Pierre Ebel*
  (CEA-CNRS-UJF)
Laboratoire de Ingéniere des Macromolécules
41 rue Jules Horowitz
F-38027 Grenoble Cedex 1
France

Mark S. WARREN
Microcide Pharmaceuticals Inc.
850 Maude Ave
Mountain View, CA 94043
USA

Yan-Qiong XIONG
Department of Medicine
Division of Infectious Diseases
St. John's Cardiovascular Research Center
LAC-Harbor-UCLA Medical Center
1000 West Carson Street
Torrence, CA 90509
USA

Lijuan ZHANG
Department of Microbiology and Immunology
University of British Colombia
#300-6174, University Boulevard
Vancouver
B.C. V6T 1Z3
Canada

1. A Global Perspective on Bacterial Infections, Antibiotic Usage, and the Antibiotic Resistance Problem

Lars G. Burman and Barbro Olsson-Liljequist
Swedish Institute for Infectious Disease Control, SE-171 82 Solna, Sweden

INTRODUCTION

Antimicrobial agents represent the greatest advance in modern curative medicine. These drugs are reckoned to have extended the life span of the average US citizen by 10 years, whereas curing all cancer would prolong life by 3 years (McDermott, 1982). Recent evidence, however, points to an inexorable increase in the prevalence of microbial drug resistance apparently paralleling the expansion of the antimicrobial usage in various fields. Therapeutic difficulties are now posed by strains of certain bacteria such as enterococci and tuberculosis bacteria, which have the ability to acquire resistance to the most useful and possibly to all agents currently in use. Thus, antimicrobial resistance particularly among bacteria has become an increasingly important problem, which has serious implications for the prevention and treatment of infectious diseases. Nevertheless, it has been estimated that the benefits in real-dollar terms of resulting from the use of developed and marketed antibiotics far outweighs costs of adverse effects, including resistance (Liss and Batchelor, 1987).

History of Antibiotic Resistance

Resistance to antimicrobial agents has existed since before they were introduced into human and veterinary medicine, probably because most of the classes of compounds used clinically are produced also by microorganisms in the environment. Evidence for this are e.g. the presence of drug resistant strains in collections of bacteria dating from before the antibiotic era (Datta and Hughes, 1983; Murray and Moellering, 1978) and in wild animals with no apparent contact with man made drugs (Gilliver et al., 1999).

The introduction and increasing clinical use of each antimicrobial agent has sooner or later been followed by an increasing isolation rate of resistant bacteria. Shortly after the introduction of penicillin after World War II penicillinase production was discovered in *Staphylococcus aureus* and *Escherichia coli* and in the early 1950's the use of the first broad spectrum agents such as chloramphenicol, streptomycin and tetracycline was rapidly followed by resistance in stapylococci and Gram negative bacteria. This was true also for the isoxazolyl penicillins and the occurrence of methicillin resistance in *S. aureus* (MRSA) apparently as a result of horizontal transfer of the *mecA* gene from *S. sciuri* (Wu et al., 1996). Also, resistance was quick to develop by mutational target alteration against e.g. quinolones, rifampicin, fusidic acid and mupirocin after their introduction into clinical use.

In some cases resistance turned out to be a complicated task and its development took three decades and included a series of sophisticated genetic events until levels of resistance of clinical relevance was obtained (penicillin betalactam resistant pneu-

1

mococci, PRP; vancomycin resistant enterococci, VRE). Also, the resurgence of tuberculosis in the 1980's was accompanied by an increasing rate of multi-resistant *Mycobacterium tuberculosis* isolates, defined as those resistant to both isoniazid and rifampicin, with or without resistance to other agents, which markedly increased the risk of therapeutic failure. On the other hand, the continuing universal susceptibility of *Streptococcus pyogenes* (group A streptococcus) to penicillin and to other betalactam agents represents an unusual example of both poor ability to acquire betalactamase genes and to create mosaic penicillin binding protein genes through the import of DNA from other streptococci, despite the continuing use of penicillin as the drug of choice for cure of streptococcal infections.

Consequences of Antibiotic Resistance

There is an embarrassing lack of studies estimating the cost of antibiotic resistance in terms of increased morbidity, mortality or cost to hospitals or the society (Liss and Batchelor, 1987). In one rare and much cited study of hospitalized patients receiving empiric antibacterial therapy, resistance of the infecting strain to the drug given resulted in prolongation of the hospital stay by about 2 weeks (Holmberg *et al.*, 1987). Applying this figure to Swedish hospitals and assuming that 5% of the cures were troubled by bacterial resistance, the resulting extra cost of the patients' care would equal that of all nosocomial infections taken together or outnumber the hospitals expenditures for antibiotics by 5 to 1.

Shortage of New Agents

The resistance problem has traditionally been addressed by the development of new antimicrobials. In recent years, however, there has been a slowing down in the introduction of such agents and no truly novel antibacterial drugs have been marketed for more than 10 years. Problems have arisen in finding effective therapy for a number of pathogens such as MRSA, VRE, PRP, non-fermentative Gram-negative rods and multiply resistant *M. tuberculosis*.

Apparently, new agents against resistant cocci will soon be clinically available. However, in the case of quinupristin-dalfopristin (Synercid®) being evaluated for use in humans, a related combination of streptogramins has already been used for many years as an animal feed additive in the EU (streptogramin A + B; Virginiamycin). This has resulted in the development of resistance to Virginiamycin but also to Synercid among enterococci from animals. Regrettably, this also applies to everninomycin (SCH 27899), an oligosaccharide and a rare example of a truly new type of antibacterial compound to be introduced into human medicine (Ziracin®). This compound has already been used for many years under the name Avilamycin in animal feeds, thereby creating enterococcal resistance in European industry reared animals.

Another truly novel agent inhibiting multi-resistant cocci and now being launched is linezolid (Zyvox®), the first representative of the oxazolidinone group of agents to be used in clinical practice. It is unique in this context as bacteria outside of laboratories have so far not seen the agent (Ford *et al.*, 1997). A possible drawback from a resistance aspect is that linezolid is almost completely absorbed after oral administration and that adverse effects are rare. This might invite a widespread use of the agent both in hospitals and in the community, which might lead to the development of resistance in Gram-positive cocci and jeopardize its future usefulness in critical clinical situations.

Current International Activities

The serious potential consequences of a continuing development of resistance to antimicrobial agents among microorganisms has been considered and debated by numerous academic, professional, industrial and Government groups worldwide. Several of these bodies, both in Europe (The Copenhagen Recommendations, 1998; European Commission, 1999) and in the US (Goldmann *et al.*, 1996) as well as the WHO (WHO, 1998) have recently reported findings and issued recommendations on complementary strategies including research efforts to better control the drug resistance problem in the future. Furthermore, several efforts to improve surveillance of antibiotic usage and antibiotic resistance and to analyze their relationship are under way in the US (Flaherty and Weinstein, 1996; McGowan and Tenover, 1997; Shlaes *et al.*, 1997) and in Europe (see further below).

INDICATIONS FOR ANTIBIOTIC USAGE

Major Community Acquired Infections

Respiratory tract infections

In western countries about 75% of all antibiotics are used to treat community acquired respiratory tract infections (Gonzales *et al.*, 1997; Guillermot *et al.*, 1998a; Pennington, 1994). The peak of antibiotic usage occurs in early childhood because of the frequent acute middle ear infections (*otitis media*) affecting half of pre-school children at least once, and some children repeatedly. Otitis is caused primarily by *Streptococcus pneumoniae* and less often by *Hemophilus influenzae, S. pyogenes* and occasionally by *Moraxella catarhalis*. Otitis becomes relatively uncommon after 6 years of age. Instead, tonsillitis mainly due to *S. pyogenes* and occasionally accompanied by facial skin infection (*impetigo*) or generalized skin rash (*scarlatina*) emerges as a major bacterial upper respiratory tract infection. Streptococcal sore throat shows a peak incidence between 5 to 10 years of age but continues to be relatively common through college age annually affecting 2–3% and totally about 25% of the population (Pennington (ed.), 1994).

A less common upper respiratory tract infection is sinusitis, 90% of affected patients being in the age group 15–65 years and with *S. pneumoniae* as the dominating causative agent followed by *H. influenzae* and *S. pyogenes*. These organisms may also contribute to acute exacerbations of chronic bronchitis whereas acute bronchitis is mainly of viral origin.

Deep infection of the lung, pneumonia, shows an incidence of about 10 episodes per 1000 inhabitants and year. It occurs in small children, less often in teenagers and middle age individuals and shows a marked increase in the elderly. *S. pneumoniae* is again the dominant causative agent followed by *Mycoplasma pneumoniae*, especially among younger patients. Other major bacterial pathogens casing pneumonia are *H. influenzae, Chlamydia pneumoniae* and occasionally *Legionella pneumophila, M. tuberculosis* or *Bordetella pertussis*, the causative agent of whooping cough.

Because of the expected spectrum of bacterial pathogens, as outlined above, penicillins, macrolides, tetracyclines and amoxicillin/clavulanic acid are the major classes of antimicrobial agents used in community acquired respiratory tract infections. Therapy is usually empiric, i.e. started without a microbiological diagnosis. This is because apart from rapid *S. pyogenes* tests there are currently no possibilities to differentiate between viruses and bacteria or to identify them and their resistance pattern in the doctors office. The drug selected in each case depends both on the organism expected and its presumed susceptibility to antibiotics. Thus, the type of infection, the age of the patient, local susceptibility patterns, previous exposure to antibiotics and other circumstances such as drug allergy have to be taken into consideration.

Urinary tract infections (UTI)

UTI is the indication for 10–20% of antibiotic prescriptions in primary health care. It shows a peak in early childhood, then occurs as the most common bacterial infection in women in the child-bearing age group, and finally increases among the elderly, both women and men. The leading causative agent is *Escherichia coli* (80%), usually originating from the patient's own faecal flora, followed by *Staphylococcus saprophyticus* (10%), *Klebsiella spp*, other Gram negative bacteria and enterococci. Therefore, broad spectrum penicillins (amoxicillin, mecillinam), oral cephalosporins, nitrofurantion, trimethoprim, and in recent years phosphomycin and particularly fluoroquinolones, have become frequently used against UTI (see below).

Helicobacter pylori

A relatively new indication for antibiotic usage is *H. pylori* infection, which is the cause of duodenal ulcer and sometimes gastric ulcer. The association between *H. pylori* and diffuse stomach discomfort such as dyspepsia is still controversial. Recommended therapy is combinations including metronidazole, amoxicillin, macrolides and sometimes bismuth. Taken together, these patient groups could represent up to 5% of those needing antimicrobial agents outside hospitals. Currently, patients with milder gastric discomfort lacking evidence of ulcer or laboratory diagnosed *H. pylori* infection are, however, not given antibiotics. It is therefore estimated that currently only about 1% of all antibiotic prescriptions but 10% of the prescriptions of macrolides (e.g. in Sweden) refer to *H. pylori* infection.

Enterocolitis

In western countries the incidence of the classical enteric infections has become relatively low. Furthermore, only severe infections due to *Salmonella, Shigella* or *Campylobacter spp.* and strains of *E. coli* causing diarrhea (ETEC, EHEC, EIEC etc.) are treated with antibiotics, usually fluoroquinolones, but erythromycin for *Campylobacter*. In contrast, milder to severe colitis due to overgrowth of toxin producing *Clostridium difficile* has become a major problem, particularly in hospitals. This is seen as a consequence of the use of modern antibiotics, disrupting the large bowel microflora by virtue of poor absorption, large biliary excretion and unfavorable antimicrobial spectrum. For example, a recent nationwide Swedish survey showed that the incidence of *C. difficile* disease, almost unknown 20 years earlier, now exceeds that of the domestic cases of *Salmonella, Shigella, Campylobacter, Yersinia* and parasitic and amoeba diarrhea taken together (Karlström *et al.*, 1998). The specific therapy of *C. difficile* diarrhea is metronidazole or oral vancomycin, but due to the current problems with VRE vancomycin is no longer recommended for *C. difficile* infections.

Sexually transmitted disease (STD)

Among these syphilis is rare and there has been a reduction of the incidence of gonorrhea in many western countries since the mid-1970's. Current isolates of *Neisseria gonorrhoeae* tend to be resistant to penicillins, and antibiotic therapy is preferably selected based on results of *in vitro* susceptibility testing of the infecting isolate. Despite the decline of gonorrhea there has been a corresponding increase in genital *Chlamydia trachomatis* infections over the last decades now affecting about 0.5% of the sexually active population every year. The drugs used for genital chlamydial infection are tetracyclines, and in recent years also modern macrolides like azithromycin and sometimes fluoroquinolones.

Other infections

Bone infection or *osteomyelitis* may occur either spontaneously via the blood stream or after penetrating trauma. The causative agent, *S. aureus*, requires weeks or months of therapy with e.g. an isoxazolylpenicillin for eradication.

Developing countries

The current panorama of community acquired bacterial infections described above is typical of the modern western society. In poor or developing countries inadequate housing conditions, different social behavioral patterns, less developed veterinary medicine, contaminated food, poor water quality etc. contribute to much higher incidences of e.g. purulent skin infections due to *S. aureus* and *S. pyogenes*, tuberculosis and other respiratory infections, STD, *H. pylori* disease as well as bacterial diarrheal disease and a spill over of these infections from the community to hospitals. It has been estimated that the impact on human health of infections, including parasitic diseases, occurring in the developing countries is more than twice as great as the impact of all human diseases occurring in the developed countries (Table 1).

Major Hospital Acquired Infections (HAI)

The definition of HAI is that the infection started during or as a result of hospitalization. The incidence of HAI depends on the type of hospital and varies between 3 and 9% of the admitted patients (Mayhall, 1996; Wenzel, 1997).

Table 1. Human impact of diseases in developing and developed countries expressed as Disability Adjusted Life Years (DALY). DALY summarizes the total number of years of life lost or affected by poor health in the respective populations (20–21).

Diseases	DALYs (millions)	
	Developing countries	Developed countries
Infectious/parasitic	377	6
Other	833	146

Urinary tract infections (UTI)

The HAIs are dominated by UTI which regularly account for 30–35% of the total incidence and represent a major source of nosocomial septicemia and associated mortality (Mayhall, 1996; Wenzel, 1997). UTI acquired in hospital is usually due to indwelling urinary catheters or introduction of other instruments into the urinary tract and are mainly caused by *E. coli* (50%), followed by enterococci, *Pseudomonas aeruginosa, Klebsiella, Enterobacter, Proteus, Morganella* and *Providencia spp., Serratia marcescens, Stenotrophomonas maltophilia* and yeasts. The oral antibiotics used are the same as those used in other UTI, but sometimes parenteral drugs such as aminoglycosides and newer cephalosporins are used.

Surgical site infections (SSI)

SSI represent 20–30% of HAI. *S. aureus* is the major pathogen in so-called clean surgery (70%), which means procedures where, after the skin incision, the site of operation is normally sterile. In clean procedures involving foreign implants (hip or knee joint prosthesis, osteosynthesis material, pacemaker, ocular lens, breast implant etc) coagulase-negative staphylococci, dominated by *S. epidermidis*, also become major pathogens and outnumber *S. aureus* as a cause of SSI by 3- to 4-fold. The antibiotics used for therapy and, in selected clean procedures, prophylaxis (major orthopedic, thoracic and neurosurgical operations) are therefore directed against staphylococci and include isoxazolylpenicillins, first generation cephalosporins, clindamycin and, in case of MRSA or MRSE, vancomycin. Should postoperative staphylococcal bone infection (*osteomyelitis*, prosthesis infection) occur, any foreign material usually has to be removed and

antibiotic therapy, based on *in vitro* susceptibility testing of the pathogen, must continue for weeks or months.

In abdominal surgery opening of the gastro-intestinal, biliary, genital or urinary tracts may lead to contamination with endogenous bacteria dramatically increasing the risk of SSI from 2–3%, typical of clean operations, to up to 40%. Thus, wound infection, peritonitis or encapsulated infection (abscess) after abdominal surgery tend to be mixed infections and may involve not only *S. aureus* but more often bacteria from the intestinal flora such as *E. coli*, other Gram-negative enterobacteria, enterococci or anaerobes such as *Bacteriodes* and *Clostridium spp.* Antibiotic perioperative prophylaxis and therapy for SSI associated with abdominal surgery therefore should have a broad antibacterial coverage and often includes a combination of agents, e.g. a cephalosporin plus metronidazole against the anaerobic flora.

Pneumonia

Nosocomial lung infection also accounts for 20–25% of HAI and is usually associated with ventilator (respirator) therapy that invalidates the natural first line of defense of the respiratory tract. The causative organisms may be *S. pneumoniae* or *H. influenzae* during the first days of hospitalization. Later *S. aureus*, Gram negative enterobacteria or *P. aeruginosa* dominate as pathogens. The often weak state of the patient and a massive tissue involvement accounts for a high mortality in nosocomial pneumonia (about 30%). Antibiotic therapy is complicated both by the complex microbiology usually involving multiply drug resistant hospital organisms and the difficulties in avoiding contamination from the upper respiratory tract and identifying the causative agent without dangerous invasive procedures for sampling of deep respiratory tract secretions for culture (broncho-scopy). Consequently, the choice of antimicrobial therapy is often empiric and has a broad spectrum such as a carbapenem or a newer cephalosporin sometimes combined with an aminoglycoside.

Due to the dismal prognosis in nosocomial pneumonia European hospitals have evaluated a much-debated antimicrobial prophylaxis for venti-lated patients named selective digestive deconta-mination (SDD) during the last 15 years. SDD involves applying a paste containing e.g. tobramy-cin, vancomycin and nystatin into the mouth and stomach of the patient in order to eliminate potentially pathogenic Gram-negative bacteria, cocci and yeasts without disturbing the dominating anaerobic flora of the upper respiratory and gastrointestinal tracts. In addition, the patient is given intravenous antibiotic therapy, usually cefo-taxime. A recent meta-analysis of a large number of local SDD studies performed by the Cochrane organization showed that this controversial proce-dure reduced both morbidity and mortality in ventilator associated pneumonia but also pointed out that the ecological consequences need to be further followed up (D'Amico *et al.*, 1998). Thus, hospitals are still reluctant to use SDD for prophylaxis of nosocomial pneumonia in Europe and even more so in the US.

Bacteremia

No. 4 among HAI is blood stream infection or bacteremia (15–20%), defined as growth of bacteria from cultures of a patient's blood. Bacteremia is often accompanied by fever and other general symptoms and then named septicemia or septic shock, and carrying a mortality of about 20%. The focus of this generalized infection may be any of the above localized infections. Another common portal of entry of bacteria is via the inside or outside of an intravascular device like a central venous catheter associated particularly with infec-tion due to skin staphylococci. The causative organisms in blood stream infections may be either staphylococci, enterococci or Gram negative bac-teria. When bacteremia is suspected, empiric therapy with a parenteral broad spectrum agent or with a combination covering these bacteria is

started as soon as blood has been drawn for culture. If necessary, the therapy is later changed according to results of susceptibility testing of the strain(s) isolated.

Other HAI

The bacterial HAI also include *C. difficile* colitis, skin infections, infections in mothers or infants during or after delivery, and infections after other special hospital procedures such as peritoneal dialysis.

USAGE OF ANTIBACTERIAL AGENTS

Total Consumption

If the assumption that increased antimicrobial use is associated with increased prevalence of bacter-ial drug resistance holds true, then data on consumption of antimicrobials in human and veterinary medicine, as feed additives or for plant protection, should provide insight into the relative contribution of the four areas to the overall problem. The amount of accurate information available, however, is limited. Furthermore, it is not known whether the key factor is the total weight of active substance consumed per capita or the way in which antibiotics are administered (e.g. dosage, duration of therapy) in hospitals and in the community.

Antimicrobial agents represent about 10% of the market value of all pharmaceuticals and 10% of prescriptions (Col and O'Connor, 1987). The world-wide annual production has been estimated to about 15 g per person, but with wide variation between countries, and may now exceed a total of 100,000 tons. Data have recently been provided indicating that in the EU 10,493 tons of active substances were consumed in 1997, 52% of which was used in human medicine, 33% in veterinary medicine and 15% in animal production (FEDESA, 1997). Similar proportions were found when perhaps more accurate local estimates were made, e.g. in Finland (European Commission, 1999). For certain drugs the relative contribution from animal feed additives may be considerably larger. Thus, in Denmark 24 kg of vancomycin was used for human therapy whereas 24,000 kg of the related glycopeptide avoparcin was used as a feed additive for animals in 1994 (Wegener et al., 1997).

In some EU countries, where the requirement for a physician's prescription in order to obtain antibiotics is not strictly enforced, consumption could be twice as high as in others. Furthermore, factors such as access to medical care, willingness of physicians to prescribe antibiotics, and to what extent the drug cost is subsidized, also affect usage. On the other hand, in countries in e.g. south east Asia, where counterfeit drugs are made by small local companies and dispensed at a low price, antibiotic consumption probably is much higher than the low levels indicated by official sales data. Concerted efforts in the EU and cooperation with US authorities are now being started in order to obtain more precise information on the usage of antibiotics as well as the prevalence of drug resistant bacteria in different countries (see below).

Methods to Measure Consumption in Human Medicine

There is currently no consensus on the unit of measurement to compare the consumption of antibiotics between hospitals or communities. Most of the commercially prepared data give information based on costs, with limited or no information on the way antimicrobials are prescribed by indication, dose, dosing regimen, duration of treatment, compliance, or treatment outcome. As the single dose on a weight basis may differ up to 100-fold between antibiotics, data released by governmental authorities is sometimes given as defined daily doses (DDD) providing a better idea of the number of individuals or microfloras exposed. The DDD is defined for each drug by the WHO.

The ATC (Anatomical and Therapeutical Classification) classification of drugs combined with the DDD methodology has been used by several investigators to compare the use of antimicrobial agents and has been used for all drug consumption data in countries with a national drug delivery system such as in the Nordic countries. There is no agreement as yet on the denominator to be used for drug consumption in hospitals, but DDD/100 bed days has been suggested and used. Although there is also no international consensus on the methodol-

ogy for comparison of antibiotic consumption outside hospitals or on a national basis it is currently expressed as DDD/1000 inhabitants/day in e.g. the Nordic countries.

Consumption in Human Medicine

Although there is a lack of comprehensive data on antibiotic consumption for most countries certain information is available, e.g. from limited studies and from countries with state controlled distribution of pharmaceuticals. In the UK (Davey et al., 1996) and in France (Guillermot et al., 1998b) there are indications of a continuing increase of antibiotic consumption during the last decades. This was the case also in the Nordic countries but national programs to curb this development now seem to have had an effect. For example, in Sweden efforts by a nation-wide network dedicated to optimizing antibiotic therapy halted the increase and have been followed by decreasing antibiotic usage during the period 1993–1997, mainly in the community and in total by 23% (Mölstad and Cars, 1999). It has been possible to keep the consumption level relatively constant since then.

In England and Wales sufficient oral antibiotics are prescribed annually in the community to treat every member of the population for five days (Prescribing Pricing Authority, 1997). This is true also for the Nordic countries. Comparisons between the UK, Germany and France indicate that the annual number of consultations for the major cause of prescribing antibiotics, upper respiratory tract infections, per 1,000 inhabitants is greater in France than in the other two countries as is the proportion of patients receiving a prescription (European Commission, 1999). In a recent comparison of three teaching hospitals each in Estonia, Spain and Sweden the overall antibiotic consumption was similar (41, 51 and 47 DDD/100 bed days, respectively). In other words 25–50% of the patients were receiving antibiotics although the predominant prescription preferences differed between the hospitals (Kiivet et al., 1998). As comprehensive and precise nation-wide or regional data on antibiotic consumption e.g. Europe, the US or Eastern regions of the world currently is not available, we have chosen for the discussion below 1998 data from Sweden to represent the situation in

Table 2. Antibiotic consumption in Sweden 1998. Other agents includes aminoglycosides, carbapenems, clindamycin, fusidic acid, glycopeptides, imidazoles, phosphomycin, rifampicin and other agents against mycobacteria. Data from Apoteket AB, Stockholm.

ATC code	Antibiotic or group	Consumption (DDD/1000 inhabitants/day)	
		Community	Hospitals
J01CE	Penicillin V	5.12	0.16
J01A	Tetracyclines	3.34	0.16
J01CA	Ampicillins	1.23	0.11
J01M	Quinolones	1.03	0.15
J01F	Macrolides	1.03	0.06
J01CF	Isoxazolylpenicillins	0.99	0.17
J01DA	Cephalosporins	0.60	0.20
J01EA	Trimethoprim	0.57	0.03
J01EE	Trimethoprim/sulfa	0.18	0.03
J01CR	Amoxycillin/clavulanic acid	0.26	0.01
	Other	0.03	0.09
	Total	14.4	1.2

northern Europe. In Sweden and the other Nordic countries all drugs are distributed through a national network of hospital and community pharmacies, which enables a precise follow up of the amounts of different antibiotics dispensed to humans.

Antibiotic usage in the community

Prescriptions used to account for 80% of the antibiotics 20 years ago but due to the steady increase until 1993 (Mölstad and Cars, 1999) they now account for 92.5% or 14.4 DDD/1000 inhabitants/day (5.1 DDD/inhabitant/year). As expected, agents used mainly against respiratory pathogens dominate, phenoximethylpenicillin (penicillin V) being the most frequently prescribed drug followed by tetracyclines, ampicillin derivatives, macrolides, cephalosporins and amoxicillin/clavulanic acid and together comprising 80% of the total antibiotic usage (Table 2). Other countries may have a less conservative consumption pattern than in the Nordic countries and thus, a lower proportion of penicillin V prescriptions. Although penicillin is the classical drug of choice in streptococcal tonsillitis, cephalosporins are preferred in recurrent cases due to their lower risk of relapse, and in countries with a high prevalence of PRP agents other than betalactams have to be considered in e.g. otitis and pneumonia. Agents

used mainly in UTI, such as fluoroquinolones, trimethoprim and trimethoprim/sulfa represented the second largest group of antimicrobials used in the community (12%) followed by the antistaphylococcal isoxazolylpenicillins (7%).

Antibiotic usage in hospitals

Although only about 1% of the population is hospitalized at any given time it is receiving 7.5% of all antimicrobials used in Sweden, and in teaching hospitals the antibiotic pressure per capita may be more than 20 times higher than that in the community. The generally broader antibacterial spectrum and the greater ecological impact on the human microflora of agents used in hospitals than of those used in the community further contribute to a high risk of selection of resistant strains, which also are given ample opportunities to become transmitted between patients. Various biological response modifiers improving the immune defense in severe acute bacterial infections that could complement or replace antibiotics and thus diminish their ecological pressure have so far had little or no clinical success.

As antibiotic resistant staphylococci, enterococci and Gram negative bacteria dominate as causative agents in HAI, penicillin, macrolides and tetracyclines are used mainly for patients hospitalised due to certain community acquired infections. Instead, fluoroquinolones, isoxazolylpenicillins, and newer injectable cephalosporins are more prominent in hospitals together with other parenteral drugs such as glycopeptides, carbapenems and aminoglycosides (Table 2). Most of the antimicrobials used are for therapy, but in surgical departments up to 30% may be used for peroperative prophylaxis in selected procedures.

ANTIBIOTIC USAGE *VERSUS* ANTIBIOTIC RESISTANCE

The So-called Ecological Impact of Antibiotics

Definition

Ecological impact is often defined in microbiological terms and relates to the ability of drugs to inhibit components of the skin, upper respiratory, gastrointestinal and genital tract microflora of patients and also its consequences. These may include symptoms such as vaginitis or diarrhea but are mostly silent, while room is made for expansion of endogenous drug resistant or *C. difficile* strains and colonization with new ones. As several body sites and numerous different bacterial species may be involved and the ecological changes are gradual and dose and time dependent, there is no generally accepted standardized method to test and grade antibiotics according to their ecological impact on the human microflora. However, systematic studies of the short time impact of antibiotics on the salivary and fecal flora have been conducted (Edlund and Nord, 1993; Edlund and Nord, 1999).

An alternative definition is even more multifaceted and complex (see below) and relates to the ability of antibiotics to increase the occurrence of drug resistant strains or *C. difficile*. These end points of antibiotic usage can occur and be studied at the patient, hospital ward, hospital, community, national, or continental level.

Drug factors

Perhaps the most important property of an antibiotic in deciding its ecological potential is its

ability to reach the microflora via e.g. the saliva, mucous membrane secretions, biliary excretion or for oral agents, the proportion of unabsorbed drug excreted by the fecal route. To what extent antibiotics reach the skin flora depends on their ability to be secreted via hair follicles or sweat glands and particularly their intracellular concentration as epidermis cells move to the skin surface and degenerate to become the horny cell layer. Another drug specific factor influencing the ecological impact of an antibacterial agent is its antimicrobial spectrum and particularly, whether it can inhibit the anaerobic bacteria that constitute about 99% of the flora at all sites.

Whether selection of drug resistant strains may occur also depends on the daily dose, the duration of therapy, the microorganism, and the type of resistance mechanism studied or rather the level of resistance conferred. Thus, if the potential resistance level to the drug studied is high, the risk of selection of resistant strains probably increases both with dosage and duration of treatment. On the other hand, for selection of low level resistant mutants or e.g. pneumococcal strains with a moderate degree of betalactam resistance, prolonged exposure to a low dose may be optimal for resistance selection. This was suggested to be the case with PRP by a recent French study, where only low aminopenicillin or cephalosporin doses combined with therapy for at least 6 days was associated with a PRP carriage rate significantly higher than that among untreated control children (Guillermot *et al.*, 1998c).

Patient factors

These include pharmacokinetic factors such as rate of intestinal absorption and renal or biliary excretion that vary between individuals and with age and influence drug concentrations in body secretions and the gut lumen. Finally, components of the individual patient's microflora may counteract an ecological impact of a drug by destroying it, e.g. by β-lactamase production, that further varies considerably between throat and large bowel microfloras from different persons. Why only a subset of *C. difficile* carriers are susceptible to antibiotic associated overgrowth and diarrheal disease is currently not known, but may be related to properties of their individual colonic microflora.

Bacterial factors

To what extent antibiotic usage will lead to selection of resistant bacterial strains will also be influenced by the ability of each bacterial species to develop and retain resistance and the prevalence of such strains in the individual or population studied. In the absence of bacteria able to become resistant such selection should theoretically not occur. The further spread of a resistant strain depends partly on its individual "epidemic potential", that is based on mostly unknown factors facilitating its survival outside the human body and capacity to colonize and be transmitted between individuals in hospital and community environments. For resistant strains the cost of carrying and expressing the resistance phenotype will probably also influence its "epidemic potential" (see Andersson and Hughes, Chapter 2, this book). For example, MRSA strains differ considerably with regard to local (nosocomial) or world-wide transmission risk although the underlying factors are not understood (Musser and Kapur, 1992).

Human population factors

As most pathogenic bacteria have evolved into many distinct serotypes, the actual "epidemic potential" of a drug resistant strain is further dependent on herd immunity, i.e. the current immunity status of the population against the serotype of the strain in question. Thus, a resistant strain may come and go in a human population depending on its "epidemic potential" including serotype/herd immunity and under constant antibiotic pressure or independently of actual increases and decreases of antibiotic usage. As humans seem to carry only one or a few strains of a certain species at a time apparently competing within their own ecological niche in the microflora, competition with closely related strains may also affect the fate of a particular drug resistant strain.

Environmental factors

Hospitals represent an environment characterized not only by high antibiotic pressure per capita, but also rapidly changing populations and numerous situations involving direct or indirect person to person contact, that is the major mode of transmission of bacteria. Therefore, the nosocomial prevalence of antibiotic resistant bacterial strains depends not only on antibiotic usage but also on hospital structure, population density, and hospital hygiene, notably staff hand hygiene. Large units, overcrowding of patients and understaffing are classical risk factors for transmission of bacteria and outbreaks of colonization or infection involving drug resistant bacteria (Wenzel, 1997; Mayhall, 1996; Phillips, 1983; Fryklund et al., 1993). It has recently been estimated that the influence of person to person transmission of resistant strains on their nosocomial occurrence almost equals that of the ecological pressure of antibiotics (Lipsitch et al., 2000). It is likely that hygienic factors together with inappropriate use also contribute to the apparent paradox that bacteria resistant to older antimicrobial drugs are more prevalent in developing countries than in developed countries despite the fact that antibiotic sales per capita are 10-fold lower in the former countries (Col and O'Connor, 1987; Kunin, 1993).

Problems in Studies of the Relation between Antibiotic Usage and Antibiotic Resistance

Given the above mentioned complexity of the ecological impact of antibiotics there are many problems and pitfalls when trying to clarify antibiotic usage *versus* resistance rates, that tend to make such studies fragmentary and weak (McGowan, 1994). The most important of these difficulties are summarized in Table 3.

A major problem is the magnitude and complexity of the question to be addressed. Suppose that we are interested in knowing the impact of each of 10 drugs on resistance in 10 bacterial species. That amounts to asking 100 questions. We might like to answer these 100 questions for both colonizing and infecting isolates, among three different populations in the community (e.g. the general public, day care attendees and their families, and nursing home residents), and a minimum of nine major populations of hospitalized patients (e.g. adult and neonatal intensive care, adult general and orthopedic surgery, internal medicine, hematology, oncol-

Table 3. Methodological problems in studies of the relation between antibiotic usage and antibiotic resistance. See also text.

Study design	
	Selection bias (results fragmentary: specific for one drug dosage, length of therapy, type of hospital unit or patient population, but perhaps not generally applicable to these and the many other possible clinical situations)
	Studies too small (insufficient statistical power, risk of chance results)
	Variable definition and identification of resistance
Confounding factors[a]	
	Co-selection of resistant bacteria by related and, for multiply resistant strains, by unrelated drugs
	Bacterium specific factors
	Host factors
	Population factors
	Environmental factors
	Others

[a] Factors other than the consumption of a particular antibiotic that may affect the prevalence of bacteria resistant to the drug.

ogy, geriatrics, pediatrics) in the three types of hospitals (primary, secondary, tertiary) each of two sizes (large, small). Furthermore, we are not only interested in the general ecological impact of each drug but also the impact for a high and a low dose, for a short and a long duration of therapy, and for three levels of consumption (DDD/population or bed day). This is in order to define a threshold level when selection and elimination of resistant strains are balancing each other at a low resistance rate. Below this consumption level resistance is uncommon and the drug can be safely used from an ecological point of view, but if it is exceeded resistant strains will increase in prevalence. Taken together, we are now confronted with over 20,000 questions concerning a drug, a bug, and a common and relevant clinical situation for an antibiotic prescribing physician! Thus, numerous well-designed studies will be required in order to address all of these questions. A further problem is that the results obtained apply to a particular microflora, hospital unit, hospital, population or time period. In order to know to what extent the results are generally applicable, these situation studies have to be multicenter or reproduced, and give similar results.

Still another problem is that if studies are too small there is an increased risk of a false negative or positive correlation. Furthermore, a number of confounding factors, i.e. factors other than the properties and consumption of each antibiotic, may also affect the prevalence of bacterial strains resistant to the drug, especially in small studies (Table 3). Confounding bacterial factors include species and strain specific ability to acquire and keep the resistance genotype and phenotype, colonize and compete in the human microflora, survive in community and hospital environments and be transmitted directly or indirectly between individuals. An increased use of modern typing methods allowing tracing of individual antibiotic-resistant strains should help to clarify their complex epidemiology. Other confounding bacterial factors are cross-resistance and multiple resistance enabling related and unrelated antibiotics, respectively, to select for the same resistant strain.

A little studied confounding host factor, also affecting the epidemiology of drug resistant bacterial strains, is herd immunity that influences the susceptibility of individuals or a whole human population to colonization by a particular strain. Finally, among environmental confounding factors influencing drug resistance rates are crowding in the community and in hospitals, compliance with hospital infection control procedures (hand hygiene/disinfection, barrier nursing, isolation or cohort nursing of patients with resistant strains) and staffing. In order to iron out confounding factors, randomized, complex and costly multicenter or reproduced studies, not likely to be financed by the pharmaceutical industry, are needed. Taking all methodological difficulties together we may never be able to fully understand the connection between antibiotic usage and bacterial antibiotic resistance.

Relation between Drug Consumption and Drug Resistance

A large number of studies, particularly of hospital units or whole hospitals, have to a variable degree documented a correlation between the clinical use of antibiotics and the development of bacterial resistance (Mouton *et al.*, 1976; McGowan, 1983; Wiedemann, 1986; Kling *et al.*, 1989; Moellering, 1990; Chow *et al.*, 1991; Fujita *et al.*, 1994; Seppälä *et al.*, 1995; Arason *et al.*, 1996; Tenover and McGowan, 1996; Gaynes and Monnet, 1997; Swartz, 1997; Austin *et al.*, 1999; Kristinsson, 1999; Struelens *et al.*, 1999; Torell *et al.*, 1999; DeMan *et al.*, 2000). The associations were both temporally and exposure or dosage related and the types of evidence can be marshalled into six categories:

1. Introducing new antimicrobial agents into clinical use was followed by an increase of patient isolates with preexisting resistance mechanisms or resistant to man-made drugs such as trimethoprim or quinolones.
2. Unplanned or policy-based decreases of antibiotic usage in hospital units, hospitals or the community was followed by decreased resistance and *vice versa*.
3. Development of antibiotic resistance among repeat isolates from hospitalized and non-hospitalized patients during antibiotic therapy.
4. Intensive care units, other hospital wards and the community, represent different levels of antibiotic usage and bacterial resistance rates among clinical isolates.
5. During hospital outbreaks, patients infected by the resistant strains were more likely than others to have received antibiotic therapy.
6. Increasing dosage of antimicrobials to hospitalized patients lead to a greater likelihood of colonization or superinfection with resistant organisms.

As most of these results were not obtained from randomized, multicenter or large trials and sometimes from rather small studies they can be criticized on the basis that the results may be accidental (due to chance) or due to any of the confounding factors mentioned above. There is also a risk of publication bias (negative results not submitted nor accepted for publication). For example, the much publicized rise and fall of PRP on Iceland and of erythromycin resistant *S. pyogenes* in Finland paralleling antibiotic consumption may partly be a result of changing herd immunity. Nevertheless, data from Sweden shows that Malmöhus county having the highest penicillin and antibiotic consumption rate also has the nation's highest isolation rate of a serotype 9 PRP clone with a typical DNA fingerprint, and that consumption and prevalence of PRP correlate also when comparing communities within that county (Melander *et al.*, 1998). On the other hand, although the consequences of the above mentioned general decrease of antibiotic consumption in Sweden by nearly 25% starting in 1993 has not yet been fully analyzed, studies of nation-wide random samples of isolates (n = 3000 isolates per species and year) performed so far do not indicate any following decrease of antibiotic resistance (Swedish reference group for antibiotics, unpublished data).

Other types of criticism are that most hospitals studies were from specific units in large teaching (tertiary) hospitals and may not be representative of other similar or different units or smaller hospitals, that some studies depicted outbreak situations so that the results my not be applicable to the typical endemic situation, and that some study periods were short so that later ecological problems due to the new regimen had not yet occurred. Furthermore, in some outbreaks involving resistant strains also other measures, mentioned or not in the publications, may have contributed to the disappearance of the strain. A rare exception was a recent Dutch study of neonates showing a much higher isolation rate of bacteria resistant to the empiric regimen when cefotaxime/amoxicillin was used than when tobramycin/penicillin G was the standard (DeMan *et al.*, 2000). Its design was unusual in that two neonatal intensive care units were compared during one year in a cross-over study (switch of regimens after six months). This study also showed that the microflora change may be very rapid in a hospital environment (a few weeks) as recently predicted (Lipsitch *et al.*, 2000).

In summary, despite weaknesses in some studies, and the absence or low level of correlation in others, it can be concluded that both a temporal and a dose-relationship between antibiotic usage and bacterial drug resistance has been documented enough to support a causal association. This notion is also biologically plausible and further supported by numerous *in vitro* experiments. Nevertheless, a major remaining problem is that the information on the ecological impact of different dosages and consumption rates available for each antibacterial agent still is too fragmentary to answer the overwhelmingly complex question about the relation between antibiotic consumption and bacterial resistance. Thus, current knowledge does not allow precise recommendations for the optimized use of various agents in different clinical settings from a resistance development point of view. It is also possible that cross-selection between different drugs and bacterial mechanisms compensating for the biological cost of antibiotic resistance tend to slow and even prevent the disappearance of drug resistant strains following efforts decreasing antibiotic usage. More randomized or multicenter studies comparing countries, communities, hospitals and hospital units with regard to consumption and resistance as well as intervention studies are urgently needed. The only safe recommendations to date seem to be local surveillance of antibiotic resistance and usage, minimizing usage, enforcement of hospital hygiene and spreading the risk of resistance over many antibacterial agents. The activities on a local as well as on an international basis regarding surveillance of antibiotic resistant bacteria will be discussed in the following section.

BACTERIAL SUSCEPTIBILITY AND RESISTANCE TO ANTIBIOTICS

Chemical compounds which are synthesized or, more commonly, derived from natural sources and further modified, and which exert a certain action on the growth of bacteria, are named antibiotics or antibacterial agents. Bacteria which are inhibited by such substances are defined as susceptible, whereas those which are to a lesser extent or not at all affected by an antibacterial agent are named resistant. As mentioned, resistance to antibiotics has become an increasingly common phenomenon encountered all over the world and in all clinically important bacterial species. It has been observed ever since antimicrobial agents began to be used for the treatment of infectious diseases caused by bacteria more than 60 years ago, and has evolved against all different classes of antibiotics that have

been used in human and veterinary medical practice. In contrast, clinical isolates of bacteria from the "pre-antibiotic" era have rarely been shown to carry plasmid-mediated or other resistance genes (Datta and Hughes, 1983).

Defining Resistance

The MIC

Bacterial resistance to an antibiotic is best described and defined in comparison to its contrary, bacterial susceptibility. The basis of both entities is quantitative measurements of the efficiency (activity measured as weight per volume, i.e. concentration) of an antibiotic against a specific organism. Available methodologies for the *in vitro* measurement of antibacterial activity are based on testing increasing concentrations (as series of doubling dilutions, or as continuous gradients in diffusion techniques) of an antibiotic against a bacterial isolate to find out at which concentration a measurable effect on bacterial growth occurs. This growth-inhibiting concentration is known as the minimum inhibitory concentration, MIC, and is generally defined as the concentration giving inhibition of visible growth in liquid medium or radically decreased growth on solid medium (light growth or a barely visible haze at the spot of inoculation normally being disregarded).

The MIC is, however, a relative measure of the lowest amount of drug required to inhibit the growth (cell division) of a bacterium. This is because for most combinations of drug and microbe any change of the conditions under which the test is made (inoculum size, medium, pH, temperature, other incubation conditions etc.) will produce a shift in the obtained MIC value. Standardized conditions are therefore of utmost importance, and there are several national guidelines emphasizing this (Courvalin and Soussy, 1996; NCCLS document M7–A4, 1997; Olsson-Liljequist *et al.*, 1997).

Breakpoints

Breakpoints are nationally or internationally defined values (expressed as concentrations in mg/L, or as inhibition zones in mm for the diffusion techniques) which aim at discriminating resistant bacteria from susceptible ones among clinical isolates. The breakpoints are to be used as tools for translating the results of susceptibility testing into categories of probable therapeutic success or failure. Breakpoints are also necessary for expressing and reporting drug resistance rates (% of isolates). There are two different ways of defining breakpoints, both currently in use in clinical laboratories.

Traditionally, the breakpoints discriminating between susceptibility and resistance were mainly based on pharmacokinetic data of the antibiotics, but also considered microbiological *in vitro* data and aimed only at predicting whether or not therapy would be successful in the individual infected patient (so called pharmacological breakpoints) (BSAC, 1991; Courvalin and Soussy, 1996; NCCLS document M7-A4, 1997). The pharmacokinetics of any chosen drug describes the fate of that drug in the human body (interaction drug – man) and has no relation to the interaction between drug and bacteria. Accordingly, pharmacokinetic data are the same regardless of the infecting microorganism, and therefore the pharmacological breakpoint of a certain drug is equal for all bacterial species. As a consequence, the breakpoint may for some species hit the normal population of isolates (Figure 1) making the results of *in vitro* testing ambiguous (varying between S and I/R for the same strain). Pharmacological breakpoints are thus mainly theoretical and based on the assumption that therapy will be successful if the drug concentration(s) achieved in body fluids exceed the *in vitro* determined MIC for the clinical isolate. Defining these breakpoints is, however, difficult because the drug levels may fluctuate dramatically during the dosage interval for agents with rapid elimination (short serum half-life, e.g. the β-lactams) and differ between body sites, but also because of the embarrassing shortage of clinical studies trying to relate clinical outcome with MIC of the infecting strain for different drug dosages and sites of infection.

In recent years interpretations of susceptibility testing have become more microbiologically oriented, focusing on each antibiotic/bacterium combination. This has led to so-called species-related or microbiologal breakpoints (Ringertz *et al.*, 1997) aiming at separating the natural (often susceptible) population of clinical isolates from those with any degree of acquired resistance to the tested anti-

biotic. Species-related breakpoints are generally lower than pharmacological breakpoints, providing an early warning of the emerging appearance of less susceptible bacterial strains. This striving to identify bacterial strains that deviate from the supposedly virgin (S = susceptible) bacteria by naming them I (Intermediate or Indeterminant) or R (Resistant, despite moderate MICs) already at the clinical laboratory level and disregarding pharmacokinetics is thought to have several advantages. The results of *in vitro* testing focus antibiotic therapy on the fully susceptible strains, increasing therapeutic safety and decreasing the risk of fueling further development of resistance that may occur when moderately susceptible strains are treated with a specific antibiotic. Furthermore, in contrast to pharmacokinetic breakpoints that currently differ between countries, the microbiological ones are less disputed. Thus, they may lead to worldwide definitions of S, I and R and enable a better system to compare resistance rates between countries.

A drawback with species specific microbiological breakpoints is that in some cases, the number of alternative agents recommended for therapy will be more limited than when using pharmacological ones. For example, unless the bacteria are located behind the blood brain barrier (in meningitis) infections due to low-level penicillin (often multiply) resistant pneumococci are considered treatable with β-lactams. Using microbiological breakpoints, penicillin will not be recommended by the laboratory in such cases and the choice of therapy may be difficult unless the treating physician is familiar with this discrepancy between *in vitro* and *in vivo*.

Measuring Resistance

When any of the methodologies for MIC determination of a drug (broth dilution or agar dilution) is applied to a large enough set of strains of a defined bacterial species, the result will almost invariably be a normal (Gaussian) distribution of MICs over a range of 3–5 dilution steps (Figure 1). Some bacterial species possess a natural (intrinsic) resistance to certain antibiotics by virtue of general factors such as inaccessibility of the drug to its target or natural resistance of the target. This type of resistance will not be further discussed in this chapter. In most species, however, the MIC

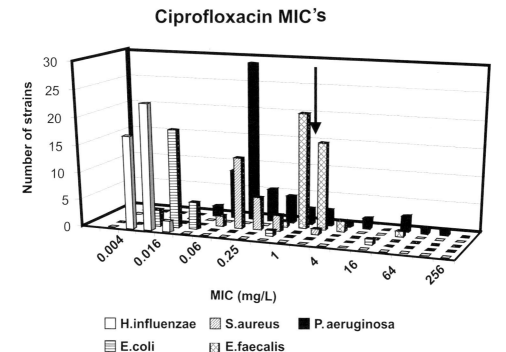

Figure 1. MICs of ciprofloxcin for a reference collection of strains of *Haemophilus influenzae*, *Escherichia coli*, *Staphylococcus aureus*, *Enterococcus faecalis* and *Pseudomonas aeruginosa*, illustrating that mic populations of different species are distinct but overlapping. The arrow indicates the pharmacological breakpoint for susceptibility ($S \leq 1$ mg/L), cutting through the population of *E. faecalis*, while there is substantial space for the evolution of unrecognized resistance in *H. influenzae* and *E. coli*.

distribution is in the low range of the concentration scale representing the normal (virgin) susceptible population of the tested bacterial species. Strains that require higher concentrations of the antibiotic for inhibition of growth are subsequently considered to be microbiologically resistant, having acquired a specific mechanism of resistance, and are often clearly separated from the susceptible ones on the concentration scale. There are however exceptions to this rule, one of the best examples being the activity of penicillin and other β-lactams against *S. pneumoniae*. In this species strains with reduced susceptibility or resistance to penicillin are not clearly separated from susceptible ones, as they represent many different subpopulations of strains exhibiting increasing levels of resistance. The MIC distribution reflects the epidemiology of pneumococci at any given time and place (Figure 2).

Bacteria may acquire resistance to antibiotics by genetic changes of the ancestral gene in the

susceptible strain by mutation or transformation altering the target and decreasing its ability to interact with the drug. Alternatively, resistant bacteria produce novel proteins which are able to counteract the effect of the antibacterial agents. Such proteins are often expressed by extra bacterial genes acquired through DNA transfer from a strain of the same or a different bacterial species. Both these processes are described as emergence of bacterial resistance. Phenotypic methods for measuring and quantifying resistance do not give any information about the actual mechanism of resistance in a bacterium, nor of its origin. Many resistance mechanisms are however well characterized, and for those mechanisms genotypic methods have also been developed. Phenotypic methods are nevertheless currently essential for the rapid detection of antibiotic resistance in clinical isolates of bacteria, and also for detection of bacteria with as yet unknown mechanisms of resistance. Geno-

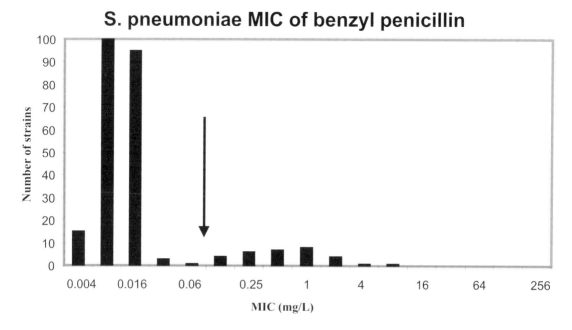

Figure 2.   Distribution of MICs of penicillin for strains of *Streptococcus pneumoniae*.

typic methods are routinely used only when phenotypic methods are slow, as for mycobacteria, or not fully reliable discriminators, such as in the case of detecting methicillin resistance. This is best done by demonstrating the presence of the *mecA* gene in staphylococci (Olsson-Liljequist *et al.*, 1993). It is possible that modern microchip technology will allow rapid genotypic detection of resistance mechanisms in clinical isolates in the future. Given the immense variety of mechanisms, such a task currently seems insurmountable and very expensive. Furthermore, as gene expression is variable and resistance mechanisms may operate together, the end result, measured phenotypically as MIC or inhibition zone, will remain a cornerstone when measuring antibiotic resistance in clinical laboratories in the near future.

Monitoring Resistance Rates

There would be no need for bacteria to accumulate mutations or acquire extrachromosomal DNA specifying resistance mechanisms if it were not for the use of antibiotics. The world-wide use, and often misuse, of antibiotics represents a constant selective pressure on bacteria, allowing only the fittest, those with acquired resistance mechanisms, to survive. Therefore, surveillance of antibiotic resistance in human pathogens, which has long been performed by hospital and reference laboratories, and in national and sometimes international cooperations for various bacterial species and classes of antibiotics, is becoming even more important to perform. Such surveys should be done regularly and preferably in international networks using standardized test methods and definitions of resistance. Point-prevalence studies are the hitherto most common sources of information on rates of resistance to antibiotics, and if they are performed repeatedly using the same hospitals and the same inclusion criteria for specimens and patients, they can mirror trends in rates of resistance, usually expressed as a percentage of the total number of isolates tested. Longitudinal studies are, however, preferable in terms of following trends of resistance. They are readily performed at the local level, provided a hospital or a microbiology laboratory has a computerized system for registration of quantitative data on clinical isolates, which can then be easily accessible for analysis. Efforts are now being made by the WHO, both to gather available information on antibiotic resistance from reliable sources to form a global drug resistance map, and also to provide a user-friendly software program for handling results of routine susceptibility testing in clinical laboratories (O'Brien and Stelling, 1995).

National and international networks

National networks are now being set up, the oldest being CDC's NNIS (National Nosocomial Infection Surveillance), which follows trends in the infection and microbiological panorama (including drug resistance) in about 200 US hospitals (Emori *et al.*, 1991). In recent years many new networks have been formed, funded both by public health institutions and by the pharmaceutical industry, to collect data and analyse trends in the occurrence of antibiotic resistance. One example of the first is the EARSS project funded by the European Union and involving all EU countries and also Iceland and Norway (Bronzwaer *et al.*, 1999). This project aims at creating a European network of national surveillance systems to aggregate comparable and reliable data for public health purposes. Initially it has focused on two pathogens, *S. aureus* and *S. pneumoniae* isolated from blood to be a pilot study representing one hospital acquired and one community acquired infection, respectively. An example of a project funded by the pharmaceutical industry is the Alexander project concentrating on community-acquired bacterial lower respiratory tract pathogens (Finch *et al.*, 1996). The purpose of such projects is not merely to gather data on resistance but to better understand the epidemiology of drug resistant bacteria and of antibiotic resistance genes among bacteria in relation to antibiotic consumption, with the ultimate goal to reduce antibiotic resistance or at least prevent a further increase. To this end, several additional aspects must be investigated, such as the clonality or diversity of bacterial strains and of their resistance genes. Detailed clinical data on patients from different clinical settings are also needed, as well as more precise data on the local and nation-wide consumption of antibiotics by different patient populations in hospitals and in the community (see above).

Examples of Clinically Important Current Antibiotic Resistance Problems Acquired by Common Bacterial Pathogens

There are many publications on resistance rates and describing specific mechanisms of antibiotic resistance, many of which have created clinical problems when their prevalence have reached 10% or more. In Table 4 are shown some examples of the current problems around the world. The interested reader is referred to the cited literature and numerous other papers published.

SUMMARY

The origin and the precise mode of acquisition and transfer of drug resistance remains unidentified in some cases and the mechanisms by which bacteria minimize the metabolic cost and thus, manage to retain and stabilize the mutations, genes and plasmids conferring resistance are also poorly understood. Therefore, research efforts need to be directed towards resolving the scientific basis of these problems. This applies also to the role of person to person transmission and the nature and the impact of the epidemiological potential of resistant bacterial strains and host herd immunity for the emergence of antimicrobial resistance in hospitals and in the community. Also, the consequences of antibiotic resistance in terms of morbidity, mortality and cost needs to be much better defined. Finally, improved systems for surveillance of antibiotic resistance and antibiotic usage and many more large scale comparative and intervention studies relating antibiotic usage and resistance are needed until the use of each antimicrobial agent in each clinical setting can be optimized from a resistance point of view.

REFERENCES

Aarestrup, F.M., Ahrens, P., Madsen, M., Pallesen, L.V., Poulsen, R.L. and Westh, H. (1996) Glycopeptide susceptibility among Danish *Enterococcus faecium* and *Enterococcus faecalis* isolates of animal and human origin and PCR identification of genes within the vanA cluster. *Antimicrob. Agents Chemother.* **40**, 1938–1940.
Arason, V.A., Kristinsson, K.G., Sigurdsson, J.A., Stefansdottir, G., Mölstad, S. and Gudmundsson, S. (1996) Do antimicro-

Table 4. Examples of current antibiotic resistance rates among major pathogens.

Bacterial pathogen	Antibiotic resistance	Prevalence
Escherichia coli	Betalactam resistance due to plasmid-mediated betalactamases of narrow (TEM-1) or extended-spectrum activity (ESBL of TEM- or SHV-origin) with or without resistance to betalactamase inhibitors (IRT)	TEM-1 present in 20–90% of E. coli (Nguyen Thi et al., 1991; Lipermore, 1995) ESBL of TEM-origin or SHV-origin < 10% (Livermore, 1995; Aubry-Damon and Courvalin, 1999; Pai et al., 1999)
	Trimethoprim resistance due to DHFR enzymes	Type I DHFR 10–20% (Nguyen Thi et al., 1991; Heikkolä et al., 1990) 8% in Spain, 2–3% in Sweden (Pena et al., 1995; Österlund and Olsson-Liljequist, 1999)
	Quinolone resistance due to mutations in *gyrA*	
Klebsiella spp	Betalactam resistance due to plasmid-mediated betalactamases with extended-spectrum activity (ESBL of SHV-, TEM- or other origin)	15% in France (Aubry-Damon and Courvalin, 1999)
	Quinolone resistance	1–20% in Europe (Hanberger et al., 1999)
	Aminoglycoside resistance	1–30% in Europé (Hanberger et al., 1999)
Enterobacter and other enterobacteria	Constitutive high-level production of potent chromosomal cephalosporinases	25–50% in Europé (Hanberger et al., 1999)
	Quinolone resistance	5–30% (Hanberger et al., 1999)
Salmonella, Shigella	Multidrug resistance, plasmid- and chromosomally encoded	>50% (Nguyen Thi et al., 1991)
Campylobacter	Quinolone resistance	20–50% (Piddock, 1999)
Vibrio cholerae	Trimethoprim/sulfa and streptomycin resistance (O139)	100% (Waldor et al., 1996)
Helicobacter pylori	Macrolide and metronidazole resistance	3% and 24% (Megraud, 1999)
Pseudomonas aeruginosa, Stenotrophomonas maltophilia	Carbapenem resistance; aminoglycoside resistance, quinolone resistance	Carbapenems 15–20%, Gentamicin 5–50%, Amikacin 5–15%, Ciprofloxacin 10–40% (Hanberger et al., 1999)
Staphylococcus aureus	Methicillin resistance (MRSA) often combined with multidrug resistance (aminoglycosides, macrolides, tetracyclines)	20–40% (Ronveaux et al., 1998; Pfaller et al., 1999; Spencer, 1999)
Enterococci	Glycopeptide resistance; aminoglycoside resistance; quinolone resistance; penicillin/carbapenem resistance (mainly in *E. faecium*)	Glycopeptides 1–50% (Aarestrup et al., 1996; Torell et al., 1999)
Streptococcus pneumoniae	Penicillin resistance combined with multidrug resistance (tetracycline, macrolide, chloramphenicol and/or trimethoprim/sulfa resistance)	3–30% (Baquero et al., 1991)
Streptococcus pyogenes	Macrolide resistance: tetracycline resistance	10–20% in Finland (Seppälä et al., 1995)
Haemophilus influenzae	Betalactam resistance (mediated by betalactamase or by target modification); chloramphenicol resistance	10–20% (Finch et al., 1996)
Mycobacteria	Multidrug resistance (rifampin, isoniazid and others)	Crofton et al. (2000)

bials increase the carriage rate of penicillin resistant pneumococci in children? Cross sectional prevalence study. *Br. Med. J.* **313**, 387–391.

Aubry-Damon, H. and Courvalin, P. (1999) Bacterial resistance to antimicrobial agents: Selected problems in France, 1996 to 1998. *Emerging Infect. Dis.* **5**, 315–320.

Austin, D.J., Kristinsson, K.G. and Anderson, R.M. (1999) The relationship between the volume of antimicrobial consumption in human communities and the frequency of resistance. *Proc. Natl. Acad. Sci. USA* **96**, 1152–1156.

Baquero, F., Martinez-Beltran, J. and Loza, E. (1991) A review of antibiotic resistance patterns of *Streptococcus pneumoniae* in Europe. *J. Antimicrob. Chemother.* **28 (Suppl C)**, 31–38.

Bronzwaer, S.L., Goettsch, W., Olsson-Liljequist, B., Wale, M.C.J., Vatopoulos, A.C. and Sprenger, M.J.W. (1999) European antimicrobial resistance surveillance system (EARSS): objectives and organisation. *Eurosurveillance* **4**, 41–44.

BSAC, British Society for Antimicrobial Chemotherapy (1991) A guide to sensitivity testing. Report of the working party on antibiotic sensitivity testing of the BSAC. *J. Antimicrob. Chemother.* **27, Suppl D**.

Chow, J.W., Fine, M.J., Shlaes, D.M., Quinn, J.P., Hooper, D.C., Johnson, M.P., *et al.* (1991) *Enterobacter* bacteremia: clinical features and emergence of antibiotic resistance during therapy. *Ann. Intern. Med.* **115**, 585–590.

Col, N.F. and O'Connor, R.W. (1987) Estimating worldwide current antibiotic usage: Report of Task Force 1. *Rev. Infect. Dis.* **9 (Suppl 3)**, S232–S243.

Courvalin, P. and Soussy J-J (eds.) (1996) Report of the Comité de lÁntibiogramme de la Société Francaise de Microbiologie. *Clin. Microbiol. Infect.* **2 (Suppl 1)**.

Crofton, J., Chaulet, P. and Maher, D. (2000) Guidelines for the management of drug-resistant tuberculosis. WHO Global Tuberculosis Programme, WHO, Geneva.

D'Amico, R., Pifferi, S., Leonetti, C., Torri, V., Tinazzi, A. and Liberati, A. (1998) Effectiveness of antibiotic prophylaxis in critically ill adult patients: Systematic review of randomized controlled trials. *Br. Med. J.* **316**, 1275–1285.

Datta, N. and Hughes, V.M. (1983) Plasmids of the same Inc groups in Enterobacteria before and after the medical use of antibiotics. *Nature* **306**, 616–617.

Davey, P.G., Bax, R.P., Newey, J., Reeves, D., Lutherford, D., Slack, R., *et al.* (1996) Growth in the use of antibiotics in the community in England and Scotland in 1980–93. *Br. Med. J.* **312**, 613.

De Man, P., Verhoeven, B., Verbrugh, H., Vos, M.C. and van den Anker, J.N. (2000) An antibiotic policy preventing the emergence of resistant bacilli: A prospective crossover study in neonatal intensive care. *Lancet* **355** (9208), 973–978.

Edlund, C. and Nord, C.E. (1993) Ecological impact of antimicrobial agents on human intestinal microflora. *Alpe Adria Microbiol. J.* **3**, 137–164.

Edlund, C. and Nord, C.E. (1999) Effect of quinolones on intestinal ecology. *Drugs* **58** Suppl. 2, 65–70.

Emori, T.G., Culver, D.H., Horan, T.C., *et al.* (1991) The National Nosocomial Infections Surveillance (NNIS) System: description of surveillance methods. *Am. J. Infect. Control* **19**, 19–35.

European Commission. Opinion of the Scientific Steering Committee on Antimicrobial Resistance (28 May, 1999).

FEDESA (1997) Animal Health Dossier 15: European Federation of Animal Health.

Finch, R.G., Wilcox, M.H. and Wood, M.J. (eds) (1996) An international collaborative surveillance study of community-acquired bacterial LRTI pathogens: The Alexander project. *J. Antimicrob. Chemother.* **38, Suppl A**.

Flaherty, J.P. and Weinstein, R.A. (1996) Nosocomial infection caused by antibiotic-resistant organisms in the intensive-care unit. *Infect. Control Hosp. Epidemiol.* **17**, 236–248.

Ford, C.W., Hamel, J.C., Stapert, D., Moerman, J.K., Hutchinson, D.K., Barbachyn, M.R., *et al.* (1997) Oxazolidinones: new antibacterial agents. *Trends Microbiol.* **5**, 196–200.

Fryklund, B., Tullus, K. and Burman, L.G. (1993) Relation between nursing procedures, other local characteristics and transmission of enteric bacteria in neonatal wards. *J. Hosp. Infect.* **23**, 199–210.

Fujita, K., Murono, K., Yoshikawa, M. and Murai, T. (1994) Decline of erythromycin resistance of group A streptococci in Japan. *Pediatr. Infect. Dis.* **13**, 1075–1078.

Gaynes, R. and Monnet, D. (1997) The contribution of antibiotic use on the frequency of antibiotic resistance in hospitals. *CIBA Found. Symp.* **207**, 47–60.

Gilliver, M.A., Bennett, M., Begon, M., Hazel, S.M. and Hart, C.A. (1999) Antibiotic resistance found in wild rodents. *Nature* **40**, 233–234.

Goldmann, D.A., Weinstein, R.A., Wenzel, R.P., Tablan, O.C., Duma, R.J., Gaynes, R.P., *et al.* (1996) Strategies to prevent and control the emergence and spread of antimicrobial-resistant microorganisms in hospitals. A challenge to hospital leadership. *JAMA* **275**, 234–240.

Gonzales, R., Steiner, J.F. and Sande, M.A. (1997) Antibiotic prescribing for adults with colds, upper respiratory tract infections, and bronchitis by ambulatory care physicians. *J. Am. Med. Assoc.* **278**, 901–904.

Guillermot, D., Carbon, C., Vauzelle-Kervroedran, F., Balkau, B., Maison, P., Bouvenot, G., *et al.* (1998a) Inappropriateness and variability of antibiotic prescribing among French office-based physicians. *J. Clin. Epidemiol.* **51**, 61–68.

Guillermot, D., Maison, P., Carbon, C., Balkan, B., Vauzelle-Kervroedan, F., Sermat, C., *et al.* (1998b) Trends in antimicrobial drug use in the community between 1981 and 1992. *J. Infect. Dis.* **177**, 492–497.

Guillermot, D., Carbon, C., Balkan, B., Geslin, P., Lecoeur, H., Vauzelle-Kervroedan, F., *et al.* (1998c) Low dosage and long treatment duration of β-lactams: risk factors for carriage of penicillin-resistant *Streptococcus pneumoniae*. *JAMA* **279**, 365–370.

Hanberger, H., Garcia-Rodrigues, J.A., Gobernado, M., Goossens, H., Nilsson, L.E. and Struelens, M.J. (1999) Antibiotic susceptibility among aerobic gram-negative bacilli in intensive care units in 5 European countries. French and Portugese ICU Study Groups. *JAMA* **281**, 67–71.

Heikkilä, E., Renkonen, O-V., Sunila, R., Uurasmaa, P. and Huovinen, P. (1990) The emergence and mechanisms of trimethoprim resistance in Escherichia coli isolated from outpatients in Finland. *J. Antimicrob. Chemother.* **25**, 275–283.

Holmberg, S.D., Solomon, S.L. and Blake, P.A. (1987) Health and economic impacts of antimicrobial resistance. *Rev. Infect. Dis.* **9**, 1065–1078.

Karlström, O., Fryklund, B., Tullus, K., Burman, L.G. and the Swedish *C. difficile* Study Group (1998) A prospective nationwide study of *Clostridium difficile*-associated diarrhea in Sweden. *Clin. Infect. Dis.* **26**, 141–145.

Kiivet, R-A., Dahl, M-L., Llerena, A., Maimets, M., Wettermark, B. and Berecz, R. (1998) Antibiotic use in 3 European university hospitals. *Scand. J. Infect. Dis.* **30**, 277–280.

Kling, P., Östensson, R., Granström, S. and Burman, L.G. (1989) A 7-year survey of drug resistance in aerobic and anaerobic faecal bacteria of surgical inpatients: clinical relevance and relation to local antibiotic consumption. *Scand. J. Infect. Dis.* **21**, 598–596.

Kristinsson, K.G. (1999) Modification of prescribers behavior: the Icelandic approach. *Clin. Microbiol. Infect.* **5**, 4S43–4S47.

Kunin, C.M. (1993) Resistance to antimicrobial drugs – A worldwide calamity. *Ann. Int. Med.* **118**, 557–561.

Lipsitch, M., Bergstrom, C.T. and Levin, B.R. (2000) The epidemiology of antibiotic resistance in hospitals: Paradoxes and prescriptions. *Proc. Natl. Acad. Sci. USA* **97**, 1938–1943.

Liss, R.H. and Batchelor, F.R. (1987) Economic evaluation of antibiotic use and resistance – A perspective: Report of Task Force 6. *Rev. Infect. Dis.* **9 (Suppl 3)**, S297–S312.

Listorti, J.A. (1999) Is environment health really a part of economic development – or only an afterthought. *Environment and urbanization* **11**, 89–100.

Livermore, D.M. (1995) β-lactamases in laboratory and clinical resistance. *Clin. Microbiol. Rev.* **8**, 557–584.

Mayhall, C.G. (ed.) (1996) Hospital epidemiology and infection control. Williams & Wilkins.

McDermott, W. (1982) Social ramifications of control of microbial disease. *Johns Hopkins Med. J.* **151**, 302–312.

McGowan, J.E. (1994) Do intensive hospital control programs prevent the spread of antibiotic resistance? *Infect. Control Hosp. Epidemiol.* **15**, 478–483.

McGowan, J.E. Jr. (1983) Antimicrobial resistance in hospital organisms and its relation to antibiotic use. *Rev. Infect. Dis.* **5**, 1033–1048.

McGowan, J.E. Jr. and Tenover, F.C. (1997) Control of antimicrobial resistance in the health care system. *Inf. Dis. Clin. North Am.* **11**, 297–311.

Megraud, F. (1999) Resistance of *Helicobacter pylori* to antibiotics: the main limitation of curent proton-pump inhibitor triple therapy. *Eur. J. Gastroenterol. Hepatol.* **11 Suppl 2**, S35–S37.

Melander, E., Ekdahl, K., Hansson, H.B., Kamme, C., Laurell, M., Nilsson, P., et al. (1998) Introduction and clonal spread of penicillin- and trimethoprim/sulfamethoxazole-resistant *Streptococcus pneumoniae*, serotype 9V, in southern Sweden. *Microb. Drug Resist.* **4**, 71–78.

Moellering, R.C. Jr. (1990) Interaction between antimicrobial consumption and selection of resistant bacterial strains. *Scand. J. Infect. Dis. Suppl.* **70**, 18–24.

Mouton, R.P., Glerum, J.H. and van Loenen, A.C. (1976) Relationship between antibiotic consumption and frequency of antibiotic resistance of four pathogens – a seven-year study. *J. Antimicrob. Chemother.* **2**, 9–19.

Murray, B.E. and Moellering, R.C. Jr. (1978) Patterns and mechanisms of antibiotic resistance. *Med. Clin. North Am.* **62**, 899–923.

Murray, C.J. and Lopez, A.D. (eds) (1994) Global comparative assessments in the health sector. WHO, Geneva.

Musser, J. and Kapur, V. (1992) Clonal analysis of methicillin-resistant *Staphylococcus aureus* strains from intercontinental sources: Association of the *mec* gene with divergent phylogenetic lineages implies dissemination by horizontal transfer and recombination. *J. Clin. Microbiol.* **30**, 2058–2063.

Mölstad, S. and Cars, O. (1999) Major change in the use of antibiotics following a national programme: Swedish Strategic Programme for the Rational Use of Antimicrobial Agents and Surveillance of Resistance (STRAMA). *Scand. J. Infect. Dis.* **31**, 191–195.

NCCLS (1997) Methods for dilution antimicrobial susceptibility tests for bacteria that grow aerobically – Fourth edition; Approved standard. NCCLS document M7-A4. NCCLS, 940 West Valley Road, Suite 1400, Wayne, Pennsylvania 19087.

Nguyen, T., Thanh, H., Hoang, T.L., Nguyen, T.H., Nguyen, T.T., Lundström, G., Olsson-Liljequist, B. and Kallings, I. (1991) Antibiotic sensitivity of enteric pathogens in Vietnam. *Int. J. Antimicrob. Agents* **1**, 121–126.

O'Brien, T.F. and Stelling, J. (1995) WHONET: An information system for monitoring antimicrobial resistance. *Emerging Infect. Dis.* **1**, 66.

Olsson-Liljequist, B., Larsson, P., Ringertz, S. and Löfdahl, S. (1993) Screening methods for methicillin resistance in staphylococci and verification by an oligonucleotide probe. *Eur. J. Clin. Microbiol. Infect. Dis.* **12**, 527–533.

Olsson-Liljequist, B., Larsson, P., Walder, M. and Miörner, H. (1997) Antimicrobial susceptibility testing in Sweden. III. Methodology for susceptibility testing. *Scand. J. Infect. Dis. Suppl.* **105**, 13–23.

Pai, H., Lyu, S., Lee, J.H., Kim, J., Kwon, Y., Kim, J.W., et al. (1999) Survey of extended-spectrum β-lactamases in clinical isolates of *Escherichia coli* and *Klebsiella pneumoniae*: prevalence of TEM-52 in Korea. *J. Clin. Microbiol.* **37**, 1758–1763.

Pena, C., Albareda, J.M., Pallares, R., Pujol, M., Tubau, F. and Ariza, J. (1995) Relationship between quinolone use and emergence of ciprofloxacin-resistant *Escherichia coli* in bloodstream infections. *Antimicrob. Agents Chemother.* **39**, 520–524.

Pennington, J.E. (ed.) (1994) Respiratory infections – diagnosis and treatment. Raven Press.

Pfaller, M.A., Jones, R.N., Doern, G.V., Sader, H.S., Kugler, K.C. and Beach, M.I. (1999) Survey of blood stream infections attributable to gram-positive cocci: frequency of occurrence and antimicrobial susceptibility of isolates collected in 1997 in the United States, Canada, and Latin America from the SENTRY Antimicrobial Surveillance Program. SENTRY Participants Group. *Diagn. Microbiol. Infect. Dis.* **33**, 283–297.

Phillips, I. (1983) Environmental factors contributing to antibiotic resistance. *Infect. Control* **4**, 448–451.

Piddock, L. (1999) Quinolone resistance and *Campylobacter*. *Clin. Microbiol. Infect.* **5**, 239–243.

Prescribing Pricing Authority (1997) Annual Report. 1st April 1996 – 31st March 1997. Newcastle upon Tyne.

Ringertz, S., Olsson-Liljequist, B., Kahlmeter, G. and Kronvall, G. (1997) Antimicrobial susceptibility testing in Sweden. II. Species-related zone diameter breakpoints to avoid interpretive errors and guard against unrecognized evolution of resistance. *Scand. J. Infect. Dis. Suppl.* **105**, 8–12.

Ronveaux, O., Jans, B., Suetens, C. and Carsauw, H. (1998) Epidemiology of nosocomial bloodstream infections in Belgium, 1992–1996. *Eur. J. Clin. Microbiol. Infect. Dis.* **17**, 695–700.

Seppälä, H., Klaukka, T., Lehtonen, R., Nenonen, E., Finnish Study Group for Antimicrobial Resistance, and Huovinen, P. (1995) Outpatient erythromycin use – link to increased erythromycin resistance in group A streptococci. *Clin. Infect. Dis.* **21**, 1378–1385.

Shlaes, D.M., Gerding, D.N., John, J.F. Jr., Craig, W.A., Bornstein, D.L., Duncan, R.A., et al. (1997) Society for Healthcare Epidemiology of America Joint Committee on the Prevention of Antimicrobial Resistance: guidelines for the prevention of antimicrobial resistance in hospitals. *Clin. Infect. Dis.* **25**, 584–599.

Spencer, R.C. (1999) Bacteremia caused by multi-resistant Gram-positive micrroorganisms. *Clin. Microbiol. Infect.* **5 (Suppl 2)**, 2S17–2S28.

Struelens, M.J., Baudouin, B. and Vincent, J-L. (1999) Antibiotic policy: a tool for controlling resistance of hospital pathogens. *Clin. Microbiol. Infect.* **5**, S19–S24.

Swartz, M.N. (1997) Use of antimicrobial agents and drug resistance. *New Engl. J. Med.* **337**, 491–492.

Tenover, F.C. and McGowan, J.E. (1996) Reasons for the emergence of antibiotic resistance. *Ann. J. Med. Sci.* **311**, 9–16.

The Copenhagen Recommendations. Report from the Microbial Threat EU Conference (1998). Copenhagen: Ministry of Health, Ministry of Food, Agriculture and Fisheries of Denmark.

Torell, E., Olsson-Liljequist, B., Hoffman, B-M., Cars, O., Lindbäck, J, and Burman, L.G. (1999) Near absence of vancomycin-resistant enterococci but high carriage rates of quinolone-resistant ampicillin-resistant enterococci among hospitalized patients and nonhospitalized individuals in Sweden. *J. Clin. Microbiol.* **37**, 3509–3513.

Waldor, M.K., Tschape, H. and Mekalanos, J.J. (1996) A new type of conjugative transposon encodes resistance to sulfa-methoxazole, trimethoprim, and streptomycin in *Vibrio cholerae O139. J. Bacteriol.* **178**, 4157–4165.

Wegener, H.C., Aarestrup, F.M., Jensen, L.B., Hammerum, A.M. and Bager, F. (1999) Use of antimicrobial growth promoters in food animals and *Enterococcus faecium* resistance to therapeutic antimicrobial drugs in Europe. *Emerging Infect. Dis.* **5**, 329–335.

Wenzel, R.P. (ed.) (1997) Prevention and control of nosocomial infections. Williams & Wilkins.

WHO, World Health Organization, Emerging and other Communicable Diseases, Surveillance and Control (1998). Use of quinolones in food animals and potential impact on human health. Report of a WHO meeting. Geneva, Switzerland.

Wiedemann, B. (1986). Selection of β-lactamase producers during cephalosporin and penicillin therapy. *Scand. J. Infect. Dis. Suppl.* **49**, 100–105.

Wu, S., Piscitelli, C., de Lencastre, H. and Tomasz, A. (1996) Tracking the evolutionary origin of the methicillin resistance gene: Cloning and sequencing of a homologue of *mecA* from a methicillin susceptible strain of *Staphylococcus sciuri. Microbial. Drug Res.* **2**, 435–441.

Österlund, A. and Olsson-Liljeuqist, B. (1999) Fluoroquinolone-resistance in human pathogenic bacteria; resistant bacteria now appearing in Sweden. *Läkartidningen* **96**, 1965–1966 (in Swedish).

2. Target Alterations Mediating Antibiotic Resistance

Diarmaid Hughes[1] and Dan I. Andersson[2]
[1]Department of Cell and Molecular Biology, The Biomedical Center, Box 596, Uppsala University, S-751 24 Uppsala, Sweden
[2]Swedish Institute for Infectious Disease Control, S-171 82 Solna, Sweden

INTRODUCTION

For several clinically important combinations of pathogen and antibiotic, the primary cause of antibiotic resistance is mutation or gene acquisition resulting in target alteration. These altered antibiotic targets are often the products of normal housekeeping genes on the chromosome, but they can also arise from horizontal genetic transfer and recombination, resulting in the creation of novel genetic mosaics. Our choice of examples covers the main antibiotic targets and is focussed on clinically important pathogens such as *Streptococcus pneumoniae*, *Mycobacterium tuberculosis*, *Staphylococcus aureus*, and *Escherichia coli*. In each case we relate the antibiotic target to clinically relevant antibiotics and pathogens. The extent of the resistance problem and the genetic basis of the alterations to the antibiotic target are discussed with reference to information from both *in vitro* selections and clinical isolates.

CLINICALLY IMPORTANT PATHOGENS

Mycobacterium tuberculosis

The causative agent of tuberculosis, *Mycobacterium tuberculosis*, currently infects about one-third of the world's population and was estimated to be responsible for killing almost 3 million people in 1990 (Bloom and Murray, 1992). Tuberculosis has been treated successfully with a succession of antibiotics, streptomycin from the 1940's, isoniazid from the 1950's, rifampicin from the 1970's, and fluoroquinolones from the 1980's (reviewed in Blanchard, 1996). The resurgence of tuberculosis infections in the USA and Europe since the mid-1980's is associated with HIV infections, and accompanied by the appearance of multidrug-resistant *M. tuberculosis*. Resistance to antituberculosis drugs is a global problem (Pablos-Méndez *et al.*, 1998). In order to develop improved strategies to combat multidrug-resistant tuberculosis great efforts have recently been invested in understanding the genetic basis of drug resistance in *M. tuberculosis* (reviewed in Blanchard, 1996).

Streptococcus pneumoniae

Pneumonia causes about three million deaths a year in young children, nearly all of which are in developing countries. *Streptococcus pneumoniae* is the most important bacterial cause of pneumonia in young children and so is likely to be responsible for a high proportion of these deaths. The pneumococcus is also responsible for a substantial proportion of the 100,000–500,000 deaths that occur from meningitis in children each year (Greenwood, 1999) and is a common cause of middle ear infection (otitis media). Infection with human immunodeficiency virus (HIV) also predisposes to invasive pneumococcal disease. The recommended therapy for nonmeningeal pneumococcal infections (e.g., pneumonia, sepsis, acute otitis media) includes a beta-lactam antibiotic. The recommended therapy for meningitis is a cephalosporin (cefotaxime or ceftriaxone), with the addition of vancomycin until susceptibility is known. The incidence of penicillin resistance in strains of *S. pneumoniae* approaches 40% in some areas of the United States, and the incidence of high-level resistance has increased by 60-fold during the past 10 years (Jacobs, 1999).

Staphylococcus aureus

The most common habitat of *S. aureus* is the upper respiratory tract, especially the nose and throat, as well as the surface of the skin. *S. aureus* is associated with pathological conditions such as boils, pimples, impetigo, pneumonia, osteomyelitis, carditis, meningitis, and arthritis. *S. aureus* is a common cause of nosocomial infections of the respiratory tract, of surgical sites, of burn wounds, and of the blood stream. Serious staphylococcal infections usually occur only when the resistance of the host is low, typically because of debilitating illness or because of wounds to the skin surface. Appropriate antibiotic therapy for *S. aureus* infections is a problem. Although some community-acquired infections are treatable with penicillin, nosocomial (hospital-acquired) infections are often drug-resistant.

Escherichia coli

E. coli is important because it is the most thoroughly explored experimental organism, but also because it is the causative agent of some common human infections. Strains of *E. coli* cause diarrhoea and gastroenteritis. In addition, *E. coli* is frequently the cause of urinary tract infections, the most common type of nosocomial infection, typically related to urinary catheterization. Treatment for *E. coli* infections is with fluoroquinolones, sulfonamides, cephalosporins, or ampicillin.

TARGETS AND MECHANISMS OF RESISTANCE

The examples of antibiotic resistance considered in this chapter are the result of genetic alterations to the genes encoding antibiotic target molecules. In the simplest cases clinical resistance is caused by single amino acid substitutions in a target molecule, while in other cases resistance requires the successive mutational alteration of several target molecules. More complex cases involve the horizontal transfer of genetic information encoding resistant targets into the host organism. In these cases the resistant target may recombine into the host chromosome at a novel site, or it may recombine with a related host target gene resulting in a genetic mosaic. In some of these cases mutation is the only known mechanism of resistance in clinical isolates, while in other cases several different mechanisms are known to contribute to clinical resistance. Here we consider each of the main antibiotic targets in relation to clinically relevant and problematic pathogens and antibiotics. Information on causes of resistance obtained from *in vitro* selections is compared with the available information from clinical isolates.

DNA Topoisomerases

The bacterial DNA gyrase, an ATP-dependent type II DNA topoisomerase that catalyzes negative supercoiling of DNA, is a target of a clinically important class of antibiotic, the quilolones (reviewed in Maxwell, 1997). The gyrase enzyme is a heterotetramer composed of two GyrA and two GyrB subunits, encoded by the *gyrA* and *gyrB* genes respectively. The active site of DNA gyrase includes residue Y122 which is bound to DNA in the gyrase-DNA complex (Horowitz and Wang, 1987). Quinolones act by binding to the complex of DNA and DNA gyrase, thus interfering with processes dependent on DNA topology such as DNA replication and transcription. The quinolone binding affinities of the complex are probably determined by both GyrA and GyrB subunits acting in concert (Yoshida *et al.*, 1993).

Bacteria also contain a second type II topoisomerase, named topoisomerase IV, which is essential for chromosome partitioning. Unlike gyrase, topoisomerase IV can relax, but cannot supercoil, DNA (Kato *et al.*, 1990). Topoisomerase IV is also composed of two subunits, ParC and ParE (known in many species as GrlA and GrlB) and is also a target for quinolone antibiotics (Ferrero *et al.*, 1994; Yamagishi *et al.*, 1996). GrlA and GrlB share significant sequence similarity with the GyrA and GyrB subunits of gyrase.

The first quinolone drugs to be synthesized, nalidixic acid and oxolonic acid, have been superceded by the fluoroquinolones (e.g. norfloxacin, ciprofloxacin and sparfloxacin) which have a broader antibacterial spectrum in addition to being effective against many strains of bacteria that are multiresistant to beta-lactam antibiotics and aminoglycosides (reviewed in von Rosenstiel and Adams, 1994). Fluoroquinolones are used in the treatment of urinary tract infections (e.g. *E. coli*), in cases of gastroenteritis (*Campylobacter jejuni* and *C. coli*) and in the treatment of patients with multidrug-resistant tuberculosis.

Quinolone resistance mutations

Mutations in GyrA that confer resistance to quinolones have been identified as single amino acid substitutions in the "quinolone resistance-determining region (QRDR) located between amino acid residues 81–87 on the *E. coli* GyrA sequence, with additional lower-level resistance mutations at residues 67 and 106 (Yoshida *et al.*, 1990). Individual substitutions in the QRDR increase the MIC of quinolones by a factor of at least 15 (Yoshida *et al.*, 1990). By comparison with *E. coli* it has been possible to identify the quinolone binding site and map a series of mutations in the same region of GyrA conferring resistance on other clinically important pathogens such as *M. tuberculosis* (reviewed in Blanchard, 1996), *S. aureus* (Sreedharan *et al.*, 1990; Goswitz *et al.*, 1992), and *S. pneumoniae* (Pan *et al.*, 1996).

In *E. coli* stepwise *in vitro* selections for resistance to increasing concentrations of nalidixic acid and ciprofloxacin lead to mutants with multiple changes. Three rounds of selection resulted first in S83L in GyrA, second an altered outer membrane protein, and third D87G in GyrA (Heisig and Tschorny, 1994). Similar stepwise

selections in *M. smegmatis* for ofloxacin resistance resulted in strains with equivalent individual substitutions in GyrA after the first step, and double mutants in GyrA after the second step (Revel *et al.*, 1994). Clinical isolates of ciprofloxacin resistant *S. aureus* (Sreedharan *et al.*, 1990; Goswitz *et al.*, 1992), and *E. coli* (Conrad *et al.*, 1996; Ozeki *et al.*, 1997) also have a high incidence of double mutations in the GyrA QRDR and a good correlation between the number of mutations and the MIC for fluoroquinolones. This shows that an accumulation of substitutions near the presumptive quinolone binding in GyrA site can have cumulative effects on MIC's for fluoroquinolones.

Mutations conferring quinolone resistance also occur in GyrB in *E. coli*. Only two different mutations have been identified, the single amino acid substitutions D426N and K447E (Yoshida *et al.*, 1991). Although GyrB mutations occur *in vitro* as frequently as GyrA mutations they are associated with lower MIC's and are infrequent in clinical isolates of fluoroquinolone-resistant *E. coli* (Nakamura *et al.*, 1989).

Quinolone resistance mutations have also been mapped to the GrlA/ParC subunit of topoisomeraes IV (functionally equivalent to GyrA in DNA gyrase) in several bacterial species. Of eight *S. aureus* clinical isolates resistant to ciprofloxacin all carry an amino acid substitution in ParC (GrlA), while six also carry a substitution in GyrA (Ferrero *et al.*, 1994). The highest MIC's are associated with the GrlA GyrA double mutants. When resistant mutants of *S. aureus* are selected in a stepwise manner on increasing concentrations of ciprofloxacin, amino acid substitutions occur first in GrlA, and in a second or third step in GyrA (Ferrero *et al.*, 1995; Ng *et al.*, 1996). In genetic reconstruction experiments it was shown that the mutant form of GyrA only expresses a fluoroquinolone resistance phenotype in the presence of a mutant GrlA (Ng *et al.*, 1996). This suggests that topoisomerase IV is the primary target of ciprofloxacin in *S. aureus*, and that multiple mutations contribute to high MIC's for ciprofloxacin. The resistance mutations occur at equivalent positions in the related QRDR sequences of GrlA and GyrA (Ferrero *et al.*, 1994; Ferrero *et al.*, 1995). Similar results have been obtained with both laboratory-selected and clinical isolates of ciprofloxacin-resistant *S. pneumoniae* (Pax and Fisher, 1996; Pax *et al.*, 1996; Munoz and De La

Campa, 1996; Pax and Fisher, 1999). In these cases resistance mutants to ciprofloxacin first occur in either transport functions or in GrlA, and subsequently in GyrA or GyrB. Ciprofloxacin resistance in clinical isolates of *Campylobacter jejuni* is also associated with mutations in the QRDR of both GyrA and GrlA (Gibreel *et al.*, 1998). In contrast, when the anti-pneumococcal fluoroquinolone sparfloxacin is used to stepwise select resistant mutants of *S. pneumoniae*, GyrA alterations arise first, and GrlA mutations only appear in second-step mutants (Pan and Fisher, 1997). This suggests that structural differences between ciprofloxacin and sparfloxacin may be responsible for focusing primary target preference on either topoisomerase IV or DNA gyrase. In *E. coli* the substitution L445H in ParE (GrlB) enhances norfloxacin resistance only in a strain already carrying a mutant GyrA, suggesting that in *E. coli* topoisomerase IV is a secondary target for the antibiotic (Breines *et al.*, 1997). In clinical isolates of *Pseudomonas aeruginosa* GyrA, ParC (GrlA) and efflux system mutations are all found in various combinations, with GyrA most common but apparently not a prerequisite for expression of resistance (Jalal and Wretlind, 1998). Thus, in all bacteria tested, both DNA gyrase and topoisomerase IV are targets of fluoroquinolones, but the order of target preference depends both of the particular quinolone and on the bacterial species. The development of new fluoroquinolones with altered preferences for attacking each target may assist in reducing the rate of resistance development (Zhao *et al.*, 1997).

Multidrug-resistance systems also have an important role in fluoroquinolone resistance in clinically relevant organisms (reviewed in Martínez *et al.*, 1998). Thus, overexpression of the efflux transporter protein NorA in *S. aureus*, often as the result of a promoter mutation, confers resistance to fluoroquinolones and other antibiotics and can act individually or in combination with mutations in the target molecules GrlA (ParC) and/or GyrA to increase MIC (Kaatz and Seo, 1997). Efflux systems are also important in resistance to fluoroquinolones and other antibiotics in *S. pneumoniae* (Zellor *et al.*, 1997), *P. aeruginosa* (Kohler *et al.*, 1997) and *C. jejuni* (Charvalos *et al.*, 1995). In *E. coli* mutations affecting OmpF modification or expression increase resistance levels (Heisig and Tschorny, 1994) as does overexpression of the *mar* (multiple antibiotic resistance) locus (Goldman

et al., 1996). In *E. coli* an active AcrAB efflux pump is required so that gyrase mutations can produce clinically relevant levels of fluoroquinolone resistance (Oethinger *et al.*, 2000).

RNA Polymerase

DNA-dependent RNA transcription is inhibited by binding of the antibiotic rifampicin to RNA polymerase, both in *E. coli* (Hartmann *et al.*, 1967) and in the mycobacteria (Levin and Hatfull, 1993). Rifampicin is an important component of multidrug therapies for combating mycobacterial diseases such as tuberculosis and leprosy (reviewed in Blanchard, 1996). Rifampicin resistance in *E. coli* (Jin and Gross, 1988) and in clinical isolates of mycobacteria (Telenti *et al.*, 1993; Kapur *et al.*, 1994; Williams *et al.*, 1994; Ohno *et al.*, 1996) correlates strongly with individual amino acid substitutions, and a few small deletions and insertions, in a limited region of the *rpoB* gene encoding the β subunit of RNA polymerase (residues 509–533 in *E. coli*, and 432–458 in *M. tuberculosis*). In approximately 5% of cases mutations have not been found in this region but where more extensive sequencing has been made additional sites of substitution within *rpoB* have been found (Taniguchi *et al.*, 1996; Nash *et al.*, 1997). The correlation between individual amino acid substitutions and the MIC's of clinical isolates in imperfect, suggesting that additional unidentified mutations may be involved in at least some cases (Ohno *et al.*, 1996; Yang *et al.*, 1998). Currently there is no evidence that any mutations outside of *rpoB* are involved in rifampicin resistance in clinical isolates of *M. tuberculosis*. This is also true for *M. tuberculosis* clinical isolates with multiple drug resistance where the evidence supports the accumulation of mutations in each of the individual drug target genes rather than the expression of a novel broad spectrum resistance determinant (Morris *et al.*, 1995).

The Ribosome

Ribosomes are the factories of protein synthesis. They are large complexes of protein and RNA, composed of two subunits and interacting with dozens of different ligands including mRNA, aminoacyl-tRNA's, and translation factors. Ribosomes are targeted by many antibiotics but we shall limit our discussion to streptomycin, which has historical and clinical importance. Streptomycin, (reviewed in Waksman, 1965) is an aminoglycoside antibiotic and was the first drug shown to be effective in the treatment of tuberculosis. Its importance lessened after the introduction of other effective anti-tuberculosis antibiotics, isoniazid and rifampicin, but since the 1980's it has regained importance, especially in the treatment of multidrug-resistant tuberculosis. Streptomycin binds to the small subunit of the ribosome, probably interacting directly with 16S rRNA (Abad and Amils, 1994) close to the position of ribosomal protein S12. [Streptomycin on the ribosome decreases the accuracy of protein synthesis resulting in the synthesis of incomplete or defective proteins (reviewed in Kurland et al., 1996).]

Resistance to streptomycin has been extensively studied in the genetically amenable bacteria E. coli and S. typhimurium (reviewed in Kurland et al., 1996). Mutations causing high-level resistance are found exclusively as single amino acid substitutions in the ribosomal protein S12. Most SmR mutations in S12 alter amino acid 42, or one of a few amino acids in the 87–90 region, and a few alter amino acid 53. Because the 16S rRNA genes are present in seven copies in these bacteria resistance due to individual rrn mutations is generally not detected unless special genetic techniques are employed to overcome the gene dosage effect. Mutations to streptomycin resistance in 16S rRNA have been found in three regions: position 13, the nucleotide 530 stem and loop, and the region around nucleotide 912 (reviewed in Kurland et al., 1996). Ribosomal resistance mutations generally act by reducing the binding of streptomycin to the ribosome (Kurland et al., 1996). Another source of streptomycin resistance is streptomycin-inactivating enzymes such as the aminogycoside phosphotransferases encoded by strA strB genes. These are found on plasmids and within transposons in a broad range of commensal and pathogenic bacteria (Sundin and Bender, 1996) including many resistant clinical isolates of E. coli (Korfmann et al., 1983).

Effective treatment of tuberculosis, and prevention of transmission, requires rapid identification of resistance patterns among the infecting strains to allow correct drug treatment. This is particularly important since the rise in frequency of multidrug-resistant tuberculosis and has encouraged the search for molecular methods to rapidly characterize resistance patterns in clinical isolates. In M. tuberculosis, where streptomycin is used as part of a multidrug treatment regime, mutations associated with resistance are identified by comparison with the E. coli data. M. tuberculosis differs significantly from E. coli in having only one copy of the gene for 16S rRNA, thus allowing expression of a resistance phenotype due to a single mutation. Presumed streptomycin resistance mutations in clinical isolates of M. tuberculosis from many different countries have been identified by sequencing the rpsL gene (protein S12) and parts of the rrn gene corresponding to the 530 and 912 regions in E. coli 16S rRNA. The results show that approximately 47% of isolates have single amino acid substitutions in ribosomal protein S12 (K42R ~36%, K87R ~8%, other changes at residues 42 or 87 ~3%), about 17% have single nucleotide substitutions in 16S rRNA (513 region ~13%, 912 region ~4%), about 1% have two mutations. In the remaining one third of isolates no mutation was identified (Finken et al., 1993; Meier et al., 1994; Sreevatsan et al., 1996; Dobner et al., 1997; Katsukawa et al., 1997). There are several notable aspects to these findings. Firstly, about one third of the isolates, making up almost all isolates with high MIC's (Cooksey et al., 1996; Meier et al., 1996), carry the mutation K42R in ribosomal protein S12. In S. typhimurium this particular mutation is unique among rpsL mutations in not reducing the in vivo fitness of streptomycin resistant strains (Björkman et al., 1998) suggesting that it may also be selected in M. tuberculosis clinical strains for the same reason. Secondly, for approximately one third of the streptomycin-resistant clinical isolates of M. tuberculosis no mutation has yet been identified. This group of strains may be a homogenous class because they are associated with relatively low MIC's (Meier et al., 1996). So far there is no evidence that streptomycin modifying or inactivating enzymes such as strA-strB (Sundin and Bender, 1996) are involved but this has not been tested. Thirdly, strains with the same mutation can have different MIC's. This is especially true for strains carrying rrn mutations and mutations of residue 87 in S12 (Meier et al., 1996). This suggests that in these strains there are additional unidentified mutations contributing to the level of resistance. It also suggests that there may be fitness costs associated with an initial mutation which are compensated by secondary mutations leading to the observed differences in MIC values. In the search for resistance-associated mutations a lack of genetic methods restricts the search in *M. tuberculosis* to comparative studies with amenable bacteria like *E. coli*.

Translation Factors

Translation factors targeted by clinically relevant antibiotics include elongation factor G (EF-G) the ribosomal translocase, and isoleucine-tRNA synthetase. Fusidic acid, (Godtfredsen *et al.*, 1962) binds to the ribosome-EF-G complex and interferes with translocation (Tanaka *et al.*, 1968). Mupirocin (Sutherland *et al.*, 1985) binds to bacterial isoleucyl-tRNA synthetase, inhibiting tRNA charging (Hughes and Mellows, 1980; Yanagisawa *et al.*, 1994). Both antibiotics are used primarily in the topical treatment of *Staphylococcus aureus* infections.

Fusidic acid resistance

In *S. typhimurium* all *in vitro*-selected high-level fusidic acid resistance mutants are due to single amino acid substitutions in EF-G. More than twenty different amino acid substitutions individually responsible for high level resistance have been identified in EF-G, clustered in the GTP-binding domain, in domain III, and a few in domain V (Johanson and Hughes, 1994; Johansson *et al.*, 1996). In *S. aureus*, both *in vitro*-selected and resistant clinical isolates, have amino acid substitutions in the same regions of EF-G (Nagaev, Björkman, Andersson and Hughes, unpublished data). However, in contrast to *S. typhimurium*, a significant percentage of the resistant mutants in *S. aureus*, comprising about two thirds of resistant clinical isolates, do not have alterations in EF-G (Lacey and Rosdahl, 1974; Nagaev, Björkman, Andersson and Hughes, unpublished data). This group of resistant strains may carry plasmids encoding an inducible function creating a permeability barrier to fusidic acid (Chopra, 1976). There

are currently no details on the mechanisms of non-EF-G fusidic acid resistance. In spite of being in use since 1962 resistance to resistance to fusidic acid is found in only about 1% of clinical *S. aureus* strains (Faber and Rosdahl, 1990).

Mupirocin resistance

Mupirocin resistant *S. aureus* isolates fall into two classes; those showing low-level resistance (MIC = 8–256 μg/ml), and those showing high-level resistance (MIC >1,000 μg/ml). Low-level resistance in *S. aureus* is thought to be due to mutations in the chromosomally encoded IleS, although these have been argued to be of little clinical significance (Cookson, 1998). In *E. coli* resistance is caused by a F594L substitution in IleS (Yanagisawa *et al.*, 1994). High-level resistance, on the other hand, is due to the presence of a novel IleS encoded on a plasmid, which in some strains has become incorporated into the chromosome (Gilbart *et al.*, 1993; Hodgson *et al.*, 1994). This high-level resistance IleS gene may have been acquired by horizontal transfer from another species.

The Bacterial Cell Wall

β-lactam antibiotics

Penicillin-binding proteins (PBP's), minor membrane components functioning in the late stages of murein biosynthesis, are the target enzymes for *β*-lactam antibiotics. PBP's interact with *β*-lactams enzymatically by forming a covalent complex via an active-site serine. PBP's can be divided into three classes (Ghuysen, 1994), one of low molecular mass (lmm) and two of high molecular mass (class A hmm, and class B hmm), each with different enzymatic activities. In gram-negative bacteria the main mechanism of *β*-lactam resistance is the production of *β*-lactamases that are effective in opening the *β*-lactam ring, thus inactivating the antibiotic before it reaches its target (Livermore, 1995). Gram-negative bacteria can also acquire resistance by changes in their outer membrane permeability which decrease their susceptibility to *β*-lactams (Nikaido, 1989).

Most gram-positive bacteria, in contrast, gain resistance to β-lactams by changing the specificity of their PBP's or by acquiring new PBP's with a lower affinity for these antibiotics. This mechanism involves alterations of essential PBP's such that the interaction with β-lactams takes place at much higher antibiotic concentrations than with the PBP's of susceptible strains. Two important pathogens where β-lactam resistance poses a serious problem are S. pneumoniae and S. aureus. In the case of S. pneumoniae resistance has involved extensive recombination with related species generating β-lactam-resistant mosaic PBP's, whereas S. aureus which is less proficient in inter-specific transformation has acquired an entirely novel low-affinity PBP by horizontal transfer (reviewed in Chambers, 1999).

S. pneumoniae contains six PBP's, the class A hmm PBP's, 1a, 1b, 2a; the class B hmm PBP's, 2x and 2b; and the lmm PBP3. In clinical isolates, low affinity variants of PBP's 2b, 2x and 1a are usually encoded by mosaic genes that are the result of gene transfer from related species and subsequent recombination events (Dowson et al., 1989; Laible et al., 1991; Martin et al., 1992; reviewed in Hakenbeck, 1998). Mosaic genes are defined as having regions which differ by up to 20% in nucleotide sequence from the parental sequence. In clinical isolates more than one altered mosaic PBP is usually present, and the mosaic genes usually contain multiple amino acid substitutions compared with susceptible S. pneumoniae isolates. Only some of these alterations are relevant for the development of resistance, with the remainder due to the different genetic origins of the DNA. In order to distinguish which alterations are important for resistance development mutants of S. pneumoniae resistant to clinically relevant penicillins and cephalosporins have been isolated in the laboratory.

The isolation of mutants of S. pneumoniae in the laboratory has revealed that different PBP's are primary targets for different β-lactam antibiotics. Many in vitro selection studies make use of cefotaxime (a cephalosporin) and piperacillin (a typical lytic penicillin) as representatives of two clinically important classes of β-lactam antibiotics. Thus, cefotaxime-resistant mutants carry mutations in PBP 2x or PBP3 that reduce β-lactam affinity and confer resistance (Laible and Hakenbeck, 1991; Kraub and Hakenbeck, 1997), whereas

piperacillin-resistant mutants carry mutations in PBP 2b (Grebe and Hakenbeck, 1996). However, all six PBP's can occur as low-affinity variants in β-lactam-resistant laboratory mutants (Hakenbeck, 1998). In addition, a variety of non-PBP mutations arise in these selections. These include alterations of CiaH (selected with cefotaxime) or CpoA (selected with piperacillin), mutations which affect bacterial competence (Guenzi et al., 1994; Grebe et al., 1997). Additional mutations also arise but are currently unidentified. In each case the increases in MIC associated with each individual mutation are modest (typically 2–4 fold) and in the PBP cases are associated with only one or a few amino acid alterations in individual PBP's.

In clinical isolates it is typically found that three or more PBP's are altered, often in a complex manner as a result of horizontal gene transfer and recombination. Highly resistant clinical isolates with MIC's 100-fold above that of susceptible strains, are found to have mutant PBP 1a (altered at Thr371), in addition to mutant PBP 2b and 2x (Asahi and Ubukata, 1998). In these strains mutations in two of the three genes result in up to ten-fold increases in MIC, but all three mutant PBP are required for the highest level of resistance. In PBP2b mutations reducing the affinity of β-lactams occur close to the active site serine, and at Thr446. This latter mutation reduces cell lysis in the presence of β-lactams and may be clinically important in allowing survival during drug treatment (Grebe and Hakenbeck, 1996). In PBP2x mutations contributing to resistance occur at Thr338 (close to the active site Ser337), His394, Thr550 and Gln552 (Hakenbeck and Coyette, 1998).

Non-β-lactamase-mediated penicillin resistance is also a serious clinical problem with S. aureus infections. Methicillin resistant S. aureus (MRSA) with intrinsic resistance to all β-lactams are generally found in hospital settings (reviewed in Lyon and Skurray, 1987). MRSA is due to the mecA gene borne on a 40 kb element integrated into the S. aureus chromosome. The origin of the mecA acquired by MRSA is uncertain but probably involved transposition (Archer and Niemeyer, 1994). All methicillin-resistant strains of S. aureus are clonal descendants from the few ancestral strains that acquired mecA (Kreiswirth et al., 1993). MRSA strains have a reduced susceptibility to β-lactams because the *mecA* gene codes for a novel PBP2' with a low affinity for β-lactams, allowing the strains to survive when penicillin inhibits the other PBP's.

Fosfomycin resistance

Fosfomycin tromethamine is an oral antibiotic used for the treatment of uncomplicated lower urinary tract infections. It is active in the urine against common uropathogens such as *E. coli*, *Citrobacter*, *Enterobacter*, *Klebsiella*, *Serratia* and *Enterococcus spp*. Fosfomycin enters cells through the GlpT uptake system, and inhibits peptidoglycan synthesis through inactivation of the enzyme UDP-GlcNAc (pyruvyl transferase). Fosfomycin resistant mutants can have mutations in the chromosomal permease gene *glpT* reducing uptake (Kadner and Winkler, 1973), or target mutations reducing the affinity of pyruvyl transferase for fosfomycin (Venkateswaran and Wu, 1972). Fosfomycin resistance can also be plasmid-encoded by *fosA* or *fosB*, resulting in modification and inactivation of the antibiotic (Arca *et al.*, 1997). In clinical isolates the major cause of resistance is by alteration of the GlpT transport system (Arca *et al.*, 1997) with high level resistance associated with additional mutations causing over-expression of the target enzyme (Horii *et al.*, 1999).

Isoniazid resistance

Isoniazid (INH) acts by inhibiting the biosynthesis of mycolic acids, important constituents of the cell wall, and is very potent against *M. tuberculosis*. Clinical resistance to INH has been attributed to mutations in two different genes. A minority of mutations, up to 30% (Miesel *et al.*, 1998) are found in the *inhA* operon, which is required for the synthesis of mycolic acids (Mdluli *et al.*, 1996). The majority of clinical mutations are of the *katG* gene encoding catalase-peroxidase, which converts INH to an active form which then affects proteins such as InhA (Musser *et al.*, 1996; Haas *et al.*, 1997; Marttila *et al.*, 1998). Among these resistant isolates a single amino acid substitution, S315T in KatG, is found in the great majority of cases, but deletions are also found. The loss of normal KatG function in INH-resistant mutants is apparently

compensated by a second mutation resulting in overexpression of AhpC, an alternative alkyl hydroperoxidase (Sherman *et al.*, 1996).

Vancomycin resistance

Glycopeptide antibiotics such as vancomycin and teicoplanin inhibit cell wall synthesis by forming complexes with the C-terminal D-alanyl-D-alanyl (D-Ala-D-Ala) residues of peptidoglycan precursors and block their incorporation into the bacterial cell wall. Vancomycin-resistant bacteria attenuate antibiotic sensitivity by produc-ing peptidoglycan precursors that terminate in D-Ala-D-lactate, rather than D-Ala-D-Ala. Vancomycin resistance is usually associated with an alteration of the antibiotic target and is for that reason included in this chapter. However, the cause of resistance is primarily an acquired function, not a target mutation. In enterococci, inducible resistance to high levels of vancomycin and teicoplanin (VanA phenotype) is associated with the transposon Tn*1546* which encodes a ligase (VanA), a dehydrogenase (VanH) and a dipeptidase (VanX) required for the replacement of D-Ala-D-Ala with D-Ala-D-Lac in the peptidoglycan assembly pathway (reviewed in Arthur *et al.*, 1996). This substitution prevents the binding of glycopeptides to the cell wall components and allows peptidoglycan polymerization to continue in the presence of the antibiotics. Another transposon, Tn*1547*, carries a related set of genes conferring inducible low-level resistance to vancomycin but not teicoplanin (VanB phenotype) also associated with the synthesis of D-Ala-D-Lac precursors. A related set of genes expressing constitutive resistance to various levels of both vancomycin and teicoplanin in *E. faecium*, also associated with peptidoglycan precursors ending in D-Ala-D-Lac, is termed the VanD phenotype (Perichon *et al.*, 1997). Constitutive low-level resistance to vancomycin but not teicoplanin, found in several entercocci species synthesizing D-Ala-D-Ser peptidoglycan precursors is termed the VanC phenotype (Leclercq *et al.*, 1992). A similar resistance phenotype also caused by D-Ala-D-Ser precursors and constitutively expressed in *E. faecalis* is termed VanE (Fines *et al.*, 1999). A third resistance phenotype, VanC, is a constitutive low-level resistance to vancomycin but not teico-planin, and is caused by chromosomal genes in several entercocci species. The The emergence in the past decade of vancomycin-resistant entercocci (VRE) is now a clinical problem, for example in treating urinary tract infections and in immunocompromised patients (reviewed in French, 1998). The predominant outbreak organism is *E. faecicum* and the clinical resistance phenotype is nearly always VanA. However, entercocci are not highly pathogenic and the more serious problem associated with VRE is that these bacteria are now significant nosocomial pathogens and as such are a reservoir of resistant genes with the potential to spread to more pathogenic bacteria such as staphylococci and streptococci.

As discussed in an earlier section, the emergence of MRSA poses serious clinical problems. Currently the only antibiotic reliably effective against MRSA is vancomycin. Thus, there is great interest in determining the risks of vancomycin resistance emerging in *S. aureus* and in particular in MRSA. The VanA-phenotype genes can be transferred from enterococci to *S. aureus* under laboratory conditions (Noble *et al.*, 1992) but this has not yet been observed in clinical isolates. However, low to intermediate-level resistance to vancomycin has been detected in MRSA in a few hospital patients in Japan and the USA (Hiramatsu *et al.*, 1997; Smith *et al.*, 1999; Sieradzki *et al.*, 1999a). These strains are being referred to as GISA (glycopeptide-intermediate susceptible *S. aureus*). The mechanism of resistance has not been deduced but the treatment regime and cell wall analysis suggest that it is due to selected mutations arising within the resistant strains (Sieradzki and Tomasz, 1997; Smith *et al.*, 1999). In laboratory isolates of GISA strains mutational inactivation of penicillin binding protein 4 (Sieradzki *et al.*, 1999b) and overproduction of penicillin binding protein 2 (Moreira *et al.*, 1997) are associated with resistance. As yet no high-level glycopeptide resistance has been detected in clinical *S. aureus* isolates.

Folic Acid Metabolism

The coenzyme tetrahydrofolate (THF) is required for essential functions in cell growth including the synthesis of dTMP and purines, and hence the synthesis of DNA and RNA. THF is synthesized by the reduction of its precursor dihydrofolic acid

(DHF). This pathway is targeted by the synthetic antibiotics trimethoprim and the sulfonamides. Sulfonamides inhibit the enzyme dihydropteroate synthase (DHPS, required for the synthesis of DHF), while trimethoprim inhibits dihydrofolate reductase (DHFR, required for the reduction of DHF to THF). Trimethoprim is widely used clinically in combination with sulfonamides such as sulfamethoxazole in the treatment of urinary, enteric and respiratory infections, especially in developing countries.

Trimethoprim resistance

Trimethoprim resistance can be plasmid/transposon-mediated, and this is the most frequently encountered mechanism in enterobacteria (reviewed in Houvinen *et al.*, 1995). Plasmid/transposon-mediated resistance is caused by non-allelic and drug-insusceptible variants of DHFR and will not be considered here. Chromosomal mutations are of two classes. One class works by increasing the production of DHFR, and/or decreasing its affinity for the drug. Thus, high-level resistant clinical isolates of *E. coli* and *H. influenza* have been found to have promoter-up mutations resulting in overproduction of DHFR, in addition to amino acid substitutions within the protein (Flensburg and Sköld, 1987). DHFR is approximately 160 amino acid in length, depending on bacterial species, and the mutational changes observed were G30W, E69K and D77N. In clinical isolates of *S. pneumoniae* resistance is caused by a single amino acid substitution, I100L, in DHFR (Adrian and Klugman, 1997; Pikis *et al.*, 1998). The same amino acid substitution also occurs in trimethoprin resistant *H. influenza* (de Groot *et al.*, 1996), and in clinical isolates of *S. aureus* the substitution F98Y is responsible for resistance (Dale *et al.*, 1997).

A second class of trimethoprim resistance found clinically is caused by mutations in the chromosomal *thyA* gene. Such bacteria are unable to methylate deoxyuridylic acid to thymidylic acid and are thymine-requiring. They are resistant to low levels of trimethoprim because DHFR is no longer required to carry out its major task of regenerating the THF which is used in the thymine synthesis pathway (King *et al.*, 1983). Thus, effectively the ThyA defect acts by making the inhibition of

DHFR by trimethoprim less important to the bacterial cell.

Sulfonamide resistance

In laboratory mutants of *E. coli* a chromosomal mutation, P64S, in *folP* (encoding DHPS), in combination with additional mutations confers sulfonamide resistance (Vedantam *et al.*, 1998). Clinically occurring sulfonamide-resistance in gram-negative enteric bacteria is however largely plasmid-borne and is due to the presence of alternative drug-resistant variants of the DHPS enzymes (these are reviewed in Houvinen *et al.*, 1995). In *Neisseria meningitidis*, *S. pneumoniae* and *S. pyrogenes* on the other hand, most clinical resistance to sulfonamides is linked to the chromosome. Chromosomal resistance results either from single amino acid substitutions in DHPS (Dallas *et al.*, 1992), or from the addition of one or two amino acids to DHPS by small internal duplications in the region between amino acid 58–67 (Lopez *et al.*, 1987; Maskell *et al.*, 1997). These duplications and other alterations are thought to reflect the creation of a mosaic DHPS gene by recombination of foreign DNA (Rådström *et al.*, 1992; Swedberg *et al.*, 1998).

SPREAD OF RESISTANCE

The spread of antibiotic resistance caused by target alterations is a multifactorial problem. Initially the limiting factor is the rate of mutation to resistance, or in the case of the creation of mosaic genes, the rate of horizontal transfer and recombination. These rates may be low but are clearly not a limiting factor in the examples discussed in this chapter. At best, a very low rate of mutation, or a requirement for multiple mutations to achieve high-level resistance, will delay the initial appearance of resistance. In this respect the appearance of GISA strains is worrying because it may be the initial stage in the creation of a reservoir of bacteria where an additional mutation could lead to high-level vancomycin-resistance.

Once a clinically-relevant level of resistance to an antibiotic appears in a pathogen the next question is whether the mutant bacteria will survive and spread

geographically. This is essentially a question of the fitness of the resistant bacteria coupled with the level of environmental antibiotic pressure. A resistant clone suffering a consequent fitness defect is more likely to survive if it is subject to antibiotic selection pressure, or if it acquires fitness-restoring compensating mutations (Björkman *et al.*, 1998). A defect-free mutant may survive regardless of antibiotic usage, but will be assisted in spreading by antibiotic usage, which removes any competing sensitive bacteria. Thus, the stronger the selective pressure of antibiotic usage, the more likely it is that a resistant clone will spread. Once resistance is established it does not have to be re-invented by new mutations as long as the resistant pathogen survives and can move between humans. This is a particular problem in hospital environments where antibiotic pressure is likely to be greatest, and for those pathogens which are easily transmitted.

Resistance mediated through target mutations differs from plasmid- and transposon-borne resistance in one significant respect: problems arise one at a time. Thus, the problem of plasmid- and transposon-borne resistance is not that it is necessarily more frequent, or more likely to spread rapidly, but rather, that it is more likely to be associated with the simultaneous acquisition of multi-drug resistance. In this respect the problem is more serious because once it arises the alternative treatment options may be very limited. This problem is discussed in the next chapter.

ACKNOWLEDGEMENTS

This work was supported by grants from the Swedish Natural Sciences Research Council, the Swedish Medical Research Council, the Swedish Institute for Infectious Disease control, the Leo Research Foundation, and the EU Biotechnology Research Programme.

REFERENCES

Abad, J.P. and Amils, R. (1994) Location of the streptomycin ribosomal binding site explains its pleiotrophic effects on protein biosynthesis. *J. Mol. Biol.* **235**, 1251–1260.

Adrian, P.V. and Klugman, K.P. (1997) Mutations in the dihydrofolate reductase gene of trimethoprim-resistant isolates of *Streptococcus pneumoniae*. *Antimicrib. Agents Chemother.* **41**, 2406–2413.

Arca, P., Reguera, G. and Hardisson, C. (1997) Plasmid-encoded fosfomycin resistance in bacteria isolated from the urinary tract in a multicentre survey. *J. Antimicrob. Chemother.* **40**, 393–399.

Archer, G.L. and Niemeyer, D.M. (1994) Origin and evolution of DNA associated with resistance to methicillin in *staphylococci*. *Trends Microbiol.* **2**, 343–347.

Arthur, M., Reynolds, P. and Courvalin, P. (1996) Glycopeptide resistance in enterococci. *Trends Microbiol.* **4**, 401–407.

Asahi, Y. and Ubukata, K. (1998) Association of a Thr-371 substitution in a conserved amino acid motif of penicillin-binding protein 1a with penicillin resistance of *Streptococcus pneumoniae*. *Antimicrob. Agents Chemother.* **42**, 2267–2273.

Björkman, J., Hughes, D. and Andersson, D.I. (1998) Virulence of antibiotic-resistant *Salmonella typhimurium*. *Proc. Natl. Acad. Sci. USA* **95**, 3949–3953.

Blanchard, J.S. (1996) Molecular mechanisms of drug resistance in *Mycobacterium tuberculosis*. *Annu. Rev. Biochem.* **65**, 215–239.

Bloom, B.R. and Murray, C.J.L. (1992) Tuberculosis: commentary on reemergent killer. *Science* **257**, 1055–1064.

Breines, D.M., Ouabdesselam, S., Ng, E.Y., Tankovic, J., Shah, S., Soussy, C.J., *et al.* (1997) Quinolone resistance locus *nfxD* of *Escherichia coli* is a mutant allele of the *parE* gene encoding a subunit of topoisomerase IV. *Antimicrob. Agents Chemother.* **41**, 175–179.

Chambers, H.F. (1999) Penicillin-binding protein-mediated resistance in *Pneumococci* and *Staphylococci*. *J. Infect. Dis.* **179, suppl. 2**, S353–359.

Charvalos, E., Tselentis, Y., Michae-Hamzenpour, M., Kohler, T. and Pechere, J.C. (1995) Evidence for an efflux pump in multi-drug resistant *Campylobacter jejuni*. *Antimicrob. Agents Chemother.* **39**, 2019–2022.

Chopra, I. (1976) Mechanisms of resistance to fusidic acid in *Staphylococcus aureus*. *J. Gen. Microbiol.* **96**, 229–238.

Conrad, S., Oethinger, M., Kaifel, K., Klotz, G., Marre, R. and Kern, W.V. (1996) *gyrA* mutations in high-level fluoroquinolone-resistant clinical isolates of *Escherichia coli*. *J. Antimicrob. Chemother.* **38**, 443–455.

Cooksey, R.C., Morlock, G.P., McQueen, A., Glickman, S.E. and Crawford, J.T. (1996) Characterization of streptomycin resistance mechanisms among *Mycobacterium tuberculosis* isolates from patients in New York City. *Antimicrob. Agents Chemother.* **40**, 1186–1188.

Cookson, B.D. (1998) The emergence of mupirocin resistance: a challenge to infection control and antibiotic prescribing practice. *J. Antimicrob. Chemother.* **41**, 11–18.

Dale, G.E., Broger, C., D'Arcy, A., Hartman, P.G., DeHoogt, R., Jolidon, S., *et al.* (1997) A single amino acid substitution in *Staphylococcus aureus* dihydrofolate reductase determines trimethoprim resistance. *J. Mol. Biol.* **266**, 23–30.

Dallas, W.S., Gowen, J.E., Ray, P.H., Cox, M.J. and Dev, I.K. (1992) Cloning, sequencing, and enhanced expression of the dihydropteroate synthase gene of *Escherichia coli* MC4100. *J. Bacteriol.* **174**, 5961–5970.

de Groot, R., Sluijter, M., de Bruyn, A., Campos, J., Goessens, W.H.F., Smith, A.L., *et al.* (1996) Genetic characterization of trimethoprim resistance in *Haemophilus influenza*. *Antimicrob. Agents Chemother.* **40**, 2131–2136.

Dobner, P., Bretzel, G., Rüsch-Gerdes, S., Feldman, K., Rifai, M., Löscher, T., *et al.* (1997) Geographic variation of the predictive values of genomic mutations associated with streptomycin resistance in *Mycobacterium tuberculosis*. *Mol. Cell. Probes* **11**, 123–126.

Dowson, C.G., Hutchison, A., Brannigan, J.A., George, R.C., Hansman, D., Liñares, J., *et al.* (1989) Horizontal transfer of penicillin-binding protein genes in penicillin-resistant clinical isolates of *Streptococcus pneumoniae*. *Proc. Natl. Acad. Sci. USA* **86**, 8842–8846.

Faber, M. and Rosdahl, V.T. (1990) Susceptibility to fusidic acid among Danish *Staphylococcus aureus* strains and fusidic acid consumption. *J. Antimicrob. Chemother.* **25, Suppl. B**, 7–14.

Ferrero, L., Cameron, B., Manse, B., Lagneaux, D., Crouzet, J., Famechon, A., *et al.* (1994) Cloning and primary structure of *Staphylococcus aureus* DNA topoisomerase IV: a primary target of fluoroquinolones. *Mol. Microbiol.* **13**, 641–653.

Ferrero, L., Cameron, B. and Crouzet, J. (1995) Analysis of *gyrA* and *grlA* mutations in stepwise-selected ciprofloxacin-resistant mutants of *Staphylococcus aureus*. *Antimicrob. Agents Chemother.* **39**, 1554–1558.

Fines, M., Perichon, B., Reynolds, P., Sahm, D.F. and Courvalin, P. (1999) VanE, a new type of acquired glycopeptide resistance in *Enterococcus faecalis* BM4405. *Antimicro. Agents Chemother.* **43**, 2161–2164.

Finken, M., Kirschner, P., Meier, A., Wrede, A. and Böttger, E.C. (1993) Molecular basis of streptomycin resistance in the *Mycobacterium tuberculosis:* alterations of the ribosomal protein S12 gene and point mutations within a functional 16S ribosomal RNA pseudoknot. *Mol. Microbiol.* **9**, 1239–1246.

Flensburg, J. and Sköld, O. (1987) Massive overproduction of dihydrofolate reductase in bacteria as a response to the use of trimethoprim. *Eur. J. Biochem.* **162**, 473–476.

French, G.L. (1998) Enterococci and vancomycin resistance. *Clin. Infect. Dis.* **27 (Suppl 1)**, S75–83.

Ghuysen, J.M. (1994) Molecular structures of penicillin-binding proteins and β-lactamases. *Trends Microbiol.* **2**, 372–380.

Gibreel, A., Sjögren, E., Kaijser, B., Wretlind, B. and Sköld, O. (1998) Rapid development of high-level resistance to quinolones in *Campylobacter jejuni* associated with mutational changes in *gyrA* and *parC*. *Antimicrob. Agents Chemother.* **42**, 3276–3278.

Gilbart, J., Perry, C.R. and Slocombe, B. (1993) High-level mupirocin resistance in *Staphylococcus aureus*: evidence for two distinct isoleucyl-tRNA synthetases. *Antimicrob. Agents Chemother.* **37**, 32–38.

Gill, M.J., Brenwald, N.P. and Wise, R. (1999) Identification of an efflux pump gene, *pmrA*, associated with fluoroquinolone resistance in *Streptococcus pneumoniae*. *Antimicrob. Agents Chemother.* **43**, 187–189.

Godtfredsen, W., Roholt, K. and Tybring, L. (1962) Fucidin. A new orally active antibiotic. *Lancet i*, 928–931.

Goldman, J.D., White, D.G. and Levy, S.B. (1996) Multiple antibiotic resistance (*mar*) locus protects *Escherichia coli* from rapid cell killing by fluoroquinolones. *Antimicrob. Agents Chemother.* **40**, 1266–1269.

Goswitz, J.J., Willard, K.E., Fasching, C.E. and Peterson, L.R. (1992) Detection of *gyrA* gene mutations associated with ciprofloxacin resistance in methicillin-resistant *Staphylococcus aureus*: analysis by polymerase chain reaction and automated direct DNA sequencing. *Antimicrob. Agents Chemother.* **36**, 1166–1169.

Grebe, T. and Hakenbeck, R. (1996) Penicillin-binding proteins 2b and 2x of *Streptococcus pneumoniae* are primary resistance determinants for different classes of β-lactam antibiotics. *Antimicrob. Agents Chemother.* **40**, 829–834.

Grebe, T., Paik, J. and Hakenbeck, R. (1997) A novel mechanism against β-lactams in *Streptococcus pneumoniae* involves CpoA, a putative glycosyltransferase. *J. Bacteriol.* **179**, 3342–3349.

Greenwood, B. (1999) The epidemiology of pneumococcal infection in children in the developing world. *Philos. Trans. R. Soc. Lond. B Biol. Sci.* **354**, 777–785.

Guenzi, E., Gasc, A.M., Sicard, M.A. and Hakenbeck, R. (1994) A two-component signal-transducing system is involved in competence and penicillin susceptibility in laboratory mutants of *Streptococcus pneumoniae*. *Mol. Microbiol.* **12**, 505–515.

Haas, W.H., Schilke, K., Brand, J., Amthor, B., Weyer, K., Fourie, P.B., *et al.* (1997) Molecular analysis of *katG* gene mutations of *Mycobacterium tuberculosis* complex from Africa. *Antimicrob. Agents Chemother.* **41**, 1601–1603.

Hakenbeck, R. (1998) Mosaic genes and their role in penicillin-resistant *Streptococcus pneumoniae*. *Electrophoresis* **19**, 597–601.

Hakenbeck, R. and Coyette, J. (1998) Resistant penicillin-binding proteins. *Cell Mol. Life Sci.* **54**, 332–340.

Hartmann, G., Honikel, K.O., Knusel, F. and Nuesch, J. (1967) The specific inhibition of the DNA-directed RNA synthesis by rifamycin. *Biochem. Biophys. Acta* **145**, 843–844.

Heisig, P. and Tschorny, R. (1994) Characterization of fluoroquinolone-resistant mutants of *Escherichia coli* selected in vitro. *Antimicrob. Agents Chemother.* **38**, 1284–1291.

Hiramatsu, K., Hanaki, H., Ino, T., Yabuta, K., Oguri, T. and Tenover, F.C. (1997) Methicillin-resistant *Staphylococcus aureus* clinical strain with reduced vancomycin susceptibility. *J. Antimicrob. Chemother.* **40**, 135–136.

Hodgson, J.E., Curnock, S.P., Dyke, K.G.H., Morris, R., Sylvexter, D.R. and Gross, M.S. (1994) Molecular characterization of the gene encoding high-level mupirocin resistance in *Staphylococcus aureus* J2870. *Antimicrob. Agents Chemother.* **38**, 1205–1208.

Horii, T., Kimura, T., Sato, K., Shibayama, K. and Ohta, M. (1999) Emergence of fosfomycin-resistant isolates of Shiga-like toxin-producing *Escherichia coli* O26. *Antimicrob. Agents Chemother.* **43**, 789–793.

Horowitz, D.S. and Wang, J.C. (1987) Mapping the active site tyrosine of *Escherichia coli* DNA gyrase. *J. Biol. Chem.* **262**, 5339–5344.

Hughes, J. and Mellows, G. (1980) Interaction of pseudomonic acid A with *Escherichia coli* B isoleucyl tRNA synthetase. *Biochem. J.* **191**, 209–219.

Huovinen, P., Sundström, L., Swedberg, G. and Sköld, O. (1995) Trimethoprim and sulfonamide resistance. *Antimicrob. Agents Chemother.* **39**, 279–289.

Jacobs, M.R. (1999) Drug-resistant *Streptococcus pneumoniae*: rational antibiotic choices. *Am. J. Med.* **106**, 19S–25S; discussion 48S–52S.

Jalal, S. and Wretlind, B. (1998) Mechanisms of quinolone resistance in clinical strains of *Pseudomonas aeruginosa*. *Microb. Drug Resist.* **4**, 257–261.

Jin, D. and Gross, C. (1988) Mapping and sequencing of mutations in the *Escherichia coli rpoB* gene that lead to rifampicin resistance. *J. Mol. Biol.* **202**, 45–58.

Johanson, U. and Hughes, D. (1994) Fusidic acid-resistant mutants define three regions in elongation factor G of *Salmonella typhimurium*. *Gene* **143**, 55–59.

Johanson, U., Ævarsson, A., Liljas, A. and Hughes, D. (1996) The dynamic structure of EF-G studied by fusidic acid resistance and internal revertants. *J. Mol. Biol.* **258**, 420–432.

Kaatz, G.W. and Seo, S.M. (1997) Mechanisms of fluoroquinolone resistance in genetically related strains of *Staphylococcus aureus*. *Antimicrob. Agents Chemother.* **41**, 2733–2737.

Kadner, R.J. and Winkler, H.H. (1973) Isolation and characterization of mutations affecting the transport of hexose phosphates in *Escherichia coli*. *J. Bacteriol.* **113**, 895–900.

Kapur, V., Li, L-L., Iordanescu, S., Hamrick, M.R., Wanger, A., Krieswirth, B.N., *et al.* (1994) Characterization by automated DNA sequencing of mutations in the gene (*rpoB*) encoding the RNA polymerase β subunit in rifampicin-resistant *Mycobacterium tuberculosis* strains from New York City and Texas. *J. Clin. Microbiol.* **32**, 1095–1098.

Kato, J., Nishimura, Y., Imamura, R., Niki, H., Higara, S. and Suzuki, H. (1990) New topoisomerase essential for chromosome segregation in *E. coli*. *Cell* **63**, 393–404.

Katsukawa, C., Tamaru, A., Miyata, Y., Abe, C., Mikino, M. and Suzuki, Y. (1997) Characterization of the *rpsL* and *rrn* genes of streptomycin-resistant clinical isolates of *Mycobacterium tuberculosis* in Japan. *J. Appl. Microbiol.* **83**, 634–640.

King, C.H., Shlaes, D.M. and Dul, M.J. (1983) Infection caused by thymidine-requiring, trimethoprim-resistant bacteria. *J. Clin. Microbiol.* **18**, 79–83.

Kohler, T., Michae-Hamzepour, M., Plesiat, P., Kahr, A.L. and Pechere, J.C. (1997) Differential selection of multidrug efflux systems by quinolones in *Pseudomonas aeruginosa*. *Antimicrob. Agenta Chemother.* **41**, 2540–2543.

Korfmann, G., Ludtke, W., van Treeck, U. and Wiedemann, B. (1983) Dissemenation of streptomycin and sulfonamide resistance by plasmid pBP1 in *Escherichia coli*. *Eur. J. Clin. Microbiol.* **2**, 463–468.

Kraub, J. and Hakenbeck, R. (1997) A mutation in the D,D-carboxypeptidase penicillin-binding protein 3 of *Streptococcus pneumoniae* contributes to cefotaxime resistance of the laboratory mutant C604. *Antimicrob. Agents Chemother.* **41**, 936–942.

Kreiswirth, B., Kornblum, J., Arbeit, R.D., Eisner, W., Maslow, J.N., McGeer, A., *et al.* (1993) Evidence for a clonal origin of methicillin resistance in *Staphylococcus aureus*. *Science* **259**, 227–230.

Kurland, C.G., Hughes, D. and Ehrenberg, M. (1996) Limitations of translational accuracy. In *Escherichia coli* and *Salmonella*: Cellular and Molecular Biology, 2nd edn, Vol. 1. Neidhardt, F.C., *et al* (eds). Washington, DC. American Society for Microbiology Press, pp. 979–1004.

Lacey, R.W. and Rosdahl, V.T. (1974) An unusual ''penicillinase plasmid'' in *Staphylococcus aureus*; evidence for its transfer under natural conditions. *J. Med. Microbiol.* **7**, 1–9.

Laible, G. and Hakenbeck, R. (1991) Five independent combinations of mutations can result in low-affinity penicillin-binding protein 2x of *Streptococcus pneumoniae*. *J. Bacteriol.* **173**, 6986–6990.

Laible, G., Spratt, B.G. and Hakenbeck, R. (1991) Inter-species recombinational events during the evolution of altered PBP 2x genes in penicillin-resistant clinical isolates of *Streptococcus pneumoniae*. *Mol. Microbiol.* **5**, 1993–2002.

Leclerq, R., Dutka-Malen, S., Duval, J. and Courvalin, P. (1992) Vancomycin resistance gene *vanC* is specific to *Enterococcus gallinarum*. *Antimicrob. Agents Chemother.* **36**, 2005–2008.

Levin. M.E. and Hatfull, G.F. (1993) *Mycobacterium smegmatis* RNA polymerase: DNA supercoiling, action of rifampicin and mechanism of rifampicin resistance. *Mol. Microbiol.* **8**, 277–285.

Livermore, D.M. (1995) β-lactamases in laboratory and clinical resistance. *Clin. Microbiol. Rev.* **8**, 557–584.

Lopez, P., Espinosa, M., Greenberg, B. and Lacks, S.A. (1987) Sulfonamide resistance in *Streptococcus pneumoniae*: DNA sequence of the gene encoding dihydropteroate synthase and characterization of the enzyme. *J. Bacteriol.* **169**, 4320–4326.

Lyon, B.R. and Skurray, R. (1987) Antimicrobial resistance of *Staphylococcus aureus*: genetic basis. *Microbiol. Rev.* **51**, 88–134.

Martin, C., Sibold, C. and Hackenback, R. (1992) Relatedness of penicillin-binding protein 1a genes from different clones of penicillin-resistant *Streptococcus pneumoniae* isolated in South Africa and Spain. *EMBO J.* **11**, 3831–3836.

Martínez, J.L., Alonso, A., Gómez-Gómez, J.M. and Baquero, F. (1998) Quinolone resistance by mutations in chromosomal gyrase genes. Just the tip of the iceberg? *J. Antimicrib. Chemother.* **42**, 683–688.

Marttila, H.J., Soini, H., Eerola, E., Vyshnevskaya, E., Vyshnevskiy, B.I., Otten, T.F., *et al.* (1998) A Ser315Thr substitution in KatG is predominant in genetically heterogenous multidrug-resistant *mycobacterium tuberculosis* isolates originating from the St. Petersburg area in Russia. *Antimicrob. Agents Chemother.* **42**, 2443–2445.

Maskell, J.P., Sefton, A.M. and Hall, L.M.C. (1997) Mechanism of sulfonamide resistance in clinical isolates of *Streptococcus pneumoniae. Antimicrob. Agents Chemother.* **41**, 2121–2126.

Maxwell, A. (1997) DNA gyrase as a drug target. *Trends Microbiol.* **5**, 102–109.

Mdluli, K., Sherman, D.R., Hickey, M.J., Kreiswirth, B.N., Morris, S., Stover, K. and Barry III, C.E. (1996) Biochemical and genetic data sugest that InhA is not the primary target for activated isoniazid in *Mycobacterium tuberculosis. J. Infect. Dis.* **174**, 1085–1090.

Meier, A., Kirschner, A., Bange, F-C. Vogel, U. and Böttger, E.C. (1994) Genetic alterations in streptomycin-resistant *Mycobacterium tuberculosis*: Mapping of mutations conferring resistance. *Antimicrob. Agents Chemother.* **38**, 228–233.

Meier, A., Sander, P., Schaper, K-J., Scholz, M. and Böttger, E.C. (1996) Correlation of molecular resistance mechanisms and phenotypic resistance levels in streptomycin-resistant *Mycobacterium tuberculosis. Antimicrob. Agents Chemother.* **40**, 2452–2454.

Miesel, L., Rozwarski, D.A., Sacchettini, J.C. and Jacobs, W.R. Jr. (1998) Mechanisms for isoniazid action and resistance. *Novartis Found. Symp.* **217**, 209–221.

Moreira, B., Boyle-Vavra, S., deJonge, B.L. and Daum, R.S. (1997) Increased production of penicillin-binding protein 2, increased detection of other penicillin-binding proteins, and decreased coagulase activity associated with glycopeptide resistance in *Staphylococcus auseus. Antimicrob. Agents Chemother.* **41**, 1788–1793.

Morris, S., Bai, G.H., Suffys, P., Portillo-Gomez, L., Fairchok, M. and Rouse, D. (1995) Molecular mechanisms of multiple drug resistance in clinical isolates of *Mycobacterium tuberculosis. J. Infect. Dis.* **171**, 954–960.

Munoz, R. and De La Campa, A.G. (1996) ParC subunit of DNA topoisomerase IV of *Sterptococcus pneumoniae* is a primary target of fluoroquinolones and cooperates with DNA gyrase A subunit in forming resistance phenotype. *Antimicrob. Agents Chemother.* **40**, 2252–2257.

Musser, J.M., Kapur, V., Williams, D.L., Kreiswirth B.N., van Soolingen, D. and van Embden, J.D.A. (1996) Characterization of the catalase-peroxidase gene (*katG*) and *inhA* locus in isoniazid-resistant and -susceptible strains of *Mycobacterium tuberculosis* by automated DNA sequencing: restricted array of mutations associated with drug resistance. *J. Infect. Dis.* **173**, 196–202.

Nakamura, S., Nakamura, M., Kojima, T. and Yoshida, H. (1989) *gyrA* and *gyrB* mutations in quinolone-resistant strains of *E. coli. Antimicrob. Agents Chemother.* **33**, 254–255.

Nash, K.A., Gaytan, A. and Inderlied, C.B. (1997) Detection of rifampicin resistance in *Mycobacterium tuberculosis* by use of a rapid, simple, and specific RNA/RNA mismatch assay. *J. Infect. Dis.* **176**, 533–536.

Ng, E.Y., Trucksis, M. and Hooper, D.C. (1996) Quinolone resistance mutations in topoisomerase IV: relationship to the *flqA* locus and genetic evidence that topoisomerase IV is the primary target and DNA gyrase is the secondary target of fluoroquinolones in *Staphylococcus aureus. Antimicrob. Agents Chemother.* **40**, 1881–1888.

Nikaido, H. (1989) Outer membrane barrier as a mechanism of antimicrobial resistance. *Antimicrob. Agents Chemother.* **33**, 1831–1836.

Noble, W.C., Virani, Z. and Cree, R.G.A. (1992) Co-transfer of vancomycin and other resistance genes from *Entercoccus faecilis* NCTC 12201 to *Staphulococcus aureus. FEMS Microbiol. Lett.* **72**, 195–198.

Oethinger, M., Kern, W.V., Jellen-Ritter, A.S., McMurray, L.M. and Levy, S.B. (2000) Ineffectiveness of topoisomerase mutations in mediating clinically significant fluoroquinolone resistance in *Escherichia coli* in the absence of the AcrAB efflux pump. *Antimicrob. Agents Chemother.* **44**, 10–13.

Ohno, H., Koga, H., Kohno, S., Tashiro, T. and Hara, K. (1996) Relationship between rifampicin MICs for and *rpoB* mutations of *Mycobacterium tuberculosis* strains isolated in Japan. *Antimicrob. Agents Chemother.* **40**, 1053–1056.

Ozeki, S., Deguchi, T., Yasuda, M., Nakano, M., Kawamura, T., Nishino, Y., *et al.* (1997) Development of a rapid assay for detecting *gyrA* mutations in *Escherichia coli* and determination of incidence of *gyrA* mutations in clinical strains isolated from patients with complicated urinary tract infections. *J. Clin. Microbiol.* **35**, 2315–2319.

Pablos-Méndex, A., Raviglione, M.C., Laszlo, A., Binkin, N., Rieder, H.L., Bustreo, F., *et al.* (1998) Global surveillance for antituberculosis-drug resistance. *New Eng. J. Med.* **338**, 1641–1649.

Pan, X.S. and Fisher, L.M. (1996) Cloning and characterization of the *parC* and *parE* genes of *Streptococcus pneumoniae* encoding DNA topoisomerase IV: role in fluoroquinolone resistance. *J. Bacteriol.* **178**, 4060–4069.

Pan, X.S. and Fisher, L.M. (1997) Targeting of DNA gyrase in *Streptococcus pneumoniae* by sparfloxacin: selective targeting of gyrase or topoisomerase IV by quinolones. *Antimicrob. Agents Chemother.* **41**, 471–474.

Pan, X.S. and Fisher, L.M. (1999) *Streptococcus pneumoniae* DNA gyrase and topoisomerase IV: overexpression, purification, and differential inhibition by fluoroquinolones. *Antimicrib. Agents Chemother.* **43**, 1129–1136.

Pan, X.S., Ambler, J., Mehtar, S. and Fisher, L.M. (1996) Involvement of topoisomerase IV and DNA gyrase as ciprofloxacin targets in *Streptococcus pneumoniae. Antimicrob. Agents Chemother.* **40**, 2321–2326.

Perichon, B., Reynolds, P. and Courvalin, P. (1997) VanD-type glycopeptide-resistant *Enterococcus faecium* BM4339. *Antimicrob. Agents Chemother.* **41**, 2016–2018.

Pikis, A., Donkersloot, J.A., Rodriguez, W.J. and Keith, J.M. (1998) A conservative amino acid mutation in the chromosome-encoded dihydrofolate reductase confers trimethoprim resistance in *Sterptococcus pneumoniae. J. Infect. Dis.* **178**, 700–706.

Revel, V., Cambau, E., Jarlier, V. and Sougakoff, W. (1994) Characterization of mutations in Mycobacterium smegmatis involved in resistance to fluoroquinolones. *Antimicrob. Agents Chemother.* **38**, 1991–1996.

Rådström, P., Fermer, C., Kristiansen, B.-E., Jenkins, A., Sköld, O. and Swedburg, G. (1992) Transformational exchanges in the dihydropteroate synthase gene of *Neisseria meningitidis*, a novel mechanism for the acquisition of sulfonamide resistance. *J. Bacteriol.* **174**, 6386–6393.

Sherman, D.R., Mdluli, K., Hickey, M.J., Arain, T.M., Morris, S.L., Barry, C.E. III, *et al.* (1996) Compensatory *ahpC* gene expression in isoniazid-resistant *Mycobacterium tuberculosis. Science* **272**, 1641–1643.

Sieradzki, K. and Tomasz, A. (1997) Inhibition of cell wall turnover and autolysis by vancomycin in a highly vancomycin-resistant mutant of *Staphylococcus aureus. J. Bacteriol.* **179**, 2557–2566.

Sieradzki, K., Roberts, R.B., Haber, S.W. and Tomasz, A. (1999a) The development of vancomycin resistance in a patient with methicillin-resistant *Staphylococcus aureus* infection. *N. Eng. J. Med.* **340**, 517–523.

Sieradzki, K., Pihno, M.G. and Tomasz, A. (1999b) Inactivated pbp4 in highly glycopeptide-resistant laboratory mutants of *Staphylococcus aureus*. *J. Biol. Chem.* **274**, 18942–18946.

Smith, T.L., Pearson, M.L., Wilcox, K.R., Cruz, P.H.C., Lancaster, M.V., Robinson-Dunn, B., *et al.* (1999) Emergence of vancomycin resistance in *Staphylococcus aureus*. *N. Eng. J. Med.* **340**, 493–501.

Sreedharan, S., Oram, M., Jensen, B., Peterson, L.R. and Fisher, L.M. (1990) DNA gyrase *gyrA* mutations in ciprofloxacin-resistant strains of *Staphylococcus aureus*: close similarity with quinolone resistance mutations in *Escherichia coli*. *J. Bacteriol.* **172**, 7260–7262.

Sreevatsan, S., Pan, X., Stockbauer, K.E., Williams, D.L., Kreiswirth, B.N. and Musser, J.M. (1996) Characterization of *rpsL* and *rrn* mutations in streptomycin-resistant *Mycobacterium tuberculosis* isolates from diverse geographic localities. *Antimicrob. Agents Chemother.* **40**, 1024–1026.

Sundin, G.W. and Bender, C.L. (1996) Dissemenation of the *strA-strB* streptomycin-resistance genes among commensal and pathogenic bacteria from humans, animals, and plants. *Mol. Ecol.* **5**, 133–143.

Sutherland, R., Boon, R.J., Griffin, K.E., Masters, P.J., Slocombe, B. and White, A.R. (1985) Antibacterial activity of mupirocin (pseudomonic acid), a new antibiotic for topical use. *Antimicrob. Agents Chemother.* **27**, 495–498.

Swedberg, G., Ringertz, S. and Sköld, O. (1998) Sulfonamide resistance in *Streptococcus pyrogenes* is associated with differences in the amino acid sequence of its chromosomal dihydropteroate synthase. *Antimicrob. Agents Chemother.* **42**, 1062–1067.

Tanaka, N., Kinoshita, T. and Masukawa, H. (1968) Mechanism of protein synthesis inhibition by fusidic acid and related antibiotics. *Biochem. Biophys. Res. Comm.* **30**, 278–283.

Taniguchi, H., Aramaki, H., Nikaido, Y., Mizugichi, Y., Nakamura, M., Koga, T., *et al.* (1996) Rifampicin resistance and mutation of the *rpoB* gene in *Mycobacterium tuberculosis*. *FEMS Microbiol. Lett.* **144**, 103–108.

Telenti, A., Imboden, P., Marchesi, F., Lowrie, D., Cole, S., Colston, M.J., *et al.* (1993) Detection of rifampicin-resistance mutations in *Mycobacterium tuberculosis*. *Lancet* **341**, 647–650.

Vedamtan, G., Guay, G.G., Austria, N.E., Doktor, S.T. and Nichols, B.P. (1998) Characterization of mutations contributing to sulfathiazole resistance in *Escherichia coli*. *Antimicrob. Agents Chemother.* **42**, 88–93.

Venkateswaran, P.S. and Wu, H.C. (1972) Isolation and characterization of a phosphonomycin-resistant mutant of *Escherichia coli*. *J. Bacteriol.* **110**, 935–944.

von Rosenstiel, N. and Adams, D. (1994) Quinolone antibacterials. An update of their pharmacology and therapeutic use. *Drugs* **47**, 872–901.

Waksman, S.A. (1965) A quarter-century of the antibiotic era. *Antimicrob. Agents Chemother.* **5**, 9–19.

Williams, D.L., Waguespack, C., Eisenach, K., Crawford, J.T., Portaels, F., Salfinger, M., *et al.* (1994) Characterization of rifampicin resistance in pathogenic *Mycobacteria*. *Antimicrob. Agents Chemother.* **38**, 2380–2386.

Yamagishi, J-I., Kojima, T., Oyamada, Y., Fujimoto, K., Hattori, H., Nakamura, S., *et al.* (1996) Alterations in the DNA topoisomerase *grlA* gene responsible for quinolone resistance in *Staphylococcus aureus*. *Antimicrob. Agents Chemother.* **40**, 1157–1163.

Yanagisawa, T., Lee, J.T., Wu, H.C. and Kawakami, M. (1994) Relationship of protein structure of isoleucyl-tRNA synthetase with pseudomonic acid resistance of *Escherichia coli*. A proposed mode of action of pseudomonic acid as an inhibitor of isoleucyl-tRNA synthetase. *J. Biol. Chem.* **269**, 24304–24309.

Yang, B., Koga, H., Ohno, H., Ogawa, K., Fukuda, M., Hirakata, Y., *et al.* (1998) Relationship between antimycobacterial activities of rifampicin, rifabutin and KRM-1648 and *rpoB* mutations of *Mycobacterium tuberculosis*. *J. Antimicrob. Chemother.* **42**, 621–628.

Yoshida, H., Bogaki, M., Nakamura, M. and Nakamura, S. (1990) Quinolone resistance determining region in the DNA gyrase gene of *Escherichia coli*. *Antimicrob. Agents Chemother.* **34**, 1271–1272.

Yoshida, H., Bogaki, M., Nakamura, M., Yamanaka, L.M. and Nakamura, S. (1991) Quinolone resistance-determining region in the DNA gyrase *gyrB* gene of *Escherichia coli*. *Antimicrob. Agents Chemother.* **35**, 1647–1650.

Yoshida, H., Nakamura, M., Bogaki, M., Ito, H., Kojima, T., Hattori, H., *et al.* (1993) Mechanism of action of quinolones against *Escherichia coli* DNA gyrase. *Antimicrob. Agents Chemother.* **37**, 839–845.

Zeller, V., Janoir, C., Kitzis, M.D., Gutmann, L. and Moreau, N.J. (1997) Active efflux as a mechanism of resistance to ciprofloxacin in *Streptococcus pneumoniae*. *Antimicrob. Agents Chemother.* **41**, 1973–1978.

Zhao, X., Xu, C., Domagala, J. and Drlica, K. (1997) DNA topoisomerase targets of fluoroquinolones: A strategy for avoiding bacterial resistance. *Proc. Natl. Acad. Sci. USA* **94**, 13991–13996.

3. Evolution of Multiple Antibiotic Resistance by Acquisition of New Genes

Ruth M. Hall[1] **and Sally R. Partridge**[1,2]
[1] CSIRO Molecular Science, PO Box 184, North Ryde, NSW 1670, Australia
[2] School of Biological Sciences, Macquarie University Sydney, NSW 2109, Australia

INTRODUCTION

Over the past 40 years, the study of antibiotic resistance has revealed a new world of genes that are part of what can best be described as a ''floating'' or ''itinerant'' genome that is shared by a vast number of bacterial species and indeed genera. The floating genome contains genes that are not essential to the basic life of a bacterium but are ''optional extras'' that allow adaption to new conditions. Movement of these genes from one bacterium to another occurs by the process of horizontal gene transfer and the mechanisms by which this is achieved are the subject of a separate chapter (Chapter 3a). Genes that are part of mobile elements that can translocate have the capacity to board the vehicles that traffic the highways and byways of horizontal gene transfer. The main forces that drive the evolution of multidrug resistant bacteria are thus horizontal gene transfer and translocation and, though these two types of gene movement are inextricably intertwined, for clarity it is important to clearly separate them conceptually. The difference between these two fundamentally distinct types of gene movement is illustrated in Figure 1.

In the light of the steadily increasing incidence of antibiotic resistance in important human pathogens, it is critical to remember that the total antibiotic load is likely to be the most important factor

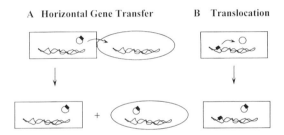

A Horizontal Gene Transfer   B Translocation

Figure 1. Gene movement.
A. Horizontal gene transfer. DNA, in this case a plasmid (circle) containing a resistance gene (black box), is transferred to a sensitive bacterial cell. Both cells are then resistant.
B. Translocation. A resistance gene contained within a mobile element (black box) moves from one replicon to another within the same cell.

driving this phenomenon. Because antibiotics select for and amplify the number of resistant bacteria, their use alters the composition of the bacterial community, favouring naturally resistant species and resistant variants of susceptible species. As a consequence, resistant bacterial strains that arise only very rarely can achieve prominence in the bacterial community. Furthermore, because other types of genes, e.g. virulence and pathogenicity determinants and determinants of resistance to heavy metals, are also present in the floating genome, they can and do become closely associated with antibiotic resistance genes. When this occurs any selective force that favors organisms containing these other genes (e.g. the ability to grow in a specific niche or exposure to heavy metals) will also favor the antibiotic-resistant bacteria.

The independently-replicating small genomes collectively known as plasmids and their non-replicating relatives the conjugative transposons (here called conjugons) are undoubtedly major forces in the intra-species, inter-species and ultimately the inter-genera movement of genes, particularly in the case of antibiotic resistance genes. The ability of plasmids to accumulate genetic material that is not essential for their primary (replication, stability/partition) or secondary (conjugation, mobilization, cointegration) functions permits them to be the freight trains and cargo ships of the floating genome. In the context of resistance, intercellular transfer, either self-transfer (conjugation) or assisted transfer (mobilization or cointegration and cotransfer), is an important property of plasmids as they evolve because plasmids that retain such properties have a strong advantage. When clinical isolates of antibiotic-resistant Enterobacteriaceae, Pseudomonads and Staphylococci are examined, the majority are found to contain plasmids, and most of the antibiotic resistance genes are located on these plasmids. While in the Staphylococci many of the plasmids are small and probably move to new host cells by processes other than conjugation (e.g. transduction), a large proportion of the plasmids found in Staphylococci and in other species can either self-transfer to a new host bacterium or be mobilized by certain conjugative plasmids (Paulsen et al., 1996). In a recent examination of the community carriage

of trimethoprim resistance in commensal gram-negative enterobacteria it was found that over 70% of isolates from human faeces were trimethoprim resistant and amongst these over 50% contained plasmids carrying a trimethoprim resistance determinant that was transferable to a new E. coli host (Adrian et al., 1995). Though relatively few detailed studies are available, it is likely that this extends to the resistance genes found in environmental bacteria, because it is self-evident that horizontal gene transfer is unlikely to be confined to bacteria that are human or animal pathogens.

Even when the antibiotic resistance genes are not found on plasmids, but instead incorporated into the bacterial chromosome, it is often the case that plasmids have played a role in bringing the new genes into the host cell. An incoming plasmid can be stably incorporated into the chromosome or the relevant genes contained in a transposon can jump onto the chromosome before the plasmid is lost. For example, in Salmonella strains with chromosomally-located resistance genes, the genes are identical to ones previously found in plasmids (Rajakumar et al., 1997; Briggs and Fratamico, 1999) and the trimethoprim resistance of Campylobacter that was thought to be intrinsic is in fact caused by acquired resistance genes that have become stably incorporated into the chromosome (Gibreel and Sköld, 1998). The mobile elements known as conjugative transposons, here called conjugons, are also found in bacterial chromosomes. Conjugons can be viewed as conjugative plasmids that have dispensed with the ability to be maintained as an independently replicating DNA molecule or replicon and replaced it with the ability to integrate into the bacterial chromosome, where they are replicated passively

Table 1. Properties of horizontal gene transfer vehicles[a]

	Stable maintenance			Transfer	
	Replication	Stability	Integration	Conjugation	Mobilization
Plasmid	+	+	o	o	o
Conjugative Plasmid	+	+	o	+	+
Mobilizable Plasmid	+	+	o	−	+
Conjugon[b]	−	−	+	+	+
NBU[c]	−	−	+	−	+

a. + = essential, − = absent, o = optional

b. conjugative transposon

c. non-replicating Bacteroides units

(Table 1; Salyers and Amábile-Cuevas, 1997). They came to light in cases where transfer of resistance occurred but plasmids were not detectable. Their role in the broad dissemination of antibiotic resistance genes in gram-positive genera is possibly as important as that of plasmids and, as in the case of plasmids, the resistance genes are effectively cargo (Rice, 1998; Clewell *et al.*, 1995; Salyers *et al.*, 1995).

The pivotal role of plasmids and conjugons in the emergence and spread of antibiotic resistance genes (particularly in human and animal pathogens) is dependent on mechanisms that enable them to acquire antibiotic resistance genes. Antibiotic resistance genes are not found in plasmids prior to antibiotic use (Datta and Hughes, 1983). Many plasmids that carry two or more genes each conferring resistance to a different antibiotic or family of antibiotics have been identified in organisms such as Enterobacteria, Pseudomonads, Vibrio, Staphylococci and Enterococci. A number of specific processes that facilitate such gene acquisition events are well studied, and these processes can collectively be described as translocation. In this chapter, the term "translocation" is used in the most general sense to describe any process that facilitates inter-replicon or inter-genome gene movement. Translocation of genes represents a specific sub-set of gene mobility in which the genes move from a location in one replicon (bacterial chromosome or plasmid) into a second replicon or other small genome (e.g. conjugon, transposon) and is illustrated in Figure 1B.

This chapter focuses on the antibiotic resistance genes that are acquired from the shared bacterial gene pool and resistance arising by alteration of genes in the bacterial chromosome is not discussed. A large number of acquired resistance determinants have now been identified and sequenced and in many instances their ability to translocate has been studied. Two common mechanisms for movement are transposition and site-specific recombination and the role of these processes in the evolution of multidrug resistance plasmids is examined. However, other processes that facilitate the movement of genes to new locations are known and more are likely to come to light in the future. The potential origins of the acquired genes, and their further evolution by mutation in response to the selective pressure of the use of particular antibiotics are also covered.

TRANSLOCATION

A number of distinct mechanisms that facilitate the movement of discrete DNA segments from one location in the DNA to another location are known. These mobile DNA segments have identifiable contents and boundaries and thus each is a definable small genome In the absence of direct experimental evidence for movement, they can be identified once identical copies are found in three or more distinct locations and their prior movement can be traced. The two groups of translocatable small genomes that frequently contain antibiotic resistance genes are transposons (Craig, 1996a; Kleckner, 1977) and gene cassettes (Hall *et al.*, 1991; Recchia and Hall, 1995a; Hall and Collis, 1998) and these are described in more detail below.

Transposons

The discrete small mobile genomes known as insertion sequences (IS) and transposons (Tn) are bounded by short inverted repeats and include one or more genes encoding the proteins required for their own movement. When they move, they create a duplication of the target site. The translocation of IS and Tn involves the recognition of and cleavage at the discrete boundaries of the mobile element, which is then spliced into a new location. The details of the transposition process are not always the same. Both direct, or "cut and paste", and "replicative" transpositional mechanisms are known (Craig, 1996a; Kleckner, 1977). However, in both cases a transposase recognizes the inverted repeats that bound IS and Tn and cleaves the DNA at the ends. With a cut and paste mechanism, the IS or Tn is cut out and spliced into a new location, which is also cleaved by the transpositional machinery. In replicative transposition, only one nick is formed in each strand of the original DNA molecule containing the IS or Tn, and after splicing into the new location, replication is required to complete the process. This creates two copies of the IS or Tn in a cointegrate of the two participating DNA molecules, and further events are required to resolve this cointegrate. The transpositional machinery of IS and Tn also catalyse other events that are important in shaping bacterial genomes (Kleckner, 1977), one of which is the deletion of adjacent DNA.

Insertion sequences carry only the information needed for transposition whereas transposons are somewhat larger and carry additional information e.g. antibiotic resistance genes. Some transposons, known as class I or composite transposons, consist of a pair of insertion sequences (IS) that encode the transpositional machinery flanking a region of DNA that contains one or more antibiotic resistance genes. Representative examples are shown in Figure 2. Such Tn presumably arise by transposition of two copies of the IS into the DNA molecule that is the source of the central region. A variation of this type of IS-assisted movement has given rise to several of the multidrug resistance arrays found in Staphylococci (Paulsen *et al.*, 1996, Skurray and Firth, 1997). In this case, it appears that the integration of small plasmids or other DNA fragments carrying antibiotic resistance genes is effected by replicative transposition of IS257 into either a plasmid or the chromosome without subsequent resolution (Needham *et al.*, 1995, Leelaporn *et al.*, 1996). Multiple events of this type or homologous recombination between two copies of IS257 in different DNA molecules leads to the formation of arrays of several antibiotic resistance genes interspersed by copies of IS257.

The second major group of transposons (the class II Tn) are bounded by only short (20–50 bp)

inverted repeats and a region at one end contains the transposition genes (Figure 3A). The most numerous members of this group belong to the Tn3 family (Sherratt, 1989). Multiple antibiotic resistance is most common in the Tn21-subfamily of the Tn3 family (Grinsted *et al.*, 1990; Brown *et al.*, 1996) largely because more than one transposon from this group (e.g. Tn21 and Tn1696) has acquired and carries within it an integron which is able to accumulate gene cassettes containing the resistance genes (see below). The Tn7 family also falls into this broad group of transposons though the mechanism of transposition is quite distinct (Craig, 1996b) and Tn7 includes the features characteristic of an integron (see below) and an array of three antibiotic resistance cassettes.

Transposons that are Integrons

The features that characterize an integron (Figure 4A) are a gene, *intI*, coding for a site-specific recombinase of the bacterial integrase superfamily, a recombination site, *attI*, that is recognised by the integrase and is the receptor site for gene cassettes, and a promoter P_c (formerly P_{ant}) that directs transcription of genes found in the integrated gene cassettes (Stokes and Hall, 1989; Collis and Hall, 1995). In gram-negative bacteria, the commonly isolated carriers of gene cassettes containing antibiotic resistance genes are integrons that are either active transposons or transposition-defective derivatives of transposons

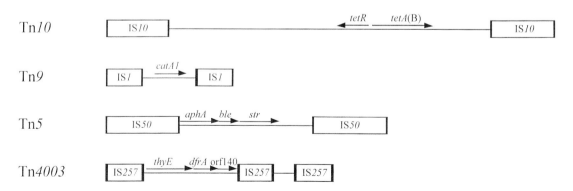

Figure 2.   Composite transposons.
Composite or class I transposons consist of two IS elements flanking a DNA sequence which includes one or more antibiotic resistance genes. The resistance genes shown confer resistance to tetracycline (*tetA* (B)), chloramphenicol (*catA1*), kanamycin (*aphA*), bleomycin (*ble*), streptomycin (*str*) and trimethoprim (*dfrA*). The extent and location of genes is indicated by an arrow. Transposition functions are encoded by the IS sequences.

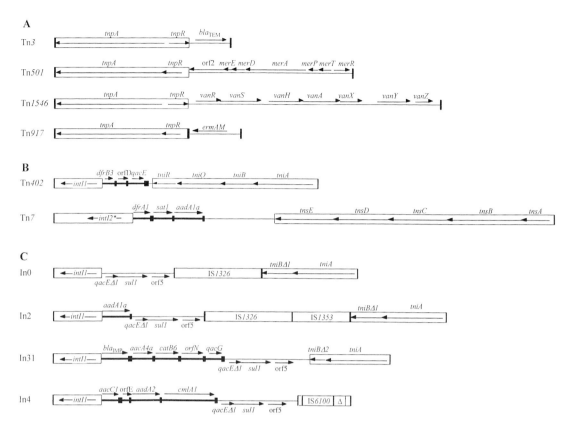

Figure 3.   Class II transposons and integrons.

A. Class II Tn are bounded by short terminal inverted repeats (black bars) and consist of a region containing genes (*tnp*) essential for transposition grouped at one end (open box) and a region containing genes that confer resistance to antibiotics (*bla*TEM, *van*, *ermAM*) or mercury (*mer*).

B. Integrons which are active transposons. Tn*402* (Tn*5090*, In28), a class 1 integron and Tn7, a class 2 integron, contain both transposition (*tni* or *tns*) and integron (*intI*) regions (open boxes) separated by an array of gene cassettes. Gene cassettes consist of a gene coding region (indicated by an arrow) and a downstream 59-base element recombination site (depicted by a small black box). The cassettes shown contain genes that determine resistance to trimethoprim (*dfrA1*, *dfrB3*), quaternary ammonium compounds (*qacE*), streptothricin (*sat*) and streptomycin and spectinomycin (*aadA1a*). The orfD cassette contains an open reading frame of unknown function.

C. Class 1 integrons that contain the 3'-conserved segment (3'-CS). The 3'-CS (thin line) containing the *sul1* sulphonamide resistance determinant, the *qacE∆1* gene and orf 5 is not found in Tn*402* and these integrons are transposition-defective due to loss of part or all of the *tni* region but are still bounded by terminal inverted repeats. IS, transposition and integrase gene regions are represented by open boxes. Further cassettes shown contain genes that determine resistance to gentamicin (*aacC1*), amikacin (*aacA4a*), streptomycin and spectinomycin (*aadA2*), imipenem (*bla*IMP), chloramphenicol (*cmlA1* and *catB6*) and quaternary ammonium compounds (*qacG*). The orfE and orfN cassettes contain open reading frames of unknown function. In0, which contains no cassettes, is from plasmid pVS1; In2 is in Tn*21* which is found in plasmid NR1; In4 is in Tn*1696* which is from R1033; In31 is from pPAM-101.

(Stokes and Hall, 1989; Hall and Collis, 1998; Hall, 1997). These integrons fall into two classes that are distinguished by the sequence of the *intI* gene. In the case of the class 2 integrons, the dual functions are most easily seen. Tn7 includes both a complete set of genes (*tns*) for the transposition functions (Craig, 1996b) and also contains, at the left end, the characteristic features of an integron, namely a gene encoding the class 2 integrase (*intI2**) and a site (*attI2*) recognized by the integrase. Tn7 contains an array of gene cassettes that includes the antibiotic resistance genes *dfrA1*, *sat* and *aadA1a* (Figure 3B).

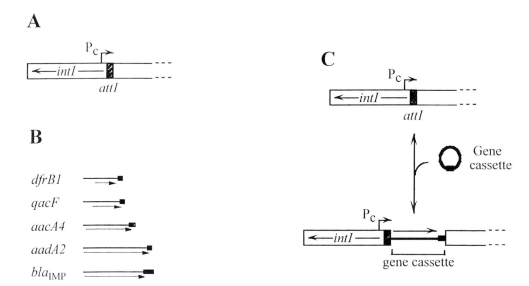

Figure 4. Gene cassettes and integrons.

A. Schematic representation of integron features. The *intI* gene encodes the integrase, which recognises the adjacent *attI* site (shaded). The cassette-encoded genes are transcribed from a promoter, P_c, in the integron.

B. Gene cassettes. Some representative gene cassettes are shown in the linear form. Each gene cassette consists of a gene coding region (indicated by an arrow) and a 59-base element recombination site (small black box). The cassettes shown all contain antibiotic resistance genes: *dfrB1* (trimethoprim resistance), *aacA4* and *aadA2* (aminoglycoside resistance), *bla*$_{IMP}$ (imipenem resistance) and *qacF* (resistance to quaternary ammonium compounds).

C. Integration of a circular gene cassette into an integron. Circular gene cassettes may be incorporated into the *attI* receptor recombination site (shaded) of the integron by site-specific recombination catalysed by the IntI integrase encoded by the integron.

In other transposons of this group the *dfrA1* cassette either has been lost or replaced by a different cassette (see Recchia and Hall, 1995a and below).

Though the class 1 integrons are more prevalent than class 2 integrons in clinical multidrug resistant bacterial isolates, only one member of class 1 that is an active transposon has been characterized to date (Rådström *et al.*, 1994). The transposon Tn*402* (referred to as Tn*5090* by Rådström *et al.*, 1994) has an overall organization that is similar to that of Tn*7* (Figure 3B). However, the integrase encoded by the *intI1* gene is distinct and there are differences in the transposition gene (*tni*) set (see Kholodii *et al.*, 1995 for details).

The more commonly isolated members of the class 1 integron group are defective transposons (Figure 3C), having lost most or all of the *tni* genes (Brown *et al.*, 1996; Hall, 1997). However, most of these class 1 integrons retain the short inverted repeats that bound Tn*402* and are thus potentially

mobile if transposition functions are supplied *in trans*. Indeed, the fact that class 1 integrons with identical backbone structures are found in several distinct locations is clear evidence of past movement (Stokes and Hall, 1989; Hall *et al.*, 1994; Brown *et al.*, 1996). With the exception of Tn*402*, known class 1 integrons all contain a region known as the 3′-conserved segment that includes the sulphonamide resistance determinant *sul1* and a truncated version (*qacEΔ1*) of the *qacE* gene that confers resistance to antiseptics and disinfectants. The 3′-conserved segment may represent an array of gene cassettes that have become fused by deletion events, as *qacE* is found in a gene cassette in Tn*402* (Recchia and Hall, 1995a). However, further characterization of this group of class 1 integrons has revealed two distinct backbone structures to the right of the 3′-CS. The first group commonly contains IS*1326* and only part of the *tni* gene module and is represented by In0. In2, which is the integron in Tn*21*, is related to In0 but contains a further IS (Brown *et al.*, 1996) and In31 is a close relative that has lost IS*1326* (Laraki *et al.*, 1999). The second group is represented by In4 and contains IS*6100* and none of the *tni* genes (Hall *et al.*, 1994; Hall, 1997). In4 is found in Tn*1696*, which has a backbone structure similar to that of Tn*21* and, like Tn*21*, contains a mercury resistance gene set.

An integron does not necessarily contain a gene cassette and In0 (Figure 3C) is an example of an integron with no gene cassettes. However, most class 1 integrons found in clinical bacterial isolates do contain one or more gene cassettes, and can contain multiple cassettes each conferring resistance to a different antibiotic or group of antibiotics. In4 and In31 both contain such a multidrug resistance array (Figure 3C). However, if the 3′-CS and hence the *sul1* gene is present, the presence of even one cassette results in multidrug resistance.

Gene Cassettes

Gene cassettes are a novel family of small mobile elements that consist of a single gene (or occasionally two genes) and a recombination site that is located downstream of the gene (Figure 4B). These recombination sites, known as 59-base elements (59-be), are recognized by integron integrases and confer mobility. Though all 59-be tested to date are functional as IntI1 recognition sites, they have a wide range of sizes and sequences. The common features are regions of 20–25 bp at the outer ends that each conform to a consensus (Collis and Hall, 1992b; Stokes *et al.*, 1997). These two consensus regions are also related, being imperfect inverted repeats of one another, and both include the features of a simple recombinase integration site (Stokes *et al.*, 1997). However, 59-be are distinct from the *attI* sites found in the integrons that are also recognized by integron-encoded integrases (Collis *et al.*, 1998).

A total of over 60 antibiotic resistance genes have now been found in a gene cassette (Partridge and Hall, unpublished) and amongst them are genes that determine resistance to trimethoprim, chloramphenicol, aminoglycosides, β-lactam family antibiotics and quaternary ammonium antiseptics and disinfectants (Recchia and Hall, 1995a), and also

rifampin (Tribuddharat and Fennewald, 1999) and erythromycin (Plante and Roy, unpublished). In some cases, genes that confer resistance to the same antibiotic by distinct mechanisms are found e.g. multiple chloramphenicol resistance genes that confer resistance by efflux (*cmlA*) or by acetylation (*catB*) have been found. For other antibiotics, though several different genes have been identified all of them confer resistance by the same mechanism e.g. *dfr* genes confer resistance to trimethoprim by supplying a trimethoprim-insensitive dihydrofolate reductase. Thus a wide variety of antibiotic resistance genes have been packaged together with a 59-be recombination site to form a gene cassette.

Cassettes can be excised by the IntI1 integrase (Collis and Hall, 1992a) and can exist transiently as free circles (Collis and Hall, 1992b). Cassettes in the circular form are integrated into an integron by IntI-mediated recombination between the 59-be and the *attI* site in the integron (Collis *et al.*, 1993; Figure 4C). However, the same integrative and excisive recombination events occurring between *attI* or 59-be sites in two different integrons can lead to the loss, gain or rearrangement of cassettes without the necessity for the free circular intermediate (Collis and Hall, 1992a; Hall and Collis, 1995) and homologous recombination within the conserved sequences flanking the cassettes can also effect the exchange of cassette arrays (see Hall and Collis, 1995).

Though the gene cassettes containing antibiotic resistance genes are generally found within an integron (class 1, 2 or 3), they can also become incorporated elsewhere (Recchia and Hall, 1995b; Segal and Elisha, 1997). This is assumed to involve a reaction between the 59-be in the cassette and a secondary site catalysed by an integron-encoded integrase, and IntI1 has been shown to catalyse such events (Recchia *et al.*, 1994; Francia *et al.*, 1993). Cassettes incorporated at secondary sites cannot be excised and the gene they contain is silent unless an appropriately-oriented promoter is present upstream of the cassette (Recchia and Hall, 1995b).

MULTIDRUG RESISTANCE PLASMIDS AND THE PROCESSES THAT SHAPE THEM

The plasmid NR1 and several other well-studied plasmids provide a paradigm for understanding the evolution of plasmids that carry multiple antibiotic resistance genes. NR1 (R100), which was recovered from a multidrug resistant *Shigella flexneri* isolated in Japan in the late 1950s, has been extensively studied (Watanabe, 1963; Womble and Rownd, 1988) and determines resistance to several antibiotics, tetracycline, chloramphenicol, fusidic acid, sulphonamides, streptomycin and spectinomycin, and also determines resistance to mercury ions (HgII). The resistance genes of NR1 also represent a broad range of mechanisms. The *tetA* gene encodes an inner membrane efflux pump that confers resistance by reducing the intracellular concentration of tetracycline (Levy, 1992). The *catA1* (*catI*) gene encodes an acetyltransferase that inactivates chloramphenicol by adding an acetyl group and also confers resistance to fusidic acid, though this effect appears to be fortuitous (Murray and Shaw, 1997). The *aadA1* gene confers resistance to both streptomycin and spectinomycin; the AAD(3″)(9) protein is an adenylyltransferase that modifies both of these antibiotics. The *sul1* (formerly *sulI*) gene codes for a sulphonamide-resistant dihydropteroate synthase that substitutes for the susceptible cellular enzyme (Sundström *et al.*, 1988).

NR1 is a conjugative plasmid belonging to the IncFII incompatibility group and contains, in addition to the essential replication (*ori, rep*) and stability (*stb*) functions, a region that is essential for conjugative transfer and includes transfer (*tra*) genes, the transfer origin (*oriT*) and the nickase (*mob*) that recognizes *oriT*. The antibiotic resistance genes are all part of mobile elements that have presumably been incorporated into an ancestral plasmid backbone. The tetracycline resistance determinant *tetA*(B) is found in the transposon Tn*10* (Figure 2). The mercury resistance determinants and the resistance determinants for all of the remaining antibiotics are clustered in a complex array (Figure 5) that includes representatives of each of the most important types of mobile elements; insertion sequences (IS), a compound class I transposon, a class II transposon, a gene cassette and an integron are found within this region. Overall, the region is organized as a series of mobile elements each located within a larger structure which is also mobile. The structure and organization of Tn*21*, which is part of this region, has recently been reviewed (Liebert *et al.*, 1999).

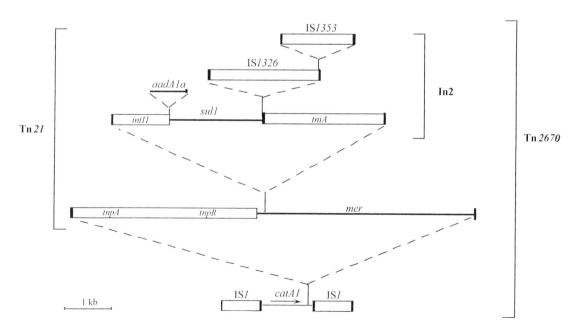

Figure 5. The complex multidrug resistance transposon Tn*2670* from plasmid NR1. Tn*2670* consists of Tn*21* inserted in a Tn that is closely related to Tn*9* (see Figure 2) and includes a number of discrete transposons and IS that are able to move independently or together in different combinations. Features of IS, integrons, transposons and *aadA1* gene cassette are depicted as in Figures 3 and 4.

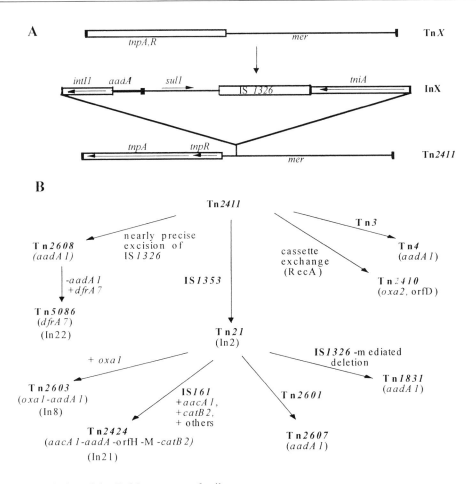

Figure 6.   Evolution of the Tn21 transposon family.

A. The progenitor Tn2411 is proposed to have been created by insertion of an integron (InX) into an unknown mercury resistance transposon (TnX).

B. The relationships of all of the transposons in this group can then be explained by subsequent events including integration, excision or exchange of gene cassettes, insertion of transposons or insertion sequences, insertion-sequence mediated deletions or insertion sequence excision e.g. In2 and Tn21 are derived from InX and Tn2411 by acquisition of IS1353.

Many very close relatives of Tn21 have been reported (see Grinsted *et al.*, 1990). They differ from the common progenitor Tn2411 mainly in that they have accumulated further IS and transposons, have added to the cassette array found in In2 or have exchanged the *aadA1a* gene cassette for one or more cassettes that determine resistance to other antibiotics (Brown *et al.*, 1996). This group of transposons illustrate that evolution of multidrug resistance plasmids and transposons is an ongoing process (Figure 6). Further moulding of these genomes has occurred by deletion and rearrangements effected by a variety of both known and currently little-understood mechanisms. A particularly influential one is the deletion of DNA immediately adjacent to one end of an IS or Tn (Kleckner, 1977). Such IS-mediated adjacent deletions are common and many examples such as the loss of the *tni* genes from transposition defective class 1 integrons (Brown *et al.*, 1996) and variations in the structure of Tn4003 that affect the level of expression of the *dfrA* gene (Leelaporn *et al.*, 1994) appear to have occurred by this mechanism.

A further way of shaping plasmid genomes is via cointegration of two distinct plasmids to form a single larger structure. For example, some of the antibiotic resistance genes in multidrug resistant Staphylococci plasmids are part of the genome of a small plasmid that carries a single antibiotic resistance gene which has become integrated into a larger, often conjugative plasmid via an IS257-mediated process (Skurray and Firth, 1997). It is likely that transposition by a replicative mechanism leads to the formation of such cointegrates, and the resulting structures can be very stable. Furthermore, even when a cointegrate formed in this way exists only transiently, a cointegrate between a self-transferable plasmid and one that is not can effect the transfer of the non-transferable plasmid into a new host in a process known as conduction. If resolution occurs within the IS or Tn that formed the cointegrate, the two original plasmid structures will be re-formed, but both will now contain the IS. Resolution can occur by either homologous recombination or by specific processes. For example, transposons of the Tn*3*/Tn*21* family transpose by a replicative mechanism, but encode a resolvase that effects the resolution of the cointegrate by a site-specific recombination event that re-forms two molecules now each containing the transposon (Sherratt, 1989). However, because these resol-vases can recognize sites that are closely related to the one at which they normally act, if a related transposon is present in the same cointegrate molecule, resolution at the alternate site can lead to a redistribution of the genomes of the two original plasmids. The same effect can be achieved by homologous recombination if the two original plasmids contained any other identical DNA segment. In this case two new plasmids will be formed if both resolution products contain a replication origin and replication functions, but only one new plasmid if one segment retains both of the original replication regions.

ORIGINS OF ANTIBIOTIC RESISTANCE GENES

Though the sequencing of antibiotic genes got off to a slow start, over the last 15 years many have been isolated and sequenced. This has revealed that the antibiotic resistance gene pool is large and often contains a number of genes that confer resistance to one or more members of the same antibiotic family. Frequently, resistance is effected by more than one

mechanism. These mechanisms include inactivation by modification of the antibiotic, efflux to reduce the intracellular concentration, target replacement and modification of the target such that it is no longer affected. For example, resistance to tetracycline is conferred by both efflux and ribosomal (target) modification and resistance to chloramphenicol is effected by both efflux and acetylation of the antibiotic. In many cases, more than one gene for each individual mechanism has been found. Thus families of genes whose products are related have been identified. This information has permitted questions to be asked about the evolutionary origins of the genes and their immediate origins to be sought.

Evolutionary Origins

The origins of resistance genes are not well understood, though speculation abounds, and there are several plausible theories for the origins of antibiotic resistance genes (Davies, 1997). Many antibiotics are natural products of bacteria or other microorganisms and both the antibiotic producers themselves and bacteria living in the same environment as the producers need strategies to protect themselves from the deleterious effects of the antibiotics and other harmful natural products. However, the chromosomes of all bacterial species also provide a potential source of resistance genes.

Many resistance genes isolated from antibiotic-producing Streptomycetes have been sequenced in the last 15 years and the mechanisms are mostly the same as those determined by antibiotic resistance genes from antibiotic resistant bacteria of human or animal origin. However, the relationships between these two groups of genes and the proteins they encode are, generally, very distant in evolutionary terms, where 1% divergence represents about 1 million years (Ochman and Wilson, 1987). For example, in Streptomycetes resistance to many aminoglycosides involves phosphorylation of the antibiotic(s) and several acquired aminoglycoside phosphotransferase (*aph*) genes have been found in both gram-positive and gram-negative bacteria. Though measurable levels of similarity are found when these proteins are aligned, the Streptomycete enzymes form a cluster that is distinct from the cluster formed by known resistance genes from both gram-positive and gram-negative bacteria (Martin *et al.*, 1988; Kirby, 1990). A similar situation applies in the case of class A β-lactamases (Kirby, 1992) and the cluster of genes that confer resistance to vancomycin (Marshall *et al.*, 1998). In other cases the relationships are even more distant; acetyltransferases from two Streptomyces species that produce antibiotics in the streptothricin family are related to one another but show only very limited similarity to the acetyltransferases found in Enterobacteriaceae that confer resistance to these antibiotics (Krügel *et al.*, 1993). In one case, a closer relationship has been demonstrated. Genes related to the oxytetracycline resistance genes of *Streptomyces rimosus* in were found in species of Mycobacteria using stringent hybridization (Pang *et al.*, 1994; Roberts, 1997). However, these genes may still be only 80% identical and thus represent inter-species transfer events that are recent in terms of evolutioary time but ancient in real time and sequences are needed to resolve this issue.

Thus, while the Streptomycetes and other antibiotic-producing organisms remain a potential source of acquired antibiotic resistance genes, most of the known relationships are distant in terms of evolutionary time, and they do not appear to be the immediate source of any of the known acquired resistance genes. While this can be interpreted as indicating that the acquired resistance genes found in clinically relevant bacteria have been evolving over very long periods, other potential sources also need to be considered.

A second likely origin for the antibiotic resistance genes is the genomes of eubacteria. For the synthetic antibiotics in the trimethoprim and sulphonamide families, acquired resistance is most commonly due to antibiotic-insensitive enzymes that perform the same function as the antibiotic-sensitive targets of the antibiotics, namely dihydrofolate reductase and dihydropteroate synthase. Many acquired trimethoprim-resistance genes (*dfr*) have been sequenced as well as the trimethoprim-sensitive chromosomal *dfr* genes from many bacteria. The products form a single family (Sundström *et al.*, 1995), now designated *dfrA*, with the exception of three closely-related acquired resistance genes (*dfrB*). The known trimethoprim-resistant DHFR are distributed on several arms of the highly branched evolutionary tree, where mostly they cluster together with one or more

chromosomally-determined bacterial DHFR. This suggests that the acquired *dfr* genes that confer resistance are as evolutionarily diverse as the chromosomal *dfr* genes, and it is possible that each has arisen from a different bacterial source by acquiring both mutations that confer resistance and a place in a plasmid, transposon or gene cassette. Whether this has, in each case, occurred since the widespread use of trimethoprim or whether these genes were already present in the shared floating genome is not known.

A more distant relationship is seen in the case of Tet proteins that confer resistance by a ribosomal protection mechanism (Taylor and Chau, 1996). These proteins appear to have an evolutionary origin in common with the protein synthetic elongation factors EF-G and EF-Tu. Likewise, chromosomal genes for a variety of acetyl, phosphoryl and adenylyl (nucleotidyl) transferases also provide a potential source of antibiotic modifying enzymes (Davies, 1997), class A and D β-lactamases have evolutionary links to the cell wall biosynthetic enzymes inhibited by β-lactam antibiotics and to other penicillin binding proteins (Massova and Mobashery, 1998), and potential progenitors of two cassette-associated antibiotic resistance genes have been found (Recchia and Hall, 1997).

Recent Resistance Gene Mobilization Events

Though the similarities between proteins that perform the same or related functions provide insights into the potential evolutionary origins of resistance genes, they in no way explain the sources of the antibiotic resistance genes currently found in transposons, gene cassettes and plasmids. However, two cases of recent gene pick-up events are known. Both of these examples are likely to represent quite recent events but they also provide insight into how other genes may have become mobilized in the more distant past.

In one case the immediate origin of a trimethoprim resistance gene has been identified. The *dfrA* gene found in Tn*4003* in *Staphylococcus aureus* plasmids that confer resistance to trimethoprim is part of a region of 2 kb that is flanked by two copies of IS*257* (Figure 2). This segment of Tn*4003* appears to have been created by pickup of a DNA fragment that contains the genes for dihydrofolate reductase (*dfrC*) and thymidylate synthase (*thyF*) from the chromosome of a strain of *S. epidermidis* (Dale, 1995). The trimethoprim-sensitive chromosomal *dfr* gene has evolved by a few mutations to create the trimethoprim-resistant *dfrA* gene found in Tn*4003* and its relatives and a promoter created using a -35 region in IS*257* leads to high levels of expression of *dfrA* (Leelaporn *et al.*, 1994).

A second example of pickup of chromosomal genes is provided by several *ampC* β-lactamase genes recently found on plasmids. Many Enterobacteria encode a class C β-lactamase (AmpC) within their genome, though its normal function is unknown. Some of the plasmid-encoded AmpC are closely related to AmpC from the chromosomes of Enterobacteriaceae, and in one case a sequence identical to the *ampC* and *ampR* genes found in the chromosome of a strain of *Morganella morganii* is found in a plasmid that was isolated from *E.coli* (Barnaud *et al.*, 1998; Poirel *et al.*, 1999).

EVOLUTION OF ACQUIRED ANTIBIOTIC RESISTANCE GENES

Each individual antibiotic resistance determinant has a particular spectrum of activity; for example a large number of β-lactamases have been found (Bush *et al.*, 1995) and each confers resistance to a specific subset of the β-lactam family of antibiotics (penicillins, cephalosporins, carbapenems and monobactams). Some of these genes are more prevalent than others, and new members of this antibiotic family have been developed in response to this. However, as each new antibiotic is used more extensively, over time strains containing lactamases that can inactivate the new antibiotic emerge (Jacoby, 1994; Bush, 1997). These represent either new resistance genes or mutant versions of the prevalent resistance gene that have an altered spectrum of activity. In the case of the widely-distributed β-lactamases known as TEM and SHV, several variants with an extended substrate range have been identified. In each case one or a few single base changes has led to subtle alterations that allow further members of this antibiotic family to be efficiently hydrolysed, or reduce the inhibitory effects of β-lactamase inhibitors, such as clavulanate (Bush, 1997; Du Bois *et al.*, 1995). A single base change is also known to alter the substrate specificity of one of the most prevalent aminoglycoside acetyltransferases (Rather *et al.*, 1992). Thus, individual acquired antibiotic resistance genes continue to evolve as a consequence of the selective pressure of antibiotic use.

CONCLUSIONS

The driving force in the development and spread of antibiotic resistance is the selective pressure of antibiotic use. Selection favors the resistant bacterium. Resistance and multi-drug resistance has arisen by the movement of antibiotic resistance genes onto the most successful vehicles for horizontal gene transfer, namely plasmids and conjugative transposons followed by the spread of these elements to other species. Antibiotic resistance genes have made their way onto the horizontal gene transfer highway by a variety of translocation mechanisms, the best understood of which are transposition and site-specific recombination. Transposons and the integron and gene cassette systems have clearly contributed extensively to the emergence of antibiotic-resistant strains of both pathogenic and commensal bacteria. The accumulation of more than one resistance determinant on the same plasmid or conjugon leads to multidrug resistance that can transfer from one bacterium to another and also has the potential to become stably incorporated into bacterial chromosomes.

There are now many known genes whose products confer resistance to each antibiotic or antibiotic family and often more than one mechanism (e.g. antibiotic modification and antibiotic efflux) is represented. This suggests that the pool of resistance genes is very large. Though many examples of families of related genes or proteins have been found, the extent of the divergence is such that they have clearly evolved over a far longer time span than the 50 years of antibiotic use in human and veterinary medicine. The evolutionary origins of these acquired antibiotic resistance genes remain the subject of much speculation. The immediate origins of resistance genes have been identified only in a few instances where known chromosomal genes have become associated with plasmids.

Though the story as it stands today is one where the identity and relative contribution of individual mobile elements is specific for each organism studied, the themes are common. Furthermore, though each individual resistance gene is generally most prevalent in one species or group of species, as more detailed studies become available, identical genes are being found in other species, indicating that the genes are continuing to spread. It seems likely that selective pressure has led to the capture and amplification of individual genes or mobile genetic elements that may move only exceedingly rarely into the populations of bacteria associated with humans and domestic and food-producing animals. If this is so, it is possible that the initial association of some of the individual genes with individual bacterial species is simply the result of chance. Subsequent clonal spread of the organism containing the resistance gene has ultimately led to the dissemination of many of the genes globally and horizontal gene transfer has distributed them into new species and genera.

REFERENCES

Adrian, P.V., Klugman, K.P. and Aymes, S.G.B. (1995) Prevalence of trimethoprim resistant dihydrofolate reductase genes identified with oligonucleotide probes in plasmids from isolates of commensal faecal flora. *J. Antimicrob. Chemother.*, **35**, 497–508.

Barnaud, G., Arlet, G., Verdet, C., Gaillot, O., Lagrange, P.H. and Philippon, A. (1998) *Salmonella enteritidis*: AmpC plasmid-mediated inducible beta-lactamase (DHA-1) with an *ampR* gene from *Morganella morganii*. *Antimicrob. Ag. Chemother.* **42**, 2352–8.

Briggs, C.E. and Fratamico, P.M. (1999) Molecular characterization of an antibiotic resistance gene cluster of *Salmonella typhimurium* DT104. *Antimicrob. Ag. Chemother.* **43**, 846–849.

Brown, H.J., Stokes, H.W. and Hall, R.M. (1996) The integrons In0, In2 and In5 are defective transposon derivatives. *J. Bact.* **178**, 4429–4437.

Bush, K. (1997) The evolution of β-lactamases. In D.J. Chadwick and J. Goode (eds.), Ciba Foundation Symposium 207 on *Antibiotic Resistance: Origins, Evolution, Selection and Spread*, John Wiley and Sons, Chichester, pp. 152–166.

Bush, K., Jacoby, G.A. and Medeiros, A.A. (1995) A functional classification scheme for β-lactamases and its correlation with molecular structure. *Antimicrob. Ag. Chemother.* **39**, 1211–1233.

Clewell, D.B., Flannagan, S.E. and Jaworski, D.D. (1995) Unconstrained bacterial promiscuity: the Tn*916*-Tn*1545* family of conjugative transposons. *Trends Microbiol.* **229**, 229–236.

Collis C.M. and Hall, R.M. (1992a) Site-specific deletion and rearrangement of integron insert genes catalysed by the integron DNA integrase. *J. Bact.* **174**, 1574–1585.

Collis, C.M. and Hall, R.M. (1992b) Gene cassettes from the insert region of integrons are excised as covalently closed circles. *Mol. Microbiol.* **6**, 2875–2885.

Collis, C.M., Grammaticopolous, G., Briton, J., Stokes, H.W. and Hall, R.M. (1993) Site-specific insertion of gene cassettes into integrons. *Mol. Microbiol.* **9**, 41–52.

Collis, C.M. and Hall, R.M. (1995) Expression of antibiotic resistance genes in the integrated cassettes of integrons. *Antimicrob. Agents. Chemother.* **39**, 155–162.

Collis, C.M., Kim, M.-J., Stokes, H.W. and Hall, R.M. (1998) Binding of the purified integron DNA integrase IntI1 to integron- and cassette-associated recombination sites. *Mol. Microbiol.* **29**, 477–490.

Craig, N. (1996a) Transposition. In F.C. Neidhart, (ed.), *Escherichia coli and Salmonella*. American Society for Microbiology, Washington, D.C., pp. 2339–2362.

Craig, N. (1996b) Transposon Tn*7*. *Curr. Top. Microbiol Immunol.* **204**, 27–48.

Dale, G.E., Broger, C., Hartman, P.G., Langen, H., Page, M.G.P., Then, R.L., *et al.* (1995) Characterization of the gene for the chromosomal dihydrofolate reductase (DHFR) of *Staphylococcus epidermidis* ATCC 14990: the origin of the trimethoprim-resistant S1 DHFR from *Staphylococcus aureus*? *J. Bact.* **177**, 2965–2970.

Datta, N. and Hughes, V.M. (1983) Plasmids of the same Inc groups in Enterobacteria before and after the medical use of antibiotics. *Nature* **306**, 616–617.

Davies, J.E. (1997) Origins, acquisition and dissemination of antibiotic resistance determinants. In D.J. Chadwick and J. Goode (eds.), Ciba Foundation Symposium 207 on *Antibiotic Resistance: Origins, Evolution, Selection and Spread*, John Wiley and Sons, Chichester, pp. 15–35.

Du Bois, S.K., Marriott, M.S. and Amyes, S.G.B. (1995) TEM- and SHV-derived extended spectrum β-lactamases: relationship between selection, structure and function. *J. Antimicrob. Chemother.* **35**, 7–22.

Francia, M.V., de la Cruz, F. and García Lobo, J.M. (1993) Secondary sites for integration mediated by the Tn*21* integrase. *Mol. Microbiol.* **10**, 823–828.

Gibreel, A. and Sköld, O. (1998) High-level resistance to trimethoprim in clinical isolates of *Campylobacter jejuni* by acquisition of foreign genes (*dfr1* and *dfr9*) expressing drug-insensitive dihydrofolate reductases. *Antimicrob. Ag. Chemother.* **42**, 3059–3064.

Grinsted, J., de la Cruz, F. and Schmitt, R. (1990) The Tn*21* subgroup of bacterial transposable elements. *Plasmid* **24**, 163–189.

Hall, R.M. (1997) Mobile gene cassettes and integrons: moving antibiotic resistance genes in gram-negative bacteria. In D.J. Chadwick and J. Goode (eds.), Ciba Foundation Symposium 207 on *Antibiotic Resistance: Origins, Evolution, Selection and Spread*, John Wiley and Sons, Chichester, pp. 192–205.

Hall, R.M. and Collis, C.M. (1995) Mobile gene cassettes and integrons: capture and spread of genes by site-specific recombination. *Mol. Microbiol.* **15**, 593–600.

Hall, R.M. and Collis, C.M. (1998) Antibiotic resistance in gram-negative bacteria: the role of gene cassettes and integrons. *Drug Resist. Updates* **1**, 109–119.

Hall, R.M., Brookes, D.E. and Stokes, H.W. (1991) Site-specific insertion of genes into integrons: role of the 59-base element and determination of the recombination cross-over point. *Mol. Microbiol.* **5**, 1941–1959.

Hall, R.M., Brown, H.J., Brookes, D.E. and Stokes, H.W. (1994) Integrons found in different locations have identical 5′ ends but variable 3′ ends. *J. Bact.* **176**, 6286–6294.

Jacoby, G.A. (1994) Extrachromosomal resistance in Gram-negative organisms: the evolution of β-lactamase. *Trends Microbiol.* **2**, 357–360.

Kholodii, G.Ya., Mindlin, S.Z., Bass, I.A., Yurieva, O.V., Minakhina, S.V. and Nikiforov, V.G. (1995) Four genes, two ends, and a *res* region are involved in transposition of Tn*5053*: a paradigm for a novel family of transposons carryingeither a *mer* operon or an integron. *Mol. Microbiol.* **17**, 1189–1200.

Kirby, R. (1990) Evolutionary origin of aminoglycoside phosphotransferase resistance genes. *J. Mol. Evol.* **30**, 489–492.

Kirby, R. (1992) Evolutionary origin of the class A and class C β-lactamases. *J. Mol. Evol.* **34**, 345–350.

Kleckner, N. (1977) Translocatable elements in procaryotes. *Cell* **11**, 11–23.

Krügel, H., Fielder, G., Smith, C. and Baumberg, S. (1993) Sequence and transcriptional analysis of the nourseothricin acetyltransferase-encoding gene *nat*1 from *Streptomyces noursei*. *Gene* **127**, 127–131.

Laraki, N., Galleni, M., Thamm, I., Riccio, M.L., Amicosante, G., Frère, J-M., et al. (1999) Structure of In31, a *bla*IMP-containing *Pseudomonas aeruginosa* integron phyletically related to In5, which carries an unusual array of gene cassettes. *Antimicrob. Ag. Chemother.* **43**, 890–901.

Leelaporn, A., Firth, N., Byrne, M.E., Roper, E. and Skurray, R.A. (1994) Possible role of insertion sequence IS*257* in dissemination and expression of high- and low-level trimethoprim resistance in Staphylococci. *Antimicrob. Ag. Chemother.* **38**, 2238–2244.

Leelaporn, A., Firth, N., Paulsen, I.T. and Skurray, R.A. (1996) IS*257*-mediated cointegration in the evolution of a family of staphylococcal trimethoprim resistance plasmids. *J. Bact.* **178**, 6070–6073.

Levy, S.B. (1992) Active efflux mechanisms for antimicrobial resistance. *Antimicrob. Ag. Chemother.* **36**, 695–703.

Liebert, C.A., Hall, R.M. and Summers, A.O. (1999) Transposon Tn*21*, flagship of the floating genome. *Microb. Mol. Biol. Rev.* **63**, 507–522.

Marshall, C.G., Lessard, A.D. Park, I.-S. and Wright, G.D. (1998) Glycopeptide antibiotic resistance genes in glycopeptide-producing organisms. *Antimicrob. Ag. Chemother.* **42**, 2215–2220.

Martin, P., Jullien, E. and Courvalin, P. (1988) Nucleotide sequence of *Acinetobacter baumannii aphA*-6 gene: evolutionary and functional implications of sequence homologies with nucleotide-binding proteins, kinases and other aminoglycoside-modifying enzymes. *Mol. Microbiol.* **2**, 615–25.

Massova, I. and Mobashery, S. (1998) Kinship and diversification of bacterial penicillin-binding proteins and β-lactamases. *Antimicrob. Ag. Chemother.* **42**, 1–17.

Murray, I.A. and Shaw, W.V. (1997) O-acetylation for chloramphenicol and other natural products. *Antimicrob. Ag. Chemother.* **41**, 1–6.

Needham, C., Noble, W.C. and Dyke, K.G.H. (1995) The Staphylococcal insertion sequence IS*257* is active. *Plasmid* **34**, 198–205.

Ochman, H. and Wilson, A.C. (1987) Evolution in bacteria: evidence for a universal substitution rate in bacterial genomes. *J. Mol. Evol.* **26**, 74–86.

Pang, Y., Brown, B.A., Steingrube, V.A., Wallace, R.J. and Roberts, M.C. (1994) Tetracycline resistance determinants in *Mycobacterium* and *Streptomyces* species. *Antimicrob. Ag. Chemother.* **38**, 1408–1412.

Paulsen, I.T., Firth, N. and Skurray, R.A. (1996) Resistance to antimicrobial agents other than β-lactams. In K.B. Crossley and G.L. Archer (eds.), *The Staphylococci in Human Disease*, Churchill Livingstone, New York. pp. 175–212.

Plante, I. and Roy, P.H. Accession No. AF099140.

Poirel, L., Guibert, M., Girlich, D., Naas, T. and Nordmann, P. (1999) Cloning, sequence analyses, expression, and distribution of ampC-ampR from *Morganella morganii* clinical isolates. *Antimicrob. Ag. Chemother.* **42**, 769–776.

Rådström, P., Sköld, O., Swedberg, G., Flensburg, J., Roy, P.H. and Sundström, L. (1994) Transposon Tn*5090* of plasmid R751, which carries an integron, is related to Tn*7*, Mu, and the retroelements. *J. Bact.* **176**, 3257–3268.

Rajakumar, K., Bulach, D., Davies, J., Ambrose, L., Sasakawa, C. and Adler, B. (1997) Identification of a chromosomal *Shigella flexneri* multi-antibiotic resistance locus which shares sequence and organizational similarity with the resistance gene region of the plasmid NR1. *Plasmid* **37**, 159–168.

Rather, P.N., Munayyer, H., Mann, P.A., Hare, R.S., Miller, G.H. and Shaw, K.J. (1992) Genetic analysis of bacterial acetyltransferases: identification of amino acids determining the specificities of the aminoglycoside 6′-N-acetyltransferase Ib and Iia proteins. *J. Bact.* **174**, 3196–3203.

Recchia, G.D. and Hall, R.M. (1995a) Gene cassettes: a new class of mobile element. *Microbiol.* **141**, 3015–3027.

Recchia, G.D. and Hall, R.M. (1995b) Plasmid evolution by acquisition of mobile gene cassettes: plasmid pIE723 contains the *aadB* gene cassette precisely inserted at a secondary site in the IncQ plasmid RSF1010. *Mol. Microbiol.* **15**, 179–187.

Recchia, G.D. and Hall, R.M. (1997) Origins of the mobile gene cassettes found in integrons. *Trends Microbiol.* **5**, 389–394.

Recchia, G.D., Stokes, H.W. and Hall, R.M. (1994) Characterisation of specific and secondary recombination sites recognised by the integron DNA integrase. *Nucleic Acids Res.* **11**, 2071–2078.

Rice, L.B. (1998) Tn*916* family conjugative transposons and dissemination of antimicrobial resistance determinants. *Antimicrob. Ag. Chemother.* **42**, 1871–1877.

Roberts, M.C. (1997) Genetic mobility and distribution of tetracycline resistance determinants. In D.J. Chadwick and J. Goode (eds.), Ciba Foundation Sympsium 207 on *Antibiotic Resistance: origins, evolution, selection and spread*, John Wiley and Sons, Chichester. pp. 206–222.

Salyers, A.A. and Amábile-Cuevas, C.F. (1997) Why are antibiotic resistance genes so resistant to elimination? *Antomicrob. Ag. Chemother.* **41**, 2321–2325.

Salyers, A.A., Shoemaker, N.B., Stevens, A.M. and Li, L-Y. (1995) Conjugative transposons: an unusual and diverse set of integrated gene transfer elements. *Microb. Rev.* **59**, 579–590.

Segal, H. and Elisha, B.G. (1997) Identification and characterization of an *aadB* gene cassette at a secondary site in a plasmid from Acinetobacter. *FEMS Lett.* **153**, 321–326.

Sherratt, D. (1989) Tn*3* and related transposable elements: site-specific recombination and transposition. In D.E. Berg and M.M. Howe (eds.), *Mobile DNA.*, American Society for Microbiology, Washington, D.C. pp. 163–184.

Skurray, R.A. and Firth, N. (1997) In D.J. Chadwick and J. Goode (eds.), Ciba Foundation Sympsium 207 on *Antibiotic Resistance: origins, evolution, selection and spread*, John Wiley and Sons, Chichester. pp. 167–191.

Stokes, H.W. and Hall, R.M. (1989) A novel family of potentially mobile DNA elements encoding site-specific gene integration functions: integrons. *Mol. Microbiol.* **3**, 1669–1683.

Stokes, H.W., O'Gorman, D.B., Recchia, G.D., Parsekhian, M. and Hall, R.M. (1997) Structure and function of 59-base element recombination sites associated with mobile genetic elements. *Mol. Microbiol.* **26**, 731–745.

Sundström, L., Radström, P., Swedberg, G. and Skuold, O. (1988) Site-specific recombination promotes linkage between trimethoprim- and sulphonamide resistance genes. Sequence characterization of *dhfrV* and *sulI* and a recombination active locus of Tn*21*. *Mol. Gen. Genet.* **213**, 191–201.

Sundström, L., Jansson, C., Bremer, K, Heikkilä, E., Olsson-Liljequist, B. and Sköld, O. (1995) A new *dhfrVIII* trimethoprim-resistance gene, flanked by IS*26*, whose product is remote from other dihydrofolate reductases in parsimony analysis. *Gene* **154**, 7–14.

Taylor, D.E. and Chau, A. (1996) Tetracycline resistance mediated by ribosomal protection. *Antimicrob. Ag. Chemother.* **40**, 1–5.

Tribuddharat C. and Fennewald, M. (1999) Integron-mediated rifampin resistance in *Pseudomonas aeruginosa*. *Antimicrob. Ag. Chemother.* **43**, 960–962.

Watanabe, T. (1963) Infective heredity of multiple drug resistance bacteria. *Bacteriol. Rev.* **27**, 87–115.

Womble, D.D. and Rownd, R.H. (1988) Genetic and physical map of plasmid NR1: comparison with other IncFII antibiotic resistance plasmids. *Microbiol Rev.* **52**, 433–451.

4. Adaptive Resistance to Antibiotics

Yan-Qiong Xiong,[1*] Arnold S. Bayer[1,2] and Gilles Potel[3]

[1.] Department of Medicine, Division of Infectious Diseases, St. John's Cardiovascular Research Center, LAC-Harbor UCLA Medical Center, 1000 West Carson Street, Torrance, CA 90509, USA

[2.] UCLA School of Medicine, Los Angeles, CA 90024, USA

[3.] Laboratoire d'Antibiologie Clinique et Expérimentale, Faculté de Médecine, Centre Hospitalier Universitaire, 44035 Nantes, France

BACKGROUND

Although newer antibiotics, such as extended-spectrum β-lactams and fluoroquinolones, have been developed and marketed, aminoglycosides are still important antibacterial agents for the prevention and treatment of serious aerobic gram-negative bacterial infection (Bayer et al., 1985; Edson et al., 1991; Karlowsky et al., 1994, 1997; MacArthur et al., 1984; Spivey et al., 1990; Zhanel et al., 1994). The aminoglycosides possess several features justifying their continued clinical use, including their rapid and potent bactericidal activity, long-lasting postantibiotic effect, synergy with other antibiotics, and low cost (Craig et al., 1991; Daikos et al., 1990; Karlowsky et al., 1994). Furthermore, progress in optimizing dosing strategies has helped circumvent nephro- and ototoxicity, the major short-comings that characterize this class of antibiotics (Edson et al., 1991; Gilbert et al., 1991).

The in vitro antimicrobial spectrum of activity of aminoglycosides includes a broad range of aerobic gram-negative bacilli, many staphylococci, and certain mycobacteria (Craig et al., 1991; Edson et al., 1991; MacArthur et al., 1984; Xiong et al., 1997). Furthermore, aminoglycosides display concentration-dependent bactericidal activity; i.e., the drug concentration is directly related to the rate of bacterial killing. Under these circumstances, the contribution of the time of exposure of bacteria to the drug is relatively unimportant to the killing process. Concentration-dependent activity of aminoglycosides has been demonstrated in vitro, and in several in vivo animal studies, predominantly those in which infection is caused by Gram-negative bacilli (Craig et al., 1991; Potel et al., 1991, 1992; Powell et al., 1983; Xiong et al., 1996, 1997).

Besides concentration-dependent bactericidal activity, aminoglycosides also exert a growth inhibitory, post-antibiotic effect (PAE) against bacteria from several genera, including S. aureus, P. aeruginosa and E. coli. PAE is defined as the continued suppression of bacterial growth after the organism is no longer in contact with the antibiotic (Craig et al., 1991, 1993; Karlowsky et al., 1994; Xiong et al., 1997). Several factors are known to influence the presence and duration of the PAE. They include type of organism, class and concentration of the antibiotic (relative to the MIC of the drug against an organism), duration of antimicrobial exposure, and antimicrobial combinations

* Corresponding author. Division of Infectious Diseases, St. John's Cardiovascular Research Center, LAC-Harbour UCLA Medical Center, 1000 West Carson Street, Bldg. RB-2, Torrance, California 90509. Phone: (310) 222-6423. Fax: (310) 782-2016. E-mail: XIONG@HUMC.EDU.

(Craig *et al.*, 1991; Xiong *et al.*, 1996). Generally, *β*-lactam antibiotics have PAEs against only Gram-positive organisms, while aminoglycosides exhibit a PAE against both Gram-positive and Gram-negative pathogens (Craig *et al.*, 1991). Drug-induced, reversible damage is the probable mechanism underlying the PAE of aminoglycosides and *β*-lactams. A direct correlation between increased concentrations of aminoglycoside and longer duration of the PAE has been reported (Craig *et al.*, 1991, 1993; Karlowsky *et al.*, 1994; Xiong *et al.*, 1996). A PAE duration range of 0.5–7.5 hrs has been reported in the literature for aminoglycosides against a variety of Gram-positive and Gram-negative pathogens (Craig *et al.*, 1991).

The mechanism(s) involved in the bactericidal action of aminoglycosides has been best delineated with streptomycin. An energy- and oxygen-dependent transport mechanism is required for aminoglycosides to penetrate the bacterial membrane of susceptible bacteria. For this reason, aminoglycosides demonstrate poor *in vitro* activity against anaerobes, and have decreased ability to kill otherwise susceptible bacteria within the relatively anaerobic environment of abscesses (Edson *et al.*, 1991). Besides the energy- and oxygen-dependent bactericidal activity of aminoglycosides, the pH of the medium also affects the activity of aminoglycosides. For example, streptomycin is 500 times as active in alkaline pH, compared with acidic medium. Other aminoglycosides show similar but less drastic shifts in activity with changes in pH (Garrod *et al.*, 1973). The mechanism for the marked pH effect on the aminoglycosides may be explained by the degree of ionization of these compounds. Aminoglycosides exert their bactericidal effects by irreversibly binding to the 30S ribosomal subunit of susceptible bacteria, resulting in the inhibition of protein synthesis (Edson *et al.*, 1991). Aminoglycosides interfere with protein synthesis in three ways; 1) interfering with the initiation complex, which prevents its transition to a chain-elongating functional ribosomal complex; 2) misreading of the mRNA template code, which causes miscoding of amino acids in the peptide; and 3) disruption of polysomes into nonfunctional monosomes.

Like many classes of antibiotics, the emergence and spread of aminoglycoside resistance determinants in pathogenic bacteria is a problem of increasing importance worldwide, particularly among nosocomial isolates. In general, stable aminoglycoside resistance in both Gram-positive and Gram-negative bacteria is based on one of three principal mechanisms (Edson *et al.*, 1991): 30S ribosomal target alteration; enzyme-induced aminoglycoside modification (e.g., acetylation); and/or intracellular transport defects. The presence of aminoglycoside-modifying enzymes is, by far, the most prevalent and important mechanism. However, anaerobiosis, high cation concentrations, acidic pH, binding to abscess exudates, and other biologic conditions may also cause functional resistance (i.e., unstable or phenotypic resistance) (Craig *et al.*, 1991; Xiong *et al.*, 1996).

In vitro and *in vivo* evidence demonstrating reproducible, unstable phenotypic resistance to aminoglycosides in bacterial populations preexposed to these agents has been provided from a number of laboratories (Barclay *et al.*, 1992; Daikos *et al.*, 1990, 1991; Karlowsky *et al.*, 1996; Xiong *et al.*, 1996, 1997). In these studies, aerobic and facultative Gram-negative bacilli, (e.g. *Pseudomonas aeruginosa*) were exposed to fractions or multiples of the MICs for various aminoglycosides for 1 to 2 hrs. Following removal of the extracellular aminoglycoside by dilution or repeated washing-centrifugation steps, aliquots from the cultures were then re-exposed to the aminoglycoside at specific time intervals after initial exposure. The relative degrees of bactericidal activities of aminoglycosides in the pre-exposed cells were then quantified at different time intervals after initial exposures, and then were compared with aminoglycoside-induced killing of a control culture. A reproducible pattern was identified in which a population of initially susceptible bacteria acquired transient resistance to aminoglycosides after an initial exposure. This phenomenon has been termed ''adaptive resistance''. Adaptive resistance in bacteria surviving the first exposure to aminoglycosides involves the development of an unstable, time-dependent refractivity to aminoglycoside-mediated killing by subsequent drug exposure. It is a reversible form of phenotypic resistance, since the intrinsic MIC for organisms in which induced adaptive resistance occurs does not change (Daikos *et al.*, 1990). Cross-resistance to other aminoglycosides, characteristic of adaptive resistance, contrasts with enzymatic modification and ribosomal mutation, which tend to affect specific aminoglycosides (Daikos *et al.*, 1990; Karlowsky *et al.*, 1996). Furthermore, adaptive resistance is distinct from PAE in several ways. First, adaptive resistance is rarely observed immediately after bacteria are removed from aminoglycoside exposure; in contrast, PAE is the greatest at this early post-exposure stage. Similarly, adaptive resistance is the greatest when growth of the exposed culture resumed, which is generally considered the end of the PAE period. Additionally, adaptive resistance is more prolonged than PAE, occurs during bacteria growth, and makes bacteria much more impervious to the presence or reintroduction of the inducing drug (Barclay *et al.*, 1992; Daikos *et al.*, 1990; Xiong *et al.*, 1996).

The majority of studies examining adaptive resistance have focused on *Pseudomonas aeruginosa*, one of the most common bacteria causing serious nosocomial infections (Reyes *et al.*, 1983). These studies have shown that adaptive resistance is a key factor in aminoglycoside killing of Gram-negative bacteria. Thus, understanding the development of adaptive resistance, and ways to circumvent it, is very important for delineating the optimal clinical use of aminoglycosides. The following discussion will focus on the *in vitro* and *in vivo* properties of adaptive resistance to aminoglycosides, and dose-regimen strategies to minimize this form of phenotypic resistance.

ADAPTIVE RESISTANCE TO AMINOGLYCOSIDES *IN VITRO*

Adaptive resistance experiments *in vitro*: Bacterial susceptibility to aminoglycosides depends on maximizing the movement of drug across the bacterial membrane (Edson *et al.*, 1991; Spivey *et al.*, 1990). In the first step of this process, the drug attaches to the bacterial membrane in an energy-independent, ionic process. The rate of this process is dependent on drug concentration. The drug then crosses the membrane and binds to ribosomes in two sequential, energy-dependent steps. If the amount of drug transported into the cell is sufficiently high, an irreversible bactericidal effect takes place. If the initial exposure to the agent is sublethal, the bacterial cell down-regulates subsequent uptake of aminoglycosides in a process

known as "adaptive resistance". Surviving bacteria minimize further internalization of the drug and become relatively refractory to ensuing drug exposure for a defined period of time. Thus, second-exposure drug uptake is mitigated, and the cell becomes transiently resistant (Barclay *et al.*, 1992; Daikos *et al.*, 1990; Karlowsky *et al.*, 1996; Xiong *et al.*, 1996).

A number of previous investigations have been carried out to examine adaptive resistance of *P. aeruginosa* to the aminoglycosides, using static *in vitro* studies (Daikos *et al.*, 1990; Xiong *et al.*, 1996). Daikos *et al.* (1990) first confirmed the phenomenon of adaptive resistance to gentamcin and netilmicin in *P. aeruginosa* (including two standard strains ATCC 27853, NCTC 10701 and two clinical isolates) following initial exposure to the drug. Netilmicin produced progressive bacterial killing over a 7 hr interval in the control culture (no drug preexposure) (Figure 1). However, bacterial cells preexposed to netilmicin for 1 hr exhibited adaptive resistance. Maximum adaptive resistance occurred between 2–5 hrs after removal of the initial aminoglycoside drug over this period. Six hours after the initial preexposure, when the bacterial densities had increased 2–3 $\log_{10}$ CFU/ml in drug-free medium, cells again became susceptible to the second aminoglycoside exposure (Figure 1). The phenomenon of adaptive resistance to aminoglycosides has been demonstrated over MIC multiples ranging from 0.25 to 4 × MIC. Moreover, adaptive resistance induced by one member of the class produced uniform adaptive cross-resistance to other aminoglycosides (e.g., netilmicin, tobramycin, amikacin, and isepamicin [Daikos *et al.*, 1990; Xiong *et al.*, 1996]). Yet, in no instances of adaptive resistance *in vitro* have stable drug-induced resistant mutants been selected.

As determined in other studies (Barclay *et al.*, 1992; Darkos *et al.*, 1990; Xiong *et al.*, 1996), adaptive resistance is influenced by several factors. Xiong *et al.* (1996) reported that: 1) the degree of adaptive resistance was greater and the duration more prolonged with higher initial aminoglycoside concentrations (above the MIC) at pH 7.4; 2) the degree of adaptive resistance was also substantially influenced by alterations in media pH *in vitro*, a result related to pH-dependent effects on the magnitude of bacterial killing by the first aminoglycoside exposure. For example, the bactericidal effect

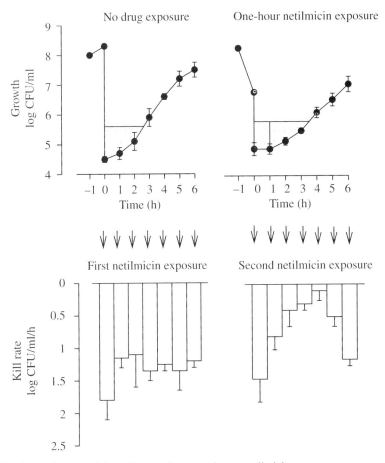

Figure 1. Adaptive resistance of three *P. aeruginosa* strains to netilmicin.
Upper graphs show number of viable bacteria (CFU) at different times in the control (no drug exposure, [left panels])
and test culture (had 1 hr prior exposure to 8MIC of netilmicin, [right panels]). The postantibiotic effect (PAE) of
neltimicin is shown in upper left graph (double crosshatched area).
Lower graphs show the bactericidal rate measured 1 hr after the addition of 4MIC netilmicin at successive hourly
intervals. Lower left panels shows the bactericidal rate of 4MIC netilmicin on control culture (first netilmicin
exposure). Lower right panels shows the development and recession of unstable adaptive resistance (less of
bactericidal action) of *P. aeruginosa* to second netilmicin exposure. The postantibiotic effect of netilmicin is shown in
upper right graph (double crosshatched area). This figure is modified from Daikos *et al.* (1990), with permission from
J. Infect. Dis.

of aminoglycoside at pH 6.5 was far less than that at
pH 7.4, but the magnitude of adaptive resistance was
more marked at the lower pH; and 3) adaptive
resistance was suppressed when rifampicin was
present (e.g., rifampicin was added after, but not
during, first exposure to aminolgycoside). The effect
of rifampicin could account for the known synergy
between rifampicin and aminoglycosides against *P.
aeruginosa in vitro* (Xiong *et al.*, 1996).

**Pharmacodynamic Model of Adaptive Re-
sistance *in vitro*:** The phenomenon of adaptive

resistance following single, first-exposure doses of
aminoglycosides has been assessed using a dynamic
in vitro model to mimic human aminoglycoside
pharmacokinetics *in vivo* (Barclay *et al.*, 1992, 1996;
Begg *et al.*, 1992). Barclay *et al.* used a computer-
controlled syringe pump to simulate a gentamicin
elimination half-life of 2.5 hrs. They showed that
following a first dose of gentamicin with a peak
concentration of 25 μg/ml (10 × MIC), a gentamicin-
susceptible strain of *P. aeruginosa* was almost
completely resistant to a second exposure to genta-

micin (at concentrations of up to 20×MIC) between six and sixteen hours of growth after the first exposure. Full recovery of susceptibility occurred approximately 43 hrs after initial exposure. They also demonstrated that higher initial gentamicin concentrations produced greater bactericidal activity, as well as a more marked and more prolonged adaptive resistance effect. In addition, they confirmed that adaptive resistance to aminoglycosides is a reversible phenotype, since the MICs of 30 individual colonies exhibiting adaptive resistance were similar to those of the control strain when such cells were retested for their MICs. Of note, up until recently aminoglycosides had been conventionally administered to patients at 8–12 hrs intervals in treating human infections, at which time surviving cells may well be adaptively resistant. Thus these results, using a dynamic *in vitro* model which simulates human pharmacokinetics, were important for reassessing the dose-regimen strategies for aminoglycosides in clinical practice.

Begg *et al.* evaluated the interaction of *Pseudomonas aeruginosa* with gentamicin at dosing regimens in a one-compartment pharmacodynamic model *in vitro*. Gentamicin was administered as a first exposure to produce peak concentrations of 8 μg/ml every 6 hrs, 13.5 μg/ml every 12 hrs, or 26 μg/ml once for over 24 hrs. In all cases, second exposure doses in all regimens produced a bactericidal effect that was comparably less than that observed with the first doses. Although MICs were not assessed in this experiment, the authors reported no changes in gentamicin MICs of adaptively resistance cells in subsequent studies (Begg *et al.*, 1992). In this study, they used a single-compartment model, which may not be applicable to infection in deeper compartments. Persistence of the antibiotic at these sites may result in a longer period of adaptive resistance. Thus, further experiments using a two-compartment model would be required.

Although not studied to the same degree as aminoglycosides, the quinolones (ciprofloxacin, enoxacin and ofloxacin) appear to induce adaptive resistance against *P. aeruginosa* cells (Blaser *et al.*, 1987; Dudley *et al.*, 1991; Madaras *et al.*, 1996). This is not surprising since this class of antibiotics also exhibits concentration-dependent bacterial killing.

Barclay *et al.* also showed the effect of aminoglycoside-induced adaptive resistance on the antibacterial activity of six other antibiotics (Barclay *et al.*, 1996). These antibiotics represented different drug classes that have in vitro activity against *P. aeruginosa*, and which might be used in combination with an aminoglycoside in clinical infections. The different modes of action of these antibiotics include inhibition of either cell wall synthesis, DNA gyrase, or DNA-dependent RNA polymerase. Adaptive resistance was then induced in *P. aeruginosa* in a dynamic *in vitro* model of infection. The presence of adaptive resistance in *P. aeruginosa* induced by aminoglycosides did not alter the bactericidal effect of subsequent exposure to ceftazidime, piperacillin, imipenem, aztreonam, or ciprofloxacin. Moreover, rifampicin had a greater bactericidal effect in the presence of adaptive resistance than other compounds. The reason for this latter effect is unclear since the mechanism of action of aminoglycosides and rifampicin are not known to be related. Because *P. aeruginosa* is only modestly sensitive to rifampicin, the concentration used in this study was higher than that achieved in clinical practice. Therefore, although this result is of interest in relation to the mechanism of adaptive resistance, it may not have direct clinical relevance. Of interest, we have been unable to induce adaptive resistance to aminoglycosides during *in vitro* studies in *S. aureus* (Xiong, Y-Q. and Bayer, A.S. unpublished observations), suggesting this phenomenon may well be restricted to Gram negative bacilli.

The presence of adaptive resistance may contribute to aminoglycoside failures when these drugs are used alone. Amingolycosides are usually given in combination with other antibiotics, which may explain why adaptive resistance has not been widely recognized earlier. The presence of adaptive resistance may provide a rationale for the success of combination antibiotic therapy versus aminoglycoside monotherapy, as well as for using these agents at longer dosing intervals (e.g. once-a-day dosing strategies).

THE MECHANISM OF ADAPTIVE RESISTANCE

The mechanisms of adaptive resistance have not been fully delineated. However, pathways responsible for aminoglycoside adaptive resistance do not appear to pose a significant metabolic disadvantage to such bacteria, since cells demonstrating peak adaptive resistance have mean generation times similar to those of control cells (Daikos *et al.*, 1990; Karlowsky *et al.*, 1996). Some studies have suggested that aminoglycoside-induced adaptive resistance is related to perturbations in drug uptake, especially during the period of accelerated, energy-dependent drug transport (EDP II) (Daikos *et al.*, 1990; Karlowsky *et al.*, 1996). For example, Daikos *et al.* (1990), evaluated uptake of labeled gentamicin by *P. aeruginosa* following exposure to non-labeled drug for 1 hr; the drug was then removed and cultures incubated in drug-free medium for 4 or 7 hrs before a second exposure to ^{14}C-labeled gentamicin. No bacterial uptake of drug was detected during the period of maximum adaptive resistance (4 hrs after removal from period drug), whereas rapid bacterial uptake was noted after 7 hrs when susceptibility to the bacterial action of gentamicin was re-established. Another experiment demonstrated that during initial exposure, cells rapidly internalized ^{3}H-tobramycin, corresponding to bacterial killing. In contrast, cells pre-exposed to tobramycin (0.5 MIC for 8 h), failed to take up ^{3}H tobramycin (4 MIC), and this concentration of tobramycin as a second exposure had no significant bactericidal effect (Daikos *et al.*, 1990). The reduction of aminoglycoside accumulation in bacteria exhibiting adaptive resistance suggests that impaired drug uptake is a major mechanism underlying the adaptive resistance phenotype, and that ribosomal resistance mechanisms are not directly operative (Daikos *et al.*, 1990; Karlowsky *et al.*, 1996).

Another study (Karlowsky *et al.*, 1996) compared the proton motive force (pmf) and cell envelope component profiles of unexposed control cells, adaptively-resistant cells, and postadaptively-resistant (drug-susceptible) cells. Aminoglycoside adaptive resistance was shown to significantly correlate with a reduction in overall pmf during the adaptive resistance period. However, no changes in outer membrane protein or lipopolysaccharide profiles (by SDS-PAGE) were noted when control, adaptively resistant, and postadaptively-resistant cells were compared (Karlowsky *et al.*, 1996). Furthermore, cytoplasmic membrane profiles of adaptively-resistant cells demonstrated six prominent protein band changes when compared with control and post-adaptively resistant

cells. Adaptively-resistant cells demonstrated increases in the intensities of three such bands at 35, 41, and 60 KDa, two new bands at 28 and 45 KDa, and a decrease in the intensity of a 36 KDa band when compared with control and postadaptively resistant cells. The precise implications of such band changes are not clear, but may represent alterations in electron transport chain components or in the proteins associated with these metablic pathways. Thus, aminoglycoside adaptive resistance in *P. aeruginosa* may coincide with cytoplasmic membrane changes, but appears to be independent of lipopolysaccharide and outer membrane protein changes (Karlowsky *et al.*, 1996). Interestingly, no change in mean generation time was noted during adaptive resistance even though pmf was reduced during this time. The importance of these pmf changes, although they were statistically significant, remains enigmatic, since the reduction in pmf observed during peak adaptive resistance was only 5% relative to the control cells.

Anaerobiosis has also been reported to reduce aminoglycoside-induced bacterial killing and to limit intracellular aminoglycoside accumulation. Therefore, a common mechanism may facilitate both the reduced bacterial killing and aminoglycoside accumulation under anaerobic conditions. Karlowsky *et al.* (1997) studied gene expression by Northen blot analysis of *P. aeruginosa* cells which were adaptively-resistant to gentamicin. In these studies, increased mRNA levels for both the *denA* (nitrite reductase) gene, which facilitates terminal electron acceptance in the anaerobic respiratory pathway, and its regulatory protein, *anr* (anaerobic regulation of arginine deaminase and nitrate reduction), when compared with wild-type and post-adaptively-resistant cell were demonstrated. Similar results were demonstrated with the reference strain ATCC 27853 and two blood culture isolates of *P. aeruginosa* tested. These observations suggested that *P. aeruginosa* may up-regulate the expression of genes in its anaerobic respiratory pathway in response to an aminoglycoside insult, and this may partially explain, *P. aeruginosa* adaptive resistance to aminoglycosides. Interestingly, identical genetic up-regulation occurred when pseudomonal cells were grown anaerobically *in vitro*. These investigators also provided data that increased *denA* and *anr* mRNA levels likely were the result of one or more regulatory events in cells exhibiting adaptive resistance. The *anr* locus has previously been demonstrated to positively regulate *den*A transcription (Karlowsky *et al.*, 1997); however, the factors regulating *anr* mRNA levels are presently unknown, as are the mechanisms of signal transduction by which adaptively-resistant *P. aeruginosa* cells experience increased *anr* and *den*A mRNA levels. Thus, it appears that the mechanism of adaptive resistance in *P. aeruginosa* is associated with enhanced expression of pseudomonal genes in its anaerobic respiratory pathway, activated either under anaerbic conditions or in response to initial aminoglycoside exposures. A relationship between increased anaerobic respiratory pathway use and *P. aeruginosa* adaptive resistance to aminoglycosides may exist for several reasons. First, anaerobically grown *P. aeruginosa* cells accumulate aminoglycosides less effectively than under aerobic conditions (Bryan *et al.*, 1979). Second, cytoplasmic aminoglycoside accumulation has been shown to be dependent upon a functional aerobic respiratory pathway, in addition to an intact pmf (Parr *et al.*, 1988).

ADAPTIVE RESISTANCE TO AMINOGLYCOSIDES IN ANIMAL MODELS AND IN HUMAN INFECTIONS

Animal models of infection represent a unique transition between *in vitro* and clinical studies, disclosing important *in vivo* relevance regarding antibacterial efficacy, pathogenesis of infection, role of host defenses, and the emergence of antibiotic resistance (Fantin *et al.*, 1992). The advantages of discriminative models of infection include close simulation of the characteristics of the same infection in humans, and definitive endpoints which allow statistical comparisons among study groups. Several studies have been carried out to examine the *in vivo* relevance of adaptive resistance in *P. aeruginosa*, using distinct animal models (e.g., the rabbit endocarditis model, and the normal or neutropenic murine thigh infection model) (Bayer *et al.*, 1987; Daikos *et al.*, 1991; Xiong *et al.*, 1997).

Rabbit endocarditis model: Experimental endocarditis in rabbits and rats has been shown to be a reliable model for evaluation of antimicrobial

efficacy. Xiong *et al.* (1997) studied adaptive resistance in *P. aeruginosa* following a single dose of amikacin in the rabbit model. Rabbits with *P. aeruginosa* endocarditis received either no therapy (control) or a single intravenous dose of amikacin (80 mg/kg of body weight) at 24 hrs postinfection, after which vegetations were removed at different time points. Excised vegetations were then exposed *ex vivo* to amikacin at different concentrations for 90 min. *In vivo* adaptive resistance was identified when amikacin-induced pseudomonal killing within excised aortic vegetations was less in animals receiving single-dose amikacin *in vivo* than in vegetations from control animals not receiving amikacin *in vivo*. They showed that the maximal adaptive resistance occurred between 8 and 16 hrs after the *in vivo* amikacin dose, with complete refractoriness to *ex vivo* killing by amikacin seen at 12 hrs postdose. By 24 hrs postdose, bacteria within excised vegetations had partially recovered their initial amikacin susceptibility. In a parallel treatment study, amikacin given once daily (but not twice daily) at a total dose of 80 mg/kg, iv for 1 day treatment significantly reduced pseudomonal densities within aortic vegetation versus those in untreated controls. When therapy was continued for 3 days with the same total daily dose (80 mg/kg/day), amikacin given once or twice daily significantly reduced intravegetation pseudomonal densities versus those in controls. However, amikacin given once daily was still more effective than the twice-daily regimen (Xiong *et al.*, 1997). These data confirm the induction of aminoglycoside adaptive resistance *in vivo*, and further supported the advantages of once-daily aminoglycoside dosing regimens in the treatment of serious pseudomonal infections.

Bayer *et al.* analyzed five amikacin-resistant variants of *P. aeruginosa* isolated from aortic valve vegetations during unsuccessful therapy of experimental endocarditis with amikacin (Bayer *et al.*, 1987). The organisms were cross-resistant to other aminoglycosides, and aminoglycoside-modifying enzymes were not identified. All five variants had significantly reduced intracellular accumulation of H^3-amikacin as compared to the amikacin-susceptible parent strain. The reductions in drug permeability were unstable *in vitro*, and returned to normal after serial passage of these pseudomonal cells in antibiotic-free media. The drug-resistant variants often grew as "small colony variants" and had *in vitro* generation times approximately 1.5–2 times longer than those of their parent strain.

Normal and neutropenic murine thigh infection models: For the past decade, Craig *et al.* have reported using a neutropenic murine thigh infection model to study the *in vivo* PAE with a much wider range of drug/organism combinations (Craig *et al.*, 1991). Recently, Daikos *et al.* (1991) demonstrated biphasic bactericidal action and induction of adaptive resistance to aminoglycosides using the *P. aeruginosa* thigh abscess model in both normal and neutropenic murine infections (Daikos *et al.*, 1991). Mice were made neutropenic to examine the role of adaptive resistance in altering drug efficacy in the absence of this aspect of host immunity. Adaptive resistance occurred when doses were given more than 2 hrs after an initial dose. For example, in neutropenic mice doses 4, 6, and 8 hrs after the initial dose showed no bactericidal effect. A drug-free interval of 8 hrs renewed bacterial susceptibility to the aminoglycoside bactericidal effect. Thus, adaptive resistance occurred when second aminoglycoside doses were given between 2 and 8 hrs after the start of treatment (Daikos *et al.*, 1991).

Human lung infections in cystic fibrosis: Patients with cystic fibrosis frequently develop lung colonization with *P. aeruginosa*. Once infection is established, it is very difficult to eradicate. Aminoglycosides remain important in the treatment of *P. aeruginosa* lung infection despite the development of newer agents with less toxicity (Craig *et al.*, 1991; MacArthur *et al.*, 1984). However, aminoglycoside bactericidal activity is reduced in cystic fibrosis sputum or any infected sputum due to a number of factors, including bacterial biofilms, the slow growth phase of established infecting bacteria (Gilbert *et al.*, 1995), and molecular binding of aminoglycosides (Bataillon *et al.*, 1992). The induction of adaptive resistance in *P. aeruginosa* by aminoglycoside treatment may be another important factor causing reduced antibacterial activity in this clinical setting. Barclay *et al.* (1996) studied patients with cystic fibrosis treated with inhaled tobramycin to determine whether adaptive resistance occurred in sputum isolates of *P. aeruginosa*. Adaptive resistance was detected in *P. aeruginosa* 1–4 hrs after

such tobramycin administrations in patients who had not recently taken antibiotics. Moderate resistance was present at 24 hrs and full susceptibility returned between 24 and 48 hrs after the initial treatment dose. In other patients on long-term twice-daily inhaled aminoglycoside treatment, adaptive resistance was present before, and 4 hrs after tobramycin administration. Thus, adaptive resistance has been shown to occur in *P. aeruginosa* in humans following a dose of tobramycin, and the adaptive resistance time-course characteristics appear to mirror those determined *in vitro*. These results further support the use of higher concentration and longer aminoglycoside dose-intervals, allowing time for adaptive resistance to abate.

AMINOGLYCOSIDE REGIMENS

As described above, aminoglycoside antibiotics are widely-used and valuable agents in the treatment of serious infections. They have been used successfully for more than 50 years. However, they have a low therapeutic index, and the choice of dosage-regimen is critical. Traditionally, aminoglycosides were administered in several divided daily doses on the assumptions that the serum concentration should be prevented from falling below the MIC, and that bactericidal action is optimal once the minimum bactericidal concentration (MBC) is reached or exceeded. For example, peak plasma concentrations of 6 to 10 μg/ml and a trough concentration of less than 2 μg/ml for gentamicin and tobramycin had been traditional goals of therapy to maximize efficacy and limit toxicity (Bayclay *et al.*, 1994; Edson *et al.*, 1991; Gilbert *et al.*, 1995; Preston *et al.*, 1995). Recent evidence from *in vitro*, animal and human studies suggests that these target concentrations and dose-regimens need revision.

A recent emphasis in the literature advocates the use of once-daily aminoglycoside dosing (as opposed to multiple-dose regimens) in order to both minimize nephro- and oto-toxicities, while maintaining clinical efficacy (Bayclay *et al.*, 1994; Preston *et al.*, 1995). The ability to utilize once-daily aminoglycoside dose-regimens with an efficacy equivalent to that of multidose regimens has several theoretical foundations. First, aminoglycosides have concentration-dependent (rather than a time-dependent) bactericidal activity against aerobic gram-negative bacilli; the higher peak levels in serum obtained by once-daily dosing regimens (versus multidose regimens) would, thus, tend to maximize killing. This concept has been supported by clinical data from Moore et al. (1987), who showed a correlation between aminoglycoside efficacy in serious human infections and the magnitude of the peak serum drug level:MIC ratio. Second, even after serum aminoglycoside levels fall below the MIC of the infecting strain, the magnitude of the post-antibiotic growth inhibition duration (PAE) is also related to the peak aminoglycoside:MIC ratios (Galloe et al., 1995; Karlowsky et al., 1994). Third, during the terminal elimination phase following a once-daily aminoglycoside dose regimen (about 16 to 24 h postdose in patients with normal renal function [Nicolau et al., 1995]), the target cells for aminoglycoside toxicity (e.g., the cochlear hair cells) are likely to be exposed to exceedingly low drug levels for a substantial time prior to subsequent doses. This drug-free duration is felt to be important in limiting end-organ toxicities (Van Der Auwera et al., 1991). Additionally, the once-daily regimen offers the potential for decreased cost, related to decreased serum level monitoring and drug administration charges.

An additional theoretical advantage for once-daily aminoglycoside dosing was recently identified by Daikos et al. (1990, 1991) and others (Barclay et al., 1992, 1996; Xiong et al., 1996, 1997) with delineation of the phenomenon of adaptive resistance both in vitro (Barclay et al., 1992, 1996; Daikos et al., 1990; Xiong et al., 1996), in vivo (animal models [Bayer et al., 1987; Daikos et al., 1991; Xiong et al., 1997] and in human infections [Barclay et al., 1996]). These investigators observed that, following initial exposure of P. aeruginosa to aminoglycosides, cells entered a prolonged period of relative refractivity to further bactericidal effects during subsequent aminoglycoside exposure. This phenotypic trait was unstable and disappeared following passage of such refractory cells in aminoglycoside-free media. In addition, the longer the time interval following the initial aminoglycoside exposure (usually at > 16 hrs in vivo), the lower the adaptive resistance response

of bacterial cells. Both in vitro and in vivo experience indicates that, when repeated doses are given at 8–12 hrs intervals, it is likely that a substantial portion of surviving bacterial cells may exhibit adaptive resistance to the antibiotic. Thus, the combined effect of low-dose and short-interval administration of aminoglycoside could both augment adaptive resistance and minimize the drug's bactericidal potential (Barclay et al., 1992, Daikos et al., 1990; Xiong et al., 1997).

A substantial number of trials have compared once-daily aminoglycoside administration with more frequent administration regimens (Kapusnik et al., 1988; Kovarik et al., 1989; Maller et al., 1988; Munchhof et al., 1996; Potel et al., 1992; Preston et al., 1995). There have been 29 clinical trials comparing once-daily administration of aminoglycosides with conventional administration (2 to 4 times daily, [review by Bayclay et al., 1994]). In general, efficacy has not been shown to be different between regimens, although one trial has shown an advantage for once-daily administration compared with administration 3 times daily (Bayclay et al., 1994). A small number of trials have shown less nephrotoxicity and ototoxicity with once-daily administration (Bayclay et al., 1994). Thus, evidence to-date would suggest that once-daily aminoglycoside administration is at least as effective as more frequent administration, and probably causes less nephrotoxicity and ototoxicity (Bayclay et al., 1994; Preston et al., 1995). These findings suggest that there is sufficient evidence to warrant a change to once-daily administration of aminoglycosides. However, once-daily administration has not been well-studied in patients with renal failure, Gram-negative endocarditis and in patients requiring long-term aminoglycosides (> 7days), and cannot be recommended in these patients as yet. Moreover, the use of once-daily regimens for synergy against Gram-negative pathogens, and for Gram-positive endocarditis remains largely up proven, except for viridans streptococcal endocarditis. It should be emphasized that for use in synergy, once-daily aminoglycoside dose-regimens should be substantially lower (2–3 mg/kg/d) then when used as primary therapy against a proven or suspected Gram-negative infection (5–7 mg/kg/d).

SUMMARY

Adaptive resistance is defined as reduced antimicrobial killing in originally susceptible bacterial populations after initial exposure to an
aminoglycoside. It has been observed *in vitro*
and *in vivo* with gram-negative bacilli, most often
P. aeruginosa. The mechanisms of adaptive
resistance of *P. aeruginosa* to aminoglycosides
appear to involve: 1) a marked reduction in intracellular aminoglycoside accumulation; 2) reduction in proton motive force; 3) emergence of six
prominent band changes in the bacterial cytoplasmic membrane; and 4) increased mRNA
levels for genes involved in facilitation of
terminal electron acceptance in the anaerobic
respiratory pathway. Understanding the process
of aminoglycoside-induced *P. aeruginosa* adaptive resistance is of particular clinical importance,
since these agents are very useful in the treatment
of pseudomonal infections. As described above,
aminoglycosides have concentration-dependent
bacterial killing, a long postantibiotic effect and
also induce adaptive resistance in Gram negative
bacteria. All these factors support the use of daily
dose regimens of aminoglycosides in which drug
administrations are given at less frequent intervals
than conventional therapy to maximize the clinical
effectiveness of these agents. However, more
studies are still necessary to define the optimal
administration regimen for aminoglycoside therapy.

ACKNOWLEDGMENTS

The efforts and insight of numerous individuals have
greatly contributed to this chapter. Dr. Henri
Drugeon, Dr. Denis Baron, and Dr. Jocelyne Caillon
were active participants in the renaissance of studying adaptive resistance to aminoglycoside *in vitro*
and *in vivo*, and provided invaluable input. Their
contributions are greatly appreciated. We also greatly
acknowledge Dr. Michael R. Yeaman for reading the
chapter and providing valuable critical comments.

REFERENCES

Barclay, M.L., Begg, E.J. and Chambers, S.T. (1992) Adaptive
    resistance following single doses of gentamicin in a dynamic
    in vitro model. *Antimicrob. Agents Chemother* **36**, 1951–
    1957.
Barclay, M.L., Begg, E.J., Chambers, S.T. and Peddie, B.A.
    (1996) The effect of aminoglycoside-induced adaptive
    resistance on the antibacterial activity of other antibiotics
    against *Pseudomonas aeruginosa in vitro. J. Antimicrob.
    Chemother.* **38**, 853–858.
Barclay, M.L, Begg, E.J., Chambers, S.T., Thornley, P.T.,
    Pattemore, P.K. and Grimwood, K. (1996) Adaptive resis-
    tance to tobramycin in *Pseudomonas aeruginosa* lung
    infection in cystic fibrosis. *J. Antimicrob. Chemother.* **37**,
    1155–1164.
Bayclay, M.L., Begg, E.J. and Hickling, K.G. (1994) What is the
    evidence for once-daily aminoglycoside therapy? *Clin.
    Pharmacokinet.* **27(1)**, 32–48.
Bataillon, V., Hermitte, M.L., Lafitte, J.J., Pommery, J. and
    Roussel, P. (1992) The binding of amikacin to macromole-
    cules from the sputum of patients suffering from respiratory
    diseases. *J. Antimicrob. Chemother.* **29**, 499–508.
Bayer, A.S., Norman, D. and Kim, K.S. (1985) Efficacy of
    amikacin and ceftazidime in experimental aortic valve
    endocarditis due to *Pseudomonas aeruginosa. Antimicrob.
    Agents Chemother.* **28**, 781–785.
Bayer, A.S., Norman, D.C. and Kim, K.S. (1987) Characteriza-
    tion of impermeability variants of *Pseudomonas aeruginosa*
    isolated during unsuccessful therapy of experimental endocar-
    ditis. *Antimicrob. Agents Chemother.* **31**, 70–75.
Begg, E.J., Peddie, B.A., Chambers, S.T. and Boswell, D.R.
    (1992) Comparison of gentamicin dosing regimens using an
    in vitro model. *J. Antimicrob. Chemother.* **29**, 427–433.
Blaser, J., Stone, B.B., Groner, M.C. and Zinner, S.H. (1987)
    Comparative study with enoxacin and netilmicin in a
    pharmacodynamic model to determine importance of ratio
    of antibiotic peak concentration to MIC for bacterial activity
    and emergence of resistance. *Antimicrob. Agents Chemother.*
    31, 1054–1060.
Bryan, L.E., Kowand, S.K. and van den Elzen, H.M. (1979)
    Mechanism of aminoglycoside antibiotic resistance in anae-
    robic bacteria: *Clostridum perfringens* and *Bacteroides
    fragilis. Antimicrob. Agents Chemother.* **15**, 7–13.
Craig, W.A., Redington, J. and Ebert, S.C. (1991) Pharmaco-
    dynamics of amikacin *in vitro* and in mouse thigh and lung
    infections. *J. Antimicrob. Chemother.* **27 (Suppl. C)**, 29–40.
Craig, W.A. and Gudmundson, S. (1991) Postantibiotic effect. In
    V. Lorian (ed), *Antibiotic in laboratory medicine*, The
    Williams & Wilkins Co., London, pp. 403–431.
Craig, W.A. (1993) Post-antibiotic effects in experimental
    infection model: relationship to *in vitro* phenomena and to
    treatment of infections in man. *J. Antimicrob. Chemother.*
    31(Suppl. D), 149–158.
Daikos, G.L., Jackson, G.G., Lolans, V.T. and Livermore, D.M.
    (1990) Adaptive resistance to aminoglycoside antibiotics from
    first-exposure down-regulation. *J. Infect. Dis.* **62**, 414–420.
Daikos, G.L., Lolans, V.T. and Jackson, G.G. (1991) First-
    exposure adaptive resistance to aminoglycoside antibiotics *in
    vivo* with meaning for optimal clinical use. *Antimicrob. Agents
    Chemother.* **35**, 117–123.
Daikos, G.L., Jackson, G.G., Livermore, D. and Papafragas, E.
    (1990) Prevention and repair of adaptive resistance from first
    exposure to an aminoglycoside by beta-lactam antibiotic. In:
    Program and abstracts of the 30th Interscience Conference on
    Antimicrobial Agents and Chemotherapy. Washington, DC.
    American Society for Microbiology, pp. 323.
Dudley, M.N., Blaser, J., Gibert, D., Mayer, K.H. and Zinner,
    S.H. (1991) Combination therapy with ciprofloxacin plus
    azlocillin against *Pseudomonas aeruginosa*: effect of simulta-

neous versus staggered administrated in an *in vitro* model. *J. Infect. Dis.* **64**, 499–506.

Edson, R.S. and Terrell, C.L. (1991) *The aminoglycosides. Mayo Clin Proc.* **66**, 1158–1164.

Fantin, B. and Carbon, C. (1992) *in vivo* antibiotic synergism: contribution of animal model. *Antimicrob. Agents Chemother.* **36**, 907–9012.

Galloe, A.M., Graudal, N., Christensen, H.R. and Kampman, J.P. (1995) Aminoglycosides: single or multiple daily dosing? A meta-analysis on efficacy and safety. *Eur J. Clin. Pharmacol.* **48**, 39–43.

Garrod, L.P., Lambert, H.P. and O'Grady, R. (eds) (1973) Antibiotics and chemotherapy. Edinburgh: Churchill Livingstone, pp. 490–531.

Gilbert, D.N. (1991) Once-daily aminoglycoside therapy. *Antimicrob. Agents Chemother.* **35**, 399–405.

Gilbert, D.N. (1995) Aminoglycosides, In G.L. Mandell, R. Douglas, and J.E. Bennett (eds), *Principles and practice of infectious diseases*, 4th edn, Churchill Livingstone, New York, pp. 279–306.

Kapusnik, J.E., Hackbarth, C.J., Chambers, H.F., Carpenter, T. and Sande, M.A. (1988) Single, large, daily dosing versus intermittent dosing of tobramycin for treating experimental *Pseudomonas pneumonia. J. Infect. Dis.* **158**, 7–12.

Karlowsky, J.A., Saunders, M.H., Harding, G.A.J., Hoban, D.J. and Zhanel, G.G. (1996) *in vitro* characterization of aminoglycoside adaptive resistance in *Pseudomonas aeruginosa. Antimicrob. Agents Chemother.* **40**, 1387–1393.

Karlowsky, J.A., Zhanel, G.G., Davidson, R.J. and Hoban, D.J. (1994) Once-daily aminoglycoside dosing assessed by MIC reversion time with *Pseudomonas aeruginosa. Antimicrob. Agents Chemother.* **38**, 1165–1168.

Karlowsky, J.A., Hoban, D.J., Zelenitsky, S.A., Kabani, A.M. and Zhanel, G.G. (1997) Altered *den*A and *anr* gene experssion in aminoglycoside adaptive resistance in *Pseudomonas aeruginosa. J. Antimicrob. Chemother.* **40**, 371–376.

Karlowsky, J.A., Zhanel, G.G., Davidson, R.J. and Hoban, D.J. (1994) Postantibiotic effect in *Pseudomonas aeruginosa* exposures *in vitro. J. Antimicrob Chemother.* **33**, 937–947.

Kovarik, J.M., Hopelman, I.M. and Verhoef, J. (1989) Once-daily aminoglycoside administration: new strategies for an old drug. *Eur. J. Clin. Microbiol. Infect. Dis.* **8**, 761–769.

MacArthur, R.F., Lolans, V.T., Zer, F.A. and Jackson, G.G. (1984) Biphasic, concentration-dependent and rate-limited, concentration-independent bacterial killing by an aminoglycoside antibiotic. *J. Infect. Dis.* **150**, 778–779.

Madaras–Kelly, K.J., Ostergard, B.E., Beaker Hovde, L. and Rotschafer, J.C. (1996) Twenty-four-hour area under the concentration-time curve/MIC ratio as a generic predictor of fluoroquinolone antimicrobial effect by using three strains of *Pseudomonas aeruginosa* and an *in vitro* pharmacodynamic model. *Antimicrob. Agents Chemother.* **40**, 627–632.

Maller, R., Isaksson, B., Nilsson, L. and Soren, L. (1988) A study of amikacin given once versus twice daily in serious infections. *J. Antimicrob. Chemother.* **22**, 75–79.

Moore, R.D., Lietman, P. and Smith, C.R. (1984) The association of aminoglycoside plasma levels with mortality in patients with gram-negative bacteremia. *J. Infect. Dis.* **149**, 443–448.

Moore, R.D., Lietman, P. and Smith, C.R. (1987) Clinical response to aminoglycoside therapy: importance of the ratio of peak concentration to minimum inhibitory concentration. *J. Infect. Dis.* **155**, 93–97.

Munchhof, W.J., Grayson, M.L. and Turnidge, L.D. (1996) A meta-analysis of studies on the safety and efficacy of aminoglycoside given either once daily or as divided doses. *J. Antimicrob Chemother.* **37**, 645–663.

Nicolau, D.P., Freeman, C.D., Belliveau, P.P., Nightingale, C.H., Ross, J.W. and Quintiliani, R. (1995) Experience with a once-daily aminoglycoside program administered to 2,184 adult patients. *Antimicrob. Agents Chemother.* **39**, 650–655.

Parr, T.R. and Bayer, A.S. (1988) Mechanism of aminoglycoside resistance in variants of *Pseudomonas aeruginosa* isolated during treatment of experimental endocarditis in rabbits. *J. Infect. Dis.* **158**, 1003–1010.

Potel, G., Caillon, J., Le Gallou, F., Bugnon, D., Le Conte, P., Raza, J., *et al.* (1992) Identification of factors affecting *in vivo* aminoglycoside activity in an experimental model of gram-negative endocarditis. *Antimicrob. Agents Chemother.* **36**, 744–750.

Potel, G., Caillon, J., Fantin, B., Raza, J., Le Gallou, F. and Lepage, J.Y. (1991) Impact of dosage schedule on the efficacy of gentamicin, tobramycin, or amikacin in an experimental model of *Serratia marcescens* endocarditis: *in vitro-in vivo* correlation. *Antimicrob Agents Chemother.* **35**, 111–116.

Potel, G., Caillon, J., Le Gallou, F., Bugnon, D., Le Conte, P. and Raza, P. (1992) Identification of factors affecting *in vivo* aminoglycoside activity in an experimental model of gram-negative endocarditis. *Antimicrob. Agents Chemother.* **36**, 744–750.

Powell, S.H., Thompson, W.L., Luthe, M.A., Stern, R.C., Grossinklaus, D.A. and Bloxham, D.D., *et al.* (1983) Once daily versus continuous aminoglycoside dosing: efficacy and toxcity in animal and clinical studies of gentamicin, netilmicin and tobramycin. *J. Infect. Dis.* **147**, 918–932.

Preston, S.L. and Briceland, L.L. (1995) Single daily dosing of aminoglycosides. *Pharmacotherapy.* **15(3)**, 297–316.

Reyes, M.P. and Lerner, A.M. (1983) Current problems in the treatment of endocarditis due to *Pseudomonas aeruginosa. Rev. Infect. Dis.* **5**, 314–321.

Spivey, J.M. and Schentag, J.J. (1990) Aminoglycosides revisited. *Curr. Opin. Infect. Dis.* **3**, 770–772.

Van Der Auwera, P., Meunier, F., Ibrahim, S., Kaufman, L., Derde, M.P. and Tulkens, P.M. (1991) Pharmacodynamic parameters and toxicity of netilmicin (6 milligram/kilogram/day) given once daily or in three divided doses to cancer patients with urinary tract infection. *Antimicrob. Agents Chemother.* **35**, 640–647.

Xiong, Y.Q., Caillon, J., Kergueris, M.F., Drugeon, H., Baron, D. and Potel, G., *et al.* (1997) Adaptive resistance of *Pseudomonas aeruginosa* induced by aminoglycosides and killing kinetics in a rabbit endocarditis model. *Antimicrob. Agents Chemother.* **41**, 823–826.

Xiong, Y.Q., Caillon, J., Drugeon, H., Potel, G. and Baron, D. (1996) Influence of pH on adaptive resistance of *Pseudomonas aeruginosa* to aminoglycosides and their postantibiotic effects. *Antimicrob. Agents Chemother.* **40**, 35–39.

Xiong, Y.Q., Caillon, J., Drugeon, H., Potel, G. and Baron, D. (1996) The effect of rifampicin on adaptive resistance of *Pseudomonas aeruginosa* to aminoglycosided. *J. Antimicrob. Chemother.* **37**, 993–998.

Zhanel, G.G. and Craig, W.A. (1994) Pharmacokinetic contributions to postantibiotic effects:focus on aminoglycosides. *Clin. Pharmacokinet.* **27**, 377–392.

5. Efflux Mechanisms: Molecular and Clinical Aspects

Olga Lomovskaya, Mark S. Warren and Ving Lee
Microcide Pharmaceuticals, Inc., 850 Maude Avenue, Mountain View, CA 94043, USA.

INTRODUCTION

The development and clinical use of antibiotics significantly decreased adverse effects from bacterial infections. In response to the pressure of antibiotics, multiple resistance mechanisms became widespread and are threatening the clinical utility of antibacterial therapy. Bacteria protect themselves from antibiotics by various mechanisms such as enzymatic degradation or modification of antibiotics, modification of antibiotic targets or effluxing antibiotics out of cells. The active efflux of antibiotics as a mechanism of resistance was discovered by S. Levy and coworkers in 1980 when they were studying the mechanism of tetracycline resistance in *Enterobacteria* (Levy, 1992). Since then, it has been demonstrated that one or multiple efflux pumps participate in the efflux of antibiotics, and resistance to almost all antibiotics could be achieved by extruding them out of cells through these pumps. Some efflux pumps selectively extrude specific antibiotics and some, so-called multi-drug resistance (MDR) pumps, efflux varieties of structurally diverse compounds. In this latter case a single protein may confer resistance to multiple antibiotics with different modes of action. Many MDR pumps are normal constituents of bacterial chromosomes and increased antibiotic resistance is a consequence of overexpression of MDR pumps. Thus, bacteria have the potential to develop multi-drug resistance without acquisition of multiple specific resistance determinants. Constitutive expression of some MDR pumps makes bacteria less susceptible to antibiotics.

While almost all antibiotics can be effluxed by MDR pumps, resistance mediated by these pumps is not equally important for all of them in clinical settings. In general, it is of particular relevance for antibiotics that are less affected by other resistance mechanisms. Which are these antibiotics? Some of these antibiotics are semi-synthetic derivatives of naturally existing antibiotics that were specifically designed to evade multiple degrading or modifying enzymes, such as novel β-lactams. Others are man-made synthetic compounds such as fluoroquinolones, for which degradation or modification mechanisms do not exist at all due to their artificial origin. In other words, resistance mediated by MDR pumps is especially relevant for the most advanced types of antibiotics.

Besides their importance in drug resistance in clinical settings, MDR pumps pose intriguing molecular biological questions. Their ability to recognize and extrude from the cells multiple structurally diverse compounds contradicts the existing principles of enzyme-substrate recognition. It is also puzzling how these non-specific proteins distinguish their exogenous substrates from intracellular molecules.

The ever-growing interest in efflux pumps is reflected in a large number of recent review articles addressing various aspects of the efflux problem, such as the origin and evolution of efflux pumps (Saier *et al.*, 1998), mechanisms of multidrug resistance (Paulsen *et al.*, 1996b) (Nikaido, 1998a,

1998b; van Veen and Konings, 1997; Lewis, 1999), regulation of expression of MDR pumps (Nikaido, 1998a), natural functions of MDR pumps (Neyfakh, 1997), and clinical aspects of MDR (Nikaido, 1998c; Lawrence and Barrett, 1998).

FAMILIES OF BACTERIAL DRUG EFFLUX PUMPS

The genome sequence analyses conducted on bacteria with completely sequenced genomes has revealed that the number of all transport proteins is somewhat proportional to the genome size and corresponds to 5 to 10% of the total encoded genes. Proven and putative drug efflux pumps constitute 6 to 18% of all transporters (Paulsen et al., 1998b), indicating the enormous capabilities of bacteria to combat assaults by deleterious compounds.

Five families of bacterial drug efflux pumps have been currently identified based on the energy source they use to export their substrates and sequence similarity. Transport can either be driven by ATP hydrolysis as in the case of pumps from the ATP-binding cassette (ABC) Superfamily, or pumps can utilize the proton motive force (PMF). The four PMF-dependent families are Small Multidrug Resistance Family (SMR), Major Facilitator Superfamily (MFS), Multidrug and Toxic Compound Extrusion Family (MATE) and Resistance/Nodulation/Cell Division Family (RND). Members of all families are ubiquitous from bacteria to man (Saier, 1998; Saier et al., 1998; Saier et al., 1999).

ABC Family

Both specific and multi-drug-transporters were identified in the ABC-family (Higgins, 1992; Fath and Kolter, 1993; Linton and Higgins, 1998). While most of these drug-specific pumps have been found in antibiotic-producing bacteria, where they participate in antibiotic export and also confer resistance on the producer organism, they are also reported in non-producers. For example, MsrA from S. epidermidis confers resistance to erythromycin in clinical settings (see below).

No MDR pumps from this family have been documented to be relevant for drug-resistance to therapeutic agents. In fact, the only bacterial multidrug ABC-type transporter has been identified in Lactococcus lactis (van Veen et al., 1996). It shows considerable sequence similarity (48%) with mammalian P-glycoprotein (Ambudkar et al., 1999; Gottesman et al., 1996). These proteins also have an overlapping spectrum of substrates that are amphiphilic cationic compounds, such as daunomycin and ethidium bromide, and inhibitors like verapamil and reserpine (van Veen et al., 1996). In elegant experiments it was demonstrated that when LmrA is expressed in human cells, it confers typical multi-drug resistance (van Veen et al., 1998). Thus, bacterial and human multi-drug resistance proteins were shown to be functionally interchangeable, suggesting a fundamental biological role for this type of ABC-transporter in prokaryotic and eukaryotic cells.

SMR Family

The name of this family (Paulsen et al., 1996c; Saier et al., 1997) owes it to the actual small size of its members, which are typically around 110 amino acid residues in length. They have four predicted transmembrane domains and for several proteins this topology was confirmed experimentally (Grinius et al., 1992; Paulsen et al., 1995). Most probably SMR proteins function as oligomeric complexes (Yerushalmi et al., 1996). Solely MDR, but not drug-specific pumps are found among SMR drug transporters. The substrates of SMR pumps are almost exclusively monovalent hydrophobic cations, such as ethidium bromide, methyl viologen, proflavine, quaternary ammonium compounds (cetrimide, benzalkonium chloride). This last group of compounds is extensively used as antiseptics and disinfectants in clinical settings as well as in the food industry, and SMR-mediated resistance is widespread among bacteria (see below).

Two SMR transporters were reconstituted in proteoliposomes: Smr, which is found on plasmids of S. aureus (Grinius and Goldberg, 1994), and EmrE, which is encoded by the E. coli chromosome (Yerushalmi et al., 1995). Both proteins have relatively high V_{max} in the range of 1.5 to 2 μM/min/mg, and a correspondingly high turnover number of 14 to 26 min^{-1}. This efficiency of

SMR explains their ability to provide significant levels of resistance by outcompeting rapid spontaneous influx of their hydrophobic substrates back into the cytoplasm.

Major Facilitator Superfamily

Both drug-specific and MDR proteins are among members of this large and diverse superfamily (Paulsen *et al.*, 1996b; Pao *et al.*, 1998), and both play significant roles in drug resistance in clinical settings. Selective examples include tetracycline and macrolide-specific efflux pumps and efflux pumps conferring resistance to fluoroquinolones in gram-positive bacteria (see below). Most are of 400–600 amino acid residues in length and possess either 12- or 14-putative transmembrane domains. In the case of MDR MFS-pumps, while hydrophobic cationic compounds are their preferred substrates (similar to SMR-proteins), the range of substrates is broader than that of SMR pumps. For example Bmr from *B. subtilus* (Neyfakh *et al.*, 1991) and MdfA from *E. coli* (Edgar and Bibi, 1997) can extrude neutral chloramphenicol in addition to various cationic compounds. MdfA was also shown to pump out an artificial substrate of β-galactosidase, isopropyl-β-D-galactoside (Bohn and Bouloc, 1998).

The Major Facilitator Superfamily contains at least 18 families (Paulsen *et al.*, 1998b). Nevertheless, drug transporters could readily be detected among them based on sequence analyses. However, it is difficult to discriminate *in silico* between drug-specific and MDR pumps (Saier *et al.*, 1998; Saier, 1998). Specific and MDR transporters are scattered randomly on the phylogenetic tree of MFS efflux pumps. It means that while "creation" of a drug transporter was a rare and ancient evolutionary event, the broadening and narrowing of specificity towards particular drugs occurred repeatedly during evolution.

Indeed, it has been shown that a single amino acid change in the *S. aureus* QacB transporter made it recognize not only monovalent but also divalent cationic compounds (Paulsen *et al.*, 1996a). In the case of MdfA, when a membrane embedded negatively charged glutamic acid was replaced with positively charged lysine, the ability to transport cationic substrates was lost, while the ability to efflux chloramphenicol was retained (Edgar and Bibi, 1999). The modulation of the substrate specificity may very well occur in response to the strong selective pressure applied by antibiotics in the clinic.

MATE Family

This family of MDR pumps was identified quite recently (Brown *et al.*, 1999). Similar to MFS, they contain 12-transmembrane domains but do not share sequence similarity with any of the members of MFS family. NorM protein from *Vibrio parahaemoliticus* was the first identified member of this family (Morita *et al.*, 1998). It mediated resistance to a rather broad range of antibacterial agents including hydrophobic cations, aminoglycosides and fluoroquinolones. NorM homologs were identified in *E. coli* and *Haemophilus*. The significance of the MATE proteins in clinical settings has not yet been established.

RND Family

Most of the pumps that mediate resistance to clinically relevant antibiotics from gram-negative bacteria belong to the RND (Saier *et al.*, 1994; Nikaido, 1998b) family. RND transporters are large proteins up to 1000 amino acids in length. They were predicted to consist of 12-transmembrane domains with two large periplasmic loops (Saier *et al.*, 1994). This topology was confirmed for the MexB transporter from *P. aeruginosa* (Guan *et al.*, 1999). Because of their complexity, these proteins appear to be the most difficult to study at the biochemical level. Recently, the AcrB protein from *E. coli* was reconstituted in proteoliposomes containing fluorescent phospholipids, presumed substrates of AcrB. To prevent reentry of extruded lipids into the bilayer of the original vesicles, an excess of "acceptor", AcrB-free, vesicles was added as a trap. These creative experiments demonstrated efflux of phospholipids, and unequivocally showed that AcrB is indeed an efflux protein working as drug:H^+ antiporter driven by ΔpH (Zgurskaya and Nikaido, 1999b).

MULTI-COMPONENT EFFLUX PUMPS FROM GRAM-NEGATIVE BACTERIA: BYPASSING THE PERIPLASM

Efflux pumps from gram-positive bacteria excrete their substrates across a single cytoplasmic membrane. It is also the case for some pumps of gram-negative bacteria and as a result, their substrates are effluxed into the periplasmic space. Other efflux pumps from gram-negative bacteria efflux their substrates directly into external medium, bypassing the periplasm and the outer membrane. These pumps are organized in complex three component structures, which traverse both inner and outer membranes (Nikaido, 1994, 1996). They consist of a transporter located in the cytoplasmic membrane, an outer membrane channel and a periplasmic "linker" protein, which brings in contact the other two components. Both the outer membrane and periplasmic components found in a variety of complex pumps were recently recognized as members of two new families: the family of special outer membrane channels (Paulsen *et al.*, 1997) and the family of membrane fusion proteins or MFP (Dinh *et al.*, 1994), respectively. Inner membrane transporters, utilized by trans-envelope pumps for the common goal of extracellular efflux, may belong to different families, namely, ABC-, MF- or RND-families. Many complex pumps containing ABC-transporters pump out proteins or peptides, such as hemolysin in *Escherichia coli*, leukotoxin produced by *Pasteurella haemolitica*, and proteases which are destined to be in the extracellular medium as opposed to a periplasm or an outer membrane from *Pseudomonas aeruginosa* and *Serratia marcescens* (Fath and Kolter, 1993). In these cells, a dedicated direct export system is used for proteases, as opposed to a non-specific Sec-mediated secretion pathway, which would not be effective for out-of-cell secretion.

Some multi-drug resistance pumps that belong to MF- or RND-families are also organized in three component complexes, which perform direct efflux.

Why it is advantageous for gram-negative bacteria to efflux drugs by bypassing the periplasm and outer membrane? In gram-negative bacteria the outer membrane significantly slows down the entry of both lipophilic and hydrophilic agents. Lipophilic agents, such as erythromycin and fusidic acid, are slowed down by lipopolysaccharides or LPS, which are lipid components of the outer leaflet of the outer membrane bilayer (Nikaido and Vaara, 1985). Hydrophilic agents, even small ones such as fluoroquinolones and some β-lactams, which cross the outer membrane through the water-filled porins, are still large enough to easily go through the porin channels (Nikaido and Vaara, 1985). This is the case for *E. coli*, which has rather efficient permanently opened porins like OmpF (Cowan *et al.*, 1995). In the case of *Pseudomonas aeruginosa*, a large fraction of its major porin oprF (Nikaido *et al.*, 1991) molecules appear to exist in a closed conformation (Nikaido *et al.*, 1991; Bellido *et al.*, 1992), which slows down uptake of hydrophilic antibiotics even further. Apparently the same is true for other nonfermentive gram-negative bacteria such as *Burkholderia cepacia, Stenotrophomonas maltophilia* or *Acinetobacter baunanii* (Hancock, 1998).

Thus, direct efflux creates the possibility for two different mechanisms to work synergistically to provide the cell with a potent defense mechanism. Because of this synergy even rather slow velocity pumps can confer significant levels of resistance. Indeed, the velocity of an endogenous *E. coli* tetracycline efflux pump, most probably AcrAB-TolC, was estimated to be ca. 0.2 nM/min/mg (Thanassi *et al.*, 1995), which could be indeed considered rather low (compare to 2 μM/min/mg for the SMR pump). Additional estimates showed that a pump which effluxed tetracycline directly into the medium could be ca. 100 fold less active than a pump extruding the drug into the periplasm, while providing the same level of resistance (Thanassi *et al.*, 1995).

Importantly, direct efflux into the medium leads to a decreased amount of drugs not only in the cytoplasmic but also in the periplasmic space. This can explain a paradoxical finding that efflux pumps protect gram-negative bacteria from β-lactam antibiotics (Li *et al.*, 1994; Li *et al.*, 1995) that attack their targets, penicillin-binding-proteins, from within the periplasm.

PERIPLASMIC AND OUTER MEMBRANE COMPONENTS OF COMPLEX EFFLUX PUMPS

Periplasmic components of complex efflux pumps comprise the protein family called membrane fusion proteins (MFP). They owe their name to the distant homology with proteins from enveloped virions (such as paramyxovirus) whose function is indeed to fuse the membranes (Dinh *et al.*, 1994). It was hypothesized that MFPs link or fuse the outer and inner membranes of gram-negative bacteria to create a drug exit channel together with the outer membrane component (Lomovskaya and Lewis, 1992; Lewis, 1994; Saier *et al.*, 1994). A typical MFP contains two hydrophobic regions near the N and C termini, which are believed to interact with cytoplasmic and outer membrane components of the pump. Indeed, alkaline phosphotase fusion studies of EmrA (Lomovskaya and Lewis, 1992) or AcrA (Ma *et al.*, 1993) with MFPs from EmrAB or AcrAB pumps, respectively, indicated that the central portion of the protein is located in the periplasm. Recent studies provided more evidence that the given name MFP appeared to be prophetic. First, it was demonstrated that AcrA is a highly asymmetric, elongated protein (Zgurskaya and Nikaido, 1999a). This is compatible with the hypothesis that AcrA connects the inner and outer membrane. Second, it was demonstrated directly that purified AcrA is capable of association with lipid bilayers and can promote the close association or even fusion of two different membranes (Zgurskaya and Nikaido, 1999b).

Phylogenetic analysis indicates that MFPs cluster according to their substrate specificity and to the type of transporter with which they interact. This result suggests both specific interactions between MFPs and corresponding inner membrane transporters, and an active role of MFPs in determining the specificity of transported substrates.

Unlike MFPs, the outer-membrane components of tripartite efflux pumps (OMFs) do not cluster according to the type of substrates and transporters with which they function (Paulsen *et al.*, 1997). This finding is not at all surprising but rather expected, since a single OMF protein can function with different pumps. For example, the protein TolC from *E. coli* (Benz *et al.*, 1993) is essential for activity of the drug-transporters AcrB (Fralick, 1996) and EmrB (Lewis, unpublished) that belong to the RND and MF-families, respectively. It also works together with the protein transporters HlyB (Holland *et al.*, 1990) and CvaB (Gilson *et al.*, 1990), which belong to the ABC-family and extrude hemolysin and colicin V, respectively.

OprM from *P. aeruginosa* is a component of at least two tripartite RND pumps, MexAB-OprM (Poole *et al.*, 1993) and MexXY-OprM (Mine *et al.*, 1999; Aires *et al.*, 1999), which have a different spectrum of substrates. MtrE from *N. honorrhoeae* is also a component of two efflux systems, MrtCDE (Hagman *et al.*, 1995) and FarAB-MtrE (Lee and Shafer, 1999). The MtrCDE pump confers resistance to hydrophobic antibiotics, fatty acids, bile salts and antibacterial peptides (Pan and Spratt, 1994; Hagman *et al.*, 1995; Shafer *et al.*, 1998). While MtrD is a RND transporter, FarB-belongs to the MF superfamily.

At the present time it is not known how the specificity of assembly of complex efflux pumps into functional units is determined. Construction of hybrid pumps may prove useful to address this issue (Srikumar *et al.*, 1997; Srikumar *et al.*, 1998; Yoneyama *et al.*, 1998). At least for one tripartite efflux pump, PrtDEF from *Erwinia chrysanthemi*, which secretes metalloprotease (Letoffe *et al.*, 1996), it was demonstrated that the complex formation between the three components is transient and dependent on the presence of substrate. It would be interesting to learn whether the same is true for the multi-drug resistance pumps.

POSSIBLE MECHANISMS OF MULTI-DRUG EXPORT AND MULTI-DRUG RECOGNITION

In spite of the fact that various multi-drug transporters belong to different families they appear to share a lot of similarities in transport mechanisms. It was shown that ABC-transporters, mammalian P-glycoprotein (Ambudkar *et al.*, 1999; Gottesman *et al.*, 1996) and LmrA from *Lactococcus lactis* (Bolhuis *et al.*, 1996b) capture their substrates within the membrane, and more specifically, within its inner leaflet. The substrates are then effluxed directly to the exterior bypassing the outer leaflet. The same was shown for multi-drug transporters LmrP (Bolhuis *et al.*, 1996a) and QacA (Mitchell *et al.*, 1999) that belong to the MFS-family.

RND-pumps, such as AcrAB from *E. coli* or *S. typhimurium* and MexAB-OprM from *P. aeruginosa*, can efflux agents such as many β-lactams that do not cross the plasma membrane. On the basis of these observations it was suggested that β-lactams are also captured and effluxed from within the lipid bilayer, however, from its outer leaflet (Nikaido *et al.*, 1998). Recently, two RND transporters, AmrAB-OprA (Moore *et al.*, 1999) and MexXY-OprM (Aires *et al.*, 1999) were reported to strongly efflux hydrophilic and positively charged aminoglycosides. It is possible that these pumps capture these compounds in the cytoplasm and not in the membrane. Interestingly, the same pumps were capable of exporting erythromycin, a typical hydrophobic substrate of many RND pumps. Thus, it is quite possible that the same protein can operate *via* different mechanisms to perform efflux of multiple drugs.

Multi-drug resistance pumps that can recognize multiple structurally diverse compounds violate a biochemical rule of strict specificity between enzymes or transporters and their substrates. The basis for this specificity is provided by stereospecific polar interactions between amino acids and a substrate, which are perfectly matched within the binding site. Apparently, substrate recognition by multi-drug transporters abides by different principles.

One can imagine some of these principles from the fact that even the most broad-spectrum efflux pumps, such as ABC and RND transporters, still discriminate between their multiple substrates and cellular components. Substrates of MDR pumps are amphiphilic or lipophilic molecules while components of the water-filled cytoplasm are hydrophilic. It is therefore possible that MDR pumps recognize their substrates based on their hydrophobic properties. In this case, the binding site of the MDR pump should be formed by amphiphilic amino acids. Recognition would then be mediated by stacking interactions between the hydrophobic groups of the binding pocket and substrate molecule. Such binding sites indeed may accommodate a variety of structurally diverse hydrophobic compounds.

Recently, the crystal structure of a complex between the multi-drug binding protein BmrR with the tetraphenylphosphonium (TPP$^+$) was solved with a 2.8Å resolution (Zheleznova *et al.*, 1999). BmrR is a cytoplasmic regulatory protein, which activates expression of the multi-drug resistance MFS pump Bmr from *Bacillus subtilis*. This activation requires binding of BmrR to some of the multiple substrates of the Bmr pump (Ahmed

et al., 1994). The X-ray structure revealed a conelike drug-binding pocket. It is formed by hydrophobic amino acids and contains a negatively charged glutamate residue at the bottom, which is inaccessible before the drug-induced unfolding and relocation of an α helix that gates the entrance to the hydrophobic pocket. Once inside the cone, TPP$^+$ makes electrostatic interactions with glutamate and stacking interactions with hydrophobic amino acids. Thus, recognition was based on both positive charge and hydrophobicity. Interesting, a cone-like structure imposes particular restrictions on the shape of interacting molecules. For example, planar molecules such as ethidium bromide do not bind BmrR.

One can easily imagine that a multi-drug transporter can also utilize the same structural principles to bind structurally diverse substrates. The BmrR example is of particular relevance for the pumps, which mostly export hydrophobic cationic molecules, such as SMR and many MF-transporters. If they happen to have slit-shaped but not cone-shaped binding sites, they will accommodate both planar and aplanar substrates as, indeed, is the case for the most of them (Lewis, 1999).

The importance of charged residues for drug recognition was demonstrated for several SMR (Grinius and Goldberg, 1994; Paulsen *et al.*, 1995) and MFS-proteins (Paulsen *et al.*, 1996a; Edgar and Bibi, 1999). Multi-drug transporters like ABC and RND-pumps, which can efflux negatively, positively, and uncharged compounds might have multiple multi-drug binding sites. At least two binding sites were found in the ABC-transporter P-glycoprotein (Ambudkar *et al.*, 1999). Also, BmrR shows that embedding of the binding site in the membrane is not a prerequisite for multi-drug recognition.

BROAD DEFENSE AND OTHER NATURAL FUNCTION OF BACTERIAL MDR PUMPS

Many MDR pumps have a relatively high level of constitutive expression ensuring fast defense response upon encountering a potential toxin. These highly expressed pumps most probably protect bacteria against noxious compounds that bacteria constantly encounter in natural environments. Examples of these pumps include AcrAB-TolC pump from *E. coli*, MexAB-OprM pump from *P. aeruginosa*, NorA pump from *S. aureus* and many others. Deletion of these pumps makes the corresponding bacteria hypersensitive to the multiple naturally occurring toxic compounds while their increased expression either due to induction or due to mutations in the regulatory genes results in increased resistance. Thus, protecting cells against noxious agents present in their natural environment may very well be a normal physiological function of some MDR pumps. In this respect involvement of MDR pumps in determining susceptibility to clinically important antibiotics might be just a fortuitous extension of their physiological functions.

What are the toxic compounds that bacteria might encounter in their natural environments? *Neisseria honorrhoeae* is a strict human pathogen that infects various mucosal sites. Most frequently it infects genitourinary mucosae but it also capable of colonizing rectum. These mucosal surfaces possess a number of different host defense agents, all with a potent anti-gonococcal activity *in vitro*. These agents include hydrophobic agents, *i.e.* free fatty acids, bile salts (in rectum) and some antimicrobial peptides such as human cathelicidin peptide LL-37 (in genitourinary mucosa). It was shown that the cells that lack the complex efflux pump MtrCDE are more sensitive to all these agents, implicating this pump in conferring some level of intrinsic resistance to these agents. Over-expression of this pump due to induction by hydrophobic agents, which requires the positive regulator MtrA (Rouquette *et al.*, 1999), or due to mutations in the negative regulator *mtrR* (Hagman *et al.*, 1995) results in the increased resistance (Hagman *et al.*, 1995; Shafer *et al.*, 1998). Indeed, *mtrR⁻* mutants are frequently found in rectal isolates of *N. honorrhoeae* (Pan and Spratt, 1994). MtrR also regulates expression of another efflux pump FarAB, with no reported MDR phenotype, which protects gonococci against toxic effects of long-chained fatty acids (Lee and Shafer, 1999). This genetic arrangement strongly indicates that toxic hydrophobic compounds of mucosal surfaces are indeed the primary substrates of MtrCDE. Recently, increased expression of the MtrCDE efflux pump was shown to cause

decreased susceptibility to the hydrophobic antibiotic azithromycin, which is used for the treatment of some gonococcal infections (Zarantonelli *et al.*, 1999). Thus, MtrCDE enables bacteria to resist various toxic compounds, those that are normal constituents of its natural environment, and those that it meets only occasionally.

S. aureus is an organism that commonly invades the bloodstream. It is highly susceptible to the antibacterial action of the small cationic peptide tPMP-1 (thrombin-induced platelet microbicidal protein 1), which is released from platelets by thrombin in response to endovascular infections. *In vitro* resistance of *S. aureus* to this peptide correlates with the survival advantage of resistant strains at sites of endothelial damage. It was demonstrated that the cells of *S. aureus* containing the plasmid-encoded MDR pump QacA are phenotypically resistant to tPMP-1 but not to other structurally distinct endogenous cationic peptides (Kupferwasser *et al.*, 1999). It is quite possible that QacA, which was earlier implicated in the resistance towards multiple hydrophobic cations, among them many widely used antiseptic and disinfectants, in fact helps *S. aureus* to survive in the toxic host environment.

E. coli (and other enteric bacteria) exists in a free living or pathogenic form. Its latter natural environment is enriched in toxic bile salts and fatty acids. Indeed, it was demonstrated that *E. coli* cells lacking the AcrAB efflux pump are more susceptible to these agents indicating AcrAB-mediated protection (Thanassi *et al.*, 1997). Moreover, it was shown that among all studied substrates of AcrAB, bile salts have the highest affinity to the transporter (Zgurskaya and Nikaido, 1999b) implicating efflux of bile salts as a possible physiological function of this pump. Interestingly, expression of AcrAB was not increased upon exposure to bile salts (Ma *et al.*, 1995). Most probably, the affinity of the transporter is so high that even non-induced constitutive level of expression produces enough AcrAB to confer protection.

In the case of *Salmonella typhimurium*, the mutant lacking the AcrAB homolog exhibited a reduced capacity to colonize the intestinal tract, presumably due to increased sensitivity to bile salts (Lacroix *et al.*, 1996). This hypersensitivity was also thought to be the reason that the AcrAB lacking mutant of *Salmonella enteritidis* is unable to invade tissue culture cells (Stone *et al.*, 1992; Stone and Miller, 1995).

As free-living organisms, bacteria like *E. coli* or *S. aureus* are also exposed to a variety of naturally occurring toxic compounds. Those are antibacterial agents produced by other bacteria, fungi and plants. Accordingly, mutants lacking *acrAB* are hypersensitive to many noxious agents. These include conventional broad-spectrum antibiotics of natural origin like tetracycline, chloramphenicol and β-lactams (Ma *et al.*, 1995), as well as compounds like doxorubicin or pacidamycin with very weak antibacterial activity against wild-type cells (Lomovskaya, unpublished).

As was discussed above, SMR and many MF MDRs almost exclusively extrude hydrophobic cations. However, this class of agents is absent among known antibiotics of natural origin. It was suggested that naturally occurring hydrophobic cations lack inhibitory activity against bacteria exactly because they are the best substrates of many MDR pumps (Lewis, 1999; Hsieh *et al.*, 1998). Indeed, it was demonstrated that the mutant of *S. aureus* that lacked the constitutively expressed pump NorA, was hypersensitive to the plant alkaloid berberine, the typical hydrophobic cation with little activity against the wild type bacteria (Hsieh *et al.*, 1998).

Functions of MDR pumps could be further understood through their regulation. For example, AcrAB (Okusu *et al.*, 1996) is a member of at least two inducible regulons, *mar* (multiple antibiotic resistance) and *sox* (superoxide response), which are regulated by the master sensors *marRAB* and *soxRS*, respectively (Demple, 1991; Alekshun and Levy, 1997). MarA and SoxS, which are negatively regulated by MarR and SoxR, are activators of corresponding regulons. They are not involved in the constitutive expression of *acrAB* (Ma *et al.*, 1995) but are essential for its inducibility.

The *sox* regulon, which contains, besides *acrAB*, a variety of genes whose products participate in the adaptation to superoxide (O_2^-)-generated oxidative stress, is induced in the presence of redox-cycling agents. The function of AcrAB as a member of the *sox*-regulon might be to reduce intracellular accumulation of naturally-occurring redox-cycling agents. It is also possible that AcrAB participates in the excretion of the molecules damaged by free radicals.

As for the *mar* regulon, besides *acrAB* and multiple genes of unknown function, it contains at least two other genes, *tolC* and *micF*, relevant to resistance to toxic compounds. *micF* encodes an antisense RNA, which is a negative regulator of one of the major *E. coli* porins, OmpF, the gate for exogenous molecules. Thus, induction of the *mar* regulon results in the increased production of AcrAB and TolC, components of the tripartite pump, and down-regulation of OmpF, which reduces permeability of the outer membrane. This setup is perfect to create a synergistic increase in resistance to toxic compounds. The fact that AcrAB pump belongs to the same operon as *micF* makes its role in protection against environmental toxins quite plausible. Several known naturally occurring inducers of the *mar* operon, and consequently, *acrAB*, include plant-derived salicylate (Cohen *et al.*, 1993), plumbagin (Seoane and Levy, 1995), and redox-cycling naphthoquinone. Based on the identity of inducers, it was suggested that the products of the genes that belong to the *mar* regulon participate in the defense against assault by living or decomposing plants (Miller and Sulavik, 1996).

The AcrAB pump also apparently plays an important but currently not well-defined role in survival of *E. coli* under general stress conditions since the *acrAB* operon is induced by such general stress signals as onset of stationary phase, 0.5M NaCl, and 4% ethanol (Ma *et al.*, 1995).

It was shown that expression of several MDR pumps is induced in the presence of their multiple substrates and the molecular mechanism of induction is based on the direct binding of structurally diverse molecules to a single regulatory protein. This discovery further supports the role of some MDR pumps in protecting against environmental toxins. The first reported example was the BmrR protein, a positive regulator of the Bmr pump from *B. subtilis* (Ahmed *et al.*, 1994; Markham *et al.*, 1997).

Two other MDR regulators, EmrR and QacR, the repressors of EmrAB from *E. coli* and QacA from *S. aureus*, were also shown to directly bind multiple inducing molecules, which are substrates of the corresponding pumps. (Grkovic *et al.*, 1998; Brooun *et al.*, 1999).

It is noteworthy that not all of the substrates of the aforementioned pumps are capable of inducing their expression. It is tempting to speculate that it is an inducer-regulator specificity that determines a primary substrate(s), or in other words, function of MDR pumps. This idea could by illustrated by the following example.

Besides Bmr, *B. subtilus* possess another MDR pump, Blt, which is highly homologous to Bmr (Ahmed *et al.*, 1995). Overexpression of Blt results in a resistance to a spectrum of toxic compounds very similar to Bmr's, however its expression is not induced in the presence of these drugs. An important finding was that Blt was capable of effluxing of intracellular polyamine spermidine out of cells. Moreover, the *blt* gene is located in a dicistronic operon together with the gene encoding spermidine acetyltransferase, an enzyme catalyzing a critical step in spermidine degradation (Woolridge *et al.*, 1997). This arrangement strongly suggested that the efflux of an endogenous substrate spermidine is indeed the primary function of Blt. At the same time, Bmr, as judged from the nature of its inducers, appear to confer protection against toxic exogenous compounds. Very recently a regulatory gene, *mta*, has been identified in *B. subtilus* (Baranova *et al.*, 1999). It was shown that *mta* is a repressor of both Bmr and Blt since the mutant lacking *mta* overexpressed both pumps simultaneously. The signals that stimulate activity of Mta are not yet identified. However, the fact that both transporters are able to respond to the same stimuli indicates that at least under these conditions they perform related functions. Apparently, MDR pumps can perform multiple functions. Depending on a specific regulatory mechanism, pumps may switch back and forth from transporting specific exogenous or endogenous substrates to providing a broad protection against multiple environmental stresses.

EFFLUX PUMPS AND DRUG RESISTANCE

Nowadays efflux mechanisms are broadly recognized as major components of resistance to many classes of antibiotics. Both antibiotic-specific and MDR pumps are known to confer clinically significant resistance. In some cases, if efflux-mediated resistance and other resistance mechanisms operate simultaneously in the same cell, this may result in synergistic effects on drug resistance (see below).

Basal levels of activity of non-specific multi-drug efflux pumps in the wild type cells contribute to decreased antibiotic susceptibility. This intrinsic resistance may be low enough for the bacteria to still be susceptible to therapy. However, the bacteria might be even more susceptible if efflux pumps were rendered non-functional. Bacteria that are more susceptible would allow lower doses of antibiotics to be used in therapy. This could be especially important for antibiotics with narrow therapeutic indices.

If not for efflux pumps, the spectrum of activity of many so-called ''gram-positive'' antibiotics could be expanded to previously non-susceptible gram-negative species. This can be applied to ''narrow-spectrum'' β-lactams, macrolides, lincosamides, streptogramines, rifamycins, fusidic acid, and oxazolidinones that have a potent antibacterial effect against engineered mutants that lack efflux pumps (Nikaido, 1998b, 1998a).

But what are those antibiotics whose current clinical utility is strongly affected by efflux-mediated resistance? What efflux pumps confer this resistance? What is the contribution of efflux-mediated resistance as compared to other resistance mechanism? These issues will be addressed below.

Tetracyclines

Tetracyclines have been used extensively against gram-positive and gram-negative bacteria, mycoplasma, rickettsiae and protozoan parasites (Sum *et al.*, 1998; Chopra *et al.*, 1992; Levy, 1992). Besides their therapeutic use in man and animals, tetracyclines have been used indiscriminately for growth promotion in animal husbandry. As a result, tetracycline resistance became widespread in both Gram-positive and Gram-negative species. Active efflux and ribosomal protection are the two major mechanisms of tetracycline resistance (Chopra *et al.*, 1992; Schnappinger and Hillen, 1996). The proteins involved in ribosomal protection, TetO-TetM, TetQ, TetS, TetB(P) and OtrA(P) are highly specific to tetracyclines and are distributed mainly amongst gram-positive bacteria. Efflux pumps involved in tetracycline efflux can be either specific or multi-drug resistant.

Tetracycline-specific efflux pumps

TetA-TetE, TetG, TetH and TetK, TetL, TetA(P), OtrB proteins are tetracycline-specific efflux pumps found in the majority of pathogenic gram-negative and gram-positive bacteria, respectively (Roberts, 1994). The most widely distributed pumps are TetA /TetB and TetK/TetL. Both gram-negative and gram-positive pumps belong to two subfamilies of the MF-superfamily, classified as 12- and 14-TMS export proteins, respectively. Inside each subfamily, there is ca. 80% homology in amino acids. However, the Tet proteins of gram-negative and gram-positive bacteria are less similar to each other than to other exporters from corresponding families (Sheridan and Chopra, 1991). Both type of pumps are drug:H^+ antiporters with the actual substrate being a divalent metal-tetracycline chelate $[M\text{-}tet]^+$ (Yamaguchi et al., 1990). It was shown that the most extensively studied Tn10-encoded TetA protein pumps out the magnesium-tetracycline complex into the periplasm and not directly through both membranes of gram-negative bacteria (Thanassi et al., 1995).

Efflux of tetracycline by MDR pumps

Historically, tetracyclines have been widely used for the treatment of gastrointestinal infections. In E. coli (and in other enteric bacteria) tetracyclines are substrates of multiple endogenous MDR systems that belong to the RND family. The AcrAB MDR pump from E. coli contributes to the intrinsic resistance to this antibiotic (Ma et al., 1995). MIC of tetracycline in the wild type cells (0.5–1 μg/ml) is 4 to 8-fold higher than in the cells lacking AcrAB (0.125–0.25 μg/ml). Overexpression of AcrAB, mainly due to regulatory mutations, results in 2 to 4 fold-decreased susceptibility as compared to the wild type (MIC of 2–4 μg/ml). These mutations occur with relatively high frequency of 10^{-7}–10^{-8}. Elevated activity of MDR pumps in E. coli results in significantly less resistance than that conferred by the tetracycline-specific TetA protein (MIC of 32 μg/ml). However even this resistance is high enough to bring E. coli to the boarder of clinical susceptibility (the MIC breakpoint for susceptibility and for resistance is 2 μg/ml and 8 μg/ml, respectively).

Macrolides

The 14-membered, (erythromycin, clarythromycin) and 16-membered (josamycin and spiramycin) macrolides are extensively used for the treatment of upper and lower respiratory tract and skin and soft tissue infections (Zuckerman and Kaye, 1995) caused by Staphylococci, Streptococci and Pneumoccoci. As such they have been called "gram-positive" antibiotics. However, the azalide azithromycin also shows good activity against common gram-negative respiratory pathogens such as Haemophilus influenzae. An MIC breakpoint for susceptibility of 2 μg/ml and $\geq$8 μg/ml for resistance was proposed for azithromycin. Macrolides account for 10% to 15% of the worldwide oral antibiotic market (Kirst, 1991). Two major mechanisms of macrolide resistance are target modification (dimethylation of bacterial 23S ribosomal RNA within the macrolide binding site by the products of erm genes) and efflux. In the former case the very same dimethylation reaction confers cross-resistance to two other structurally diverse but functionally similar classes of antibiotics, lincosamides and streptogramins B. The phenotype conferred by the erm genes was designated macrolide-lincosamide-streptogramine B (MLS$_B$) resistance.

Macrolide-Streptogramin B-specific efflux pumps

A phenotype of resistance to 14- and 15-membered macrolides and streptogramin B, but not 16-membered macrolides (MS-phenotype) was identified in several species of Staphylococci. Efflux pumps belonging to the ABC-transporters family confer this phenotype. The pump-mediated antibiotic resistance is inducible and can reach a significant level of 32 to 128 μg/ml (Ross et al., 1990). Separate but adjacent genes primarily located on various plasmids encode the ATP-binding and the transmembrane domains of these pumps. The msrA and the closely related msrB genes encode the ATP-binding domains (Ross et al., 1990; Matsuoka et al., 1995; Milton et al., 1992). Introduction of either smrA or smrB into S. aureus is sufficient to confer MS-phenotype (Ross et al., 1990; Milton et al., 1992), indicating that unknown proteins encoded in the *S. aureus* chromosome can substitute for the original membrane-components of these pumps.

While the predominant phenotype in *Staphylococci* is still MLS$_B$, MS-phenotype appears in a reasonable percentage of clinical isolates, particularly in coagulase-negative *Staphylococci* (Janosi *et al.*, 1990). It is noteworthy that the *msrA* gene is often found on the plasmids also containing the *erm* genes (Eady *et al.*, 1993). The two resistance mechanisms probably act synergistically.

Macrolide-specific efflux pumps

The *mef* genes (macrolide efflux) encode efflux pumps specific for 14- and 15-membered macrolides. The phenotype conferred by these pumps was designated as M-phenotype for macrolide resistance, with lincosamide and streptogramin sensitivity (Sutcliffe *et al.*, 1996). Mef-mediated resistance is not induced in the presence of antibiotics. Mef-pumps belong to the MF-family of transporters with 12-transmembrane spanning domains (Clancy *et al.*, 1997; Tait-Kamradt *et al.*, 1997). The level of resistance provided by the *mef*-genes is 4 to 16 μg/ml which is generally lower than the *erm*-mediated resistance of 32 to 128 μg/ml (Giovanetti *et al.*, 1999).

The *mefA* gene was first identified in *S. pyogenes* (Clancy *et al.*, 1997). It was later shown that *mefA* and its close homologue, *mefE*, are widely distributed among various *Streptoccocci* (Tait-Kamradt *et al.*, 1997; Arpin *et al.*, 1999; Kataja *et al.*, 1998; Poutanen *et al.*, 1999).

In 1997 a new macrolide-resistance gene designated *mreA* was identified in a strain of *Streptococcus agalactiae* (Clancy *et al.*, 1997). The product of this gene conferred resistance to 14-, 15- and 16-membered macrolides. MreA encodes a protein that belongs to the MF-superfamily but is clearly different from the Mef proteins. Its distribution and role in clinical resistance to macrolides remain to be determined.

Multi-drug resistance pump

Macrolides were traditionally labeled ''gram-positive'' antibiotics since most gram-negative bacteria are intrinsically non-susceptible (MIC>128 μg/ml). In recent years it was shown that macrolides were excellent substrates of many MDR pumps from gram-negative bacteria. Many of these pumps belong to the RND family of transporters. Acr-pumps in *E. coli*, *S. typhimurium*, *H. influenzae*, Mex-pumps in *P. aeruginosa*, and the Amr-pump from *B. pseudomallei* are examples of these pumps. Azithromycin (Zuckerman and Kaye, 1995) is the only macrolide with borderline activity against some gram-negative bacteria, including *H. influenzae*, which is a causative agent of respiratory infections. Increasing potency of azithromycin against *H. influenzae* will significantly enhance the utility of the azithromycin for the treatment of this common clinical syndrome.

Chloramphenicol and Florfenicol

Chloramphenicol is an effective bacteriostatic broad-spectrum antibiotic and was extensively used in human and animal health to treat infections caused by gram-positive and gram-negative bacteria, especially *Salmonella*. However, serious toxic side effects and the wide spread development of resistance severely restricted the use of chloramphenicol. Enzymatic acetylation of chloramphenicol (Shaw, 1983) by the product of *cat* and active efflux are the two mechanisms of chloramphenicol resistance.

Florfenicol is a fluorinated chloramphenicol analog that retains broad-spectrum potent antibacterial activity of chloramphenicol (Syriopoulou *et al.*, 1981), but is not a substrate of chloramphenicol acetyltransferase. In 1996 florfenicol, Nuflor, was approved for veterinary use. Importantly, it is not used in human medicine. An MIC breakpoint for susceptibility of 8 μg/ml, 16 μg/ml for intermediate resistance and $\geq$32 μg/ml for resistance was proposed for chloramphenicol and florfenicol

Chloramphenicol-specific efflux pumps

cmlA-genes encode efflux pumps that belong to the 12-TMS exporters from the Major facilitator superfamily. They confer inducible resistance of 16 to 200 μg/ml to chloramphenicol, but not to florfenicol (Dorman and Foster, 1982; Bissonnette *et al.*, 1991). These genes appear to be wide spread among gram-negative bacteria. They have been identified in various species of *Enterobacteriaceae* (Dorman *et al.*, 1986; Toro *et al.*, 1990; Ploy *et al.*, 1998), in *H. influenzae* (Burns *et al.*, 1985), and in *P. aeruginosa* (Bissonnette *et al.*, 1991). They are usually located on the plasmids as part of integrons. *cmlA*-genes from different species share up to 87% of amino acid identity.

It is possible that the efflux capability of CmlA alone is not sufficient to provide the high-level chloramphenicol resistance that is associated with failures in therapy. Several reports have implicated these proteins in the repression of outer membrane porins (Toro *et al.*, 1990; Bissonnette *et al.*, 1991; Burns *et al.*, 1985).

Chloramphenicol/Florfenicol-specific pumps

Shortly after the introduction of florfenicol, researchers in Japan identified a plasmid-located novel gene, *pp-flo*, which conferred resistance to florfenicol and chloramphenicol in the important fish pathogen, *Pasteurella piscicida* (Kim and Aoki, 1996). This gene encoded a 12-TMS member of the MF-superfamily with ca. 50% homology to CmlA. Recently, a gene with 97% identity to *pp-flo*, designated *flo$_{st}$*, was discovered among *Salmonella spp.* (Bolton *et al.*, 1999). Some of the resistant isolates included *S. typhimurium* DT104, which is an emerging multi-drug resistant strain in the UK, some other European countries and in the United States and Canada (Poppe *et al.*, 1998). In the United States the prevalence of this strain increased from 0.6% in 1979–1980 to 34% in 1996 (Glynn *et al.*, 1998). A Flo-pump was also identified among the majority of chloramphenicol and florfenicol-resistant pathogenic isolates of *Escherichia coli*. The strains harboring Flo were reported to have MIC ca. 128 μg/ml. Presently, it is not known whether the Flo pump alone is sufficient to confer this high-level resistance.

Multi-drug resistance pumps

Similar to tetracyclines and macrolides, both chloramphenicol and florfenicol are substrates of MDR pumps from gram-negative bacteria belonging to the RND-family. In *E. coli*, which is a pathogen relevant for the current use of florfenicol, overexpression of AcrAB conferred 4 to 8-fold increases in chloramphenicol and florfenicol resistance in both laboratory strains and veterinary clinical isolates, resulting in MIC up to 32 μg/ml (Lee and Lomovskaya, unpublished). Clinically, this level is considered as intermediate resistance.

Thus, various efflux pumps affecting chloramphenicol and florfenicol susceptibility can coexist in a single strain. What would be the effect of simultaneous expression of two efflux pumps on the susceptibility of antibiotics, which are substrates of both pumps? Our results (Lee and Lomovskaya, unpublished) show that when AcrAB and Flo are both present in the same strain there is a synergistic increase in florfenicol resistance. While overexpression of either AcrAB or Flo singly resulted in MIC of 16 to 32 μg/ml, simultaneous expression of both pumps confers resistance up to 256 μg/ml. The explanation for these results appears to lie in the *modus operandi* of these efflux pumps. Flo is a single component efflux pump that extrudes its substrates (florfenicol) into the periplasm. As was discussed above, multi-component AcrAB-TolC is quite different since it pumps out drugs, including florfenicol, directly to exterior. Consequently, this pump is capable of decreasing florfenicol concentration in the periplasm, which in turn will enhance the efficiency of Flo-mediated efflux, resulting in a synergistic effect on florfenicol resistance.

It is possible that the high-level florfenicol resistance seen in clinical isolates of *E. coli* and *Salmonella* is a result of synergistic interactions between various efflux pumps. If this is indeed the case than inhibiting even a single efflux pump from the synergistic pair will have a significant effect on susceptibility.

β-lactams

β-lactams are the preferred class of antimicrobials because of their bacteriocidal activity. Since the introduction of penicillin G in the 1940s, more than 50 β-lactams (expanded-spectrum penicillins, cephalosporins, monobactams and carbapenems) have been developed. Degradation of β-lactams by various chromosomal or plasmid-encoded β-lactamases is the main mechanism of resistance to these antibiotics in the majority of bacterial species (Bush *et al.*, 1995). However, particularly in the case of *P. aeruginosa*, resistance to carbapenems (imipenem, meropenem) can be due to mutational loss of the porin OprD, which is the main route of uptake of carbapenems (Trias and Nikaido, 1990). Efflux as a clinically relevant mechanism of resistance to β-lactams is also well documented.

Multi-drug resistance pumps

In *P. aeruginosa*, most of the β-lactams (with exception of carbapenem imipenem) are substrates of the MexAB-OprM RND MDR efflux pump (Li *et al.*, 1994; Morshed *et al.*, 1995; Yoneyama *et al.*, 1997). This pump is under the control of the local repressor, MexR (Poole *et al.*, 1996b). Frequently, overexpression of *mexAB-oprM* is a result of mutations in *mexR* (Poole *et al.*, 1996b; Ziha-Zarifi *et al.*, 1999), although *mexR* independent overexpression has also been reported (Ziha-Zarifi *et al.*, 1999). The level of resistance to many β-lactams that is provided by overexpressed MexAB-OprM exceeds the susceptibility breakpoint, and such mutants are considered clinically resistant. Several surveys indicate that there is a high proportion of MexAB-OprM overexpressing strains among β-lactam resistant clinical isolates of *P. aeruginosa*. They were discovered in ca. 80% of resistant mutants isolated in the UK in early 1980s and 1990s (Williams *et al.*, 1984; Chen *et al.*, 1995). They have also been found in ca. 50% recent Italian isolates (Bonfiglio *et al.*, 1998). Similar fractions of the MexAB-OprM overproducing strains were discovered in a French hospital when a limited number of post-therapy resistant isolates were investigated (Ziha-Zarifi *et al.*, 1999).

Using isogenic laboratory mutants it was demonstrated that in the case of the carbapenem meropenem, increased efflux due to overexpression of MexAB-OprM and decreased permeability due to the loss of OprD had a synergistic effect. Overexpression of MexAB-OprM conferred a further increase in resistance on the strain with OprD mutation. As expected, deletion of MexAB-OprM decreased resistance to meropenem even in the strains lacking OprD (Kohler *et al.*, 1999).

Quite a different picture emerged from studying the interplay between MexAB-OprM and chromosomal β-lactamase with respect to cephalosporins, such as ceftazidime and cefepime, and the monobactam aztreonam. Overexpression of MexAB-OprM significantly increased (16 to 64-fold) the resistance towards each of these antibiotics in strains lacking β-lactamase. Derepression of β-lactamase in the strain lacking MexAB-OprM conferred comparable or even higher increases. At the same time, overexpression or deletion of efflux pumps in the strain with derepressed enzyme had only a weak effect (Masuda *et al.*, 1999; Nakae *et al.*, 1999). An important conclusion from these experiments is that inhibition of the MexAB-OprM pump in the strains overexpressing β-lactamases is not expected to have a significant effect on resistance to β-lactams, that are affected by both resistance mechanisms.

Fluoroquinolones

Fluoroquinolones are broad-spectrum bacteriocidal agents widely used in both community and hospital settings. They have good antibacterial activity after oral administration. Older fluoroquinolones such as norfloxacin, ciprofloxacin, and enofloxacin, perform better against Gram-negative than Gram-positive pathogens, such as *S. pneumonia* or *S. aureus* (Gootz and Brighty, 1996). Newer fluoroquinolones, such as levofloxacin, sparfloxacin, and trovafloxacin have improved activity against gram-positive pathogens (Brighty and Gootz, 1997). Antibacterial activity of fluoroquinolones is based on the inhibition of type II topoisomerases, DNA gyrase (encoded by the *gyrA* and *gyrB* genes) and DNA topoisomerase IV (encoded by the *parC* and *parE* genes). Two resistance mechanisms found in gram-negative and gram-positive bacteria are target modification and efflux by MDR pumps.

Escherichia coli

Single-step mutants with 4 to 8 fold increased resistance to fluoroquinolones can be easily isolated using laboratory strains of *E. coli* (Hooper *et al.*, 1986; Hirai *et al.*, 1986). These mutants either have mutations in the so-called quinolone resistance-determining region (QRDR) (Piddock, 1994; Yoshida *et al.*, 1990a) of the target gene *gyrA*, or they possess regulatory mutations in the *mar* or *sox* loci which will decrease fluoroquinolone accumulation (Cohen *et al.*, 1989; Amabile-Cuevas and Demple, 1991). Decreased accumulation of fluoroquinolones reported in such mutants is due to overexpression of the MDR pump AcrAB (Okusu *et al.*, 1996) with concomitant down-regulation of the porin OprF (Alekshun and Levy, 1997; Alekshun and Levy, 1999). Neither type of single-step mutation will make *E. coli* clinically resistant to fluoroquinolones. However, passage of *E. coli* on increasing concentrations of fluoroquinolones can produce highly resistant strains. Analysis of these strains demonstrated acquisition of multiple mutations (Hooper *et al.*, 1986; Hirai *et al.*, 1986), both target-based and efflux-mediated.

Many reports demonstrated that fluoroquinolone-resistant clinical isolates of *E. coli* indeed harbor multiple mutations, including those resulting in decreased accumulation of fluoroquinolones (Aoyama *et al.*, 1987; Everett *et al.*, 1996), indicating involvement of efflux. It was demonstrated that *mar* mutants overexpressing AcrAB could be identified in clinical strains based on their increased tolerance to organic solvents (Oethinger *et al.*, 1998b; Oethinger *et al.*, 1998a). Strains with multiple resistance mechanisms, including increased efflux, were also identified among clinical isolates of *Proteus* (Ishii *et al.*, 1991), *Shigella* (Ghosh *et al.*, 1998), *Klebsiella* (Deguchi *et al.*, 1997) and in laboratory mutants of *Campylobacter jejuni* (Charvalos *et al.*, 1995).

Pseudomonas aeruginosa

Fluoroquinolones are among the primary therapeutic agents for this important pathogen. While *P. aeruginosa* is clinically susceptible to norfloxacin, ciprofloxacin and levofloxacin, the level of its intrinsic resistance is 10–20 fold higher as compared to *E. coli*. As a result, single step mutants with MIC of 1–2 μg/ml are already clinically resistant to fluoroquinolones.

Single-step fluoroquinolone resistant laboratory mutants contain either mutations in *gyrA* (Robillard, 1990; Zhanel *et al.*, 1995; Lomovskaya *et al.*,

1999a) or mutations resulting in increased expression of any of the three efflux pumps, all belonging to the RND family. The *nalB* mutants (Masuda and Ohya, 1992), *nfxB* mutants (Hirai *et al.*, 1987), and *nfxC* mutants (Fukuda *et al.*, 1990) overexpress MexAB-OprM (Poole *et al.*, 1993), MexCD-OprJ (Poole *et al.*, 1996a) and MexEF-OprN (Kohler *et al.*, 1997a) efflux pumps, respectively. Single-step gyrase and efflux mutants confer similar levels of fluoroquinolone resistance (Lomovskaya *et al.*, 1999a). Importantly, the frequencies of mutations resulting in pump overexpression are ca. 10-fold higher than frequencies of target-based mutations (Kohler *et al.*, 1997b; Lomovskaya *et al.*, 1999a).

Both target (Yoshida *et al.*, 1990a; Yonezawa *et al.*, 1995; Nakano *et al.*, 1997) and efflux mutants (Jakics *et al.*, 1992; Fukuda *et al.*, 1990; Fukuda *et al.*, 1995; Jalal and Wretlind, 1998; Ziha-Zarifi *et al.*, 1999) were identified among clinical isolates of *P. aeruginosa* resistant to fluoroquinolones. Strains with high levels of resistance (MIC>8 μg/ml) were shown to have multiple resistance mechanisms, both target-based and efflux mediated (Yoshida *et al.*, 1994; Fukuda *et al.*, 1995; Pumbwe *et al.*, 1996).

Using isogenic laboratory strains of *P. aeruginosa* containing different combinations of target-based and efflux-mediated mutations, it was demonstrated that these two mechanisms work independently and synergistically to confer fluoroquinolone resistance (Lomovskaya *et al.*, 1999a). Overexpression of any of the efflux pumps resulted in the same 8 to 32-fold increase in resistance to levofloxacin whether or not multiple target mutations were present in the same strain. Conversely, target mutations increased resistance to the same degree in the wild-type strain and in the strain with increased efflux.

The prevalence of efflux mutants among clinical isolates of *P. aeruginosa* is not known. However, it is expected that it is higher than in the case of enteric bacteria. First, overexpression of one of three efflux pumps can confer a clinically significant level of resistance. Second, the frequencies of efflux mutations are higher than frequencies of target mutations, at least under laboratory conditions. It is possible that the majority of clinical isolates with high-level resistance (MIC > 8 μg/ml) contain both target- and efflux-mediated mutations. Efflux pumps have also been identified in other gram-negative non-fermenting bacteria, CeoAB-OpcM in *Burkholderia cepacia* (Burns *et al.*, 1996), AmrAB-OprA in *Burkholderia pseudomallei* (Moore *et al.*, 1999). Part of an efflux operon was also identified in *Stenotrophomonas maltophilia* (Alonso and Martinez, 1997) with a high level intrinsic resistance to fluoroquinolones.

Gram-positive bacteria

In gram-positive bacteria, response to fluoroquinolones is similar to that described for *P. aeruginosa*. They are only moderately susceptible to hydrophilic fluoroquinolones, such as norfloxacin and ciprofloxacin (Piddock, 1994; Brighty and Gootz, 1997), and even single-step mutations confer clinically significant levels of resistance.

In *S. aureus*, NorA (Yoshida *et al.*, 1990b; Ng *et al.*, 1994; Neyfakh *et al.*, 1993; Kaatz *et al.*, 1993) which belongs to the 12-TMS subfamily of the Major Facilitator family (see above) is implicated in increased resistance. It was also shown that NorA, similar to AcrAB and MexAB-OprM (see above) was expressed in the wild type cells, conferring intrinsic resistance in *S. aureus*. Deletion of *norA* from the wild type resulted in a 4-fold increase in susceptibility to ciprofloxacin (Yamada *et al.*, 1997; Hsieh *et al.*, 1998). Again, multiple resistance mechanisms (target-based and efflux-mediated) synergistically enhance antibiotic resistance (Kaatz and Seo, 1997).

Direct uptake studies, as well as competition experiments, demonstrated that various fluoroquinolones differ in their apparent affinities for NorA (Ng *et al.*, 1994). The apparent K_m of norfloxacin transport was 6 μM, while the apparent K_i values for competitive inhibition by ciprofloxacin and ofloxacin were 26 μM and 50 μM, respectively. When NorA is cloned into *E. coli*, the increase in resistance to these quinolones differed: 32-fold for norfloxacin, 8-fold for ciprofloxacin and 4-fold for ofloxacin. It appears that NorA confers differential resistance to the more hydrophilic ciprofloxacin, enoxacin and norfloxacin.

Recently, a multi-drug transporter PmrA, associated with fluoroquinolone resistance, was identified in *S. pneumonia* (Gill *et al.*, 1999). Like NorA, PmrA belongs to the 12-TMS subfamily of the MF family. It also shows specificity for the more

hydrophilic fluoroquinolones. Unlike the *norA* gene, *pmrA* appears not to be expressed in the wild type strains of *S. pneumonia* since disruption of this gene did not result in increased sensitivity compared to the wild type strain (Gill *et al.*, 1999).

The presence of additional MDR pumps existing in *S. pneumonia* can be inferred from several observations. First, active efflux of fluoroquinolones was demonstrated even in the wild type strains (Zeller *et al.*, 1997) and it is not PmrA-mediated. Second, mutants with increased efflux of fluoroquinolones, which was not inhibited by the addition of carbonyl cyanide m-chlorophenylhydrazone (CCCP), were reported in *S. pneumeonia*. Since CCCP will inhibit proton motive force driven efflux, this result suggests the existence of an efflux system driven by ATP hydrolysis (Piddock *et al.*, 1997).

Active efflux of fluoroquinolones was also demonstrated in the wild type strains of *Enterococcus faecalis* and *Enterococcus faecium* (Lynch *et al.*, 1997). Among gram-positive bacteria, these organisms are the most resistant to fluoroquinolones with MIC 2–8 μg/ml. The relevant efflux pumps are still awaiting identification. The same is true for *Bacteroides fragilis*. This anaerobic pathogen is intrinsically resistant to most fluoroquinolones and active efflux of norfloxacin was demonstrated in the wild type strains (Miyamae *et al.*, 1998). A one step spontaneous mutation increases resistance to norfloxacin, ethidium bromide and puromycin, suggesting that a multidrug pump with specificity similar to that of NorA catalyzes efflux.

The study of efflux-mediated resistance has been greatly accelerated by the observation that reserpine is an inhibitor of many transporters: NorA, PmrA, and efflux in Bacteroides (Neyfakh *et al.*, 1993; Gill *et al.*, 1999; Miyamae *et al.*, 1998). If reserpine decreases fluoroquinolone resistance in a given stain, involvement of efflux is suspected. The first reports indicate a significant proportion of efflux mutants among ciprofloxacin resistant clinical isolates of *S. pneumonia* (Brenwald *et al.*, 1997; Brenwald *et al.*, 1998). The actual prevalence of efflux mechanisms can still be underestimated since it is not known whether reserpine inhibits all relevant efflux pumps. For example, it does not reverse intrinsic resistance to ciprofloxacin (Gill *et al.*, 1999). Thus, a small molecule probe to evaluate the presence and epidemiology of efflux-mediated resistance among clinical isolates of fluoroquinolone resistant bacteria is compelling.

Aminoglycosides

The aminoglycosides are rapidly acting bacteriocidal agents that impair protein synthesis by binding to bacterial ribosomes. Development of new analogs was driven by both the necessity to identify compounds that resist structural modification by multiple aminoglycoside modifying enzymes and the need to have better therapeutic indices. Two main mechanisms of aminoglycoside resistance are specific enzymatic inactivation by modifying enzymes and a more general mechanism often referred to as permeability resistance. This permeability resistance applies equally to all aminoglycosides, however the level of resistance may vary markedly from strain to strain.

It appears that the prevalence of resistance mechanisms is different for various bacterial species. For example, in *Enterobacteriaceae* and *Staphylococcus*, there is a tendency to acquire multiple aminoglycoside modifying enzymes which results in the emergence of multiple genotypes and complex phenotypes (Miller *et al.*, 1997). In contrast, in *P. aeruginosa*, the main changes were towards an increase in the frequency of the permeability type of resistance (Miller *et al.*, 1997).

Recently a MDR pump, MexXY, capable of effluxing aminoglycosides was identified in *P. aeruginosa* (Mine *et al.*, 1999; Aires *et al.*, 1999; Westbrock-Wadman *et al.*, 1999). Apparently OprM serves as the outer membrane component for this pump. It was shown that MexXY is expressed in the wild type cells of *P. aeruginosa*, since deletion of the corresponding genes rendered *P. aeruginosa* more susceptible to various aminoglycosides (ca. 8-fold), such as tetracycline and erythromycin. Thus, active efflux does contribute to the intrinsic resistance towards aminoglycosides. A gene, *mexZ*, homologous to repressors of active efflux systems was identified upstream of the *mexXY* genes. Transformation of several clinical isolates of *P. aeruginosa* resistant to aminoglycosides via a ''non enzymatic'' mechanism with the cloned *mexZ* gene restored their susceptibility to the wild-type level (Aires *et al.*,

1999). Thus, the MexXY pump appears to play a key role in the low-level resistance to aminoglycosides in *P. aeruginosa*.

The MDR pump AmrAB-OprA (with homology to the Mex pumps from *P. aeruginosa*), capable of effluxing aminoglycosides, was also identified in *Burkholderia pseudomallei* (Moore *et al.*, 1999). Successful treatment of *B. pseudomallei* infections is difficult because this organism is intrinsically resistant to a variety of antibiotics, including β-lactams and aminoglycosides. Inactivation of the *amrAB* operon rendered mutants 64 to 128-fold more susceptible to the widely used aminoglycosides tobramycin and gentamycin, with the MIC of 1–2 μg/ml, which is in a normal range as opposed to 256 μg/ml in the wild-type strains.

With the discovery of efflux pumps that contribute to the resistance to aminoglycosides, it would be possible to evaluate the prevalence of this resistance mechanism. Since decreased permeability resistance was poorly defined, this mechanism was often not characterized in the epidemiological surveys. Therefore, it is quite possible that involvement of the efflux pumps in aminoglycoside resistance is significantly underestimated.

Antiseptics and Disinfectants

Antiseptics and disinfectants based on quaternary ammonium compounds (QAC), such as cetrimide, benzalconium chloride, or biguanidines, such as chlorhexidine are purely synthetic compounds. They are widely used in clinical settings as well as in the food industry due to their surface-active and low-toxic properties. Efflux due to activity of multi-drug resistance pumps is the only known mechanism of resistance to these compounds.

Resistance to these compounds was first reported in the mid-1980s in clinical multi-resistant strains of *Staphylococci* and, subsequently has been found frequently among both clinical isolates (Skurray *et al.*, 1988) and *Staphylococci* isolated from different kinds of food processing industries (Heir *et al.*, 1995; Heir *et al.*, 1998).

Two types of MDR pumps confer resistance to antiseptics and disinfectants in gram-positive bacteria. Those are QacA/QacB (Rouch *et al.*, 1990, Paulsen *et al.*, 1996a) and Smr/QacG/QacH (Littlejohn *et al.*, 1992; Heir *et al.*, 1998; Heir *et al.*, 1999) pumps that belong to MF and SMR-families, respectively.

QacA, which confers resistance to monocationic QACs as well as to divalent cations (*e.g.* chlorhexidine and pentamidine), emerged among clinical isolates of *Staphylococci* during the 1980s. QacB, which confers resistance exclusively to monocationic compounds, was found on the plasmid that was present in a clinical strain isolated in 1951 in Australia (Paulsen *et al.*, 1998a). It was shown that a single amino acid substitution is responsible for the difference in substrate specificity (Paulsen *et al.*, 1996a). It is conceivable that QacA has evolved from QacB in response to the extensive use of dicationic antiseptics in clinical settings.

Another example of disinfectant agent is triclosan (Igrasan), a trichlorinated bisphenol broad-spectrum antibacterial agent used in hand-soaps, lotions, fabrics, plastics and toothpastes. In *E. coli*, triclosan inhibits the synthesis of lipids by inhibiting the fatty acid synthase FabI (McMurry *et al.*, 1998b) and mutations in a corresponding gene, *fabI*, render *E. coli* resistant to this agent. Under laboratory conditions, triclosan also selects for mutants overexpressing the AcrAB efflux pump (McMurry *et al.*, 1998a). Therefore, an undesirable consequence of the wild use of triclosan, which is an environmentally stable agent, might be selection for mutants resistant to multiple antibiotics.

OUTSMARTING BACTERIAL PUMPS: STRATEGIES TO COMBAT EFFLUX-MEDIATED RESISTANCE

There are numerous reasons to believe that the scope of efflux-mediated resistance is broader than is currently documented. Several approaches to combat the adverse effects of efflux on the efficacy of antimicrobial therapy could be envisioned:

1. Research in existing classes of antibiotics to identify derivatives that are effluxed minimally.
2. Development of therapeutic agents, which inhibit transport activity of efflux pumps and could be used in combination with existing antibiotics to increase their potency.

Modification of Existing Antibiotics

New semisynthetic tetracyclines

A new class of semisynthetic tetracyclines, glycylcyclines, has been developed by investigators at American Cyanamid (Sum *et al.*, 1998). These compounds exhibit potent activity against a broad-spectrum of gram-positive and gram-negative bacteria, including those that carry ribosomal protection (Tet(M)) and efflux determinants (Tet (A-D), Tet (K)). Susceptibility to these compounds for tetracycline resistant strains (MIC>32 μg/ml) ranged from 0.5 to 2 μg/ml. It was demonstrated that glycylcyclines overcome efflux-mediated-resistance because they are not recognized by the transporter protein (Someya *et al.*, 1995).

It is noteworthy that glycylcyclines are still effluxed by MDR, for example, by the MexAB-OprM and MexCD-OprJ from *P. aeruginosa*.

New macrolides and ketolides

Significant progress has been accomplished in developing new macrolides with enhanced activity. The ketolides are novel semisynthetic 14-membered-ring macrolides derived from erythromycin A, but characterized by a 3-keto function instead of cladinose moiety. Notable agents are HMR-3647 (Hoechst-Marion-Roussel), ABT-773 (Abbot), and CP-544372 (Pfizer), which are in advanced development. While HMR-3647 and ABT-773 are still somewhat affected by the Mef pumps, although to much lesser degree than the older macrolides, CP-544372 appears to not be affected at all (Brennan *et al.*, 1998). While the ketolides have enhanced activity against macrolide-resistant gram-positive bacteria, their activity against gram-negative bacteria is not superior to azythromycin (Biedenbach *et al.*, 1998; Jamjian *et al.*, 1997). These results infer that these compounds are still subject to efflux by multi-drug resistance pumps in gram-negative bacteria.

New fluoroquinolones

Research in new fluoroquinolones has been focussed on enhancing activity against gram-positive pathogens *versus* that for ciprofloxacin. These efforts resulted in the discovery of trovafloxacin (Pfizer) (Brighty and Gootz, 1997), clinafloxacin (Parke-Davis), gatifloxacin (Brystol-Myers Squibb), and moxifloxacin (Bayer) (Chopra, 1998). It appears that NorA from *S. aureus* and efflux pumps from *S. pneumonia* do not confer the same increase in resistance to these new fluoroquinolones as compared to the older ones (Gootz *et al.*, 1999; Fukuda *et al.*, 1998; Gill *et al.*, 1999; Pestova *et al.*, 1999; Piddock *et al.*, 1998). Direct transport experiments are still needed to clarify whether these pumps do not recognize new derivatives, or, whether it is the increased rate of diffusion of these more lipophilic fluoroquinolones that overcomes efflux.

However, the newer fluoroquinolones have lost some activity against gram-negative bacteria (Piddock *et al.*, 1998). It is possible that they are preferentially affected by the synergistic interactions between efflux pumps and the outer membrane of gram-negative bacteria than older, more hydrophilic fluoroquinolones.

Attempts to modify existing classes of antibiotics to overcome the gram-negative efflux pumps appear to be less successful than in the case of gram-positive bacteria. In gram-negative bacteria, particular restrictions are imposed on the structure of successful drugs: they must be amphiphilic to be able to cross both membranes. It is this amphiphilicity that makes antibiotics good substrates of multi-drug resistance efflux pumps from gram-negative bacteria (see above). But what about antibiotics which do not need to cross both membranes and work in the periplasm of gram-negative bacteria? Is it possible to modify these antibiotics to make them less prone to efflux? H. Nikaido (Nikaido *et al.*, 1998) has shown the feasibility of this approach with β-lactams. The multidrug efflux pump AcrAB of *Salmonella typhimurium* indeed does not affect hydrophilic β-lactam antibiotics that do not contain lipophilic side chains.

While empirical approaches have proven to be very useful in the identification of analogs of existing antibiotics which are less affected by efflux pumps, more efforts to study the genetics and biochemistry of drug transporters, and to identify substrate binding site(s), are encouraged and may prove invaluable in the near future.

Efflux Pump Inhibitor Approach

Inhibition of efflux pumps is an alternative approach to improve clinical efficacy of antibiotics that are the substrates of efflux pumps. However, it is important to properly identify antibiotics and bacteria for which this approach would be the most productive. For example, inhibition of macrolide-specific efflux pumps from gram-positive bacteria will reverse pump-mediated resistance, but only in a sub-population of resistant organisms. On the other hand, inhibition of a single multi-drug resistant pump in *H. influenzae*, will make this bacteria more susceptible to azithromycin and probably to ketolides (that are much less affected by efflux and target-modification in gram-positive bacteria, such as *Streptococci*). Thus, a combination of an inhibitor of MDR pumps and a macrolide (or ketolide) would be significant in the treatment of respiratory infections due to *Streptococci* and *H. influenzae*. Similarly, inhibitors of the predominant transporters TetA and TetB in gram-negative bacteria would represent new entities for reversing one type of tetracycline resistance.

Indeed, inhibitors of Tet pumps were identified among certain semisynthetic tetracycline analogs (Nelson *et al.*, 1993; Nelson *et al.*, 1994; Nelson and Levy, 1999). It was shown that these compounds inhibited tetracycline efflux in everted vesicles prepared from *E. coli* cells containing the TetB protein. The most potent analog, 13-CPTC, interfered with tetracycline transport by competitively binding to TetB, but itself was transported less efficiently than tetracycline. When combined with doxycycline, 13-CPTC exhibited synergy against *E. coli* strains expressing either TetA or TetB proteins by lowering the MIC by a factor of 2 or greater. More efforts in developing tetracycline efflux pump inhibitors are warranted.

In *P. aeruginosa*, inhibition of efflux pumps as applied to β-lactams and aminoglycosides will have limitations since other mechanisms also contribute to their resistance. Moreover, strains possessing a β-lactam hydrolyzing enzyme will not show increased susceptibility in spite of the presence of an efflux pump inhibitor (see above). In contrast, the efflux pump inhibitor approach is applicable for those β-lactams for which β-lactamase inactivation is not widespread. An efflux pump inhibitor is expected to significantly improve the efficacy of meropenem, which is a substrate of MexAB-OprM. This antibiotic is not hydrolyzed by chromosomal cephalosporinase AmpC and plasmid-mediated resistance is still rare in *P. aeruginosa*.

The efflux pump inhibitor approach is compelling when applied to fluoroquinolones. Since both increased efflux and target modification mechanisms contribute to the high levels of resistance, efflux pump inhibitors should decrease acquired resistance even in cells with target-based mutations (see above). In gram-negative bacteria, MDR pumps that confer resistance to fluoroquinolones belong to the RND-family and are similar on the protein level. This increases the possibility that a single inhibitor will be active against multiple efflux pumps.

Since resistance to fluoroquinolones in gram-positive bacteria is mainly due to the activity of pumps belonging to the Major Facilitator family, it is improbable to expect that a single inhibitor will potentiate fluoroquinolones in both gram-negative and gram-positive bacteria.

The first inhibitors active against multiple RND-transporters in gram-negative bacteria were reported by the scientists of Microcide Pharmaceuticals. Empiric screening of small molecules libraries for compounds with efflux pump inhibitory activity resulted in the identification of an inhibitor active against MexAB-OprM, MexCD-OprJ and MexEF-OprN efflux pumps, which contribute to fluoroquinolone resistance in *P. aeruginosa*. This broad-spectrum inhibitor, MC-207,110, is active against multiple efflux pumps in *P. aeruginosa* as well as RND-pumps in many representatives of *Enterobacteriaceae, Neisseria honorrhoeae*, and *Haemophilus influenzae* (Blais *et al.*, 1999; Lomovskaya *et al.*, 1999b; Renau *et al.*, 1999).

MC-207,110 (EPI) decreased intrinsic resistance to levofloxacin ca. 8-fold in the wild type strain of *P. aeruginosa*, while in the strains that overexpress efflux pumps, the susceptibility was increased maximally to 64-fold. As expected, EPI potentiated levofloxacin irrespective of the presence of target based mutations. Remarkably, EPI also dramatically affected the frequency of selection of resistant bacteria. When the wild type strain was used for selection experiments at the standard 4X MIC, the frequency was less than 10^{-11} (*versus* 10^{-7}). The EPI minimized appearance of both efflux-mediated

and target-based mutations. Since the inhibitor decreased the MexAB-OprM-mediated intrinsic resistance, a single target-based mutation (i.e. in gyrase) could not confer enough resistance to emerge under the selection conditions.

Recent clinical isolates of *P. aeruginosa* with a wide range of resistant phenotypes showed increased susceptibility with levofloxacin and EPI ENRfu (Cho *et al.*, 1999).

Finally, EPI potentiated levofloxacin in several animal models of infection (Griffith *et al.*, 1999), providing ''*in vivo* proof-of-principle'' of the applicability of efflux pump inhibitors.

Since MC-207,110 is a broad-spectrum inhibitor of RND transporters, it expectedly potentiated multiple antibiotics that are substrates of RND pumps of other gram-negative species. It potentiated macrolides in *H. influenzae* and florfenicol in *E. coli* and *S. typhimurium*. Therefore, a single compound when combined with different antibiotics may have multiple clinical applications.

Inhibitors of the NorA pump from *S. aureus* that potentiate ciprofloxacin against *S. aureus* have been reported by scientists of Influx Inc. (Markham *et al.*, 1999). As in the case of *P. aeruginosa*, these compounds also affected the frequency of emergence of resistant bacteria.

While initial reports have demonstrated the multi-factorial benefits of the efflux pump inhibitor approach in combating drug resistance, substantial efforts are needed before inhibitors can be used clinically.

CONCLUSIONS/PERSPECTIVES

Efflux-mediated resistance is widespread in clinical settings and is particularly important for antibiotics that are less affected by specific resistance mechanisms. It is clear that the use of efflux pump inhibitors is an approach that should be explored in addition to the conventional methods that focus on the discovery of novel classes of antibacterial agents. The ability of a single inhibitor to potentiate various clinically used antibiotics while minimizing the development of resistance to the co-administered antibiotics are attractive features of efflux pump inhibitors.

It is believed that novel drugs capable of inhibiting multiple targets will be particularly beneficial since lower frequencies of resistance are expected (Chopra, 1998). However, efflux-mediated resistance due to MDR pumps, which do not recognize their substrates on the basis of a specific structure and mode of action, could still affect even these future wonder drugs. Can we apply the efflux pump inhibitor approach to therapeutic agents before they are introduced into general use? The tremendous benefit from mini-mizing or at least delaying the development of resistance would be a compelling argument for such research.

REFERENCES

Ahmed, M., Borsch, C.M., Taylor, S.S., Vazquez-Laslop, N. and Neyfakh, A.A. (1994) A protein that activates expression of a multidrug efflux transporter upon binding the transporter substrates. *J. Biol. Chem.* **269**, 28506–28513.

Ahmed, M., Lyass, L., Markham, P.N., Taylor, S.S., Vazquez-Laslop, N. and Neyfakh, A.A. (1995) Two highly similar multidrug transporters of Bacillus subtilis whose expression is differentially regulated. *J. Bacteriol.* **177**, 3904–3910.

Aires, J.R., Kohler, T., Nikaido, H. and Plesiat, P. (1999) Involvement of an active efflux system in the natural resistance of Pseudomonas aeruginosa to amynoglycosides. *Antimicrob. Agents Chemother.* **43**, 2624–2628.

Alekshun, M.N. and Levy, S.B. (1997) Regulation of chromosomally mediated multiple antibiotic resistance: the mar regulon. *Antimicrob. Agents Chemother.* **41**, 2067–2075.

Alekshun, M.N. and Levy, S.B. (1999) The mar regulon: multiple resistance to antibiotics and other toxic chemicals. *Trends Microbiol.* **7**, 410–413.

Alonso, A. and Martinez, J.L. (1997) Multiple antibiotic resistance in Stenotrophomonas maltophilia. *Antimicrob. Agents Chemother.* **41**, 1140–1142.

Amabile-Cuevas, C.F. and Demple, B. (1991) Molecular characterization of the soxRS genes of Escherichia coli: two genes control a superoxide stress regulon. *Nucleic Acids Res.* **19**, 4479–4484.

Ambudkar, S.V., Dey, S., Hrycyna, C.A., Ramachandra, M., Pastan, I. and Gottesman, M.M. (1999) Biochemical, cellular, and pharmacological aspects of the multidrug transporter. *Annu. Rev. Pharmacol. Toxicol.* **39**, 361–398.

Aoyama, H., Sato, K., Kato, T., Hirai, K. and Mitsuhashi, S. (1987) Norfloxacin resistance in a clinical isolate of Escherichia coli. *Antimicrob. Agents Chemother.* **31**, 1640–1641.

Arpin, C., Canron, M.H., Noury, P. and Quentin, C. (1999) Emergence of mefA and mefE genes in beta-haemolytic streptococci and pneumococci in France [letter] [In Process Citation]. *J. Antimicrob. Chemother.* **44**, 133–134.

Baranova, N.N., Danchin, A. and Neyfakh, A.A. (1999) Mta, a global MerR-type regulator of the Bacillus subtilis multidrug-efflux transporters. *Mol. Microbiol.* **31**, 1549–1559.

Bellido, F., Martin, N.L., Siehnel, R.J. and Hancock, R.E. (1992) Reevaluation, using intact cells, of the exclusion limit and role of porin OprF in Pseudomonas aeruginosa outer membrane permeability. *J. Bacteriol.* **174**, 5196–5203.

Biedenbach, D.J., Barrett, M.S. and Jones, R.N. (1998) Comparative antimicrobial activity and kill-curve investiga-

tions of novel ketolide antimicrobial agents (HMR 3004 and HMR 3647) tested against Haemophilus influenzae and Moraxella catarrhalis strains. *Diagn. Microbiol. Infect. Dis.* **31**, 349–353.

Bissonnette, L., Champetier, S., Buisson, J.P. and Roy, P.H. (1991) Characterization of the nonenzymatic chloramphenicol resistance (cmlA) gene of the In4 integron of Tn1696: similarity of the product to transmembrane transport proteins. *J. Bacteriol.* **173**, 4493–4502.

Blais, J., Cho, D., Tangen, K., Ford, C., Lee, A., Lomovskaya, O., et al. (1999) Efflux pump inhibitors enhance the activity of antimicrobial agents against a broad selection of bacteria. *39th Interscience Conference for Antimicrobial Agents and Chemotherapy, San Francisco.*

Bohn, C. and Bouloc, P. (1998) The Escherichia coli cmlA gene encodes the multidrug efflux pump Cmr/MdfA and is responsible for isopropyl-beta-D-thiogalactopyranoside exclusion and spectinomycin sensitivity. *J. Bacteriol.* **180**, 6072–6075.

Bolhuis, H., van Veen, H.W., Brands, J.R., Putman, M., Poolman, B., Driessen, A.J.M., et al. (1996a) Energetics and mechanism of drug transport mediated by the lactococcal multidrug transporter LmrP. *J. Biol. Chem.* **271**, 24123–24128.

Bolhuis, H., van Veen, H.W., Molenaar, D., Poolman, B., Driessen, A.J. and Konings, W.N. (1996b) Multidrug resistance in Lactococcus lactis: evidence for ATP-dependent drug extrusion from the inner leaflet of the cytoplasmic membrane. *Embo. J.* **15**, 4239–4245.

Bolton, L.F., Kelley, L.C., Lee, M.D., Fedorka-Cray, P.J. and Maurer, J.J. (1999) Detection of multidrug-resistant Salmonella enterica serotype typhimurium DT104 based on a gene which confers cross-resistance to florfenicol and chloramphenicol. *J. Clin. Microbiol.* **37**, 1348–1351.

Bonfiglio, G., Laksai, Y., Franchino, L., Amicosante, G. and Nicoletti, G. (1998) Mechanisms of beta-lactam resistance amongst Pseudomonas aeruginosa isolated in an Italian survey. *J. Antimicrob. Chemother.* **42**, 697–702.

Brennan, L., Duignan, J., Petitras, J., Anderson, M., Fu, W., Retsema, J., et al. (1998) CP-544372: MIC90 studies and killing kinetics against key espiratory tract pathogens. *38th Interscience Conference for Antimicrobial Agents and Chemotherapy, San Diego.*

Brenwald, N.P., Gill, M.J. and Wise, R. (1997) The effect of reserpine, an inhibitor of multi-drug efflux pumps, on the in-vitro susceptibilities of fluoroquinolone-resistant strains of Streptococcus pneumoniae to norfloxacin [letter]. *J. Antimicrob. Chemother.* **40**, 458–460.

Brenwald, N.P., Gill, M.J. and Wise, R. (1998) Prevalence of a putative efflux mechanism among fluoroquinolone-resistant clinical isolates of Streptococcus pneumoniae. *Antimicrob. Agents Chemother.* **42**, 2032–2035.

Brighty, K.E. and Gootz, T.D. (1997) The chemistry and biological profile of trovafloxacin. *J. Antimicrob. Chemother.* **39 Suppl B**, 1–14.

Brooun, A., Tomashek, J.J. and Lewis, K. (1999) Purification and ligand binding of EmrR, a regulator of a multidrug transporter. *J. Bacteriol.* **181**, 5131–5133.

Brown, M.H., Paulsen, I.T. and Skurray, R.A. (1999) The multidrug efflux protein NorM is a prototype of a new family of transporters [letter; comment]. *Mol. Microbiol.* **31**, 394–395.

Burns, J.L., Mendelman, P.M., Levy, J., Stull, T.L. and Smith, A.L. (1985) A permeability barrier as a mechanism of chloramphenicol resistance in Haemophilus influenzae. *Antimicrob. Agents Chemother.* **27**, 46–54.

Burns, J.L., Wadsworth, C.D., Barry, J.J. and Goodall, C.P. (1996) Nucleotide sequence analysis of a gene from Burkholderia (Pseudomonas) cepacia encoding an outer membrane lipoprotein involved in multiple antibiotic resistance. *Antimicrob. Agents Chemother.* **40**, 307–313.

Bush, L.M., Calmon, J. and Johnson, C.C. (1995) Newer penicillins and beta-lactamase inhibitors. *Infect. Dis. Clin. North Am.* **9**, 653–686.

Charvalos, E., Tselentis, Y., Hamzehpour, M.M., Kohler, T. and Pechere, J. C. (1995) Evidence for an efflux pump in multidrug-resistant Campylobacter jejuni. *Antimicrob. Agents Chemother.* **39**, 2019–2022.

Chen, H.Y., Yuan, M. and Livermore, D.M. (1995) Mechanisms of resistance to beta-lactam antibiotics amongst Pseudomonas aeruginosa isolates collected in the UK in 1993. *J. Med. Microbiol.* **43**, 300–309.

Cho, D., Blais, J., Tangen, K., Ford, K., Lee, A., Lomovskaya, O., *et al.* (1999) Prevalence of efflux mechanisms among clinical isolates of fluoroquinolone resistant Pseudomonas aeruginosa. *39th Interscience Conference for Antimicrobial Agents and Chemotherapy*, San Francisco.

Chopra, I. (1998) Research and development of antibacterial agents. *Curr. Opin. Microbiol.* **1**, 495–501.

Chopra, I., Hawkey, P.M. and Hinton, M. (1992) Tetracyclines, molecular and clinical aspects. *J. Antimicrob. Chemother.* **29**, 245–277.

Clancy, J., Dib-Hajj, F., Petitpas, J.W. and Yuan, W. (1997) Cloning and characterization of a novel macrolide efflux gene, mreA, from Streptococcus agalactiae. *Antimicrob. Agents Chemother.* **41**, 2719–2723.

Cohen, S.P., Levy, S.B., Foulds, J. and Rosner, J.L. (1993) Salicylate induction of antibiotic resistance in Escherichia coli: activation of the mar operon and a mar-independent pathway. *J. Bacteriol.* **175**, 7856–7862.

Cohen, S.P., McMurry, L.M., Hooper, D.C., Wolfson, J.S. and Levy, S.B. (1989) Cross-resistance to fluoroquinolones in multiple-antibiotic-resistant (Mar) Escherichia coli selected by tetracycline or chloramphenicol: decreased drug accumulation associated with membrane changes in addition to OmpF reduction. *Antimicrob. Agents Chemother.* **33**, 1318–1325.

Cowan, S.W., Garavito, R.M., Jansonius, J.N., Jenkins, J.A., Karlsson, R., Konig, N., *et al.* (1995) The structure of OmpF porin in a tetragonal crystal form. *Structure* **3**, 1041–1050.

Deguchi, T., Kawamura, T., Yasuda, M., Nakano, M., Fukuda, H., Kato, H., *et al.* (1997) In vivo selection of Klebsiella pneumoniae strains with enhanced quinolone resistance during fluoroquinolone treatment of urinary tract infections. *Antimicrob. Agents Chemother.* **41**, 1609–1611.

Demple, B. (1991) Regulation of bacterial oxidative stress genes. *Annu. Rev. Genet.* **25**, 315–337.

Dinh, T., Paulsen, I.T. and Saier, M.H., Jr. (1994) A family of extracytoplasmic proteins that allow transport of large molecules across the outer membranes of gram-negative bacteria. *J. Bacteriol.* **176**, 3825–3831.

Dorman, C.J. and Foster, T.J. (1982) Nonenzymatic chloramphenicol resistance determinants specified by plasmids R26 and R55-1 in Escherichia coli K-12 do not confer high-level resistance to fluorinated analogs. *Antimicrob. Agents Chemother.* **22**, 912–914.

Dorman, C.J., Foster, T.J. and Shaw, W.V. (1986) Nucleotide sequence of the R26 chloramphenicol resistance determinant and identification of its gene product. *Gene* **41**, 349–353.

Eady, E.A., Ross, J.I., Tipper, J.L., Walters, C.E., Cove, J.H. and Noble, W.C. (1993) Distribution of genes encoding erythromycin ribosomal methylases and an erythromycin efflux pump in epidemiologically distinct groups of staphylococci. *J. Antimicrob. Chemother.* **31**, 211–217.

Edgar, R. and Bibi, E. (1997) MdfA, an Escherichia coli multidrug resistance protein with an extraordinarily broad spectrum of drug recognition [published erratum appears in *J. Bacteriol.* 1997 Sep;179(17):5654]. *J. Bacteriol.* **179**, 2274–2280.

Edgar, R. and Bibi, E. (1999) A single membrane-embedded negative charge is critical for recognizing positively charged drugs by the Escherichia coli multidrug resistance protein MdfA. *Embo. J.* **18**, 822–832.

Everett, M.J., Jin, Y.F., Ricci, V. and Piddock, L.J. (1996) Contributions of individual mechanisms to fluoroquinolone resistance in 36 Escherichia coli strains isolated from humans and animals. *Antimicrob. Agents Chemother.* **40**, 2380–2386.

Fath, M.J. and Kolter, R. (1993) ABC transporters: bacterial exporters. *Microbiol. Rev.* **57**, 995–1017.

Fralick, J.A. (1996) Evidence that TolC is required for functioning of the Mar/AcrAB efflux pump of Escherichia coli. *J. Bacteriol.* **178**, 5803–5805.

Fukuda, H., Hori, S. and Hiramatsu, K. (1998) Antibacterial activity of gatifloxacin (AM-1155, CG5501, BMS-206584), a newly developed fluoroquinolone, against sequentially acquired quinolone-resistant mutants and the norA transformant of Staphylococcus aureus. *Antimicrob. Agents Chemother.* **42**, 1917–1922.

Fukuda, H., Hosaka, M., Hirai, K. and Iyobe, S. (1990) New norfloxacin resistance gene in Pseudomonas aeruginosa PAO. *Antimicrob. Agents Chemother.* **34**, 1757–1761.

Fukuda, H., Hosaka, M., Iyobe, S., Gotoh, N., Nishino, T. and Hirai, K. (1995) nfxC-type quinolone resistance in a clinical isolate of Pseudomonas aeruginosa. *Antimicrob. Agents Chemother.* **39**, 790–792.

Ghosh, A.S., Ahamed, J., Chauhan, K.K. and Kundu, M. (1998) Involvement of an efflux system in high-level fluoroquinolone resistance of Shigella dysenteriae. *Biochem. Biophys. Res. Commun.* **242**, 54–56.

Gill, M.J., Brenwald, N.P. and Wise, R. (1999) Identification of an efflux pump gene, pmrA, associated with fluoroquinolone resistance in Streptococcus pneumoniae. *Antimicrob. Agents Chemother.* **43**, 187–189.

Gilson, L., Mahanty, H.K. and Kolter, R. (1990) Genetic analysis of an MDR-like export system: the secretion of colicin V. *Embo. J.* **9**, 3875–3894.

Giovanetti, E., Montanari, M.P., Mingoia, M. and Varaldo, P.E. (1999) Phenotypes and genotypes of erythromycin-resistant Streptococcus pyogenes strains in Italy and heterogeneity of inducibly resistant strains. *Antimicrob. Agents Chemother.* **43**, 1935–1940.

Glynn, M.K., Bopp, C., Dewitt, W., Dabney, P., Mokhtar, M. and Angulo, F.J. (1998) Emergence of multidrug-resistant Salmonella enterica serotype typhimurium DT104 infections in the United States [see comments]. *N. Engl. J. Med.* **338**, 1333–1338.

Gootz, T.D. and Brighty, K.E. (1996) Fluoroquinolone antibacterials: SAR mechanism of action, resistance, and clinical aspects. *Med. Res. Rev.* **16**, 433–486.

Gootz, T.D., Zaniewski, R.P., Haskell, S.L., Kaczmarek, F.S. and Maurice, A.E. (1999) Activities of trovafloxacin compared with those of other fluoroquinolones against purified topoisomerases and gyrA and grlA mutants of Staphylococcus aureus. *Antimicrob. Agents Chemother.* **43**, 1845–1855.

Gottesman, M.M., Pastan, I. and Ambudkar, S.V. (1996) P-glycoprotein and multidrug resistance. *Curr. Opin. Genet. Dev.* **6**, 610–617.

Griffith, D., Lomovskaya, O., Lee, V. and Dudley, M. (1999) Potentiation of levofloxacin by a broad-spectrum efflux pump inhibitor in mouse models of infection caused by Pseudomonas aeruginosa. *39th Interscience Conference for Antimicrobial Agents and Chemotherapy*, San Francisco.

Grinius, L., Dreguniene, G., Goldberg, E.B., Liao, C.H. and Projan, S.J. (1992) A staphylococcal multidrug resistance gene product is a member of a new protein family. *Plasmid* **27**, 119–129.

Grinius, L.L. and Goldberg, E.B. (1994) Bacterial multidrug resistance is due to a single membrane protein which functions as a drug pump. *J. Biol. Chem.* **269**, 29998–30004.

Grkovic, S., Brown, M.H., Roberts, N.J., Paulsen, I.T. and Skurray, R. A. (1998) QacR is a repressor protein that regulates expression of the Staphylococcus aureus multidrug efflux pump QacA. *J. Biol. Chem.* **273**, 18665–18673.

Guan, L., Ehrmann, M., Yoneyama, H. and Nakae, T. (1999) Membrane topology of the xenobiotic-exporting subunit, MexB, of the MexA,B-OprM extrusion pump in Pseudomonas aeruginosa. *J. Biol. Chem.* **274**, 10517–10522.

Hagman, K.E., Pan, W., Spratt, B.G., Balthazar, J.T., Judd, R.C. and Shafer, W.M. (1995) Resistance of Neisseria gonorrhoeae to antimicrobial hydrophobic agents is modulated by the mtrRCDE efflux system. *Microbiology* **141**, 611–622.

Hancock, R.E. (1998) Resistance mechanisms in Pseudomonas aeruginosa and other nonfermentative gram-negative bacteria. *Clin. Infect. Dis.* **27 Suppl 1**, S93–99.

Heir, E., Sundheim, G. and Holck, A.L. (1995) Resistance to quaternary ammonium compounds in Staphylococcus spp. isolated from the food industry and nucleotide sequence of the resistance plasmid pST827. *J. Appl. Bacteriol.* **79**, 149–156.

Heir, E., Sundheim, G. and Holck, A.L. (1998) The Staphylococcus qacH gene product: a new member of the SMR family encoding multidrug resistance. *FEMS Microbiol. Lett.* **163**, 49–56.

Heir, E., Sundheim, G. and Holck, A.L. (1999) Identification and characterization of quaternary ammonium compound resistant staphylococci from the food industry. *Int. J. Food Microbiol.* **48**, 211–219.

Higgins, C.F. (1992) ABC transporters: from microorganisms to man. *Annu. Rev. Cell. Biol.* **8**, 67–113.

Hirai, K., Aoyama, H., Suzue, S., Irikura, T., Iyobe, S. and Mitsuhashi, S. (1986) Isolation and characterization of norfloxacin-resistant mutants of Escherichia coli K-12. *Antimicrob. Agents Chemother.* **30**, 248–253.

Hirai, K., Suzue, S., Irikura, T., Iyobe, S. and Mitsuhashi, S. (1987) Mutations producing resistance to norfloxacin in Pseudomonas aeruginosa. *Antimicrob. Agents Chemother.* **31**, 582–586.

Holland, I.B., Kenny, B. and Blight, M. (1990) Haemolysin secretion from E coli. *Biochimie* **72**, 131–141.

Hooper, D.C., Wolfson, J.S., Souza, K.S., Tung, C., McHugh, G.L. and Swartz, M.N. (1986) Genetic and biochemical characterization of norfloxacin resistance in Escherichia coli. *Antimicrob. Agents Chemother.* **29**, 639–644.

Hsieh, P.C., Siegel, S.A., Rogers, B., Davis, D. and Lewis, K. (1998) Bacteria lacking a multidrug pump: a sensitive tool for drug discovery. *Proc. Natl. Acad. Sci. USA.* **95**, 6602–6606.

Ishii, H., Sato, K., Hoshino, K., Sato, M., Yamaguchi, A., Sawai, T. and Osada, Y. (1991) Active efflux of ofloxacin by a highly quinolone-resistant strain of Proteus vulgaris. *J. Antimicrob. Chemother.* **28**, 827–836.

Jakics, E.B., Iyobe, S., Hirai, K., Fukuda, H. and Hashimoto, H. (1992) Occurrence of the nfxB type mutation in clinical isolates of Pseudomonas aeruginosa. *Antimicrob. Agents Chemother.* **36**, 2562–2565.

Jalal, S. and Wretlind, B. (1998) Mechanisms of quinolone resistance in clinical strains of Pseudomonas aeruginosa. *Microb. Drug Resist.* **4**, 257–261.

Jamjian, C., Biedenbach, D.J. and Jones, R.N. (1997) In vitro evaluation of a novel ketolide antimicrobial agent, RU-64004. *Antimicrob. Agents Chemother.* **41**, 454–459.

Janosi, L., Nakajima, Y. and Hashimoto, H. (1990) Characterization of plasmids that confer inducible resistance to 14-membered macrolides and streptogramin type B antibiotics in Staphylococcus aureus. *Microbiol. Immunol.* **34**, 723–735.

Kaatz, G.W. and Seo, S.M. (1997) Mechanisms of fluoroquinolone resistance in genetically related strains of Staphylococcus aureus. *Antimicrob. Agents Chemother.* **41**, 2733–2737.

Kaatz, G.W., Seo, S.M. and Ruble, C.A. (1993) Efflux-mediated fluoroquinolone resistance in Staphylococcus aureus. *Antimicrob. Agents Chemother.* **37**, 1086–1094.

Kataja, J., Seppala, H., Skurnik, M., Sarkkinen, H. and Huovinen, P. (1998) Different erythromycin resistance mechanisms in group C and group G streptococci. *Antimicrob. Agents Chemother.* **42**, 1493–1494.

Kim, E. and Aoki, T. (1996) Sequence analysis of the florfenicol resistance gene encoded in the transferable R-plasmid of a fish pathogen, Pasteurella piscicida. *Microbiol. Immunol.* **40**, 665–669.

Kirst, H.A. (1991) New macrolides: expanded horizons for an old class of antibiotics. *J. Antimicrob. Chemother.* **28**, 787–790.

Kohler, T., Michea-Hamzehpour, M., Epp, S.F. and Pechere, J.C. (1999) Carbapenem activities against Pseudomonas aeruginosa: respective contributions of OprD and efflux systems. *Antimicrob. Agents Chemother.* **43**, 424–427.

Kohler, T., Michea-Hamzehpour, M., Henze, U., Gotoh, N., Curty, L.K. and Pechere, J.C. (1997a) Characterization of MexE-MexF-OprN, a positively regulated multidrug efflux system of Pseudomonas aeruginosa. *Mol. Microbiol.* **23**, 345–354.

Kohler, T., Michea-Hamzehpour, M., Plesiat, P., Kahr, A.L. and Pechere, J.C. (1997b) Differential selection of multidrug efflux systems by quinolones in Pseudomonas aeruginosa. *Antimicrob. Agents Chemother.* **41**, 2540–2543.

Kupferwasser, L.I., Skurray, R.A., Brown, M.H., Firth, N., Yeaman, M.R. and Bayer, A.S. (1999) Plasmid-mediated resistance to thrombin-induced platelet microbicidal protein in staphylococci: role of the qacA locus [In Process Citation]. *Antimicrob. Agents Chemother.* **43**, 2395–2399.

Lacroix, F.J., Cloeckaert, A., Grepinet, O., Pinault, C., Popoff, M.Y., Waxin, H., *et al.* (1996) Salmonella typhimurium acrB-like gene: identification and role in resistance to biliary salts and detergents and in murine infection. *FEMS Microbiol. Lett.* **135**, 161–167.

Lawrence, L.E. and Barrett, J.F. (1998) Efflux pumps in bacteria: overview, clinical relevance, and potential pharmaceutical target. *Exp. Opin. Invest. Drugs* **7**, 199–217.

Lee, E.H. and Shafer, W.M. (1999) The farAB-encoded efflux pump mediates resistance of gonococci to long-chained antibacterial fatty acids [In Process Citation]. *Mol. Microbiol.* **33**, 839–845.

Letoffe, S., Delepelaire, P. and Wandersman, C. (1996) Protein secretion in gram-negative bacteria: assembly of the three components of ABC protein-mediated exporters is ordered and promoted by substrate binding. *Embo. J.* **15**, 5804–5811.

Levy, S.B. (1992) Active efflux mechanisms for antimicrobial resistance. *Antimicrob. Agents Chemother.* **36**, 695–703.

Lewis, K. (1994) Multidrug resistance pumps in bacteria: variations on a theme. *Trends Biochem. Sci.* **19**, 119–123.

Lewis, K. (1999) Multidrug resistance: Versatile drug sensors of bacterial cells. *Curr. Biol.* **9**, R403–407.

Li, X.Z., Ma, D., Livermore, D.M. and Nikaido, H. (1994) Role of efflux pump(s) in intrinsic resistance of Pseudomonas aeruginosa: active efflux as a contributing factor to beta-lactam resistance. *Antimicrob. Agents Chemother.* **38**, 1742–1752.

Li, X.Z., Nikaido, H. and Poole, K. (1995) Role of mexA-mexB-oprM in antibiotic efflux in Pseudomonas aeruginosa. *Antimicrob. Agents Chemother.* **39**, 1948–1953.

Linton, K.J. and Higgins, C.F. (1998) The Escherichia coli ATP-binding cassette (ABC) proteins. *Mol. Microbiol.* **28**, 5–13.

Littlejohn, T.G., Paulsen, I.T., Gillespie, M.T., Tennent, J.M., Midgley, M., Jones, I.G., *et al.* (1992) Substrate specificity and energetics of antiseptic and disinfectant resistance in Staphylococcus aureus. *FEMS Microbiol. Lett.* **74**, 259–265.

Lomovskaya, O., Lee, A., Hoshino, K., Ishida, H., Mistry, A., Warren, M.S., *et al.* (1999a) Use of a genetic approach to evaluate the consequences of inhibition of efflux pumps in Pseudomonas aeruginosa. *Antimicrob. Agents Chemother.* **43**, 1340–1346.

Lomovskaya, O., Lee, A., Warren, M., Galazzo, J., Fronko, R., Lee, M., *et al.* (1999b) Targetting efflux pumps in Pseudomonas aeruginosa. *39th Interscience Conference for Antimicrobial Agents and Chemotherapy*, San Francisco.

Lomovskaya, O. and Lewis, K. (1992) Emr, an Escherichia coli locus for multidrug resistance. *Proc. Natl. Acad. Sci. USA* **89**, 8938–8942.

Lynch, C., Courvalin, P. and Nikaido, H. (1997) Active efflux of antimicrobial agents in wild-type strains of enterococci. *Antimicrob. Agents Chemother.* **41**, 869–871.

Ma, D., Cook, D.N., Alberti, M., Pon, N.G., Nikaido, H. and Hearst, J.E. (1993) Molecular cloning and characterization of acrA and acrE genes of Escherichia coli. *J. Bacteriol.* **175**, 6299–6313.

Ma, D., Cook, D.N., Alberti, M., Pon, N.G., Nikaido, H. and Hearst, J.E. (1995) Genes acrA and acrB encode a stress-induced efflux system of Escherichia coli. *Mol. Microbiol.* **16**, 45–55.

Markham, P.N., LoGuidice, J. and Neyfakh, A.A. (1997) Broad ligand specificity of the transcriptional regulator of the Bacillus subtilis multidrug transporter Bmr. *Biochem. Biophys. Res. Commun.* **239**, 269–272.

Markham, P.N., Westhaus, E., Klyachko, K., Johnson, M.E. and Neyfakh, A. A. (1999) Multiple novel inhibitors of the NorA multidrug transporter of staphylococcus aureus. *Antimicrob. Agents Chemother.* **43**, 2404–2408.

Masuda, N., Gotoh, N., Ishii, C., Sakagawa, E., Ohya, S. and Nishino, T. (1999) Interplay between chromosomal beta-lactamase and the MexAB-OprM efflux system in intrinsic resistance to beta-lactams in Pseudomonas aeruginosa. *Antimicrob. Agents Chemother.* **43**, 400–402.

Masuda, N. and Ohya, S. (1992) Cross-resistance to meropenem, cephems, and quinolones in Pseudomonas aeruginosa. *Antimicrob. Agents Chemother.* **36**, 1847–1851.

Matsuoka, M., Endou, K., Saitoh, S., Katoh, M. and Nakajima, Y. (1995) A mechanism of resistance to partial macrolide and streptogramin B antibiotics in Staphylococcus aureus clinically isolated in Hungary. *Biol. Pharm. Bull.* **18**, 1482–1486.

McMurry, L.M., Oethinger, M. and Levy, S.B. (1998a) Overexpression of marA, soxS, or acrAB produces resistance to triclosan in laboratory and clinical strains of Escherichia coli. *FEMS Microbiol. Lett.* **166**, 305–309.

McMurry, L.M., Oethinger, M. and Levy, S.B. (1998b) Triclosan targets lipid synthesis [letter]. *Nature* **394**, 531–532.

Miller, G.H., Sabatelli, F.J., Hare, R.S., Glupczynski, Y., Mackey, P., Shlaes, D., *et al.* (1997) The most frequent aminoglycoside resistance mechanisms–changes with time and geographic area: a reflection of aminoglycoside usage patterns? Aminoglycoside Resistance Study Groups. *Clin. Infect. Dis.* **24 Suppl 1**, S46–62.

Miller, P.F. and Sulavik, M.C. (1996) Overlaps and parallels in the regulation of intrinsic multiple-antibiotic resistance in Escherichia coli. *Mol. Microbiol.* **21**, 441–448.

Milton, I.D., Hewitt, C.L. and Harwood, C.R. (1992) Cloning and sequencing of a plasmid-mediated erythromycin resistance determinant from Staphylococcus xylosus. *FEMS Microbiol. Lett.* **76**, 141–147.

Mine, T., Morita, Y., Kataoka, A., Mizushima, T. and Tsuchiya, T. (1999) Expression in Escherichia coli of a new multidrug efflux pump, MexXY, from Pseudomonas aeruginosa. *Antimicrob. Agents Chemother.* **43**, 415–417.

Miyamae, S., Nikaido, H., Tanaka, Y. and Yoshimura, F. (1998) Active efflux of norfloxacin by Bacteroides fragilis. *Antimicrob. Agents Chemother.* **42**, 2119–2121.

Moore, R.A., DeShazer, D., Reckseidler, S., Weissman, A. and Woods, D.E. (1999) Efflux-mediated aminoglycoside and macrolide resistance in Burkholderia pseudomallei. *Antimicrob. Agents Chemother.* **43**, 465–470.

Morita, Y., Kodama, K., Shiota, S., Mine, T., Kataoka, A., Mizushima, T., *et al.* (1998) NorM, a putative multidrug efflux protein, of Vibrio parahaemolyticus and its homolog in Escherichia coli. *Antimicrob. Agents Chemother.* **42**, 1778–1782.

Morshed, S.R., Lei, Y., Yoneyama, H. and Nakae, T. (1995) Expression of genes associated with antibiotic extrusion in Pseudomonas aeruginosa. *Biochem. Biophys. Res. Commun.* **210**, 356–362.

Nakae, T., Nakajima, A., Ono, T., Saito, K. and Yoneyama, H. (1999) Resistance to beta-lactam antibiotics in Pseudomonas aeruginosa due to interplay between the MexAB-OprM efflux pump and beta-lactamase. *Antimicrob. Agents Chemother.* **43**, 1301–1303.

Nakano, M., Deguchi, T., Kawamura, T., Yasuda, M., Kimura, M., Okano, Y., *et al.* (1997) Mutations in the gyrA and parC genes in fluoroquinolone-resistant clinical isolates of Pseudomonas aeruginosa. *Antimicrob. Agents Chemother.* **41**, 2289–2291.

Nelson, M.L. and Levy, S.B. (1999) Reversal of tetracycline resistance mediated by different bacterial tetracycline resistance determinants by an inhibitor of the Tet(B) antiport protein. *Antimicrob. Agents Chemother.* **43**, 1719–1724.

Nelson, M.L., Park, B.H., Andrews, J.S., Georgian, V.A., Thomas, R.C. and Levy, S.B. (1993) Inhibition of the tetracycline efflux antiport protein by 13-thio-substituted 5-hydroxy-6-deoxytetracyclines. *J. Med. Chem.* **36**, 370–377.

Nelson, M.L., Park, B.H. and Levy, S.B. (1994) Molecular requirements for the inhibition of the tetracycline antiport protein and the effect of potent inhibitors on the growth of tetracycline-resistant bacteria. *J. Med. Chem.* **37**, 1355–1361.

Neyfakh, A.A. (1997) Natural functions of bacterial multidrug transporters. *Trends Microbiol.* **5**, 309–313.

Neyfakh, A.A., Bidnenko, V.E. and Chen, L.B. (1991) Efflux-mediated multidrug resistance in Bacillus subtilis: similarities and dissimilarities with the mammalian system. *Proc. Natl. Acad. Sci. USA* **88**, 4781–4785.

Neyfakh, A.A., Borsch, C.M. and Kaatz, G.W. (1993) Fluoroquinolone resistance protein NorA of Staphylococcus aureus is a multidrug efflux transporter. *Antimicrob. Agents Chemother.* **37**, 128–129.

Ng, E.Y., Trucksis, M. and Hooper, D.C. (1994) Quinolone resistance mediated by norA: physiologic characterization and relationship to flqB, a quinolone resistance locus on the Staphylococcus aureus chromosome. *Antimicrob. Agents Chemother.* **38**, 1345–1355.

Nikaido, H. (1994) Prevention of drug access to bacterial targets: permeability barriers and active efflux. *Science* **264**, 382–388.

Nikaido, H. (1996) Multidrug efflux pumps of gram-negative bacteria. *J. Bacteriol.* **178**, 5853–5859.

Nikaido, H. (1998a) Antibiotic resistance caused by gram-negative multidrug efflux pumps. *Clin. Infect. Dis.* **27 Suppl 1**, S32–41.

Nikaido, H. (1998b) Multiple antibiotic resistance and efflux. *Curr. Opin. Microbiol.* **1**, 516–523.

Nikaido, H. (1998c) The role of outer membrane and efflux pumps in the resistance of gram-negative bacteria.can we improve drug access? *Drug Resistance Updates* **1**, 93–98.

Nikaido, H., Basina, M., Nguyen, V. and Rosenberg, E.Y. (1998) Multidrug efflux pump AcrAB of Salmonella typhimurium excretes only those beta-lactam antibiotics containing lipophilic side chains. *J. Bacteriol.* **180**, 4686–4692.

Nikaido, H., Nikaido, K. and Harayama, S. (1991) Identification and characterization of porins in Pseudomonas aeruginosa. *J. Biol. Chem.* **266**, 770–779.

Nikaido, H. and Vaara, M. (1985) Molecular basis of bacterial outer membrane permeability. *Microbiol. Rev.* **49**, 1–32.

Oethinger, M., Kern, W.V., Goldman, J.D. and Levy, S.B. (1998a) Association of organic solvent tolerance and fluoroquinolone resistance in clinical isolates of Escherichia coli. *J. Antimicrob. Chemother.* **41**, 111–114.

Oethinger, M., Podglajen, I., Kern, W.V. and Levy, S.B. (1998b) Overexpression of the marA or soxS regulatory gene in clinical topoisomerase mutants of Escherichia coli. *Antimicrob. Agents Chemother.* **42**, 2089–2094.

Okusu, H., Ma, D. and Nikaido, H. (1996) AcrAB efflux pump plays a major role in the antibiotic resistance phenotype of Escherichia coli multiple-antibiotic-resistance (Mar) mutants. *J. Bacteriol.* **178**, 306–308.

Pan, W. and Spratt, B.G. (1994) Regulation of the permeability of the gonococcal cell envelope by the mtr system. *Mol. Microbiol.* **11**, 769–775.

Pao, S.S., Paulsen, I.T. and Saier, M.H., Jr. (1998) Major facilitator superfamily. *Microbiol. Mol. Biol. Rev.* **62**, 1–34.

Paulsen, I.T., Brown, M.H., Dunstan, S.J. and Skurray, R.A. (1995) Molecular characterization of the staphylococcal multidrug resistance export protein QacC. *J. Bacteriol.* **177**, 2827–2833.

Paulsen, I.T., Brown, M.H., Littlejohn, T.G., Mitchell, B.A. and Skurray, R.A. (1996a) Multidrug resistance proteins QacA and QacB from Staphylococcus aureus: membrane topology and identification of residues involved in substrate specificity. *Proc. Natl. Acad. Sci. USA* **93**, 3630–3635.

Paulsen, I.T., Brown, M.H. and Skurray, R.A. (1996b) Proton-dependent multidrug efflux systems. *Microbiol. Rev.* **60**, 575–608.

Paulsen, I.T., Brown, M.H. and Skurray, R.A. (1998a) Characterization of the earliest known Staphylococcus aureus plasmid encoding a multidrug efflux system. *J. Bacteriol.* **180**, 3477–3479.

Paulsen, I.T., Park, J.H., Choi, P.S. and Saier, M.H., Jr. (1997) A family of gram-negative bacterial outer membrane factors that function in the export of proteins, carbohydrates, drugs and heavy metals from gram-negative bacteria. *FEMS Microbiol. Lett.* **156**, 1–8.

Paulsen, I.T., Skurray, R.A., Tam, R., Saier, M.H., Jr., Turner, R.J., Weiner, J.H., *et al.* (1996c) The SMR family: a novel family of multidrug efflux proteins involved with the efflux of lipophilic drugs. *Mol. Microbiol.* **19**, 1167–1175.

Paulsen, I.T., Sliwinski, M.K. and Saier, M.H., Jr. (1998b) Microbial genome analyses: global comparisons of transport capabilities based on phylogenies, bioenergetics and substrate specificities. *J. Mol. Biol.* **277**, 573–592.

Pestova, E., Beyer, R., Cianciotto, N.P., Noskin, G.A. and Peterson, L.R. (1999) Contribution of topoisomerase IV and DNA gyrase mutations in Streptococcus pneumoniae to resistance to novel fluoroquinolones. *Antimicrob. Agents Chemother.* **43**, 2000–2004.

Piddock, L.J. (1994) Mechanisms of resistance to fluoroquinolones: state-of-the-art 1992–1994. *Drugs* **49**, 29–35.

Piddock, L.J., Jin, Y.F. and Everett, M.J. (1997) Non-gyrA-mediated ciprofloxacin resistance in laboratory mutants of Streptococcus pneumoniae. *J. Antimicrob. Chemother.* **39**, 609–615.

Piddock, L.J., Johnson, M., Ricci, V. and Hill, S.L. (1998) Activities of new fluoroquinolones against fluoroquinolone-resistant pathogens of the lower respiratory tract. *Antimicrob. Agents Chemother.* **42**, 2956–2960.

Ploy, M.C., Courvalin, P. and Lambert, T. (1998) Characterization of In40 of Enterobacter aerogenes BM2688, a class 1 integron with two new gene cassettes, cmlA2 and qacF. *Antimicrob. Agents Chemother.* **42**, 2557–2563.

Poole, K., Gotoh, N., Tsujimoto, H., Zhao, Q., Wada, A., Yamasaki, T., *et al.* (1996a) Overexpression of the mexC-mexD-oprJ efflux operon in nfxB-type multidrug-resistant strains of Pseudomonas aeruginosa. *Mol. Microbiol.* **21**, 713–724.

Poole, K., Krebes, K., McNally, C. and Neshat, S. (1993) Multiple antibiotic resistance in Pseudomonas aeruginosa: evidence for involvement of an efflux operon. *J. Bacteriol.* **175**, 7363–7372.

Poole, K., Tetro, K., Zhao, Q., Neshat, S., Heinrichs, D.E. and Bianco, N. (1996b) Expression of the multidrug resistance operon mexA-mexB-oprM in Pseudomonas aeruginosa: mexR encodes a regulator of operon expression. *Antimicrob. Agents Chemother.* **40**, 2021–2028.

Poppe, C., Smart, N., Khakhria, R., Johnson, W., Spika, J. and Prescott, J. (1998) Salmonella typhimurium DT104: a virulent and drug-resistant pathogen. *Can. Vet. J.* **39**, 559–565.

Poutanen, S.M., de Azavedo, J., Willey, B.M., Low, D.E. and MacDonald, K.S. (1999) Molecular characterization of multidrug resistance in Streptococcus mitis. *Antimicrob. Agents Chemother.* **43**, 1505–1507.

Pumbwe, L., Everett, M.J., Hancock, R.E. and Piddock, L.J. (1996) Role of gyrA mutation and loss of OprF in the multiple antibiotic resistance phenotype of Pseudomonas aeruginosa G49. *FEMS Microbiol. Lett.* **143**, 25–28.

Renau, T., Leger, R., Flamme, E., Sangalang, J., She, Y., Ford, M., *et al.* (1999) Inhibitors of efflux pumps in Pseudomonas aerruginosa potentiate the activity of the fluoroquinolone antibacterial levofloxacin. *39th Interscience Conference for Antimicrobial Agents and Chemotherapy*, San Francisco.

Roberts, M.C. (1994) Epidemiology of tetracycline-resistance determinants. *Trends Microbiol.* **2**, 353–357.

Robillard, N.J. (1990) Broad-host-range gyrase A gene probe. *Antimicrob. Agents Chemother.* **34**, 1889–1894.

Ross, J.I., Eady, E.A., Cove, J.H., Cunliffe, W.J., Baumberg, S. and Wootton, J.C. (1990) Inducible erythromycin resistance in staphylococci is encoded by a member of the ATP-binding transport super-gene family. *Mol. Microbiol.* **4**, 1207–1214.

Rouch, D.A., Cram, D.S., DiBerardino, D., Littlejohn, T.G. and Skurray, R.A. (1990) Efflux-mediated antiseptic resistance gene qacA from Staphylococcus aureus: common ancestry with tetracycline- and sugar-transport proteins. *Mol. Microbiol.* **4**, 2051–2062.

Rouquette, C., Harmon, J.B. and Shafer, W.M. (1999) Induction of the mtrCDE-encoded efflux pump system of neisseria gonorrhoeae requires MtrA, an AraC-like protein [In Process Citation]. *Mol. Microbiol.* **33**, 651–658.

Saier, M.H., Jr. (1998) Molecular phylogeny as a basis for the classification of transport proteins from bacteria, archaea and eukarya. *Adv. Microb. Physiol.* **40**, 81–136.

Saier, M.H., Jr., Eng, B.H., Fard, S., Garg, J., Haggerty, D.A., Hutchinson, W.J., *et al.* (1999) Phylogenetic characterization of novel transport protein families revealed by genome analyses. *Biochim. Biophys. Acta* **1422**, 1–56.

Saier, M.H., Jr., Paulsen, I.T. and Matin, A. (1997) A bacterial model system for understanding multi-drug resistance. *Microb. Drug. Resist.* **3**, 289–295.

Saier, M.H., Jr., Paulsen, I.T., Sliwinski, M.K., Pao, S.S., Skurray, R.A. and Nikaido, H. (1998) Evolutionary origins of multidrug and drug-specific efflux pumps in bacteria. *Faseb. J.* **12**, 265–274.

Saier, M.H., Jr., Tam, R., Reizer, A. and Reizer, J. (1994) Two novel families of bacterial membrane proteins concerned with nodulation, cell division and transport. *Mol. Microbiol.* **11**, 841–847.

Schnappinger, D. and Hillen, W. (1996) Tetracyclines: antibiotic action, uptake, and resistance mechanisms. *Arch. Microbiol.* **165**, 359–369.

Seoane, A.S. and Levy, S.B. (1995) Characterization of MarR, the repressor of the multiple antibiotic resistance (mar) operon in Escherichia coli. *J. Bacteriol.* **177**, 3414–3419.

Shafer, W.M., Qu, X., Waring, A.J. and Lehrer, R.I. (1998) Modulation of Neisseria gonorrhoeae susceptibility to vertebrate antibacterial peptides due to a member of the resistance/nodulation/division efflux pump family. *Proc. Natl. Acad. Sci. USA* **95**, 1829–1833.

Shaw, W.V. (1983) Chloramphenicol acetyltransferase: enzymology and molecular biology. *CRC Crit. Rev. Biochem.* **14**, 1–46.

Sheridan, R.P. and Chopra, I. (1991) Origin of tetracycline efflux proteins: conclusions from nucleotide sequence analysis. *Mol. Microbiol.* **5**, 895–900.

Skurray, R.A., Rouch, D.A., Lyon, B.R., Gillespie, M.T., Tennent, J.M., Byrne, M.E., *et al.* (1988) Multiresistant Staphylococcus aureus: genetics and evolution of epidemic Australian strains. *J. Antimicrob. Chemother.* **21 Suppl C**, 19–39.

Someya, Y., Yamaguchi, A. and Sawai, T. (1995) A novel glycylcycline, 9-(N,N-dimethylglycylamido)-6-demethyl-6-deoxytetracycline, is neither transported nor recognized by the transposon Tn10-encoded metal-tetracycline/H+ antiporter. *Antimicrob. Agents Chemother.* **39**, 247–249.

Srikumar, R., Kon, T., Gotoh, N. and Poole, K. (1998) Expression of Pseudomonas aeruginosa multidrug efflux pumps MexA-MexB-OprM and MexC-MexD-OprJ in a multidrug-sensitive Escherichia coli strain. *Antimicrob. Agents Chemother.* **42**, 65–71.

Srikumar, R., Li, X.Z. and Poole, K. (1997) Inner membrane efflux components are responsible for beta-lactam specificity of multidrug efflux pumps in Pseudomonas aeruginosa. *J. Bacteriol.* **179**, 7875–7881.

Stone, B.J., Garcia, C.M., Badger, J.L., Hassett, T., Smith, R.I. and Miller, V.L. (1992) Identification of novel loci affecting entry of Salmonella enteritidis into eukaryotic cells. *J. Bacteriol.* **174**, 3945–3952.

Stone, B.J. and Miller, V.L. (1995) Salmonella enteritidis has a homologue of tolC that is required for virulence in BALB/c mice. *Mol. Microbiol.* **17**, 701–712.

Sum, P.E., Sum, F.W. and Projan, S.J. (1998) Recent developments in tetracycline antibiotics. *Curr. Pharm. Des.* **4**, 119–132.

Sutcliffe, J., Tait-Kamradt, A. and Wondrack, L. (1996) Streptococcus pneumoniae and Streptococcus pyogenes resistant to macrolides but sensitive to clindamycin: a common resistance pattern mediated by an efflux system. *Antimicrob. Agents Chemother.* **40**, 1817–1824.

Syriopoulou, V.P., Harding, A.L., Goldmann, D.A. and Smith, A.L. (1981) *In vitro* antibacterial activity of fluorinated analogs of chloramphenicol and thiamphenicol. *Antimicrob. Agents Chemother.* **19**, 294–297.

Tait-Kamradt, A., Clancy, J., Cronan, M., Dib-Hajj, F., Wondrack, L., Yuan, W., *et al.* (1997) mefE is necessary for the erythromycin-resistant M phenotype in Streptococcus pneumoniae. *Antimicrob. Agents Chemother.* **41**, 2251–2255.

Thanassi, D.G., Cheng, L.W. and Nikaido, H. (1997) Active efflux of bile salts by Escherichia coli. *J. Bacteriol.* **179**, 2512–2518.

Thanassi, D.G., Suh, G.S. and Nikaido, H. (1995) Role of outer membrane barrier in efflux-mediated tetracycline resistance of Escherichia coli. *J. Bacteriol.* **177**, 998–1007.

Toro, C.S., Lobos, S.R., Calderon, I., Rodriguez, M. and Mora, G.C. (1990) Clinical isolate of a porinless Salmonella typhi resistant to high levels of chloramphenicol. *Antimicrob. Agents Chemother.* **34**, 1715–1719.

Trias, J. and Nikaido, H. (1990) Outer membrane protein D2 catalyzes facilitated diffusion of carbapenems and penems through the outer membrane of Pseudomonas aeruginosa. *Antimicrob. Agents Chemother.* **34**, 52–57.

van Veen, H.W., Callaghan, R., Soceneantu, L., Sardini, A., Konings, W.N. and Higgins, C.F. (1998) A bacterial antibiotic-resistance gene that complements the human multidrug-resistance P-glycoprotein gene. *Nature* **391**, 291–295.

van Veen, H.W. and Konings, W.N. (1997) Drug efflux proteins in multidrug resistant bacteria. *Biol. Chem.* **378**, 769–777.

van Veen, H.W., Venema, K., Bolhuis, H., Oussenko, I., Kok, J., Poolman, B., *et al.* (1996) Multidrug resistance mediated by a bacterial homolog of the human multidrug transporter MDR1. *Proc. Natl. Acad. Sci. USA* **93**, 10668–10672.

Westbrock-Wadman, S., Sherman, D., Hickey, M., Coulter, S., Zhu, Y., Warrener, P., Nguyen, L., Shavar, R., Folgee, K. and Stover, K. (1999) Characterization of a Pseudomonas aeruginosa efflux pump contributing to aminoglyoside impermeability. *Antimicrob. Agents Chemother.* **43**: 2975–2983.

Williams, R.J., Livermore, D.M., Lindridge, M.A., Said, A.A. and Williams, J.D. (1984) Mechanisms of beta-lactam resistance in British isolates of Pseudomonas aeruginosa. *J. Med. Microbiol.* **17**, 283–293.

Woolridge, D.P., Vazquez-Laslop, N., Markham, P.N., Chevalier, M.S., Gerner, E.W. and Neyfakh, A.A. (1997) Efflux of the natural polyamine spermidine facilitated by the Bacillus subtilis multidrug transporter Blt. *J. Biol. Chem.* **272**, 8864–8866.

Yamada, H., Kurose-Hamada, S., Fukuda, Y., Mitsuyama, J., Takahata, M., Minami, S., *et al.* (1997) Quinolone susceptibility of norA-disrupted Staphylococcus aureus. *Antimicrob. Agents Chemother.* **41**, 2308–2309.

Yamaguchi, A., Ono, N., Akasaka, T., Noumi, T. and Sawai, T. (1990) Metal-tetracycline/H+ antiporter of Escherichia coli encoded by a transposon, Tn10. The role of the conserved dipeptide, Ser65-Asp66, in tetracycline transport. *J. Biol. Chem.* **265**, 15525–15530.

Yerushalmi, H., Lebendiker, M. and Schuldiner, S. (1995) EmrE, an Escherichia coli 12-kDa multidrug transporter, exchanges toxic cations and H+ and is soluble in organic solvents. *J. Biol. Chem.* **270**, 6856–6863.

Yerushalmi, H., Lebendiker, M. and Schuldiner, S. (1996) Negative dominance studies demonstrate the oligomeric structure of EmrE, a multidrug antiporter from Escherichia coli. *J. Biol. Chem.* **271**, 31044–31048.

Yoneyama, H., Ocaktan, A., Gotoh, N., Nishino, T. and Nakae, T. (1998) Subunit swapping in the Mex-extrusion pumps in Pseudomonas aeruginosa. *Biochem. Biophys. Res. Commun.* **244**, 898–902.

Yoneyama, H., Ocaktan, A., Tsuda, M. and Nakae, T. (1997) The role of mex-gene products in antibiotic extrusion in Pseudomonas aeruginosa. *Biochem. Biophys. Res. Commun.* **233**, 611–618.

Yonezawa, M., Takahata, M., Matsubara, N., Watanabe, Y. and Narita, H. (1995) DNA gyrase gyrA mutations in quinolone-resistant clinical isolates of Pseudomonas aeruginosa. *Antimicrob. Agents Chemother.* **39**, 1970–1972.

Yoshida, H., Bogaki, M., Nakamura, M. and Nakamura, S. (1990a) Quinolone resistance-determining region in the DNA gyrase gyrA gene of Escherichia coli. *Antimicrob. Agents Chemother.* **34**, 1271–1272.

Yoshida, H., Bogaki, M., Nakamura, S., Ubukata, K. and Konno, M. (1990b) Nucleotide sequence and characterization of the Staphylococcus aureus norA gene, which confers resistance to quinolones. *J. Bacteriol.* **172**, 6942–6949.

Yoshida, T., Muratani, T., Iyobe, S. and Mitsuhashi, S. (1994) Mechanisms of high-level resistance to quinolones in urinary tract isolates of Pseudomonas aeruginosa. *Antimicrob. Agents Chemother.* **38**, 1466–1469.

Zarantonelli, L., Borthagaray, G., Lee, E.H. and Shafer, W.M. (1999) Decreased azithromycin susceptibility of neisseria gonorrhoeae due to mtrR mutations. *Antimicrob. Agents Chemother.* **43**, 2468–2472.

Zeller, V., Janoir, C., Kitzis, M.D., Gutmann, L. and Moreau, N.J. (1997) Active efflux as a mechanism of resistance to ciprofloxacin in Streptococcus pneumoniae. *Antimicrob. Agents Chemother.* **41**, 1973–1978.

Zgurskaya, H.I. and Nikaido, H. (1999a) AcrA is a highly asymmetric protein capable of spanning the periplasm. *J. Mol. Biol.* **285**, 409–420.

Zgurskaya, H.I. and Nikaido, H. (1999b) Bypassing the periplasm: reconstitution of the AcrAB multidrug efflux pump of Escherichia coli. *Proc. Natl. Acad. Sci. USA* **96**, 7190–7195.

Zhanel, G.G., Karlowsky, J.A., Saunders, M.H., Davidson, R.J., Hoban, D.J., Hancock, R.E., *et al.* (1995) Development of multiple-antibiotic-resistant (Mar) mutants of Pseudomonas aeruginosa after serial exposure to fluoroquinolones. *Antimicrob. Agents Chemother.* **39**, 489–495.

Zheleznova, E.E., Markham, P.N., Neyfakh, A.A. and Brennan, R.G. (1999) Structural basis of multidrug recognition by BmrR, a transcription activator of a multidrug transporter. *Cell* **96**, 353–362.

Ziha-Zarifi, I., Llanes, C., Kohler, T., Pechere, J.C. and Plesiat, P. (1999) *In vivo* emergence of multidrug-resistant mutants of Pseudomonas aeruginosa overexpressing the active efflux system MexA-MexB-OprM. *Antimicrob. Agents Chemother.* **43**, 287–291.

Zuckerman, J.M. and Kaye, K.M. (1995) The newer macrolides. Azithromycin and clarithromycin. *Infect. Dis. Clin. North. Am.* **9**, 731–745.

6. Bacterial Genetics and Antibiotic Resistance Dissemination

Didier Mazel

UPMTG, Département des Biotechnologies, Institut Pasteur, 25 rue du Dr Roux, 75724 Paris, France.

INTRODUCTION

There is mounting evidence that gene transfer plays a major role in the evolution of bacteria. The recent accumulation of bacterial genome sequences has shown that bacterial chromosomes have a mosaic structure, mostly due to exogenous gene acquisition (see for example (Blattner *et al.*, 1997)). In many cases the evolution of adaptive functions, including the virulence determinants of pathogenic strains, are encoded on DNA fragments which have a heterologous origin and are carried on mobile genetic structures (for review see (Lee, 1996)). The incidence of such events can be seen in the recent and dramatic emergence of the *V. cholerae* strain O139 Bengal, the cause of the large cholera epidemic in Southeast Asia in 1992 (Albert *et al.*, 1993). Indeed, the new pathogenic characteristics of this strain seem to be entirely due to the acquisition of a complete set of cell-wall poly-saccharide genes by an O1 El tor strain, which was responsible for an earlier epidemic (Mooi and Bik, 1997). However, the most striking incidence of gene transfer can be seen in the extraordinarily rapid evolution of antibiotic resistance among bacterial pathogens. All available evidence suggests that the acquisition and dissemination of antibiotic resistance genes in these strains have occurred during the past fifty years. The best support for this recent invasion comes from studies of Hughes and Datta (Datta and Hughes, 1983; Hughes and Datta, 1983) who examined the "Murray Collection", a collection of (mainly) Gram-negative pathogens obtained from clinical specimens in the pre-antibiotic era, for antibiotic resistance and conjugative plasmid content. None of these strains, isolated between 1917 and 1952, was resistant to any of the antibiotics in current use.

The mechanisms conferring antibiotic resistance are varied and may result from the mutation of endogenous chromosomal genes, the acquisition of foreign intrinsically resistant genes, or combination of these events. Non-transferable resistance traits arise primarily through point mutations in the gene encoding the antibiotic target (e.g. rifampin, fluoroquinolones) or by deregulated expression of a regular cellular process (e.g. multiple antibiotic resistance efflux systems, inducible β-lactamases). However, the activity of many anti-bacterial compounds cannot be circumvented by alteration of a pre-existing system and resistance genes must be imported. Given the critical role of gene exchange in bacterial evolution, it is self-evident that extensive inter-species and inter-genus gene transfer must have occurred during the golden age of antibiotics. Moreover, it is very likely that the potential reservoir for resistance gene recruitment by bacteria extends beyond the limit of the bacterial kingdom. Indeed, in a recent work (Brown *et al.*, 1998) Brown and collaborators have provided strong support for an eucaryotic origin of the Ile-tRNA synthetase gene which confers a high level of mupirocin resistance in *S. aureus* (Hodgson *et al.*, 1994) and *M. tuberculosis* (Sassanfar *et al.*, 1996).

The link between gene transfer and antibiotic resistance has been reviewed many times in the past

Table 1.   Natural gene transfer processes

Process	Components	Required Genetic Determinants		Host Range	Role in Antibiotic Resistance	Comments
		Donor	Recipient*			
Transduction	bacteriophage/cell	phage receptor	phage receptor	limited to closely related species	never demonstrated	very efficient
Conjugation	cell/cell	transfer genes	may require receptor	– inter specific – inter generic – very broad	primordial in most species	some are of very high efficiency
Transformation	DNA/cell		competence determinants	very broad	essential in *Nesseiria* and *Streptococcus* sp.	variable efficiency

* All DNA exchange processes are subject to negative effects of restriction (endonuclease action) in the recipient.

and a variety of mechanisms of gene transfer have been invoked in the process of resistance determinant acquisition and dispersion (see Table 1). These processes have been extensively characterized in laboratory studies and it can be assumed that all of these (and probably others) occur in nature.

We present here an overview of the three known avenues for bacteria to acquire exogenous DNA: transduction, conjugation and transformation, and discuss their pivotal roles in the spread of antibiotic resistance among the bacterial genera. To acknowledge their contribution to this global phenomenon, a section on the mechanisms of gene capture by DNA mobile elements is also included.

TRANSDUCTION

Transduction is the exchange of bacterial genes mediated by bacteriophage, or phage. Upon infection, the phage encoded proteins direct the takeover of the host DNA and protein synthesizing machinery for the propagation of new phage particles. Transducing particles are formed when plasmid DNA or fragments of host chromosomal DNA are erroneously packaged into phage particles during the replication process. Transducing particles (those carrying non-phage DNA) are included when phage are liberated from the infected cell to encounter another host and begin the next round of infection.

Transduction has been documented in at least 60 species of bacteria found in a wide variety of environments (Kokjohn, 1989). This mechanism is suspected to play a major role in the considerable

genetic diversity observed among non naturally transformable group A *streptococci* (Kehoe *et al.*, 1996). In general the actual levels of intraspecies and interspecies transduction are unknown but, the potential for transductional gene exchange is likely to be universal among the eubacteria. A study of the presence of bacteriophage in aquatic environments has shown that there may be as many as 10^8 phage particles per milliliter (Bergh *et al.*, 1989; Wichels *et al.*, 1998)]; Bergh and collaborators calculate that at such concentrations one third of the total bacterial population is subject to phage attack every 24 hours. DNA packaged into the transducing particle is protected from the environment and may survive for relatively long periods of time (Zeph *et al.*, 1988). Stotzky has suggested that transduction could represent an evolutionary survival strategy for the genetic material and this mechanism of gene transfer may be as important as conjugation or transformation in natural habitats (Stotzky, 1989).

There are many laboratory studies of transduction of antibiotic resistance (see for example (Novick and Morse, 1967; Inoue and Mitsuhashi, 1975; Lacey, 1975; Hyder and Streitfeld, 1978; Morrison *et al.*, 1978; Brau and Piepersberg, 1983; Berg *et al.*, 1983; Germida and Khachatourians, 1988; Zeph *et al.*, 1988; Stotzky, 1989; Zeph and Stotzky, 1989; Blahova *et al.*, 1993; Blahova *et al.*, 1994; Willi *et al.*, 1997)) and this process has been demonstrated in controlled conditions mimicking the different biotopes in which pathogenic bacteria could thrive or survive, such as blood (Hyder and Streitfeld, 1978), fresh water (Morrison *et al.*, 1978) or soil (Germida and Khachatourians, 1988;

Zeph *et al.*, 1988; Zeph and Stotzky, 1989). However, this mechanism has been considered less important in the dissemination of antibiotic resistance genes because phage generally have limited host ranges; they can infect only members of the same or closely related species and the size of DNA transferred does not usually exceed 50 kb. However, phages of extraordinarily broad host specificity have been described. For example, phages PRR1 and PRD1 will infect any Gram-negative bacterium containing the resistance plasmid RP1 (Olsen and Shipley, 1973; Olsen *et al.*, 1974) and could, in principle, transfer genetic information between unrelated bacterial species. Even if there is no direct evidence to implicate phage in the dissemination of antibiotic resistance genes, their role as vehicles for the dissemination of virulence functions in many pathogenicity islands (see for example (Lee, 1996; Waldor and Mekalanos, 1996; Karaolis *et al.*, 1999)) renders their contribution most probable. Moreover, the integration machinery of several conjugative transposons which are known to carry resistance determinants, is structurally and functionally similar to that of Lambdoid bacteriophages (see below).

CONJUGATION

Conjugative DNA transfer occurs during cell-to-cell contact. It is the principal mechanism for the dissemination of antibiotic resistance genes (see for example (Waters, 1999)). First discovered in 1946 by Lederberg and Tatum through genetic recombination experiments in *E. coli* (Lederberg and Tatum, 1946), the impact of this type of antibiotic resistance gene movement soon became apparent. During an epidemic of dysentery in Japan in the late 1950's, increasing numbers of *Shigella dysenteriae* strains were isolated that were resistant to up to four antibiotics simultaneously. It became clear that the emergence of multiple resistant strains could not be attributed to mutation alone. Furthermore, both sensitive and resistant *Shigella* could be isolated from a single patient, and that *Shigella sp.* and *E. coli* obtained from the same patients often exhibited the same multiple resistance patterns (Akido *et al.*, 1960; Watanabe, 1963). These findings led to the discovery of resistance transfer factors, and were also an early indication of the

contribution of conjugative transfer to the natural evolution of new bacterial phenotypes. The first conjugative elements identified were the conjugative plasmids, which encode their own transfer functions. These studies also led to the discovery of mobilizable plasmids which do not specify the functions for their transfer, but could use the conjugative functions of other plasmids. These plasmids could either be transferred through co-integrate formation with a conjugative plasmid (a phenomenon also referred as plasmid conduction) or, more commonly, by using the conjugation apparatus of a self-conjugative plasmid (if they carry an origin of transfer recognized by the conjugation machinery). In addition to plasmid-mediated conjugal transfer, another form of conjugation, conjugative transposition, has been reported in an increasing number of cases.

Conjugative Plasmids

The list of conjugative plasmids carrying antibiotic resistance determinants which have been described over the last 40 years is extensive. In 1977, Shapiro and collaborators listed more than 500 different plasmids which had been studied in enteric bacteria and Pseudomonads, as well as about 60 plasmids characterized in Gram-positive species, mainly in *S. aureus* and *E. faecalis*; most of these plasmids were conjugative (Jacob *et al.*, 1977). More than 25 different groups of plasmids have been defined on the basis of incompatibility (Inc) properties in Gram-negative species (Couturier *et al.*, 1988). Among these, several (especially those from incompatibility groups N, P, Q and W (Thomas, 1989)) are known to have a broad host range and, therefore, are able to transfer and replicate in remote bacterial species. The host range spectrum depends on many traits, including the replication and maintenance functions, the ability of plasmid-encoded markers to be expressed in the new host and the specificity of the conjugative apparatus. The different types of conjugative plasmids use different and distinct transfer systems. They do however share several properties. First, conjugation always occurs through the transfer of a single-stranded DNA after nicking at a specific site called *oriT*. Second, usually the expression of the plasmids encoded genes needed for conjugative transfer is tightly regulated so as to minimize the burden on the host (Zatyka and Thomas, 1998).

Three distinct bacteria-to-bacteria transfer systems have been the focus of study, those found in F and IncP plasmids of Gram-negative bacteria, and the pheromone regulated plasmids of *E. faecalis*, a gram-positive bacteria. The broad host range plasmids found in *Streptococci* and *Staphylococci* have also been studied to a certain extent.

Gram-negative plasmids

The bacterial conjugation system encoded by the F sex factor was the first to be described and subjected to detailed analysis. F belongs to a large group of plasmids that encode a common transfer (*tra*) mechanism and includes several Inc sub-groups (Ippen-Ihler and Skurray, 1993). Many of the F-like plasmids carry determinants for antibiotic resistance or for toxin and hemolysin production (see (Ippen-Ihler and Skurray, 1993) for a review). Where examined, molecular and genetic analyses, such as complementation studies, have confirmed the relatedness of the *tra* systems encoded by the F-like group of plasmids. On the F-plasmid, which is about

100 kb, the 37 genes necessary for the conjugation are gathered in the 34-kb *tra* region shown in Figure 1 (Frost *et al.*, 1994). Among these genes, 15 are functionally involved in pilus synthesis and assembly. The others are either implicated in conjugative DNA metabolism for the single strand transfer, the stabilisation of cell aggregates, surface exclusion, or they have an unknown function (Ippen-Ihler and Skurray, 1993; Frost *et al.*, 1994). The majority of the *tra* loci are expressed from a single promoter, P_Y, which is positively controlled by the product of *traJ*, as well as by TraY in an autostimulatory circuit. A fertility inhibition system, the FinOP system, exists in this family of conjugative plasmids (with the exception of F itself, in which an IS3 insertion inactivates *finO*) and controls the expression of *traJ* through an antisense RNA regulation mechanism (van Biesen and Frost, 1994). This control results in repression of the conjugative functions in most of the cells so that only 0.1% of the bacteria promote plasmid transfer to recipients. In new recipients there will also be a burst of gene expression because of the initial lack of FinO and P. Thus, the state of transfer proficiency oscillates, spreading through a population of plasmid free bacteria as a wave.

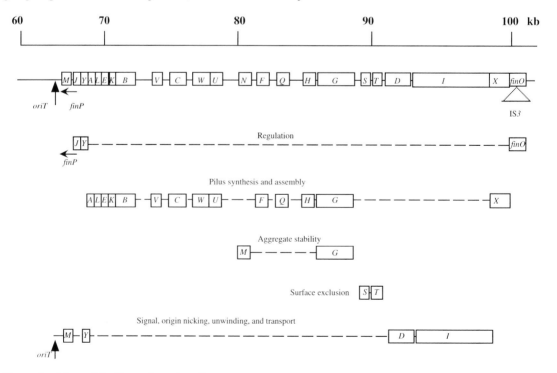

Figure 1. View of the F transfer region (Frost *et al.*, 1994). The top line gives the coordinates of the transfer region on the F map. The last five lines represent the functions of the *tra* genes identified to date (see text). *finP* is a transcript in the anti-orientation; *oriT* is the origin of transfer.

IncP plasmids, such as the R plasmids RK2 and RP4, have been studied widely because of their ability to transfer between and stably replicate themselves in almost all Gram-negative bacterial species (Thomas, 1989; Guiney, 1993). In laboratory conditions they can also promote conjugative transfer of DNA from Gram-negative bacteria to Gram-positive bacteria (Trieu-Cuot et al., 1987) as well as to yeast (Heinemann and Sprague, 1989). Furthermore, there is evidence of a genetic relationship between IncP conjugation and the bacteria-to-plant DNA transfer system of *Agrobacterium tumefaciens* (T-DNA transfer) (Waters et al., 1991; Guiney, 1993). The genes for the conjugative apparatus of the IncPα plasmid RK2 are organized in two clusters, the Tra1 and Tra2 regions, which together with the genes for essential replication functions represent about two-thirds of the 60 kb of the plasmid genome (Pansegrau et al., 1994). The complete sequence of the IncPβ plasmid R751 has recently been compiled and comparison of its sequence with that of the IncPα plasmid confirmed the conservation of the IncP backbone of replication, conjugative transfer and stable inheritance functions between the two branches of this plasmid family (Thorsted et al., 1998). In contrast to the F plasmid *tra* genes, the *tra* and *trb* genes (from the Tra1 and Tra2 regions, respectively) are not organized as a single operon. Transfer frequencies of IncP plasmids are very high (upwards of one per donor per hour); however, transfer genes are not constituvely expressed and a complex regulatory system has evolved to coordinate the expression of so many genes (for review see Zatyka and Thomas, 1998).

Gram-positive plasmids

Several Enterococcal plasmids encoding virulence factors and/or resistance genes have been shown to transfer very efficiently in liquid cultures among *Enterococci*. As for the Gram-negative plasmids described above, their transfer is tightly regulated, and a sex-pheromone recognition mechanism controls their dissemination (Clewell, 1993a; Maqueda et al., 1997). The most extensively studied plasmids from this group are pAD1, encoding hemolysin-bacteriocin production and UV resistance; pCF10, encoding tetracycline resistance; and pPD1 encoding bacteriocin production (Clewell, 1993b). Plasmid-free cells secrete multiple sex pheromones that trigger the donor cells to express transfer functions. Thus pheromone accumulation in the medium indicates to donors that recipients are in close vicinity. These pheromones are short linear hydrophobic peptides excreted in tiny amounts. As few as 1 to 10 molecules per donor are sufficient to initiate the mating process. A plasmid-free strain will secrete several pheromones, at least 5, which are specific for the corresponding conjugative plasmid. When such a strain receives a conjugative plasmid, the production of the corresponding pheromone is stopped, whereas secretion of the remaining pheromones continues. Pheromones induce the coordinated expression of different genes and gene sets involved in mating and DNA transfer functions (for review see Maqueda et al., 1997). In the Tc resistance encoding plasmid pCF10, these genes represent more than one third of the coding capacity of its 58 kb and encode the proteins participating in cell aggregation, surface exclusion, synthesis of the specific pheromone inhibitor, regulation, and plasmid transfer. The genes involved in the regulation of the pheromone response are clustered in a 7 kb region, whose organization and nucleotide sequence is conserved in each plasmid. Regulation of the transfer functions is complex and involves both positive and negative regulatory circuits which may differ slightly in the different plasmids (Maqueda et al., 1997; Zatyka and Thomas, 1998). This system of recipient recognition through a pheromone mediated pathway is very sophisticated and unique among plasmids studied to date; however, this specialization may also be a factor which limits the host range of these plasmids.

The two most extensively studied examples of conjugative broad host range streptococcal plasmids are pIP501 and pAMβ1, which are both about 30 kb in size. pIP501 confers inducible resistance to macrolide, lincosamide and streprogramin (MLS) and chloramphenicol. pAMβ1 confers constitutively expressed MLS resistance. Both plasmids are able to transfer and replicate in many Gram-positive species (reviewed in Macrina and Archer, 1993) and pAMβ1 conjugative functions have been demonstrated to work in Gram-positive to Gram-negative transfer (Trieu-Cuot et al., 1988). The conjugation apparatus and process are still poorly understood;

however, the number of genes necessary for the conjugal transfer is apparently fewer than in the broad host range Gram-negative plasmids such as the IncP members described above. Indeed, two regions representing about 11 kb implicated in conjugation proficiency have been identified by transposon mutagenesis in pIP501 (Krah and Macrina, 1989, 1991; Macrina and Archer, 1993). One of these regions has been further characterized and shown to carry a functional *oriT*, a gene whose product is related to the relaxases of some gram-negative plasmids and five contiguous unique ORFs (Wang and Macrina, 1995a, 1995b).

Another type of broad host range plasmid found in Gram-positive bacteria was identified after the emergence of gentamicin resistance among *Staphylococci* in the mid-1970s (Schaberg *et al.*, 1985; Schaberg and Zervos, 1986), for which plasmid pGO1 is the prototype. pGO1 is a 52 kb plasmid which carries a gene for a bifunctional AAC (6′) APH (2″) enzyme that confers resistance to several aminoglycosides, a *qac* gene encoding resistance to quaternary ammonium compounds and a dehydrofolate reductase gene conferring trimethoprim resistance. The transfer region (*trs*) spans 14 kb and encodes 14 ORFs (Morton *et al.*, 1993). An additional gene, *nes*, whose product shows similarity to known relaxases, is located 13.5 kb away from the *trs* region (Climo *et al.*, 1996). The *oriT* lies 100 bp upstream of *nes* and is identical to the *oriT* of the conjugative or mobilizable plasmids RSF1010, pSC101 and pIP501. In laboratory experiments, pGO1 transfers at low frequency (10^4–10^6 transconjugants per donor) and the signals that modulate or trigger the conjugation are not known. It is very likely that the product of *trsN* plays a role in the regulation of the expression of transfer functions as it has been shown to bind to several trs promoters (Sharma *et al.*, 1994). Thus, despite some similarities, the uniqueness of most of the genes implicated in conjugative functions suggest that there may be fundamental differences between plasmid transfer systems in Gram-positive and Gram-negative bacteria.

Conjugative Transposons

Conjugative transposons were first found in Gram-positive cocci but are now known to be present in a variety of clinically important groups of gram-positive and gram-negative bacteria. They were first identified when the transfer of antibiotic resistance determinants occurred in the absence of plasmids (Shoemaker *et al.*, 1980; Franke and Clewell, 1981; Gawron-Burke and Clewell, 1982). Conjugative transposons are discrete elements that are normally integrated into a bacterial genome. Their transposition and transfer is thought to start by an excision event and the formation of a covalently closed circular intermediate, which can either integrate elsewhere in the same cell or into the genome of a recipient cell following self-transfer by conjugation. In contrast to conjugative plasmids, their propagation through integration in the host genome emancipates transposons from the constraint of a compatible replication system. This property certainly contributes to the broad host range which is observed for several of them and led Salyers and collaborators to suggest that conjugative transposons may be as important as conjugative plasmids in broad host-range gene transfer between some species of bacteria (Salyers *et al.*, 1995).

Conjugative transposons are not considered typical transposons as they have a covalently closed circular transposition intermediate and they do not duplicate the target site when they integrate (Scott, 1992; Salyers *et al.*, 1995). At the present time, five different families of conjugative transposons have been established: a) the *Tn916* family (originally found in *Streptococci* but now known also to occur in Gram-negative bacteria such as *Campylobacter*); b) the *S. pneumoniae* family (*Tn5253*); c) the *Bacteroides* family; d) CTnscr94, a conjugative transposon found in enterobacteria, and d) the SXT element found in *V. cholerae* (Salyers *et al.*, 1995; Waldor *et al.*, 1996; Hochhut *et al.*, 1997; Salyers and Shoemaker, 1997; Hochhut and Waldor, 1999). Two other mobile elements are also conjugative elements that integrate rather than replicate and behave like conjugative transposons (Murphy and Pembroke, 1995; Ravatn *et al.*, 1998a; Ravatn *et al.*, 1998b). The size of the different conjugative transposons range from 15 kb to 150 kb and all but Ctnscr94 have been identified through their capacity to transfer resistance genes. Indeed, in addition to other resistance genes, most encode tetracycline resistance determinants (e.g. *Tn916* encodes the

TetM determinant, and the TetQ determinant is found on the conjugative transposons from the *Bacteroides* group (Nikolich *et al.*, 1994b; Salyers *et al.*, 1995)), while the SXT element encodes resistance to sulfamethoxazole, trimethoprim, chloramphenicol and streptomycin (Waldor *et al.*, 1996). These transposons use an integration machinery similar to that of Lambdoid phages, and all characterized elements carry an *int* gene. However, the target specificity for their integration varies considerably, *Tn916* behaves more like a real transposon, i.e. with a poor specificity, while other elements behave more like Lambdoid phages and have a higher specificity. For example, *Bacteroides* elements show a preference for three to seven sites (Salyers *et al.*, 1995), Ctnscr94 integration is restricted to two sites (Hochhut *et al.*, 1997) and one unique site of integration has been identified for SXT (Hochhut and Waldor, 1999).

In all cases studied, the transfer has been shown to occur after circularization of the element and results obtained for *Tn916* suggest that, as for conjugative plasmids, only one strand is transferred to the recipient bacteria (Scott *et al.*, 1994). Furthermore, *oriT* sequences have been identified in several elements and an homologue to the mobilization protein, MbeE, of ColE1 has been found in *Tn916* (Scott *et al.*, 1994; Jaworski and Clewell, 1995; Li *et al.*, 1995). By contrast, little is known about the mating machinery. *Tn916* has been completely sequenced (Flannagan *et al.*, 1994) and the region essential for transfer has been mapped. None of the predicted products of this region has significant similarity to sex pilus proteins of conjugative plasmids and since the number of genes is small, this transfer system may be simpler than that of F or RP4 plasmids. The region necessary and sufficient for conjugal transfer of one of the *Bacteroides* transposons (Tc^r Em^r DOT) has been localized to a 16 kb central segment (Li *et al.*, 1995), but has not been characterized further.

By contrast to most conjugative plasmids, elements of the *Tn916* or the *Bacteroides* family do not inhibit transfer of further copies or further transposition events (Shoemaker *et al.*, 1989; Shoemaker and Salyers, 1990; Norgren and Scott, 1991). Thus a strain can accumulate more than one conjugative transposon, the only limitation being their site-specificity. Flannagan and Clewell have observed that the presence of two copies of *Tn916* in the same strain results in a stimulation of transposition, a phenomenon termed transactivation (Flannagan and Clewell, 1991).

Although the *Tn916* and *Bacteroides* elements are structurally very different, *Tn916* and a *Bacteroides* transposon, Tc^r Em^r DOT, share an interesting property – their transfer is induced by the presence of low amounts of tetracycline (Stewart, 1989; Showsh and Andrews, 1992; Stevens *et al.*, 1993), to which they encode resistance. In both cases, enhanced transfer is not due to stress caused by tetracycline inhibition of protein synthesis. However, the regulatory elements governing this effect are markedly different for the two elements. A two-component regulatory system is apparently involved in the activation phenomenon for Tc^r Em^r DOT (Stevens *et al.*, 1993; Salyers *et al.*, 1995). In *Tn916* the transcription of the Tc resistance *tetM* gene is induced by Tc and this activation leads to a transcriptional read-through that allows the expression of two regulatory genes, located downstream of the *tetM* gene, which trigger circularisation and transfer (Celli and Trieu-Cuot, 1998).

Where does conjugation take place in the environment? It is a difficult question to answer, but an ideal site for gene transfer is the warm, wet, nutrient-rich environment of the mammalian intestinal tract with its associated high concentration of bacteria. The resident microflora are believed to serve as a reservoir for genes encoding antibiotic resistance which could be transferred not only to other members of this diverse bacterial population, but also to transient colonisers of the intestine, such as soil or water microbes or human pathogens. There is clear evidences that identical R plasmids can be identified in pathogen isolates of *Salmonella* as well as in native intestinal flora such as *E. coli* (see for example (Balis *et al.*, 1996). Several sets of experiments have shown that transfer could occur efficiently even between distantly related bacteria in such environment, both for conjugative plasmids (Watanabe, 1963; Sansonetti *et al.*, 1980; Doucet-Populaire *et al.*, 1992; Scott and Flint, 1995) and transposons (Doucet-Populaire *et al.*, 1991). Using oligonucleotide probes having DNA sequence homology to the hypervariable regions of the TetQ determinant, Salyers and co-workers have provided evidence that gene transfer between species of

Bacteroides, one of the predominant genera of the human intestine, and *Prevotella* sp., one of the predominant genera of livestock rumen, has taken place under physiological conditions (Nikolich *et al.*, 1994a).

TRANSFORMATION

Natural transformation is a physiological process characteristic of many bacterial species in which the cell takes up and expresses exogenous DNA. Its potential for gene transfer in the environment (terrestrial and aquatic) has been extensively reviewed (Lorenz and Wackernagel, 1994; Day, 1998), but there are only a few identified examples. While natural transformation has been reported to occur only in a limited number of genera, these include many pathogenic taxa such as *Haemophilus, Mycobacterium, Streptococcus, Neisseria, Pseudomonas*, and *Vibrio* (Stewart, 1989). Initial studies suggested that natural transformation in some of these genera was limited to DNA from that particular species. For example, an eleven-basepair recognition sequence permits *Haemophilus influenzae* to take up its own DNA preferentially compared to heterologous DNAs (Goodgal and Mitchell, 1990). Moreover, the induction of competence is tightly controlled in bacterial species such as *Streptococci*, this state being induced through a quorum-sensing mechanism mediated by peptide pheromones produced by the *Streptococci* themselves (reviewed in (Morrison, 1997; Mortier-Barriere *et al.*, 1997; Claverys *et al.*, 1997)). Given such specificity one could ask if natural transformation is an important mechanism in the transfer of antibiotic resistance genes? There are only few examples for which transformation has been implicated, mainly for the penicillin resistance in *Streptococcus pneumoniae, Neisseria gonorrhoeae* and *Neisseria meningitidis*, for the sulfonamide resistance in *N. meningitidis, S. pneumoniae* and *Streptococcus pyogenes*, and recently for the chloramphenicol resistance in *N. meningitidis*. As transformation is also suspected to play a role for other variable phenotypic traits of *Streptococcus* species (Mortier-Barriere *et al.*, 1997), it has to be considered as a general means of adaptation in these species.

In these species, β-lactam resistance is due to the development of altered penicillin-binding proteins (PBPs) with a decreased affinity for β-lactam antibiotics. Comparisons of the PBP genes from resistant and sensitive strains have shown that resistant PBPs had multiple gene segment replacements leading to mosaic structures. In most of the case, the "new gene segments" were highly similar or identical to the corresponding part of the PBP gene from a closely related commensal species (Dowson *et al.*, 1989; Martin *et al.*, 1992; Spratt *et al.*, 1992; Dowson *et al.*, 1993; Laible *et al.*, 1993; Bowler *et al.*, 1994; Dowson *et al.*, 1994). For example, in *N. meningitidis* penicillin-resistant strains, the *penA* gene contains two segments which show 22% divergence from the *penA* allele found in penicillin sensitive strains, but which are identical to the corresponding regions from the *penA* gene of *N. flavescens*, a commensal species (Spratt *et al.*, 1989). The fact that more than 30 different mosaic genes have been found in 78 isolates of penicillin-resistant *N. meningitidis* exemplifies to what extent this mechanism is common in naturally transformable species (Zhang *et al.*, 1990; Campos *et al.*, 1992).

Resistance to sulfonamide is always due either to the development or to the acquisition of an altered dehydropteroate synthase (DHPS) with a decreased affinity for these compounds. In *N. meningitidis, S. pneumoniae* and *S. pyogenes* sulfonamide resistance is mediated by mutations in the chromosomal DHPS encoding gene, *folP*. Comparisons of *folP* from resistant and sensitive strains have shown that in resistant isolates *folP* genes diverged by 8 to 14% from their counterpart in sensitive isolates (Radstrom *et al.*, 1992; Fermer *et al.*, 1995; Maskell *et al.*, 1997; Swedberg *et al.*, 1998), some of them having a mosaic structure (Radstrom *et al.*, 1992). As for the *penA* mosaic genes, the number of different *folP* mosaic genes suggest that resistance has arisen on many independent occasions in these species.

Transformation is also very likely involved in the recent dissemination of a chloramphenicol acetyl transferase gene of exogenous origin among *N. meningitidis* strains (Galimand *et al.*, 1998). In a study of 12 chloramphenicol resistant isolates of *N. meningitidis*, it has been observed that all carried a chromosomal *catP* gene. The nucleotide sequence of this gene and its flanking region was identical to an internal portion of transposon Tn*4451* characterized in *Clostridium perfringens*. In *Neisseria*, this truncated transposon corresponded almost exclusively to the *catP* gene and had lost its mobility. The resistant isolates showed a high degree of diversity; however, the location of the *catP* gene was invariant and the authors have demonstrated that this gene could be efficiently propagated to different susceptible strains trough transformation with genomic DNA from chloramphenicol resistant strains.

Another example for which transformation is suspected to be the mechanism of gene transfer is the propagation of the chromosomal non conjugative TetB determinant among naturally transformable *Haemophilus* species and in *Moraxella* catarrhalis (Roberts *et al.*, 1990; Roberts *et al.*, 1991; Roberts, 1994). However, data from molecular analysis are still forthcoming and another mechanism cannot be firmly rule out.

GENE CAPTURE IN MOBILE ELEMENTS

As mentioned in the introduction, bacteria isolated during the pre-antibiotic era did not carry antibiotic resistance determinants, yet they harboured conjugative plasmids from the same compatibility group as the R plasmids (Datta and Hughes, 1983; Hughes and Datta, 1983). Therefore, the multiresistance plasmids found in pathogens must have arisen within the past five decades. What really takes place when a new antimicrobial agent is introduced and plasmid-determined resistance develops within a few years? The most significant component in the process of antibiotic resistance flux in the microbial population is gene capture by way of transposons, which can move from replicon to replicon, be they chromosomes or episomes.

The crucial question is how do transposons acquire these resistance genes? The answer has been found for two types of transposons: composite transposons, found in both Gram positive and negative species, and transposons harbouring an integron, which are mainly observed in enterobacteriaceae (for a review see Rowe-Magnus and Mazel, 1999). Composite transposons can be constructed by random insertion of an insertion sequence (IS) on either side of any gene or cluster of genes. Most of these mobile elements carry only one or two resistance loci (e.g., Tn*9*, Tn*10*, Tn*903*, Tn*1546*, Tn*4001*, Tn*4003*, Tn*4351*, Tn*4400*) and rarely more (e.g., three in Tn*5*).

Conversely, integrons do not rely on random processes and therefore offer more flexibility. They were discovered in the 1980's when restriction mapping and heteroduplex analysis of various plasmids and transposons revealed that different resistance genes or arrays of resistances genes were flanked by identical sequences with no obvious relationship to ISs. Nucleotide sequence analysis confirmed that resistance genes integrated at a specific site (Cameron *et al.*, 1986; Ouellette *et al.*, 1987; Hall and Vockler, 1987; Sundström *et al.*, 1988). This process has been dissected genetically and biochemically (Hall and Stokes, 1993; Recchia and Hall, 1997; Hall, 1997; Sundstrom, 1998). All known integrons are composed of three key elements: a gene coding for an integrase (*intI*), a primary recombination site (*attI*) and a strong promoter. Based on the homology of the integrase genes, three classes of multi-resistant integrons have been defined. Integron integrases belong to the site-specific recombinase family and are able to recombine discrete units of DNA, now known as gene cassettes, providing them with a promoter for expression of their encoded proteins. Most of the resistance cassettes contain a single gene associated with a specific recombination sequence, the 59-base element (or *attC*). More than fifty different cassettes have been characterized, which allow Gram-negative bacteria to resist most classes of antibiotics (Mazel and Davies, 1999). The name 59-base element for the recombination sequences of the cassettes was historically given in 1986 from the comparison of the first few cassettes characterized at that time (Cameron *et al.*, 1986), they are now known to vary from 57 bp to 141 bp in length, and their nucleotide sequences similarities are primarily restricted to their extremities. Integrons belonging to the three classes appear to be able to acquire the same gene cassettes. Multi-resistant integrons harboring up to five different cassettes have been characterized. Nucleotide sequence analysis of multiresistant integrons showed that the inserted resistance gene cassettes differ markedly in codon usage, indicating that the antibiotic resistance determinants are of diverse origins.

Integrase-catalyzed insertion of resistance gene cassettes into resident integrons has been demonstrated (Martinez and de la Cruz, 1990; Collis et al., 1993). In addition, site-specific deletion and rearrangement of the inserted resistance gene cassettes can result from integrase-catalyzed events (Collis and Hall, 1992). Francia and co-workers have expanded our understanding of the role of integrons in gene mobilization (Francia et al., 1993). They have shown that the class 1 integrase can act on secondary target sites at significant frequencies; two R plasmids can be fused by interaction between the recombination hotspot of one plasmid and a secondary integrase target site on the other. However, even if the mechanisms of cassette acquisition and loss are well understood, many important details remain to be elucidated. For example, what is the origin

of the integrases and the 59-base elements? What are the 59-base elements and how do they become attached to the individual open reading frames to form the cassettes?

Recent work by Mazel and collaborators has shed some light on these matters (Mazel et al., 1998). Studies of the relationship between anti-biotic-resistance integrons of transposons and the Vibrio cholerae repeated sequences (VCRs) cluster have demonstrated striking structural similarities. The two compound genetic elements are shown in Figure 2 and the commonalties are obvious. Both the VCR cluster and antibiotic-resistance integrons possess specific integrases that are responsible for the insertion of coding sequences (ORFs) into a unique chromosomal attachment site, leading to the formation of tandem arrays of genes (albeit mostly of unidentified function). In the case of V. cholerae,

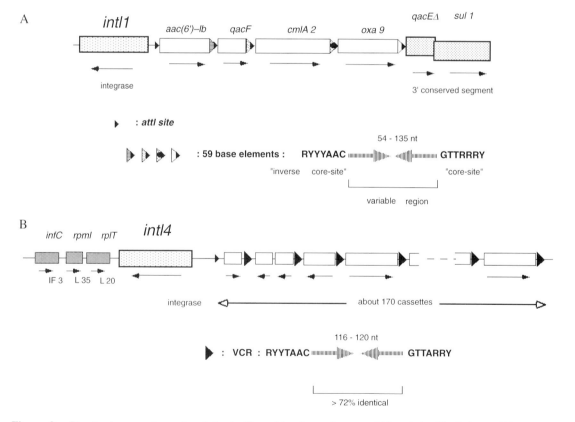

Figure 2. Structural comparison of a "classical" multi-resistant integron [A] and the V. cholerae 569B super-integron [B]. [A] Schematic representation of In40 (Ploy et al., 1998); the various resistance genes are associated with different 59-be (see text) whose consensus sequence is shown. [B] Data from the analysis of the V. cholerae super-integron (Rowe-Magnus et al., 1999); the ORFs are separated by highly homologous sequences, the VCRs, for which the consensus sequence is shown. Recombination occurs between the G and the first T of the core-site sequence.

the cluster of VCR-associated ORFs represents about 100 unidentified genes in more than 170 cassettes and occupies about 3% of the 2.5 Mb genome (Barker *et al.*, 1994; Rowe-Magnus *et al.*, 1999). Integrase-related structures are found in a number of different *Vibrio* sp. (e.g. *V. mimicus* and *V. metschnikovii*) and they share the same general characteristics (D. Mazel and collaborators, unpublished). The nature of the integrase-associated attachment sites has not been demonstrated in all cases, nor is it known if they are integrase-specific. The *Vibrio* super-integron integrases are related to each other and form a separate but closely related branch of the multi-resistance integron integrases. This suggests that multi-resistant integrons have evolved from super-integrons, through entrapment in highly mobile structures like transposons. The ancestral bacterial origin of these integrons remains to be identified.

The VCR structures differ from integrons in that (i) a certain number of ORFs possess their own promoter; (ii) an almost identical repeat (VCR) is associated with each ORF; (iii) the ORFs encode largely unknown functions. The ORF identified to date appear to be most related to virulence functions and no well-defined antibiotic-resistance genes reside within the clusters, although it should be noted that the *blaP4* and *dfrVI* cassettes of integrons are VCR-associated. What is intriguing is that many *Vibrio sp.* obviously possess a large number of clustered genes of unknown function that have been captured by a highly efficient and specific process. This genetic bank is assumed to be readily available to other *Vibrio sp.* and related bacterial species. If each *Vibrio* species proves to have cassette clusters of hundreds of unidentified genes, as in *V. cholerae*, a veritable treasure trove of genes will be available for functional genomic studies, since they clearly represent sequences that are not essential for host bacterial growth and maintenance under normal conditions. Presumably, the effects of integron-driven gene capture extend beyond the dissemination of antibiotic resistance and pathogenicity determinants and this phenomenon is likely to be an important factor in the more general process of horizontal (lateral) gene transfer in the evolution of bacterial genomes. While these tandem arrays of independent genes have been identified so far only in enteric bacteria, *M. tuberculosis* (Martin *et al.*, 1990) and *Coryne-*

bacterium glutamicum (Nesvera *et al.*, 1998), it should be noted that tandem arrays of resistance genes have been characterized in gram-positive bacteria such as *Staphylococcus aureus* (Allignet *et al.*, 1996), although no associated intergenic structural element or integrase can be identified.

Analyses of the integron-type transposons provide a good model to understand the way (i) in which antibiotic resistance genes from various (unknown) sources may be incorporated into an integron by recombination events; (ii) integrons may subsequently become associated to mobile elements such as transposons and (iii) these determinants become relocated onto a plethora of bacterial replicons. These succession of events could produce the R plasmids that we know today. However, in bacterial pathogens a variety of transposable elements have been found that undergo different processes of recombinational excision and insertion. It is not known what evolutionary mechanisms are implicated or whether some form of integron-related structure is present in all cases. For the type of integron found in transposons, we have plausible models, supported by *in vivo* and *in vitro* studies, to provide a *modus operandi* by which antibiotic resistance genes were (and are) molecularly cloned in the evolution of R plasmids. A large number of transposable elements carrying virtually all possible combinations of antibiotic resistance genes have been identified (Berg and Howe, 1989), and nucleotide sequence analysis of multiresistant integrons shows that the inserted resistance gene cassettes differ markedly in codon usage and GC%, indicating that the antibiotic resistance determinants are of diverse origins.

CONCLUSIONS

When antibiotic resistant bacteria are isolated from diseased tissue and identified as the responsible pathogen, this is the identification of the final product of a complex and poorly understood environmental system. While gene exchange may be rare under normal circumstances in stable microbial microcosms, the intense selective pressure of antibiotic usage is likely to have provoked cascades of antibiotic resistance gene transfer between unrelated microbes. These transfers must involve different biochemical mechanisms during which efficacy is not a critical factor since the survival and multiplication of a small number of resistant progeny suffices to create a clinically-problematic situation.

REFERENCES

Akido, T., Koyama, K., Ishiki, Y., Kimura, S. and Fukushima, T. (1960) On the mechanism of the development of multiple-drug-resistant clones of *Shigella*. *Japanese Journal of Microbiology*, **4**, 219–227.

Albert, M.J., Siddique, A.K., Islam, M.S., Faruque, A.S., Ansaruzzaman, M., Faruque, S.M., *et al.* (1993) Large outbreak of clinical cholera due to *Vibrio cholerae* non-O1 in Bangladesh. *Lancet*, **341**, 704.

Allignet, J., Aubert, S., Morvan, A. and el Solh, N. (1996) Distribution of genes encoding resistance to streptogramin A and related compounds among staphylococci resistant to these antibiotics. *Antimicrobial Agents & Chemotherapy*, **40**, 2523–2528.

Balis, E., Vatopoulos, A.C., Kanelopoulou, M., Mainas, E., Hatzoudis, G., Kontogianni, V., *et al.* (1996) Indications of *in vivo* transfer of an epidemic R plasmid from *Salmonella enteritidis* to *Escherichia coli* of the normal human gut flora. *Journal of Clinical Microbiology*, **34**, 977–979.

Barker, A., Clark, C.A. and Manning, P.A. (1994) Identification of VCR, a repeated sequence associated with a locus encoding a hemagglutini in *Vibrio cholerae* O1. *Journal of Bacteriology*, **176**, 5450–5458.

Berg, C.M., Grullon, C.A., Wang, A., Whalen, W. and Berg, D.E. (1983) Transductional instability of Tn5-induced mutations: generalized and specialized transduction of Tn5 by bacteriophage P1. *Genetics*, **105**, 259–263.

Berg, D.E. and Howe, M.M. (eds.) (1989) *Mobile DNA*, American Society for Microbiology, Washington, DC.

Bergh, O., Borsheim, K.Y., Bratbak, G. and Heldal, M. (1989) High abundance of viruses found in aquatic environments. *Nature*, **340**, 467–468.

Blahova, J., Hupkova, M., Babalova, M., Kremery, V. and Schafer, V. (1993) Transduction of resistance to Imipenem, Aztreonam and Ceftazidime in nosocomial strains of *Pseudomonas aeruginosa* by wild-type phages. *Acta Virologica*, **37**, 429–436.

Blahova, J., Hupkova, M. and Kremery, V., , Sr. (1994) Phage F-116 transduction of antibiotic resistance from a clinical isolate of *Pseudomonas aeruginosa*. *Journal of Chemotherapy*, **6**, 184–188.

Blattner, F.R., Plunkett, G.r., Bloch, C.A., Perna, N.T., Burland, V., Riley, M., *et al.* (1997) The complete genome sequence of *Escherichia coli* K-12. *Science* **277**, 1454–1474.

Bowler, L.D., Zhang, Q.Y., Riou, J.Y. and Spratt, B.G. (1994) Interspecies recombination between the *penA* genes of *Neisseria meningitidis* and commensal *Neisseria* species during the emergence of penicillin resistance in *N. meningitidis*: natural events and laboratory simulation. *Journal of Bacteriology*, **176**, 333–337.

Brau, B. and Piepersberg, W. (1983) Cointegrational transduction and mobilization of gentamicin resistance plasmid pWP14a is mediated by IS140. *Molecular & General Genetics*, **189**, 298–303.

Brown, J.R., Zhang, J.Z. and Hodgson, J.E. (1998) A Bacterial Antibiotic Resistance Gene With Eukaryotic Origins. *Current Biology*, **8**, R365–R367.

Cameron, F.H., Groot Obbink, D.J., Ackerman, V.P. and R.M., H. (1986) Nucleotide sequence of the AAD(2″) aminoglycoside adenylyltransferase determinant *aadB*. Evolutionary relationship of this region with those surrounding *aadA* in R538–1 and *dhfrII* in R388. *Nucleic Acids Research*, **14**, 8625–8635.

Campos, J., Fuste, M.C., Trujillo, G., Saez-Nieto, J., Vazquez, J., Loren, J.G., *et al.* (1992) Genetic diversity of penicillin-resistant *Neisseria meningitidis*. *Journal of Infectious Diseases*, **166**, 173–177.

Celli, J. and Trieu-Cuot, P. (1998) Circularization of Tn916 is required for expression of the transposon-encoded transfer functions: characterization of long tetracycline-inducible transcripts reading through the attachment site. *Molecular Microbiology*, **28**, 103–117.

Claverys, J.P., Dintilhac, A., Mortier-Barriere, I., Martin, B. and Alloing, G. (1997) Regulation of competence for genetic transformation in *Streptococcus pneumoniae*. *Society for Applied Bacteriology Symposium Series*, **26**.

Clewell, D.B. (1993a) Bacterial sex pheromone-induced plasmid transfer. *Cell*, **73**, 9–12.

Clewell, D.B. (1993b) In *Bacterial conjugation* (Ed, Clewell, D.B.) Plenum Press, New York and London, pp. 349–367.

Climo, M.W., Sharma, V.K. and Archer, G.L. (1996) Identification and characterization of the origin of conjugative transfer (oriT) and a gene (*nes*) encoding a single-stranded endonuclease on the staphylococcal plasmid pGO1. *Journal of Bacteriology*, **178**, 4975–4983.

Collis, C.M., Grammaticopoulos, G., Briton, J., Stokes, H.W. and Hall, R.M. (1993) Site-specific insertion of gene cassettes into integrons. *Molecular Microbiology*, **9**, 41–52.

Collis, C.M. and Hall, R.M. (1992) Site-specific deletion and rearrangement of integron insert genes catalyzed by the integron DNA integrase. *Journal of Bacteriology*, **174**, 1574–1585.

Couturier, M., Bex, F., Bergquist, P.L. and Maas, W.K. (1988) Identification and classification of bacterial plasmids. *Microbiological Reviews*, **52**, 375–395.

Datta, N. and Hughes, V. (1983) Plasmids of the same Inc groups in Enterobacteria before and after the medical use of antibiotics. *Nature*, **306**, 616–617.

Day, M. (1998) In *Horizontal gene transfer* sp (Eds, Syvanen, M. and Kado, C.I.) Chapman & Hall, London, pp. 144–167.

Doucet-Populaire, F., Trieu-Cuot, P., Andremont, A. and Courvalin, P. (1992) Conjugal transfer of plasmid DNA from *Enterococcus faecalis* to *Escherichia coli* in digestive tracts of gnotobiotic mice. *Antimicrobial Agents & Chemotherapy*, **36**, 502–504.

Doucet-Populaire, F., Trieu-Cuot, P., Dosbaa, I., Andremont, A. and Courvalin, P. (1991) Inducible transfer of conjugative transposon Tn1545 from *Enterococcus faecalis* to *Listeria monocytogenes* in the digestive tracts of gnotobiotic mice. *Antimicrobial Agents & Chemotherapy*, **35**, 185–187.

Dowson, C.G., Coffey, T.J., Kell, C. and Whiley, R.A. (1993) Evolution of penicillin resistance in *Streptococcus pneumoniae*; the role of *Streptococcus mitis* in the formation of a low affinity PBP2B in *S. pneumoniae*. *Molecular Microbiology*, **9**, 635–643.

Dowson, C.G., Coffey, T.J. and Spratt, B.G. (1994) Origin and molecular epidemiology of penicillin-binding-protein-mediated resistance to beta-lactam antibiotics. *Trends in Microbiology*, **2**, 361–366.

Dowson, C.G., Hutchison, A., Brannigan, J.A., George, R.C., Hansman, D., Linares, J., *et al.* (1989) Horizontal transfer of penicillin-binding protein genes in penicillin-resistant clinical isolates of *Streptococcus pneumoniae*. *Proceedings of the*

National Academy of Sciences of the United States of America, **86**, 8842–8846.

Fermer, C., Kristiansen, B.E., Skold, O. and Swedberg, G. (1995) Sulfonamide resistance in *Neisseria meningitidis* as defined by site-directed mutagenesis could have its origin in other species. *Journal of Bacteriology*, **177**, 4669–4675.

Flannagan, S.E. and Clewell, D.B. (1991) Conjugative transfer of Tn916 in *Enterococcus faecalis*: trans activation of homologous transposons. *Journal of Bacteriology*, **173**, 7136–7141.

Flannagan, S.E., Zitzow, L.A., Su, Y.A. and Clewell, D.B. (1994) Nucleotide sequence of the 18-kb conjugative transposon Tn916 from *Enterococcus faecalis*. *Plasmid*, **32**, 350–354.

Francia, M.V., de la Cruz, F. and Garcia Lobo, J.M. (1993) Secondary-sites for integration mediated by the Tn21 integrase. *Molecular Microbiology*, **10**, 823–828.

Franke, A.E. and Clewell, D.B. (1981) Evidence for a chromosome-borne resistance transposon (Tn916) in *Streptococcus faecalis* that is capable of "conjugal" transfer in the absence of a conjugative plasmid. *Journal of Bacteriology*, **145**, 494–502.

Frost, L.S., Ippen-Ihler, K. and Skurray, R.A. (1994) Analysis of the sequence and gene products of the transfer region of the F sex factor. *Microbiological Reviews*, **58**, 162–210.

Galimand, M., Gerbaud, G., Guibourdenche, M., Riou, J.Y. and Courvalin, P. (1998) High-level chloramphenicol resistance in *Neisseria meningitidis*. *New England Journal of Medicine*, **339**, 868–874.

Gawron-Burke, C. and Clewell, D.B. (1982) A transposon in *Streptococcus faecalis* with fertility properties. *Nature*, **300**, 281–284.

Germida, J.J. and Khachatourians, G.G. (1988) Transduction of *Escherichia coli* in soil. *Canadian Journal of Microbiology*, **34**, 190–193.

Goodgal, S.H. and Mitchell, M.A. (1990) Sequence and uptake specificity of cloned sonicated fragments of *Haemophilus influenzae* DNA. *Journal of Bacteriology*, **172**, 5924–5928.

Guiney, D.G. (1993) In *Bacterial conjugation* (Ed, Clewell, D.B.) Plenum Press, New York and London, pp. 75–103.

Hall, R.M. (1997) Mobile gene cassettes and integrons: moving antibiotic resistance genes in gram-negative bacteria. *Ciba Foundation Symposium*, **207**, 192–202; discussion 202–195.

Hall, R.M. and Stokes, H.W. (1993) Integrons: novel DNA elements which capture genes by site-specific recombination. *Genetica*, **90**, 115–132.

Hall, R.M. and Vockler, C. (1987) The region of the IncN plasmid R46 coding for resistance to beta-lactam antibiotics, streptomycin/spectinomycin and sulphonamides is closely related to antibiotic resistance segments found in IncW plasmids and in Tn21-like transposons. *Nucleic Acids Research*, **15**, 7491–7501.

Heinemann, J.A. and Sprague, G.F., Jr. (1989) Bacterial conjugative plasmids mobilize DNA transfer between bacteria and yeast. *Nature*, **340**, 205–209.

Hochhut, B., Jahreis, K., Lengeler, J.W. and Schmid, K. (1997) CTnscr94, a conjugative transposon found in enterobacteria. *Journal of Bacteriology*, **179**, 2097–2102.

Hochhut, B. and Waldor, M.K. (1999) Site-specific integration of the conjugal *Vibrio cholerae* SXT element into *prfC*. *Molecular Microbiology*, **32**, 99–110.

Hodgson, J.E., Curnock, S.P., Dyke, K.G., Morris, R., Sylvester, D.R. and Gross, M.S. (1994) Molecular characterization of the gene encoding high-level mupirocin resistance in *Staphylococcus aureus* J2870. *Antimicrobial Agents & Chemotherapy*, **38**, 1205–1208.

Hughes, V.M. and Datta, N. (1983) Conjugative plasmids in bacteria of the "pre-antibiotic" era. *Nature*, **302**, 725–726.

Hyder, S.L. and Streitfeld, M.M. (1978) Transfer of erythromycin resistance from clinically isolated lysogenic strains of *Streptococcus pyogenes* via their endogenous phage. *Journal of Infectious Diseases*, **138**, 281–286.

Inoue, M. and Mitsuhashi, S. (1975) A bacteriophage S1 derivative that transduces tetracycline resistance to *Staphylococcus aureus*. *Virology*, **68**, 544–546.

Ippen-Ihler, K. and Skurray, R.A. (1993) In *Bacterial conjugation*(Ed, Clewell, D.B.) Plenum Press, New York and London, pp. 23–52.

Jacob, A.E., Shapiro, J.A., Yamamoto, L., Smith, D.I., Cohen, S.N. and Berg, D.E. (1977) In *DNA insertion elements, plasmids and episomes*. (Eds, Bukhari, A.I., Shapiro, J.A. and Adhya, S.L.) Cold Spring Harbor Laboratory, pp. 601–670.

Jaworski, D.D. and Clewell, D.B. (1995) A functional origin of transfer (oriT) on the conjugative transposon Tn916. *Journal of Bacteriology*, **177**, 6644–6651.

Karaolis, D.K., Somara, S., Maneval, D.R., Jr., Johnson, J.A. and Kaper, J.B. (1999) A bacteriophage encoding a pathogenicity island, a type-IV pilus and a phage receptor in cholera bacteria. *Nature*, **399**, 375–379.

Kehoe, M.A., Kapur, V., Whatmore, A.M. and Musser, J.M. (1996) Horizontal gene transfer among group A streptococci: implications for pathogenesis and epidemiology. *Trends in Microbiology*, **4**, 436–443.

Kokjohn, T.A. (1989) In *Gene transfer in the environment* (Ed, Levy Miller, S.B. and R.V. Miller) Mac Graw-Hill, New York, pp. 73–97.

Krah, E.R.d. and Macrina, F.L. (1989) Genetic analysis of the conjugal transfer determinants encoded by the streptococcal broad-host-range plasmid pIP501. *Journal of Bacteriology*, **171**, 6005–6012.

Krah, E.R.d. and Macrina, F.L. (1991) Identification of a region that influences host range of the streptococcal conjugative plasmid pIP501. *Plasmid*, **25**, 64–69.

Lacey, R.W. (1975) Antibiotic resistance plasmids of *Staphylococcus aureus* and their clinical importance. *Bacteriological Reviews*, **39**, 1–32.

Laible, G., Spratt, B.G. and Hakenbeck, R. (1993) Interspecies recombinational events during the evolution of altered PBP 2x genes in penicillin-resistant clinical isolates of *Streptococcus pneumoniae*. *Molecular Microbiology*, **5**, 1993–2002.

Lederberg, J. and Tatum, E.L. (1946) Gene recombination in *Escherichia coli*. *Nature*, **158**, 558.

Lee, C.A. (1996) Pathogenicity islands and the evolution of bacterial pathogens. *Infectious Agents and Disease*, **5**, 1–7.

Li, L.Y., Shoemaker, N.B. and Salyers, A.A. (1995) Location and characteristics of the transfer region of a *Bacteroides* conjugative transposon and regulation of transfer genes. *Journal of Bacteriology*, **177**, 4992–4999.

Lorenz, M.G. and Wackernagel, W. (1994) Bacterial gene transfer by natural genetic transformation in the environment. *Microbiological Reviews*, **58**, 563–602.

Macrina, F.L. and Archer, G.L. (1993) In *Bacterial conjugation* (Ed, Clewell, D.B.) Plenum Press, New York and London, pp. 313–329.

Maqueda, M., Quirants, R., Martin, I., Galvez, A., Martinez-Bueno, M. and Valdivia, E. (1997) Chemical signals in gram-positive bacteria: the sex-pheromone system in *Enterococcus faecalis*. *Microbiologia*, **13**, 23–36.

Martin, C., Sibold, C. and Hakenbeck, R. (1992) Relatedness of penicillin-binding protein 1a genes from different clones of penicillin-resistant *Streptococcus pneumoniae* isolated in South Africa and Spain. *EMBO Journal*, **11**, 3831–3836.

Martin, C., Timm, J., Rauzier, J., Gomez-Lus, R., Davies, J. and Gicquel, B. (1990) Transposition of an antibiotic resistance element in mycobacteria. *Nature*, **345**, 739–743.

Martinez, E. and de la Cruz, F. (1990) Genetic elements involved in Tn*21* site-specific integration, a novel mechanism for the dissemination of antibiotic resistance genes. *The EMBO Journal*, **9**, 1275–1281.

Maskell, J.P., Sefton, A.M. and Hall, L.M. (1997) Mechanism of sulfonamide resistance in clinical isolates of *Streptococcus pneumoniae*. *Antimicrobial Agents & Chemotherapy*, **41**, 2121–2126.

Mazel, D. and Davies, J. (1999) Antibiotic resistance in microbes. *Cellular and Molecular Life Sciences*, **56**, 742–754.

Mazel, D., Dychinco, B., Webb, V.A. and Davies, J. (1998) A Distinctive Class of Integron In the *Vibrio Cholerae* Genome. *Science*, **280**, 605–608.

Mooi, F.R. and Bik, E.M. (1997) The evolution of epidemic *Vibrio cholerae* strains. *Trends in Microbiology*, **5**, 161–165.

Morrison, D.A. (1997) Streptococcal competence for genetic transformation: regulation by peptide pheromones. *Microbial Drug Resistance*, **3**, 27–37.

Morrison, W.D., Miller, R.V. and Sayler, G.S. (1978) Frequency of F116-mediated transduction of *Pseudomonas aeruginosa* in a freshwater environment. *Applied & Environmental Microbiology*, **36**, 724–730.

Mortier-Barriere, I., Humbert, O., Martin, B., Prudhomme, M. and Claverys, J.P. (1997) Control of recombination rate during transformation of *Streptococcus pneumoniae*: an overview. *Microbial Drug Resistance*, **3**, 233–242.

Morton, T.M., Eaton, D.M., Johnston, J.L. and Archer, G.L. (1993) DNA sequence and units of transcription of the conjugative transfer gene complex (trs) of *Staphylococcus aureus* plasmid pGO1. *Journal of Bacteriology*, **175**, 4436–4447.

Murphy, D.B. and Pembroke, J.T. (1995) Transfer of the IncJ plasmid R391 to recombination deficient *Escherichia coli* K12: evidence that R391 behaves as a conjugal transposon. *FEMS Microbiology Letters*, **134**, 153–158.

Nesvera, J., Hochmannova, J. and Patek, M. (1998) An integron of class 1 is present on the plasmid pCG4 from gram-positive bacterium *Corynebacterium glutamicum*. *FEMS Microbiology Letters*, **169**, 391–395.

Nikolich, M.P., Hong, G., Shoemaker, N.B. and Salyers, A.A. (1994a) Evidence for natural horizontal transfer of *tetQ* between bacteria that normally colonize humans and bacteria that normally colonize livestock. *Applied & Environmental Microbiology*, **60**, 3255–3260.

Nikolich, M.P., Shoemaker, N.B., Wang, G.R. and Salyers, A.A. (1994b) Characterization of a new type of *Bacteroides* conjugative transposon, Tcr Emr 7853. *Journal of Bacteriology*, **176**, 6606–6612.

Norgren, M. and Scott, J.R. (1991) The presence of conjugative transposon Tn916 in the recipient strain does not impede transfer of a second copy of the element. *Journal of Bacteriology*, **173**, 319–324.

Novick, R.P. and Morse, S.I. (1967) In vivo transmission of drug resistance factors between strains of *Staphylococcus aureus*. *Journal of Experimental Medicine*, **125**, 45–59.

Olsen, R.H. and Shipley, P. (1973) Host range and properties of the *Pseudomonas aeruginosa* R factor R1822. *Journal of Bacteriology*, **113**, 772–780.

Olsen, R.H., Siak, J.S. and Gray, R.H. (1974) Characteristics of PRD1, a plasmid-dependent broad host range DNA bacteriophage. *Journal of Virology*, **14**, 689–699.

Ouellette, M., Bissonnette, L. and Roy, P.H. (1987) Precise insertion of antibiotic resistance determinants into Tn21-like

transposons: nucleotide sequence of the OXA-1 beta-lactamase gene. *Proceedings of the National Academy of Sciences of the United States of America*, **84**, 7378–7382.

Pansegrau, W., Lanka, E., Barth, P.T., Figurski, D.H., Guiney, D.G., Haas, D., *et al.* (1994) Complete nucleotide sequence of Birmingham IncP alpha plasmids. Compilation and comparative analysis. *Journal of Molecular Biology*, **239**, 623–663.

Radstrom, P., Fermer, C., Kristiansen, B.E., Jenkins, A., Skold, O. and Swedberg, G. (1992) Transformational exchanges in the dihydropteroate synthase gene of *Neisseria meningitidis*: a novel mechanism for acquisition of sulfonamide resistance. *Journal of Bacteriology*, **174**, 6386–6393.

Ravatn, R., Studer, S., Springael, D., Zehnder, A.J. and van der Meer, J.R. (1998a) Chromosomal integration, tandem amplification, and deamplification in *Pseudomonas putida* F1 of a 105-kilobase genetic element containing the chlorocatechol degradative genes from *Pseudomonas* sp. Strain B13. *Journal of Bacteriology*, **180**, 4360–4369.

Ravatn, R., Studer, S., Zehnder, A.J. and van der Meer, J.R. (1998b) Int-B13, an unusual site-specific recombinase of the bacteriophage P4 integrase family, is responsible for chromosomal insertion of the 105-kilobase clc element of *Pseudomonas* sp. Strain B13. *Journal of Bacteriology*, **180**, 5505–5514.

Recchia, G.D. and Hall, R.M. (1997) Origins of the mobile gene cassettes found in integrons. *Trends in Microbiology*, **5**, 389–394.

Roberts, M.C. (1994) Epidemiology of tetracycline-resistance determinants. *Trends in Microbiology*, **2**, 353–357.

Roberts, M.C., Brown, B.A., Steingrube, V.A. and Wallace, R., Jr. (1990) Genetic basis of tetracycline resistance in *Moraxella (Branhamella) catarrhalis*. *Antimicrobial Agents & Chemotherapy*, **34**, 1816–1818.

Roberts, M.C., Pang, Y.J., Spencer, R.C., Winstanley, T.G., Brown, B.A. and Wallace, R.J.J. (1991) Tetracycline resistance in *Moraxella (Branhamella) catarrhalis*: demonstration of two clonal outbreaks by using pulsed-field gel electrophoresis. *Antimicrobial Agents & Chemotherapy*, **35**, 2453–2455.

Rowe-Magnus, D.A., Guerout, A.-M. and Mazel, D. (1999) Super-Integrons. *Research in Microbiology*, **150**, 641–651.

Rowe-Magnus, D.A. and Mazel, D. (1999) Resistance gene capture. *Current Opinion in Microbiology*, **2**, 483–488.

Salyers, A.A. and Shoemaker, N.B. (1997) Conjugative transposons. *Genetic Engineering (New York)*, **19**, 89–100.

Salyers, A.A. and Shoemaker, N.B., Stevens, A.M. and Li, L.Y. (1995) Conjugative transposons: an unusual and diverse set of integrated gene transfer elements. *Microbiological Reviews*, **59**, 579–590.

Sansonetti, P., Lafont, J.P., Jaffe-Brachet, A., Guillot, J.F. and Chaslus-Dancla, E. (1980) Parameters controlling interbacterial plasmid spreading in a gnotoxenic chicken gut system: influence of plasmid and bacterial mutations. *Antimicrobial Agents & Chemotherapy*, **17**, 327–333.

Sassanfar, M., Kranz, J.E., Gallant, P., Schimmel, P. and Shiba, K. (1996) A eubacterial *Mycobacterium tuberculosis* tRNA synthetase is eukaryote-like and resistant to a eubacterial-specific antisynthetase drug. *Biochemistry*, **35**, 9995–10003.

Schaberg, D.R., Power, G., Betzold, J. and Forbes, B.A. (1985) Conjugative R plasmids in antimicrobial resistance of *Staphylococcus aureus* causing nosocomial infections. *Journal of Infectious Diseases*, **152**, 43–49.

Schaberg, D.R. and Zervos, M.J. (1986) Intergeneric and interspecies gene exchange in gram-positive cocci. *Antimicrobial Agents & Chemotherapy*, **30**, 817–822.

Scott, J.R. (1992) Sex and the single circle: conjugative transposition. *Journal of Bacteriology*, **174**, 6005–6010.

Scott, J.R., Bringel, F., Marra, D., Van Alstine, G. and Rudy, C.K. (1994) Conjugative transposition of Tn916: preferred targets and evidence for conjugative transfer of a single strand and for a double-stranded circular intermediate. *Molecular Microbiology*, **11**, 1099–1108.

Scott, K.P. and Flint, H.J. (1995) Transfer of plasmids between strains of *Escherichia coli* under rumen conditions. *Journal of Applied Bacteriology*, **78**, 189–193.

Sharma, V.K., Johnston, J.L., Morton, T.M. and Archer, G.L. (1994) Transcriptional regulation by TrsN of conjugative transfer genes on staphylococcal plasmid pGO1. *Journal of Bacteriology*, **176**, 3445–3454.

Shoemaker, N.B., Barber, R.D. and Salyers, A.A. (1989) Cloning and characterization of a *Bacteroides* conjugal tetracycline-erythromycin resistance element by using a shuttle cosmid vector. *Journal of Bacteriology*, **171**, 1294–1302.

Shoemaker, N.B. and Salyers, A.A. (1990) A cryptic 65-kilobase-pair transposonlike element isolated from *Bacteroides uniformis* has homology with *Bacteroides* conjugal tetracycline resistance elements. *Journal of Bacteriology*, **172**, 1694–1702.

Shoemaker, N.B., Smith, M.D. and Guild, W.R. (1980) DNase-resistant transfer of chromosomal *cat* and *tet* insertions by filter mating in *Pneumococcus*. *Plasmid*, **3**, 80–87.

Showsh, S.A. and Andrews, R.E.J. (1992) Tetracycline enhances Tn916-mediated conjugal transfer. *Plasmid*, **28**, 213–224.

Spratt, B.G., Bowler, L.D., Zhang, Q.Y., Zhou, J. and Smith, J.M. (1992) Role of interspecies transfer of chromosomal genes in the evolution of penicillin resistance in pathogenic and commensal *Neisseria* species. *Journal of Molecular Evolution*, **34**, 115–125.

Spratt, B.G., Zhang, Q.Y., Jones, D.M., Hutchison, A., Brannigan, J.A. and Dowson, C.G. (1989) Recruitment of a penicillin-binding protein gene from *Neisseria flavescens* during the emergence of penicillin resistance in *Neisseria meningitidis*. *Proceedings of the National Academy of Sciences of the United States of America*, **86**, 8988–8992.

Stevens, A.M., Shoemaker, N.B., Li, L.Y. and Salyers, A.A. (1993) Tetracycline regulation of genes on *Bacteroides* conjugative transposons. *Journal of Bacteriology*, **175**, 6134–6141.

Stewart, G.J. (1989) In *Gene transfer in the environment* (Ed, Levy, S.B. and R.V. Miller) Mac Graw-Hill, New York, pp. 139–163.

Stotzky, G. (1989) In *Gene transfer in the environment* (Ed, Levy, S.B. and R.V. Miller) Mac Graw-Hill, New York, pp. 165–222.

Sundstrom, L. (1998) The potential of integrons and connected programmed rearrangements for mediating horizontal gene transfer. *APMIS. Supplementum*, **84**, 37–42.

Sundström, L., Radström, P., Swedberg, G. and Sköld, O. (1988) Site-specific recombination promotes linkage between trimethoprim- and sulfonamide resistance genes. Sequence characterization of *dhfrV* and *sull* and a recombination active locus of Tn21. *Molecular and General Genetics*, **213**, 191–201.

Swedberg, G., Ringertz, S. and Skold, O. (1998) Sulfonamide resistance in *Streptococcus pyogenes* is associated with differences in the amino acid sequence of its chromosomal dihydropteroate synthase. *Antimicrobial Agents & Chemotherapy*, **42**, 1062–1067.

Thomas, C.M. (1989) *Promiscuous plasmids of Gram-negative bacteria*, Academic Press, London.

Thorsted, P.B., Macartney, D.P., Akhtar, P., Haines, A.S., Ali, N., Davidson, P., *et al.* (1998) Complete sequence of the IncPbeta plasmid R751: implications for evolution and organisation of the IncP backbone. *Journal of Molecular Biology*, **282**, 969–990.

Trieu-Cuot, P., Carlier, C. and Courvalin, P. (1988) Conjugative plasmid transfer from *Enterococcus faecalis* to *Escherichia coli*. *Journal of Bacteriology*, **170**, 4388–4391.

Trieu-Cuot, P., Carlier, C., Martin, P. and Courvalin, P. (1987) Plasmid transfer by conjugation from *Escherichia coli* to gram-positive bacteria. *FEMS Microbiology Letters*, **48**, 289–294.

van Biesen, T. and Frost, L.S. (1994) The FinO protein of IncF plasmids binds FinP antisense RNA and its target, *traJ* mRNA, and promotes duplex formation. *Molecular Microbiology*, **14**, 427–436.

Waldor, M.K. and Mekalanos, J.J. (1996) Lysogenic conversion by a filamentous phage encoding cholera toxin. *Science*, **272**, 1910–1914.

Waldor, M.K., Tschape, H. and Mekalanos, J.J. (1996) A new type of conjugative transposon encodes resistance to sulfamethoxazole, trimethoprim, and streptomycin in *Vibrio cholerae* O139. *Journal of Bacteriology*, **178**, 4157–4165.

Wang, A. and Macrina, F.L. (1995a) Characterization of six linked open reading frames necessary for pIP501-mediated conjugation. *Plasmid*, **34**, 206–210.

Wang, A. and Macrina, F.L. (1995b) Streptococcal plasmid pIP501 has a functional oriT site. *Journal of Bacteriology*, **177**, 4199–4206.

Watanabe, T. (1963) Infective heredity of multiple resistance in bacteria. *Bacteriological Reviews*, **27**, 87–115.

Waters, V.L. (1999) Conjugative transfer in the dissemination of beta-lactam and aminoglycoside resistance. *Frontiers in Bioscience*, **4**, D433–456.

Waters, V.L., Hirata, K.H., Pansegrau, W., Lanka, E. and Guiney, D.G. (1991) Sequence identity in the nick regions of IncP plasmid transfer origins and T-DNA borders of *Agrobacterium* Ti plasmids [published erratum appears in *Proc. Natl. Acad. Sci. USA* 1991 Jul 15; 88(14): 6388]. *Proceedings of the National Academy of Sciences of the United States of America*, **88**, 1456–1460.

Wichels, A., Biel, S.S., Gelderblom, H.R., Brinkhoff, T., Muyzer, G. and Schutt, C. (1998) Bacteriophage diversity in the North Sea. *Applied & Environmental Microbiology*, **64**, 4128–4133.

Willi, K., Sandmeier, H., Kulik, E.M. and Meyer, J. (1997) Transduction of antibiotic resistance markers among *Actinobacillus actinomycetemcomitans* strains by temperate bacteriophages Aa phi 23. *Cellular & Molecular Life Sciences*, **53**, 904–910.

Zatyka, M. and Thomas, C.M. (1998) Control of genes for conjugative transfer of plasmids and other mobile elements. *FEMS Microbiology Reviews*, **21**, 291–319.

Zeph, L., Onaga Ma and Stotzky G. (1988) Transduction of *Escherichia coli* by bacteriophage P1 in soil. *Applied & Environmental Microbiology*, **54**, 1731–1737.

Zeph, L.R. and Stotzky, G. (1989) Use of a biotinylated DNA probe to detect bacteria transduced by bacteriophage P1 in soil. *Applied & Environmental Microbiology*, **55**, 661–665.

Zhang, Q.Y., Jones, D.M., Saez Nieto, J.A., Perez Trallero, E. and Spratt, B.G. (1990) Genetic diversity of penicillin-binding protein 2 genes of penicillin-resistant strains of *Neisseria meningitidis* revealed by fingerprinting of amplified DNA. *Antimicrobial Agents & Chemotherapy*, **34**, 1523–1528.

7. Mutator Bacteria and Resistance Development

Thomas A. Cebula,* Dan D. Levy and J. Eugene LeClerc
Division of Molecular Biological Research and Evaluation [HFS-235],
Center for Food Safety and Applied Nutrition, Food and Drug
Administration, 200 C Street S.W., Washington, DC 20204

PROLOGUE

Bacterial populations are constantly facing barriers that restrict their growth. Clinical and veterinary uses of antibiotics are but two barriers to growth, as host defenses and adverse environmental conditions usually keep microbial populations in check. How successful a microbe is at surviving the diverse challenges of ever-changing environments ultimately rests upon the relative diversity within the microbial population at large. Subpopulations of mutators, which exist in all bacterial populations, are a prolific source of such diversity. We discuss here how a particular set of mutators that are defective in methyl-directed mismatch repair provides multiple mechanisms for bacteria to evade the manifold barriers they face on the course to successful infection.

THE ANTIBIOTIC DILEMMA

Antibiotics are natural or synthetic products that are used to impede the growth of or kill bacteria. Unfortunately, the microbe has developed numerous ways to confound the action of many of the various antibiotics. Although the mechanisms are

many, they generally can be grouped into two broad classes with respect to the genetics engendering the resistance – those originating as a result of chromosomal mutation, and those that are acquired due to the inheritance of genes encoding information that allow bacteria to impede the action of the antibiotic. Such genes may reside on bacterial plasmids or they may be integrated into the chromosome as part of specialized elements like transposons and integrons. Consequently, mechanisms exist both for the vertical and horizontal dissemination of antibiotic resistance.

In recounting the steady increase of antibiotic resistance among human pathogens, both the scientific and lay press have made us keenly aware of this important public health issue. Antibiotic resistance among familiar (and not so familiar) bacterial pathogens has impacted negatively the outcomes of these infections and has contributed to our mounting health costs (Cohen, 1992). Emergence of multiple-drug resistant bacteria – *Mycobacterium tuberculosis* (Bloom and Murray, 1992; Greenwood, 1998; Watterson *et al.*, 1998; Segura *et al.*, 1998; Rattan *et al.*, 1998), *Enterococcus faecalis* and *Enterococcus faecium* (Brady *et al.*, 1998; Dennesen *et al.*, 1998; Murray, 1998; Huycke *et al.*, 1998), *Streptococcus pneumoniae*

* To whom reprint requests should be addressed.
Phone: (202) 205-4217   Fax: (202) 401-1105   E-mail: *tacebula@cfsan.fda.gov*

(Nuorti *et al.*, 1998; Musher, 1998; Tomasz, 1998), *Staphylococcus aureus* (Tomasz, 1998; de Sousa *et al.*, 1998; Paulsen *et al.*, 1998), *Shigella dysenteriae* (Cheasty *et al.*, 1998; Jahan and Hossain, 1997; Hughes and Tenover, 1997), and *Salmonella typhimurium* [Definitive Type 104] (Glynn *et al.*, 1998; Sandvang *et al.*, 1998) – has heightened concerns that our current antibiotic arsenal may soon be rendered useless. These concerns are sobering indeed when one considers that before the dawning of the antibiotic era, the leading cause of premature death in the United States was microbial disease. Despite the fact that there are countless antibiotics available, with various modes of action, a person is still at risk of dying because of infection by an antibiotic resistant microbe. Indeed, the very therapeutic agents that are being used to treat infections effectively today may be helping to create the more formidable microbe of tomorrow (Neu, 1992).

Many factors contribute to the complex problem of antibiotic resistance, obviating any simple solution to the problem. For example, some have argued that the emergence of antibiotic resistance is due to the over or inappropriate prescription of these agents in human and animal health arenas (Cohen, 1992; Bloom and Murray, 1992; Neu, 1992). Others have focused on use of antibiotics among agriculture animals as the primary cause for this increased incidence (Glynn *et al.*, 1998). Although such selective pressures clearly contribute, in part, to the antibiotic dilemma, they sound almost Lamarckian in abstraction. Darwinian logic reminds us that the evolutionary process will cull from the population at large, due to intrinsic diversity within that population, the fittest of the population proscribed by the nature of selective pressures applied.

What then are the selective pressures that are applied? Clearly, the antibiotic itself is one such pressure. However, it is not the sole selective pressure acting on bacteria in their feral settings for, if it were, the elusive linear tie of antibiotic use to emerging resistance would have long ago been made. Undoubtedly, selection pressure must be more globally defined to include the natural and anthropogenic challenges that the bacterium witnesses and endures each day to survive (Tenover and McGowan, 1996). For example, recent out-

breaks of *Escherichia coli* O157:H7 traced to apple cider and venison jerky serve as one such reminder that bacteria are adapting to and surviving conditions of low pH (Besser *et al.*, 1993) and high salt and temperature (Keene *et al.*, 1997) that formerly safeguarded the food supply from *E. coli* contamination. At the same time, the prevalence of antibiotic resistance among *E. coli* O157:H7 is increasing. Are these phenomena linked?

Unquestionably, bacterial pathogens are subjected to a wide variety of stresses that they must withstand be they in the food-processing milieu or within an animal or human host. For instance, they must face and survive the acid onslaughts of foods such as mayonnaise, apple cider, and fermented foods and the human body (i.e., stomach, small intestine, colon, and phagosomes) alike. So, too, must they endure temperature shifts whether external (cooking, sous vide processing, refrigeration) or associated with the host (intracellular vs. excreted pathogen). Likewise, the pathogen must weather osmolarity challenges of food processes such as brining, marinating, and jerky processing as well as the stomach of the host. Finally, the successful pathogen must persevere in the feast-and-famine environments that it experiences whether on the counters and equipment of food processing plants or in the macrophages of the host. The complex regulation and numbers of genes involved in each of these stress responses are just beginning to be appreciated. As the genetics for bacterial responses to stress and survival are often linked to the genetics of bacterial virulence (Miller *et al.*, 1989; Wright, 1990; Archer, 1990), regulation of these gene networks must be better understood. The recognition that $\sigma 38$, the gene product of *rpoS*, controls the expression of many stress response genes; regulates virulence; and modulates DNA repair processes underscore the convergence of studies involving mutagenesis, stress responses, and pathogenesis. There is little doubt that the influence that each of these pressures has in a particular niche ultimately hones the microbe that will emerge. Thus, in charting the emergence of antibiotic resistance, we must understand both the source of diversity within bacterial populations and the extent and kinds of selection pressures exerted in the feral landscape.

The role that antibiotics play in the emergence of antibiotic resistance has received ample attention in formulating strategies for prudent use of antibiotics (Cohen, 1992; Bloom and Murray, 1992; Glynn et al., 1998). Yet, neither the role that the intrinsic diversity of a bacterial population itself plays nor the effects that ancillary selective pressures have on the emergence of antibiotic resistance have been adequately addressed. Without doubt, microbes will continue to emerge and evolve rapidly. The bewildering numbers ($4-6 \times 10^{30}$ cells) are on their side (Whitman et al., 1998). Therefore, the question is not whether a new pathogen will evolve but rather when and where will it emerge. We must expect, as an unavoidable reality, that from time-to-time new pathogens will emerge – be it in feral, community, clinical, or agricultural settings – some will be truly new, some will be old with newly acquired characteristics. Recognizing this inevitability, public health initiatives must be aimed at an understanding of how these pathogens arise, propagate, and emerge. With such knowledge, we perhaps can contain an emerging pathogen long before it can gain a global distribution.

MUTATORS IN NATURAL POPULATIONS

Usually, when the genetic plasticity of a bacterium is discussed, we are reminded of the low and constant mutation rates that most bacteria possess (Drake, 1991). Recently, however, we (LeClerc et al., 1996) demonstrated that there was a high incidence of mutators among natural populations of Escherichia coli and Salmonella. That is, we found greater than 1% of the pathogenic E. coli strains and Salmonella outbreak strains that we examined mutated at frequencies at least 50-fold greater than did their laboratory-attenuated counterparts. This finding was unexpected, because mutator alleles are expected to be selected against in the population since most mutations are deleterious. At equilibrium, the frequency of deleterious alleles within a population is defined as μ/s, where μ is the mutation rate and s is the selection coefficient against the mutant (Hartl and Clark, 1989). For mutation rates around 10^{-6} per gene per replication (Drake, 1991), a mutator frequency above 1% gives an s value on the order of 10^{-4} or less, which is atypically weak and suggests there may be positive selection for these mutators. Recently, the frequency

of spontaneously arising mutator cells in unselected clones of non- mutator cells was actually determined to be $1-10 \times 10^{-6}$ (Mao et al., 1997; LeClerc et al., 1998). Such data have bolstered our contention that there are strong selection pressures operating to maintain a mutator phenotype in natural populations (LeClerc et al., 1996).

It is important to realize that in screening for the hypermutable phenotype, we selected strains that sported either rifampicin, spectinomycin, or nalidixic acid resistance. The β-subunit of RNA polymerase, the S5 protein of the 30S ribosomal subunit, and DNA gyrase, the targets for rifampicin, spectinomycin, and nalidixic acid, are encoded by rpoB (89 min), rpsE (71 min), and gyrA (48 min) in Salmonella. As these three resistances are due to different modes of action – affecting different chromosomal determinants and chromosomal locations – the hypermutable strains that we isolated were not only strong but general mutators as well. We reasoned that such mutators were at a selective advantage because they were a prolific source of genetic variation. In the changing and hostile environment of the host, the evolution of an invading mutator pathogen could be accelerated because of the increased numbers of variants sported (LeClerc et al., 1996; Cebula and LeClerc, 1997). Upon characterizing these mutators, however, we quickly suspected, as discussed below, that a hypermutable phenotype was only one part that these mutators played in evolution (LeClerc et al., 1996; Cebula and LeClerc, 1997). Most surprising was the fact that all of the mutators were found to be defective in methyl-directed mismatch repair (MMR), a repair system that helps govern ultimate fidelity of the DNA replication process by correcting base-pairing errors in newly synthesized DNA (see Modrich and Lahue, 1996). By complementation, eight of ten mutators mapped to the mutS region of the chromosome; another mutS mutator, this a multidrug-resistant S. typhimurium isolate, has recently been identified. A closer examination of these mutators revealed that deletions (ranging in size up to 17257 bp) of part or all of the mutS gene were responsible for the mutator phenotype in several of these strains. Northern and Western analysis showed that mutS message was neither expressed nor translated in these mutants (Li et al., manuscript in preparation).

Briefly, the MMR system takes advantage of the transient hemimethylation of newly replicated DNA. Dam methylase catalyzes the methylation of adenine residues within GATC sequences. However, as this is a post-replication process, particular GATC sites are left for a time unmethylated. MutS protein monitors newly replicated DNA and, upon encountering a mismatched base pair within the DNA, initiates MMR by binding to the mismatch. In the presence of ATP and MutL protein, it effects a conformational change in the DNA, bringing into close proximity both the mismatch and the nearest GATC hemimethylated sequence. The complex activates MutH protein, which incises the unmethylated strand – the nascent strand of DNA replication – at this GATC site. DNA helicase ll, the UvrD protein, unwinds the DNA and specific nucleases, depending whether the closest GATC site was $5'$ or $3'$ to the mismatch, digest the DNA from the point of incision past the mismatch, degrading as many as 1000 base pairs in the process. Repair is completed by DNA polymerase gap-filling function and the ultimate nick sealed by DNA ligase. The enzymology of MMR explains adequately why defects in any component of MMR lead to a mutator phenotype. What remained to be explained was, though there are tens of mutator loci scattered about the bacterial genome, why MMR mutators were only recovered in both our and other experiments (LeClerc et al., 1996; Cebula and LeClerc, 1997; Matic et al., 1997; Mao et al., 1997; Sniegowski et al., 1997; LeClerc et al., 1998; Rosenberg et al., 1998). As alluded to above, that explanation rests in the fact that MMR mutants are not only hypermutable but promiscuous as well.

MUTATORS AND PROMISCUITY

Biochemical studies have shown that MutS protein blocks strand transfer in response to mispairs in heteroduplex DNA, an inhibition enhanced by MutL protein (Worth et al., 1994). These studies explain the early observations that hex mutants of Streptococcus pneumoniae do not discriminate between the integration of high-efficiency and low-efficiency genetic markers during pneumococcal transformation (Claverys and Lacks, 1986). We now know from direct sequencing that the S. pneumoniae HexA protein is a homologue of MutS (Haber et al.; 1988; Priebe et al., 1988). In more recent experiments on conjugational and transductional crosses between E. coli K-12 and Salmonella typhimurium LT2, recipients mutant in mutS or mutL had recombination frequencies up to three orders of magnitude greater than mut^+ strains (Rayssiguier et al., 1989). Although a single base pair mismatch is sufficient to abort recombination if MMR is operating, sequences 30% disparate are able to recombine when MMR is defective (Matic et al., 1995). The effect of MMR gene defects is to relax recombination barriers between species that normally do not mate, leading Matic et al. (1996) to propose that this mismatch repair system is the major genetic barrier that controls speciation in bacteria. As MMR is the major gatekeeper that monitors and keeps foreign DNA sequences from being incorporated into the genome, an MMR mutant is poised to evolve and evolve quickly. It is able to mutate rapidly so that a favorable mutant can be more readily spawned and inherited vertically, or it can recombine with other bacterial species, thereby begetting the useful variation laterally. This curious pleiotropy would seem to be of especial benefit and have selective advantage for a pathogen in the adverse environments during host infection. The promiscuous phenotype affords the pathogen the opportunity to acquire new sequence elements (new traits) from similar or disparate genomes, extending the pathogen's potential for rapid emergence, in one genetic event. Indeed, it is reasoned that MMR mutators should play a critical role in adaptive evolution, especially under adverse conditions (LeClerc et al., 1996; Cebula and LeClerc, 1997; Arjan et al., 1999).

SELECTION FROM THE MUTATOR SUBPOPULATION

The question, then, is why the benefits of such mutators have not been appreciated before now. Many thought, because of the high mutation rates of MMR-defective cells, deleterious mutations should be spawned much more frequently, thus placing mutators at a long-term disadvantage within a stable environment. The error in logic, however, is that bacteria do not likely experience a stable environment for very long anywhere except in the test tube. This solecism obscured the powerful advantages that MMR mutants would have within a changing and hostile environment. In situations where strong selection is at work – e.g., when a pathogen is trying to escape both the immune surveillance of the host and the on-slaughts of antibiotic therapy – and bacteria are following a "survive or die" dictum, there clearly are advantages proffered by an MMR mutant.

The demonstration that mutators allow the bacterium to counter changes in the environment in a clinically meaningful way has been provided by mouse infection experiments using isogenic wild type, *mutS*, and *mutL* strains of *Salmonella typhimurium* (Bierman *et al.*, 1998). The mutator strains were capable of developing antibiotic resistance in greater frequency and shorter time than the wild-type strain, both after antibiotic challenge following infection and during serial passage of the isogenic strains. Since the infecting pool of genetic variants, the size of which only microorganisms can provide, likely contained beneficial, neutral, and deleterious mutants, these results showed that strong selection culls the pathogens carrying the phenotype required for successful infection in particular environments. As mutators can generate beneficial variants more often and more promptly, selective enrichment of the rare but useful mutant (or recombinant) spawned by the mutator is sufficient to explain the frequent occurrence of mutators among pathogens.

Direct selection for useful phenotypes has been used to enrich for mutators in laboratory cultures of *Escherichia coli* K-12 and *Salmonella typhimurium* LT2, experiments demonstrating that sponta-neously arising subpopulations of mutators reside in normal populations of these cells. Enrichment for mutators involved selection for a phenotype required for growth of a mutant population in circumstances where reversion of mutant cells is known to be stimulated in mismatch repair-deficient strains. In the *E. coli* case, Mao *et al.* (1997) used a Lac⁻ strain that reverts to Lac⁺ by the addition of a G in a G-G-G-G-G-G sequence, a change that is enhanced in a *mutH* strain roughly 500-fold over the *mut⁺* background. Selection for the Lac⁺ phenotype resulted in an increase in the proportion of mutators to 1 per 200 cells in the

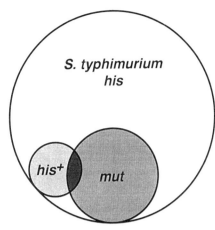

Figure 1. A subpopulation of mutator cells (*mut*) in a population of *S. typhimurium* His⁻ auxotrophs (*his*) can be enriched and detected by selection for the His⁺ prototrophs (*his⁺*) that the mutators helped to spawn.

selected population, from a level estimated to be less than 1 per 100,000 cells (1×10^{-5}) in the starting culture. Mutagenesis of the cells prior to selection, followed by two or more successive selections (e.g., Lac⁺ and resistance to antibiotics) resulted in populations that were nearly all mutator (Mao *et al.*, 1997). These results established that selection for a mutant phenotype in a population can greatly increase the proportion of mutators in the selected population.

We (LeClerc *et al.*, 1998) extended this analysis to detect the mutator subpopulation in *S. typhimurium* and to measure the frequency of mutators in unselected populations of cells. The logic of the experiment is depicted in Figure 1. Actual analysis involved screening for mutators among revertants of *S. typhimurium* histidine auxotrophs selected for the His⁺ phenotype. The increases in spontaneous reversion of the histidine mutations, an ochre mutation (CAA→TAA) in *hisG428* or a +1 frameshift (CCCC→CCCCC) in *hisC3076*, were measured in isogenic strains carrying mismatch repair-defective *mutH*, *mutL*, *mutS*, or *uvrD* alleles; they averaged 372-fold (*hisG428*) and 191-fold (*hisC3076*) greater than their mismatch repair-proficient counterparts. Screening for the mutator phenotype in nearly 12,000 revertants of the repair-proficient *his* strains yielded five isolates carrying defective *mut* alleles, either in *mutL* (4) or in *mutH* (1). Knowing the frequency of mutators in the selected population, F_S (5 per 12,000 *his⁺* revertants, or 0.04%), we then calculated the frequency of mutators in the unselected population, F_U, by:

$$F_U = F_S/E.F.$$

where E.F. is the mutator enhancement factor, that is, the fold increase in His$^+$ revertants that we measured in the isogenic mutator strains. The calculation gave a frequency of $1-4 \times 10^{-6}$, meaning that any population of reasonable size contains subpopulations of spontaneously arising mutator cells possessing increased abilities to acquire new phenotypes. The nature of selective forces then likely governs whether the proportion of mutators increases, or overtakes the population, if the new phenotype increases fitness in the population.

Long-term chemostat experiments on competition between characterized mutator and non-mutator strains showed that mutators, or the adaptive mutations they spawn, can indeed increase fitness in the population (Nestman and Hill, 1973; Cox and Gibson, 1974; Chao and Cox, 1983). Experiments by Chao and Cox (1983) showed, however, that the growth advantage for mutators was frequency-dependent. Mutators emerged, these authors concluded, only when a rare beneficial mutation was more likely to arise in a mutator cell than in one of the vast majority of non-mutators in the culture. Seeming to conflict with these studies, when Sniegowski et al. (1997) monitored the mutation frequencies of cultures started as a single (non-mutator) cell and propagated for 10,000 generations, they found three of twelve such cultures had been overtaken by mutators. Bringing these results into accord with those of Chao and Cox may be instructive on the nature of selective forces that carry the mutator to prominence in the population. Why is it that a high threshold of mutator cells was required to ensure that a rare adaptive mutation raised the mutator to high frequency under the favorable conditions of continual growth in chemostats? Yet, single mutators reached high frequency during the 10,000 generations of growth in limited glucose entailing continual adaptation to stationary phase. It may be that the latter conditions impose stronger selection – a harsher and varying environment – that may have favored adaptation of a unique genetic variant spawned in the mutator subpopulation, allowing it to survive, propagate, and successfully invade the population of non-mutator

cells. In conditions of yet stronger selection, such as an antibiotic environment, mutators may quickly become the majority population.

MMR MUTATORS AND THE EMERGENCE OF ANTIBIOTIC RESISTANCE

The role that mutators play in the emergence of antibiotic resistance has already served as a laboratory paradigm for adaptive evolution and may intimate the types of mechanisms that operate in nature. Early experiments indeed showed that when antibiotics were used to select resistance in bacterial cultures, the resistant populations were enriched for mutator cells (Siegel and Bryson, 1963; Helling, 1968). We would now say that strong selection for genetic variants had raised the frequency of the mutator allele as a requisite hitchhiker in the adaptive population it spawned. When antibiotics were used as the selective force in culture experiments to cull a bacterium carrying the required resistance from the population at large (Mao et al., 1997), or in in vivo infection experiments where resistance to antibiotic challenge was required for successful infection (Bierman et al., 1998), strong selection acted on the genetic variant that circumvented the antibiotic treatment. In each case, the frequency of the hitch-hiking mutator allele was also raised to prominence in the population.

The selection landscape – antibiotic use and the other attendant selection pressures imposed by a bacterium's stressful life style – dictates the emergence of resistant pathogens by selection from a pool of genetic variants generated by mutation and recombination. MMR mutators can influence the development, maintenance, and emergence of anti-biotic resistance in several ways. Chromosomal mutation is the most obvious route to resistance, and the 10^2- to 10^3-fold increases in mutant frequency for resistance genes seen in natural mutator strains in culture (LeClerc et al., 1996) indicates the magnitude of this hypermutability effect. The promiscuity of MMR-defective strains offers another route, permitting more ready exchange of resistance factors. Moreover, divergent sequences from different species can be assembled efficiently in MMR mutators, by homeologous recombination, to establish the reservoir of resistance sequences that are then disseminated. These sequences can be assembled either in the genome of the mutator or on the transmissible elements – plasmids, phages, or integrons – that reside there. Quite apart from generating the original antibiotic resistance, the mutator phenotype may play a substantive role in the emergence of antibiotic resistance. For example, although rapid adaptation to an antibiotic environment may serve as one driving force for maintaining mutators in natural populations, the antibiotic resistant mutator has the capacity for rapidly adapting as well. This may be seen in the development of multiple drug resistances, the expression of resistance at higher levels, and changes in the catalytic activities of resistance determinants. Finally, the MMR-defective phenotype may play a seminal role in maintaining an antibiotic resistant population within an environment devoid of antibiotic selection. From a Darwinian perspective, any antibiotic-resistant bacterium is, at first, less fit relative to its antibiotic-sensitive counterparts, engendering a greater biological burden due to either the mutation (Schrag and Perrot, 1996) or extra gene(s) it bears (Lenski *et al.*, 1994 and refs. therein). This is why one might expect to see newly resistant bacteria decline in the population when the antibiotic challenge is removed. Given time for adaptation, however, compensatory mutations arise that restore fitness without a significant loss in the level of antibiotic resistance (Schrag and Perrot, 1996). It is important to note that the effect of mutators on generating cells carrying these types of multiple adaptive mutations is exponentially greater than that of repair- proficient strains for, in the latter, each of the individual mutations is likely to be distributed into separate cells (Moxon and Thaler, 1997). Since compensatory mutations may allow fully resistant strains to compete successfully with sensitive strains in an antibiotic-free environment (Schrag and Perrot, 1996; Bjorkman *et al.*, 1998), the results are a forewarning that reducing antibiotic will not always lead to a decreased incidence of resistant bacteria. Moreover, if we are to tackle the antibiotic resistance problem intelligently, we must understand the selection landscape totally. For example, it is important to note that the compensatory mutations derived from chemostat experiments (Schrag and Perrot, 1996) are not the same as those derived from murine experiments (Bjorkman *et al.*, 1998), even though the antibiotic used to exert the pressure was the same.

MUTATORS IN CLINICAL ASSAYS FOR ANTIBIOTIC RESISTANCE

As we initially identified mutator strains by screening for the hypermutability phenotype on antibiotic plates (LeClerc *et al.*, 1996), this led us to question whether mutators might impact the routine assays that are used to assess antibiotic efficacy in clinical settings. The two most common antibiotic assays are the disc diffusion method and microtiter dilution MIC (minimum inhibitory concentration) assays. As anticipated, we found that mutator strains provoked anomalies in both of these assays.

In the disc diffusion method, an exponential or early stationary phase culture of the strain being tested is diluted to a standard turbidity and spread on a plate of rich agar. A small disk, impregnated with antibiotic, is placed on the plate. As the antibiotic diffuses out of the disk, it inhibits bacterial growth, resulting in a concentric zone of growth inhibition. The diameter of the zone is taken as a quantitative measure of the sensitivity of the strain to the antibiotic. Figure 2 (left) depicts the expected result of such an assay, shown in the cases of the antibiotics rifampin and nalidixic acid. No growth is visible in

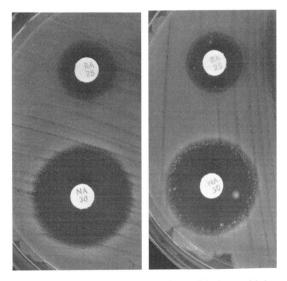

Figure 2. Disc diffusion assay for antibiotic sensitivity of two *Salmonella enteritidis* strains. (Left) A non-mutator strain shows clear zones with defined borders for inhibition of growth by rifampin (top) and nalidixic acid (bottom). (Right) A mutator strain generates both large and small satellite colonies in the zones of inhibition, especially near the border of the zones.

the zones of inhibition and the borders of the zones are remarkably sharp and even, making measurement of the diameter straightforward.

Using the disc diffusion method, we tested eight paired strains of *E. coli* and *Salmonella* pathogens against a panel of antibiotics. One member of each pair was a mutator. When mutators were tested, satellite colonies were sometimes observed in an otherwise clear zone of inhibition. Figure 2 (right) shows the results when one such mutator was tested. Instead of the smooth, distinct zone of inhibition observed for the non-mutator strain, individual colonies appeared in the zones of inhibition when the mutator strain was analyzed. The satellite colonies were due to mutants, sported by the mutator phenotype, that were resistant to the antibiotics. The patterns displayed by mutators in this assay were not always predictable. Sometimes, satellite colonies formed directly adjacent to the sharp border of the zone. Since the satellites are of fairly even diameter and distribution, the effect of this pattern was to decrease the diameter of the zone. As such, this would underestimate the potency of the antibiotic under test. At other times, mutators gave rise to satellite colonies that were distributed away from the border of the zone. The NCCLS standard (National Committee for Clinical Laboratory Standards, 1997) is that the existence of such colonies clearly denotes resistance. However, this statement is contained in notes appended to the standard in "Q and A" format. Informal discussion with clinical microbiologists suggests that in practice these guidelines are not always followed.

The second major way of determining the effectiveness of an antibiotic is by determining the MIC of the antibiotic in liquid cultures. Broth dilution assays consist of a row of tubes or microtiter plate wells containing bacteria incubated in media supplemented with serial dilutions of antibiotic. The lowest concentration at which bacteria produce a contiguous set of non-turbid wells is defined as the MIC of the antibiotic. We found that carrying out broth dilution assays using characterized mutator bacteria produced an anomaly in which resistant bacteria are propagated at higher concentrations, but, as shown in Figure 3, in an apparent stochastic fashion. The appearance of "extra" turbid wells has been observed on occasion in antibiotic assays performed in clinical settings, and has been described as the "skipped

rifampin concentration

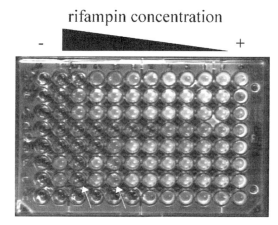

Figure 3. The MIC for rifampin as determined for eight *E. coli* and *Salmonella* strains. Wells in each row contain a series of two-fold dilutions of rifampin, flanked by wells with no cells (−) or no antibiotic (+). Of the four mutator strains tested (bottom four rows), two strains contained skipped wells (indicated by arrows).

well phenomenon" (Woods and Washington, 1995). Although skipped wells are observed frequently enough to be described in clinical manuals (Woods and Washington, 1995), NCCLS did not include this topic in the national standard (presumably because the incidence at which "lack of performance" occurred was much less frequent than that of the disk assay). However, it is not unusual to find treatment failure rates as high as 10% in clinical cases even though the strains have been shown to be sensitive to the prescribed antibiotic *in vitro* (Murray, 1991). Our estimate that 1–5% of natural bacterial isolates are defective in MMR is consistent with the thesis that a significant – if not predominant – cause of clinical treatment failures is due to mutators among these pathogenic isolates.

It is not surprising that the skipped well phenomenon has gone, largely, unnoticed and unappreciated. By direct reconstruction assays, we have demonstrated that, because so relatively few determinant dilutions are used in any one assay, the likelihood of finding a skipped well is low unless the probability of inoculating each well with a resistant mutant is close to 0.5. As expected, such reconstructions follow a simplified Poisson distribution. Although they may escape notice in routine analysis for the reasons we offer, pathogenic mutators must not be overlooked clinically, for they may seriously impact the course and outcome of antibiotic therapy.

CONCLUSION

Steady increases in antibiotic resistance among human pathogens and the emergence of multidrug resistant bacteria threaten the efficacy of antibiotics presently used to treat infections. While the overuse of antibiotics is acknowledged to be responsible, in part, for the current antibiotic resistance problem, little attention has been paid to ancillary pressures in the environment that may favor selection of resistant organisms, nor to the intrinsic genetic diversity in natural bacterial populations that is the very origin of antimicrobial resistances. Bacterial mutators, mutants exhibiting high mutation rates, are a prolific source of such diversity. We have reviewed evidence that particular mutators, those carrying defects in the methyl-directed mismatch repair, persist in natural populations and may play a prominent role in the emergence of new resistances and the evolution of new pathogens. These mutators have special benefit for pathogenesis, for they offer the bacterial pathogen the means for acquiring antimicrobial resistances both by vertical transmission (i.e., increased mutation) and by horizontal transfer of resistance elements.

REFERENCES

Archer, D.L. (1990) The need for flexibility in HACCP. *Food Technology* 44, 174–178.

Arjan, J.A., Visser, M., Zeyl, C.W., Gerrish, P.J., Blanchard, J.L. and Lenski, R.E. (1999) Diminishing returns from mutation supply rate in asexual populations. *Science* 283, 404–406.

Besser, R.E., Lett, S.M., Weber, J.T., Doyle, M.P., Barrett, T.J., Wells, J.G., et al. (1993) An outbreak of diarrhea and hemolytic uremic syndrome from *Escherichia coli* O157:H7 in fresh-pressed apple cider. *JAMA* 269, 2217–2220.

Bierman, J.C., Thaler, D.S. and Parker, C.N. (1998) Mutator strains of *Salmonella typhimurium* are not more virulent than wild type strains. *ASM 98th General Meeting Abstracts* (Atlanta, May 1998) p. 75.

Bjorkman, J., Hughes, D. and Andersson, D.I. (1998) Virulence of antibiotic-resistant *Salmonella typhimurium*. *Proc. Natl. Acad. Sci. USA* 95, 3949–3953.

Bloom, B.R. and Murray, C.J.L. (1992) Tuberculosis: commentary on a reemergent killer. *Science* 257, 1055–1064.

Brady, J.P., Snyder, J.W. and Hasbargen, J.A. (1998) Vancomycin-resistant enterococcus in end-stage renal disease. *Am. J. Kidney Dis.* 32, 415–418.

Cebula, T.A. and LeClerc, J.E. (1997) Hypermutability and homeologous recombination: ingredients for rapid evolution. *Bull. Inst. Pasteur* 95, 97–106.

Cebula, T.A. and LeClerc, J.E. (1997) To be a mutator, or how pathogenic and commensal bacteria can evolve rapidly (Discussion). *Trends Microbiol.* 5, 428–429.

Chao, L. and Cox, E.C. (1983) Competition between high and low mutating strains of *Escherichia coli*. *Evolution* 37, 125–134.

Cheasty, T., Skinner, J.A., Rowe, B. and Threlfall, E.J. (1998) Increasing incidence of antibiotic resistance in shigellas from humans in England and Wales: recommendations for therapy. *Microb. Drug Res.* 4, 57–60.

Claverys, J.P. and Lacks, S.A. (1986) Heteroduplex deoxyribonucleic acid base mismatch repair in bacteria. *Microbiol. Rev.* 50, 133–165.

Cohen, M.L. (1992) Epidemiology of drug resistance: implications for a post-antimicrobial era. *Science* 257, 1050–1055.

Cox, E.C. and Gibson, T.C. (1974) Selection for high mutation rates in chemostats. *Genetics* 77, 169–184.

de Sousa, M.A., Sanches, I.S., Ferro, M.L., Vaz, M.J., Saraiva, Z., Tendeiro, T., et al. (1998) Intercontinental spread of a multidrug-resistant methicillin-resistant *Staphylococcus aureus* clone. *J. Clin. Microbiol.* 36, 2590–2596.

Dennesen, P.J., Bonten, M.J. and Weinstein, R.A. (1998) Multiresistant bacteria as a hospital epidemic problem. *Ann. Med.* 30, 176–185.

Drake, J.W. (1991) A constant rate of spontaneous mutation in DNA-based microbes. *Proc. Natl. Acad. Sci. USA* 88, 7160–7164.

Glynn, M.K., Bopp, C., Dewitt, W., Dabney, P., Mokhtar, M. and Angulo, F.J. (1998) Emergence of multidrug-resistant *Salmonella enterica* serotype typhimurium DT104 infections in the United States. *N. Engl. J. Med.* 338, 1333–1338.

Greenwood, D. (1998) Resistance to antimicrobial agents: a personal view. *J. Med. Microbiol.* 47, 751–755.

Haber, L.T., Pang, P.P., Sobell, D.I., Mankovich, J.A. and Walker, G.C. (1988) Nucleotide sequence of the *Salmonella typhimurium mutS* gene required for mismatch repair: homology of MutS and HexA of *Streptococcus pneumoniae*. *J. Bacteriol.* 170, 197–202.

Hartl, D.L. and Clark, A.G. (1989) *Principles of population genetics*. Sunderland, MA, Sinauer, pp.198–201.

Helling, R.B. (1968) Selection of a mutant of *Escherichia coli* which has high mutation rates. *J. Bacteriol.* 96, 975–980.

Hughes, J.M. and Tenover, F.C. (1997) Approaches to limiting emergence of antimicrobial resistance in bacteria in human populations. *Clin. Infect. Dis.* 24 (Suppl 1), S131–135.

Huycke, M.M., Sahm, D.F. and Gilmore, M.S. (1998) Multiple-drug resistant enterococci: the nature of the problem and an agenda for the future. *Emerg. Infect. Dis.* 4, 239–249.

Jahan, Y. and Hossain, A. (1997) Multiple drug-resistant *Shigella dysenteriae* type 1 in Rajbari district, Bangladesh. *J. Diarrhoeal Dis. Res.* 15, 17–20.

Keene, W.E., Sazie, E., Kok, J., Rice, D.H., Hancock, D.D., Balan, V.K., et al. (1997) An outbreak of *Escherichia coli* O157:H7 infections traced to jerky made from deer meat. *JAMA* 277, 1229–1231.

LeClerc, J.E. and Cebula, T.A. (1997) Highly variable mutation rates in commensal and pathogenic *Escherichia coli* (Discussion). *Science* 227, 1834.

LeClerc, J.E., Li, B., Payne, W.L. and Cebula, T.A. (1996) High mutation frequencies among *Escherichia coli* and *Salmonella* pathogens. *Science* 274, 1208–1211.

LeClerc, J.E., Payne, W.L., Kupchella, E. and Cebula, T.A. (1998) Detection of mutator subpopulations in *Salmonella typhimurium* LT2 by reversion of *his* alleles. *Mutation Res.* **400**, 89–97.

Lenski, R.E., Simpson, S.C. and Nguyen, T.T. (1994) Genetic analysis of a plasmid-encoded, host genotype-specific enhancement of bacterial fitness. *J. Bacteriol.* **176**, 3140–3147.

Mao, E.F., Lane, L., Lee, J. and Miller, J.H. (1997) Proliferation of mutators in a cell population. *J. Bacteriol.* **179**, 417–422.

Matic, I., Radman, M., Taddei, F., Picard, B., Doit, C., Bingen, E., *et al.* (1997) Highly variable mutation rates in commensal and pathogenic *Escherichia coli. Science* **227**, 1833.

Matic, I., Rayssiguier, C. and Radman, M. (1995) Interspecies gene exchange in bacteria: the role of SOS and mismatch repair systems in evolution of species. *Cell* **80**, 507–515.

Matic, I., Taddei, F. and Radman, M. (1996) Genetic barriers among bacteria. *Trends Microbiol.* **4**, 69–72.

Miller, J.F., Mekalanos, J.J. and Falkow, S. (1989) Coordinate regulation and sensory transduction in the control of bacterial virulence. *Science* **243**, 916–922.

Modrich, P. and Lahue, R. (1996) Mismatch repair in replication fidelity, genetic recombination, and cancer biology. *Annu. Rev. Biochem.* **65**, 101–133.

Moxon, E.R. and Thaler, D.S. (1997) Microbial genetics. The tinkerer's evolving tool-box. *Nature* **387**, 659–662.

Murray, B.E. (1991) New aspects of antimicrobial resistance and the resulting therapeutic dilemmas. *J. Infect. Dis.* **163**, 1184–1194.

Murray, B.E. (1998) Diversity among multidrug-resistant enterococci. *Emerg. Infect. Dis.* **4**, 37–47.

Musher, D.M. (1998) Pneumococcal outbreaks in nursing homes. *N. Engl. J. Med.* **338**, 1915–1916.

National Committee for Clinical Laboratory Standards. (1997) *Performance standards for antimicrobial disk susceptibility tests – Sixth edition.* Approved Standard M2-A6. J.H. Jorgensen, 17–1.

Nestman, E.R. and Hill, R.F. (1973) Population changes in continuously growing mutator cultures of *Escherichia coli. Genetics* [suppl] **73**, 41–44.

Neu, H.C. (1992) The crisis in antibiotic resistance. *Science* **257**, 1064–1073.

Nuorti, J.P., Butler, J.C., Crutcher, J.M., Guevara, R., Welch, D., Holder, P., *et al.* (1998) An outbreak of multidrug-resistant pneumococcal pneumonia and bacteremia among unvaccinated nursing home residents. *N. Engl. J. Med.* **338**, 1861–1868.

Paulsen, I.T., Brown, M.H. and Skurray, R.A. (1998) Characterization of the earliest known *Staphylococcus aureus* plasmid encoding a multidrug efflux system. *J. Bacteriol.* **180**, 3477–3479.

Priebe, S.D., Hadi, S.M., Greenberg, B. and Lacks, S.A. (1988) Nucleotide sequence of the *hexA* gene for DNA mismatch repair in *Streptococcus pneumoniae* and homology of *hexA* to *mutS* of *Escherichia coli* and *Salmonella typhimurium. J. Bacteriol.* **170**, 190–196.

Rattan, A., Kalia, A. and Ahmad, N. (1998) Multidrug-resistant *Mycobacterium tuberculosis*: molecular perspectives. *Emerg. Infect. Dis.* **4**, 195–209.

Rayssiguier, C., Thaler, D.S. and Radman, M. (1989) The barrier to recombination between *Escherichia coli* and *Salmonella typhimurium* is disrupted in mismatch-repair mutants. *Nature* **342**, 396–401.

Rosenberg, S.M., Thulin, C. and Harris, R.S. (1998) Transient and heritable mutators in adaptive evolution in the lab and in nature. *Genetics* **148**, 1559–1566.

Sandvang, D., Aarestrup, F.M. and Jensen, L.B. (1998) Characterisation of integrons and antibiotic resistance genes in Danish multiresistant *Salmonella enterica* Typhimurium DT104. *FEMS Microbiol. Lett.* **160**, 37–41.

Schrag, S.J. and Perrot, V. (1996) Reducing antibiotic resistance. *Nature* **381**, 120–121.

Segura, C., Salvado, M., Collado, I., Chaves, J. and Coira, A. (1998) Contribution of beta-lactamases to beta-lactam susceptibilities of susceptible and multidrug-resistant *Mycobacterium tuberculosis* clinical isolates. *Antimicrob. Agents Chemother.* **42**, 1524–1526.

Siegel, E.C. and Bryson, V. (1963) Selection of resistant strains of *Escherichia coli* by antibiotics and antibacterial agents: role of normal and mutator strains. *Antimicrob. Agents Chemother.* **1963**, 629–634.

Sniegowski, P.D., Gerrish, P.J. and Lenski, R.E. (1997) Evolution of high mutation rates in experimental populations of *E. coli. Nature* **387**, 703–705.

Taddei, F., Matic, I., Godelle, B. and Radman, M. (1997) To be a mutator, or how pathogenic and commensal bacteria can evolve rapidly. *Trends Microbiol.* **5**, 427–428.

Tenover, F.C. and McGowan, J.E. (1996) Reasons for the emergence of antibiotic resistance. *Amer. J. Med. Sci.* **311**, 9–16.

Tomasz, A. (1998) Accelerated evolution: emergence of multidrug resistant gram-positive bacterial pathogens in the 1990's. *Neth. J. Med.* **52**, 219–227.

Watterson, S.A., Wilson, S.M., Yates, M.D. and Drobniewski, F.A. (1998) Comparison of three molecular assays for rapid detection of rifampin resistance in *Mycobacterium tuberculosis. J. Clin. Microbiol.* **36**, 1969–1973.

Whitman, W.B., Coleman, D.C. and Wiebe, W.J. (1998) Prokaryotes: The unseen majority. *Proc. Natl. Acad. Sci. USA* **95**, 6578–6583.

Worth, L., Clark, S., Radman, M. and Modrich, P. (1994) Mismatch repair proteins MutS and MutL inhibit RecA-catalyzed strand transfer between diverged DNAs. *Proc. Natl. Acad. Sci. USA* **91**, 3238–3241.

Woods, G.L. and Washington, J.A. (1995) Antibacterial susceptibility tests: Dilution and disc diffusion methods. In P.R. Murray, E.J. Baron, M.A. Pfaller, F.C. Tenover and R.H. Stokes (eds.), *Manual of Cinical Microbiology*, ASM, Washington, D.C., pp. 1327–1341.

Wright, K. (1990) Bad news bacteria. *Science* **249**, 22–24.

8. Low-level Antibiotic Resistance

Fernando Baquero

Department of Microbiology, Ramón y Cajal Hospital, National Institute of Health (INSALUD), 28034 Madrid, Spain

INTRODUCTION

The process of selection of antibiotic resistance in microorganisms essentially belongs to the category of directional selection. Antibiotic-mediated selection acts as a progressive force that favors genetic variability, leading to the continuous restoration of the fit of bacterial populations confronted with the antibiotic challenge. In most cases, an antibiotic exerts its selective pressure in a gradual way, both at the human individual level (gradients of antibiotic concentrations are formed in the body), and at the human populational level (gradients of antibiotic consumption in the society). Effective directional selection of resistant bacterial populations will occur in these gradually changing antibiotic fields. Gradual environmental change leads to gradual evolutionary change. Under gradual antibiotic pressure, and if the cost of selection is tolerable, low-level resistance may evolve to high-level resistance. As proposed by F. Jacob, adaptation typically progresses through small changes involving a local search in the space of possibilities (Jacob, 1977). In an antibiotic-polluted world, the various bacterial adaptive responses form a field of resistance genotypes, and to each one of them a fitness value can be assigned. That fitness essentially reflects the efficacy (adequacy) of the mechanism of resistance. The distributions of fitness values over the space of resistance genotypes constitute a resistance fitness landscape (Kauffman, 1993). Little hills or hillsides may correspond to low-level mechanisms of resistance; big mountains, to very efficacious high-level mechanisms. Adaptive evolution is frequently a hill-climbing process; to reach a high peak, successive slopes should be frequently climbed. From the point of view of the surveillance of the process of antibiotic resistance, the emergence of low-level antibiotic resistance should be considered as a signal of alarm, a hallmark of a possible evolutionary trend leading to the heights of therapy-resistant bacterial organisms.

LOW-LEVEL ANTIBIOTIC RESISTANCE: THE CLINICAL PERSPECTIVE

The term "antibiotic resistance" is certainly equivocal. The expression was coined by clinical microbiologists in order to communicate to clinicians in charge of infected patients that the microorganisms isolated in the infective site needed an exceedingly high quantity of a given antibiotic to be inhibited. Probably, this high amount of antibiotic could not be reached at the site of infection, and therefore the infective process would remain "resistant" to the treatment with such a drug. Note that the message is understood by clinicians as "resistant infection"; and an antibiotic-resistant microorganism tends to be consequently conceived as the one involved in untreatable infections. Indeed the results of quantitative determinations of the bacterial susceptibility to antibiotics (normally expressed as the "minimal inhibitory concentration", or MIC of the antibiotic) are translated into practical "categories". A given bacterial strain was considered to belong to the "resistant" category if it is not inhibited by the usually achievable systemic concentrations of the agent with normal dosage schedules and clinical efficacy has not been reliable in treatment studies. The more recent refinements of such concept consider other pharmacological features of the drug than the maximal systemic concentration, for instance the time during which concentrations of certain agents stay above the MIC of the offending organism. This parameter has been demonstrated to be predictive for a successful therapeutic response in animal models and clinical observations. It is important to understand that in this context "resistance" does not necessarily imply "high-level MIC value". For instance, accordingly to the NCCLS criteria (NCCLS, 1999), a member of the coagulase-negative *Staphylococcus* group of organisms will be considered resistant to cefotaxime only if its MIC exceeds a high-level MIC value of 64 μg/ml. The same strain is defined as resistant to another β-lactam drug, oxacillin, if the MIC now exceeds a low-level MIC value of 0.25 μg/ml. By similar reasoning, "low-level resistance" should not be confused with the clinical category of "intermediate resistance". According to the same widely accepted criteria, a strain has an "intermediate resistance" to a given antibiotic when the MIC approaches levels usually attainable blood and tissue, and for which clinical response rates may be lower than for susceptible isolates. Note that from the academic clinical perspective, dominant in the mind of many prescribing physicians, the aim is to improve the health of the individual patient. Only in recent years, is the interest of the community being considered at the moment of selecting strategies, agents and schedules to minimize the emergence of low-level resistance.

LOW-LEVEL ANTIBIOTIC RESISTANCE: THE POPULATIONAL PERSPECTIVE

If antibiotics were conceived as antimicrobial substances, but unrelated with therapeutic applications, it would still be possible to classify individual organisms accordingly to their various degrees of susceptibility. That would lead to the establishment of a "natural" (as opposed to "clinical") categorisation of bacterial strains. It is only in this context that the concept of "low-level resistance" can be applied. The populational definition of low-level resistance is based on one

"a priori" concept: that resistance is always "acquired resistance", meaning that any "resistant" organism is more resistant than the original "susceptible" ancestor population. That leads to the need for a definition of "susceptibility". From the population point of view, a strain can be considered susceptible to a given antibiotic if it belongs to the "population with the lowest MIC" as determined in a comprehensive collection of microorganisms of the same species. If the strain has a slightly higher MIC than that common for the susceptible population, it can be considered to have "low-level resistance".

Some caveats to this definition should be taken into account. Note that the definition is relative (established by comparison with the MIC of the susceptible strains), and its accuracy depends on the appropriateness of defining the population with the lowest MIC as fully "susceptible". It may happen, for example, that the intense effect of antibiotic selection on microbial populations for more than half a century has driven the original susceptible population to extinction, and thus the existing "most susceptible" is in fact a low-level resistant population (a situation analogous to Müller's ratchet). The available collections of pre-antibiotic bacterial isolates do not give a clear answer to this point, as prolonged storage may have shifted these bacterial populations to MIC values different from the original one. Nevertheless, they can serve to genetically detect the genes involved in possible mechanisms of resistance, for instance genes encoding the AcrAB multi-drug resistant pump in currently "susceptible" *E. coli*, or *H. influenzae* (Okusu 1996; Sánchez *et al.*, 1997). Should one consider as "basal MICs" those resulting from the inactivation of the natural (?) mechanisms of resistance? That is probably an impractical approach, but at least the concept should be taken into account. What is certain is that in the absence of these basal "low-level resistance mechanisms", the development of high-level resistance is hampered (Sánchez *et al.*, 1997), and because of that, they may be considered as potential chemotherapeutic targets.

Population analysis is generally performed accordingly to MIC results, rather than minimal bactericidal concentrations (MBCs). Therefore, the more susceptible peak in a distribution may contain strains that in spite of having identical MIC's, could differ strongly in MBC values; that is, the strains may be tolerant to the antibiotic. In the absence of distributions based on MBC's, tolerance is an invisible "low-level resistance" phenomenon. Nevertheless, tolerance may be considered as a first-step mechanism for the evolution of a high-level mechanism; this has been postulated for *Streptococcus pneumoniae* and penicillin (Moreillon and Tomasz, 1988) and more recently for vanco-mycin-resistance (Novak *et al.*, 1999). Are the antibiotic-selectable mutations in the self cell-killing mechanisms (for instance addiction mod-ules) under antibiotic stress? On the other hand, low-level resistance mechanisms may only be detectable after a prolonged bacterial-antibiotic interaction in time? For instance, we can imagine a strain producing a small quantity of β-lactamase, or a β-lactamase of poor affinity for the substrate, and therefore with a low-level resistance mechan-ism. If the cell survives the antibiotic challenge, each molecule of the enzyme will hydrolyse a molecule of the substrate, and then a new molecule, coming from the environment, will enter in the active site. After a prolonged bacteria-drug ex-posure, this "*sink effect*" may still be able to effectively deplete the surrounding environment of the antibiotic.

Similarly, very little has been done about resistance to the sub-inhibitory effects of antibiotics on bacterial cells. Sub-inhibitory concentrations of antibiotics exert important alterations in the func-tion of the bacterial cell (Lorian, 1986; Gould and MacKenzie, 1997). To explore the area of sub-inhibitory concentrations, the ill-defined concept of "minimal antibiotic concentration" (MAC) is sometimes used (Lorian, 1986), to define the minimal concentration of antibiotic producing an observable effect on the bacterial culture. The question is, what effect? Possibilities include, alterations in cell shape; reductions in ATP content; decreases in growth rate; modifications in DNA supercoiling; reductions in toxin production, or in adhesion rate (Shibl, 1983; Molinari *et al.*, 1993). The approach is of interest and certainly deserves of further attention. A well defined procedure to evaluate one or other of these parameters (adapted to the mechanism of action of each antibiotic and the biology of each bacterial organism) may be useful to understand "the very first step" in low-

level antibiotic resistance: resistance to sub-inhibi-tory effects of antimicrobial agents. The clarifica-tion of this phenomenon should be relevant in ascertaining the importance of antibiotic residues in wastewater, in the soil, in food, and in general in shaping our view of the environmental impact of antimicrobial agents. Clearly, the possibility of selecting resistant variants with very small differ-ences is the operational limit of this approach. Experiments of sustained sequential selection at precise sub-inhibitory concentrations will be re-quired to explore this field. In some cases, the so-called post-antibiotic sub-MIC effect may be related to sub-inhibitory effects; nevertheless, it can be explained in some cases by the prolongation in time of the deleterious consequences of the inhibitory concentrations; for instance, reduction in protein synthesis after macrolide treatment (Scott-Champney and Tober, 1999).

Returning to practicalities, we should accept that, in operational terms, the currently most susceptible population should be considered to possess the "basal" MIC level, raising the question of how to determine such a level. The most frequently applied method is to challenge a collection of microorgan-isms of the same species with a fixed range of antibiotic concentrations. The size, as well as the variety of individual strains forming the collection, are critical factors. For example, a collection formed by a large number of strains obtained in various countries, having different patterns of use of the antibiotic, and where a variety of naturally occurring phenotypes ranging from susceptibility to resistance are expected to be present, may be considered as appropriate. The range of challenging antibiotic concentrations chosen is also essential. In Figure 1, population distributions are shown, obtained by plotting the number of inhibited strains at each one of the sixteen antibiotic concentrations tested. In the upper part of the figure, the left peak is considered to be composed of the "susceptible population" of strains (with a modal MIC of 0.01 μg/ml). Other peaks appear with modal MICs of 0.12 (that can be considered as the low-level resistant population) and 1 μg/ml; finally, the peak on the right is composed of the population of strains with high-level resistance (modal MIC 64 μg/ml). The lower part of Figure 1 shows the detectable populations if another range of only seven antibiotic concentrations is tested on the

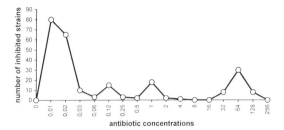

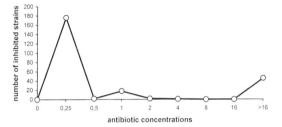

Figure 1. The detection of low-level bacterial populations is highly dependent on the methodology for antibiotic susceptibility testing. The use of a reduced number of antibiotic concentrations to determine the minimal inhibitory concentration (bottom) may be insufficient to detect peaks of low-level sub-populations present in the strain collection (top of the Figure).

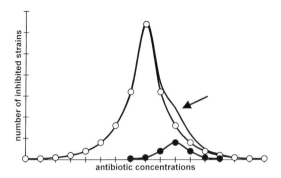

Figure 2. Apparently monomodal distribution of a susceptible collection of antibiotic "susceptible" strains (white circles). Note that a minority sub-population with a low-level resistance mechanism (black circles), despite its higher modal MIC, may be hidden by the predominance of the susceptible isolates. Sometimes its presence may be suspected from a shoulder in the susceptible distribution (arrow).

same collection of organisms. The low-level resistant population appears at 1 μg/ml; and the formerly low-level resistant population with MIC 0.12 μg/ml is no longer detectable as it is now included in the susceptible peak. The main conclusion is that a broad range of antibiotic concentrations, particularly at the lower end of the concentration range (ideally, a continuous gradient of concentrations) is needed to detect and define a low-level resistant population. Finally, apparently monomodal distributions may indeed hide the presence of low-level resistant populations (Figure 2). Typically that occurs in distributions spread over a long range of antibiotic concentrations, due to a high bacterial variability in the phenotypic response (for instance, a high dependence of the antibiotic effect on growth rate) or to uncontrolled technical variability in the procedures applied in standard MIC determinations. In Figure 2, the presence of a low-level resistant population is only detectable as a "shoulder" in the right arm of an apparently monomodal distribution. Obviously the shoulder is only apparent when the low-level resistant populations reaches a certain frequency. Surveillance analysis should be based on the

accurate recording during time of eventual cumulative increases of MICs at particular low antibiotic concentrations in order to detect "invisible" shoulders. In conclusion, population analysis provides insights into the interpretation that each one of the peaks in a multimodal distribution appearing to the right of the more susceptible one, or as a shoulder of it, corresponds to a particular type of resistant variant, presumptively harboring a given mechanism of acquired resistance.

LOW-LEVEL RESISTANCE: THE GENETIC PERSPECTIVE

From the genetic point of view, low-level resistance could be defined as a property of those bacterial cells harboring gene/s encoding a potential low-level resistance mechanism. This MIC-independent definition overcomes the problems related with the expression of the resistance, at the cost of acceptance that the presence of a potential resistance gene provides to the bacterial organism a higher possibility to survive the antibiotic challenge. The discussion about genetic-based definitions of antibiotic resistance is becoming a critical issue, due to the recent availability of oligonucleotide-array microchip technology potentially able to detect an important number of genes (or gene mutations) involved in antibiotic resistance (Cockerill, 1999). The real problem is what should be considered as a "resistance gene"; in particular, how to differentiate them from house-keeping "natural" genes able to decrease the activity of antibiotic on the bacterial cell. For instance, the physiological expression of the AcrAB pump of *E. coli* is responsible for the "typical" MIC levels of the so-called "susceptible" *E. coli* population; without the pump, the bacterial organism would be more susceptible to a wide variety of antibiotics, including β-lactams, but not aminoglycosides. For some antibiotics, this mechanism acts as a high-level resistance mechanism. Thus, the MIC to cloxacillin, a "typical" anti-Gram positive penicillin, is reduced from >256 μg/ml to only 2 μg/ml by AcrAB pump genetic disruption in *E. coli*. Conversely, mutations in several genes that are known to be involved in antibiotic resistance, as antibiotic targets, or involved in antibiotic access to the target pathways, or in antibiotic detoxifying mechanisms (see later), have no detectable phenotype by conventional methods (no MIC change). In some cases, the resistant phenotype only occurs after overproduction of the detoxifying mechanism. Overproduction of a phosphotransferase with very low detoxifying activity against tobramycin lead to a resistant phenotype (Menard *et al.*, 1993). In many cases this occurs by gene amplification (gene-copy effect). That is also the case for multiple copies of a β-lactamase *bla* gene in many cephalosporin and amoxycillin-clavulanate resistant *E. coli* strains (Reguera *et al.*, 1988). Amikacin-resistance may appear in strains harboring several copies of the gene encoding a low-efficiency phosphotransferase (Bongarts and Kaptijn, 1981). Low-level fluoroquinolone resistance may occur by multiple copies of a recessive mutant *gyrA* gene encoding a quinolone resistance gyrase (Gómez-Gómez *et al.*, 1997a). This type of adaptive mutability has been also associated with gene amplification in other circumstances, such as nutritional limitation (Andersson *et al.*, 1998). In other cases, a resistant phenotype is only expressed in combination with other genes or mutations, that may be silent by themselves (see later, "the combinatorial approach"). A typical case of these so-called "pre-resistant" mutations occurs in the early events of fluoroquinolone resistance in some Gram-positive cocci (*Streptococcus pneumoniae* or *Staphylococcus aureus*). Primary mutants in topoisomerase genes are

in some cases hardly differentiated from the susceptible population (Deplano *et al.*, 1997; Pestova *et al.*, 1999; Muñoz-Bellido *et al.*, 1999). Nevertheless, these variants are probably selected *in vivo*, indicating a real (even if non-detectable) low-level resistance. As a result of this selection, double mutants appear, now with a patent resistant phenotype. In other cases, the expression of a silent mechanism of resistance depends on an endogenous genomic arrangement, for instance mediated by an insertion event, as occurs for carbapenem resistance in *Bacteroides* (Podglajen *et al.*, 1994). The possibility of emergence of a previously cryptic resistance may also depend on contingency genes, as those involved in phase changes (Moxon *et al.*, 1994; Moxon and Thaler, 1997). Specific DNA rearrangements affecting regulatory proteins or the promoters of the genes are involved in phase change. DNA regions involved in switching present then a high mutation rate, which is specific for these regions and is tightly regulated. Few of these systems have so far been analyzed in relation to antibiotic resistance, but it is known to occur in the case of chloramphenicol acetyl transferase-mediated chloramphenicol resistance in *Proteus mirabilis* strain PM13 and probably also in *Agrobacterium radiobacter* (Charles *et al.*, 1985; Martínez *et al.*, 1989).

In vitro mutagenesis and/or *in vitro* selection and identification of bacterial mutants at low antibiotic concentrations (slightly over the MIC of the susceptible population) may be useful for predicting the frequency of emergence of bacterial variants with low-level antibiotic resistance. A number of observations support the hypothesis that low-level antibiotic resistance may occur as a result of a much broader variety of biochemical mechanisms than high-level resistance. The rate of emergence of tetracycline resistant *P. aeruginosa* mutants increased 20-fold when the concentration of antibiotic used for selection was $2\times$ instead of $4\times$ the MIC value (J.L. Martínez, personal communication). Similarly, low-level selection by nalidixic acid for resistance in *Salmonella* yielded much higher number of mutants than selection with high nalidixic concentrations (J.L. Martínez, and A.S. Sörensen, personal communication, 1999). In both studies, the mutants obtained after high-level antibiotic challenge were much more homogeneous that when selection was with a low concentration of the drug. In this context, high-level resistance can be presumptively viewed as depending on more specific mechanisms than low-level resistance. The relatively low specificity of several low-level mechanisms of resistance has important consequences. For instance, low level mechanisms of resistance can emerge and evolve under the pressure of non-antibiotic challenges. These will include toxic chemicals, organic solvents, household disinfectants, and non-antibiotic drugs (Alekshun and Levy, 1999). Thus, Mar mutants (over-expressing MarA) in *E. coli* are selected following growth on sub-inhibitory concentrations of tetracycline or chloramphenicol, but the mutant strain has increased resistance to many other antibiotics. These include β-lactams and fluoroquinolones (Goldman *et al.*, 1996), but the strain has also acquired resistance to oxidative stress agents such as menadione and paraquat, to uncoupling agents such as 2,4-dinitrophenol, to household and hospital disinfectants such as chloroxylenol and quaternary ammonium compounds, and to other chemicals such as cyclohexane. Another example is growth in the presence of salicylate, which increases antibiotic resistance in *E. coli* (Cohen *et al.*, 1993) and fluoroquinolone resistance in *Staphylococcus aureus* (Gustafson *et al.*, 1999). Mechanisms like *mar*, as well as mar-independent pathways are probably involved in the above cases. The essential message here is that many non-antibiotic compounds are potentially able to select low-level antibiotic resistance.

The concept of the *efficiency of the mechanisms of resistance* applies here. The acquisition of low-level antibiotic resistance is by definition a process of low efficacy (in terms of resistance). Nevertheless, it may be very efficient, because ability to survive and multiply when the cell is confronted with a broad range of environmental stresses (including antibiotics) may be achieved with a variety of available tools. Acquisition of high-level resistance has a high efficacy, but the process it is not always efficient, as it frequently depends on the availability and possibility of acquisition of a relatively limited number of pre-existing refined mechanisms. Because of that, most acquired low-level resistances are the result of mutational events in house-keeping genes, whereas on the contrary, the emergence of high-level resistance frequently depends on the acquisition of specialized mechanisms encoded by genes frequently located in integrons, transposons and plasmids, or resulting from extended recombination events.

For many antibiotics, mutations in house-keeping genes provide a wide range of possibilities for low-level antibiotic resistance (Baquero and Blázquez, 1997a). The possibility of appearance of a resistant mutation in a bacterial organism clearly depends on the peculiarities of the mechanism of action of each antibiotic challenging the bacterial population. Examples are discussed below.

First, the development of a resistance mutation depends on the characteristics of the antibiotic target/s, that is, the bacterial molecule/s whose interaction with the antibiotic triggers the static or lethal event. Mutational changes in the gene/s encoding the production of such target may modify the interaction with the drug, which could result in a decrease in the activity of the antibiotic. The frequency and the efficacy of these "*target gene mutations*" in producing different increases in the MIC of the antibiotics depends on a number of properties of the target/s themselves. Different target structures will tolerate different rates of variation without major functional impairments. If several of these changes are compatible with the preservation of the function, but each decreases the affinity with the antibiotic, the number of expected events leading to resistance should be higher. This is probably the case with changes in the sequences of PBP genes, reducing the affinity of β-lactams and several of them producing low-level resistance. Another important case concerns target gene redundancy. If a single copy gene is involved in an important function (e.g. *rrn*, encoding 23SrRNA, in *Mycobacterium* or *Helicobacter*), a mutation in that gene can significantly increase the MIC for antibiotics involved in ribosomal function, such as macrolides (Nash and Inderlied, 1995; Taylor *et al.*, 1997; García-Arata *et al.*, 1999). Conversely, in the case of multiple copies (*rrn* in many other microorganisms) the same event could result in very-low-level resistance, or in total susceptibility (Sander *et al.*, 1997; Prammananan *et al.*, 1999). A third case concerns recessive target genes. For example, a mutation in a gene encoding components of topoisomerases II or IV, will, if another gene encoding the same component is present in the cell, result in very-low level resistance, because of the frequent dominance of the wild-type alleles (Gómez-Gómez *et al.*, 1997a). Finally, there are cases of target co-operativity. If bacteria present multiple targets for antibiotic action, but interaction with all of them is needed for the full effect of the antibiotic, mutation in the gene encoding one of them will probably lead to low-level resistance; that may occur in interactions of β-lactams and PBPs.

Second, low-level antibiotic resistance may result from mutations in house-keeping genes involved in the entrance (and eventually activation) of an antibiotic before it reaches its cellular targets. The frequency and consequences of these "*target access gene mutations*" also depends on the characteristics of the genes encoding or regulating the pathway molecules. In the case of multiple pathways enabling the access of the antibiotic, mutations in the genes involved in one of them may slightly reduce access, and in such a way produce low-level resistance. Mutations in the outer membrane protein OmpC produce a low-level β-lactam-resistance phenotype in *E. coli* (Yoshimura *et al.*, 1985). Reductions in both OmpC and OmpF result in a slightly higher, but still low-level, β-lactam resistance in *E. coli*, suggesting the existence of alternative access routes for the antibiotic (Jaffe *et al.*, 1982). Low-level antibiotic resistance can also result from mutations in house-keeping genes with the ability to protect the target from the antibiotic. These "*target protective mutations*" increase the basal low rate of detoxification, or the active efflux of the drug, or act by modification of the target site. Mutations may also occur that increase the expression of "normally present" chromosomal β-lactamases, or result in the hyper-expression of "enormally present" efflux pumps. That leads to an increase in the basal intrinsic low-level antibiotic resistance that these mechanisms provide to the microorganism (Ma *et al.*, 1994; Li *et al.*, 1995; Nikaido, 1998; Masuda *et al.*, 1999; Ziha-Zarifi *et al.*, 1999).

One important point to be considered during any evaluation of the rate of low-level resistance variants based on mutations of house-keeping genes in a bacterial population is the *biological cost* associated with the mutational event (Lenski, 1997). In a simple way, such biological costs can be defined as the reduction in the fitness (essentially reproductive rate) of the mutant population with respect to the ancestor population, in the absence of the selector. In an intuitive way, it could be suspected that low-level resistance mutations

should be less costly than high-level resistances. That is not always the case. The fact that the "low-level" mutation is not very effective in determining resistance does not means that the bacterial cell is not severely impaired in other functions. In practice, frequently, some of the resistant mutants recovered from plates containing low antibiotic concentrations are slow growing organisms. These mutants have a chance to be selected by low antibiotic concentrations during therapy; nevertheless, if the ancestor population is not driven to extinction, they should have a low possibility of being fixed during the re-colonization process following drug elimination. Nevertheless, fixation may occur in highly compartmentalized habitats, where the competition with the residual susceptible population is diminished. If that were the case, the low-level resistant mutant could increase in number and give rise to secondary mutants. Some of these mutants could be "compensatory mutants" of the cost imposed by the first mutation (Schrag and Perrot, 1997a; Schrag *et al.*, 1997b; Bottger *et al.*, 1998). Being now able to compete with the former susceptible population, the antibiotic selective process during new antibiotic challenges will be more effective, and even considering the periods of antibiotic depletion, the low-level resistant population will predominate. Confronted with higher antibiotic concentrations, this population will have a statistical chance to give rise to high-level antibiotic resistance (Sörensen and Andersson, personal communication; Figure 3).

Low-level antibiotic resistance does not emerge exclusively from house-keeping genes. Indeed it emerges frequently as a consequence of the inadequacy of a pre-existing mechanism of resistance, having high efficacy in promoting resistance to a given antibiotic, in dealing with another compound of the same family. In other words, the mechanism producing high-level resistance to antibiotic A1 produces low-level resistance (including tolerance) to the related antibiotic A2. For instance, the same *gyrA* mutation producing high-level nalidixic-acid resistance *E. coli* produces low-level antibiotic resistance to ciprofloxacin. The same methyl-transferase promoting high-level resistance to 14C-macrolides produces low-level resistance to synergistins or ketolides (Malathum *et al.*, 1999). VanA and VanB, mediating high-level resistance to vancomycin in *Enterococcus*, cause only low-level

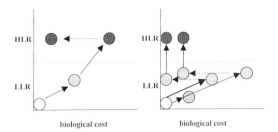

Figure 3. The acquisition of a certain level of antibiotic resistance (white circle moves to grey circle) may have a biological cost for a particular strain. A further increase (to black circle) may reduce even more the ability of the strain to survive in particular environments or to compete with wild strains. Compensatory mutations restore (at least partially) the original performance of the organism (left part). Recent experiments suggest that, under *in vivo* circumstances, early cost-compensatory mutations may facilitate the progress of the strain to higher levels of antibiotic resistance (Sörensen and Andersson, unpublished results).

resistance to new related peptide compounds such as LY333328 (Arthur *et al.*, 1999). The aminoglycoside-modifying enzyme APH3′II produces high-level resistance to kanamycin and low-level resistance to amikacin (Perlin and Lerner, 1979). Frequently, mutations in genes involved in high-level resistance to a given antibiotic, result in low-level resistance to a second related antibiotic. Thus the TEM-1 β-lactamase is extremely active in inactivating amoxycillin, but not cefotaxime; however, a single mutation in the TEM-1 gene leads to the TEM-12 β-lactamase, which provides low-level resistance to cefotaxime.

The expression of the genes involved in acquired antibiotic resistance may strongly differ among *different bacterial hosts*; in such a way, the same resistance determinant may produce high-level resistance in some organisms, but low-level resistance in others. This can be due purely to genetic factors such as differences in rates of transcription, or increased gene-copy numbers. It may also occur because of the presence of a "natural" house-keeping mechanism of resistance, such as a particular influx or efflux mechanism, increasing the effectiveness of a low-level mechanism of resistance. Indeed, the co-operation between different low-level mechanisms of resistance may produce a high-level resistance phenotype (see later). For instance, mechanisms leading to anti-

biotic extrusion, possibly in interplay with reduced outer membrane permeability, will reduce the intracellular concentration of an antibiotic, in such a way increasing the efficacy of low level-mechanisms present in the cell, such as low-level (repressed) β-lactamase production.

There is obviously a difficulty in the definition of the genetic events involved in low-level resistance, particularly when the mechanism involved can be considered as "intrinsic" or "natural" in a particular bacterial population. Knowing this difficulty, and for operative reasons, from the genetic perspective, the definition of "low-level resistance" to an antibiotic should be confined to low-level *acquired* resistance. That implies the presence, in a bacterial population with slightly higher MIC than the ancestor population, of a heritable mechanism of resistance responsible for such a phenotype, either acquired by mutation of pre-existing genes or by genetic horizontal transfer. The MIC of every one of these mutants or recipients will define the precise concentrations corresponding to low-level resistance. Before the introduction of a new antimicrobial agent in clinical practice, it could be advisable to know in advance the possibilities for development of low-level resistance in different bacterial pathogens. Analysis of both natural and hypermutable populations (with alterations in the mismatch-repair system, or mutagenised), under antibiotic challenge (using closely-spaced concentrations starting from very near the MIC), and preferably using compartmentalized (solid) media, will provide important data to predict the frequency and types of resistant variants that may arise during therapy.

LOW-LEVEL RESISTANCE: THE ECOLOGICAL PERSPECTIVE

The bacterial environment, determining the physiological conditions of the cell, may strongly modulate the expression of antibiotic resistance. The antimicrobial effect of some drugs is dependent on the bacterial growth rate (Gilbert *et al.*, 1990). For instance, β-lactams have greater effects on actively growing microorganisms. At low replication rates (frequently the case in established infections, or in normal flora) a mechanism of β-lactam resistance conferring only low-level resistance in fast-growing cells may act as a very efficient high-level mechanism. It is known that quinolone compounds (including most fluoroquinolones) and aminoglycosides have reduced effects under anaerobic conditions, an environment typical of infected tissues. Under *in-vivo* conditions, it is possible that low-level resistance to these compounds may, on some occasions, be as effective as high-level resistance. When bacterial populations reach high cell densities, the antimicrobial effects of most antibiotics are also severely reduced, a phenomenon generally known as the ''inoculum effect''. This effect, characteristic of the widespread bacterial biofilms in nature, may result from the additive effect of low-level mechanisms of resistance. For instance, a reduced single-cell production of β-lactamase may be insufficient to protect the organism against a given antibiotic, but the total β-lactamase released by a high-density colonial form of the population may be sufficient to protect the whole multicellular organization (Baquero *et al.*, 1985). Indeed the same number of TEM-1-expressing *E. coli* (10^8) under the same antibiotic pressure (50 mg/l) will be killed more rapidly by ampicillin if they are in one liter than if they are in ten microliters volume. In biofilms (or colonies) the reduced growth rate of stationary-phase populations may decrease the expression of antibiotic-target molecules, such as PBPs for β-lactams, in such a way contributing to the observed effect (Gilbert *et al.*, 1990, Stevens *et al.*, 1993, Gilbert and Brown, 1995). The expression of some low-level mechanisms of resistance (including tolerance; see Novak, 1999) is regulated in some cases by two-component regulatory systems of signal transduction, that may be eventually responsive to high cell density signals. It is also known that the expression of many influx-efflux mechanisms involved in the drug-cell interaction (from porins to pumps) is environmentally regulated, and that certainly should influence the local expression of antibiotic resistance. In general, these examples show that low-level resistance may be very effective under particular environmental conditions. A poorly explored field of research is whether antibiotics are influencing the selection in nature of bacterial sub-populations with low replication rates, or prone to biofilm formation, or more anaerobic, or with higher sporulation rates. If that were the case, then one could argue for these conditions to be considered as acquired resistance mechanisms.

The microbial environment also influences low-level resistance by triggering particular physiological features of microorganisms. Highly structured environments can be considered as continuous areas of space possessing a maximum of discrete surfaces linked by a minimum of connections. In these compartmentalized environments, different bacterial variants will occupy different niches (Rainey and Travisano, 1998). Indeed the number of different low-level resistant variants that can be recovered after antibiotic selection is much higher in solid (surfaced) medium than in liquid (non-structurated) medium. As Korona *et al.* (1994) have proposed, bacteria evolving in structured habitats present multiple adaptive peaks. Therefore, the physical structure of the environment is expected to influence the emergence and the evolution of low-level resistance.

In summary, low-level resistance is not a property associated with a given mechanism, but with a given mechanism in a particular bacterial host, located in a particular environment.

LOW-LEVEL RESISTANCE: THE COMBINATORIAL PERSPECTIVE

The combinatorial approach that bacteria apply to climb the mountains in the antibiotic selective landscape provides an important perspective on the phenomenon of low-level resistance. This topic was recently reviewed by Courvalin (1999), who proposed some of the following concepts and examples. It is an interesting fact that several mechanisms of resistance directed to protect the cell against a particular antibiotic are frequently collected in a single bacterial population, suggesting an additive step-wise evolution of resistance from low to high resistance levels. For example, a quarter of methicillin-susceptible *S. aureus* strains harbor tetracycline-resistance determinants of three different types: *tet(K), tet(L)*, and *tet(M)*. In other cases, different mechanisms can co-operate synergistically, so that the influence of two genes on the phenotype is greater than that expected from the sum of their effects, as in the combination of esterase and 23SrRNA methyltransferase in *Enterococcus faecalis*. Quantitative co-operation between antibiotic resistance genes is a widespread phenomenon, probably more frequent among the most susceptible organisms. Thus, while the *Streptococci* have intrinsic mechanisms for aminoglycoside resistance, the presence of more than one aminoglycoside-resistance enzyme is more frequent among *Staphylococci*. In *Staphylococci* the presence of a low efficiency aminoglycoside phosphotransferase acting on amikacin, in combination with impaired uptake, leads to significant drug resistance (Perlin and Lerner, 1986). Single-mutations in the genes encoding topoisomerase in *Campylobacter* may be sufficient to make a high-level resistance phenotype because the presence of a natural efflux pump assures a relatively high basal MIC to quinolones (Charvalos, 1995).

It is important to mention here that the presence of a first mechanism of resistance may facilitate the acquisition of a second mechanism. For instance, it has been shown that an inhibitor of MDR determinants from Gram-positive bacteria drastically reduces the mutation rate leading to quinolone resistant mutants in *Staphylococcus aureus* (Markham and Neyfack, 1996) and *Streptococcus pneumoniae* (Markham, 1999). Possibly, the presence of a low-level resistance mechanism allows the population to survive long enough in the antibiotic environment to be able to receive by horizontal transfer a more efficacious mechanism. Another possibility is that prolonged survival under stress conditions increases the mutability of the endangered strain, which may allow the emergence of a more favorable mutation. In general it can be predicted that the mechanism of higher efficacy will prevail in the challenged population. In the former examples concerning *S. aureus*, or *S. pneumoniae*, the mechanism of lower efficacy (MDR) will tend (particularly if there is an associated cost) to disappear. Once the high-level mechanism has been acquired at an acceptable cost, there is no apparent reason to maintain the low-level mechanism. In some cases, the ''low-level mechanism'' is an intrinsic property of the bacterial cell, and in that case, the emergence of high-level resistance appears as a primary phenomenon.

LOW-LEVEL RESISTANCE: THE EPIDEMIOLOGICAL PERSPECTIVE

The emergence of low-level antibiotic resistance may be detected in bacterial populations challenged by antibiotics *in vitro*, or in animal models of infection, or in the treated patient. But the problem of development of low-level antibiotic resistances as stepping stones for the ascent of high-level antibiotic resistance is also detectable in epidemiological studies, where the prevalence of resistance at different MIC levels in bacterial pathogens is monitored over extended periods of time. Such evolution can be particularly well traced in those geographical regions where the problem of resistance is emerging. In Figures 4 and 5, two examples are provided strongly suggesting the involvement of low-level resistance in the build-up to high-level resistance at the community level. Figure 4 corresponds to the emergence of fluoroquinolone resistance among *E. coli* strains isolated in the community in Spain (Aguiar *et al.*, 1992). The therapeutic use of fluoroquinolones from 1985 was immediately followed by a sudden increase in nalidixic acid resistance (a quinolone compound, but not a fluoroquinolone). Nalidixic acid resistance can be viewed as a marker of the first mutation in the topoisomerase II encoding gene, providing only very low-level resistance to norfloxacin, ciprofloxacin, and ofloxacin, the first three fluoroquinolones on the market. Only after a time did high-level resistance to fluoroquinolones

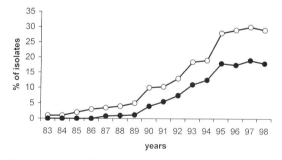

Figure 4. Evolution of the rates of nalidixic acid resistance (white circles) and ciprofloxacin resistance (black circles) in *Escherichia coli* isolates recovered in the Ramón y Cajal Hospital in Madrid, Spain, from the time of introduction of fluoroquinololes (norfloxacin, 1985) to the present day. Over several years the new drugs selected for primary mutants that had high-level resistance to nalidixic acid, but low-level resistance to fluoroquinolones. Only from 1990 onwards did double mutants appear with high-level resistance to the newer compounds. The selection of primary low-level fluoroquinolone resistant strains has continued under heavy drug usage.

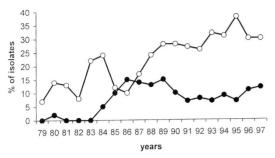

Figure 5. Yearly evolution of the rates of low-level (white circles) and high-level penicillin resistance (black circles) in *Streptococcus pneumoniae* isolates studied at the National *Streptococcus* Reference Center in Madrid, Spain (modified from Fenoll *et al.*, 1998). Note that the development of low-level resistance preceded by several years the emergence of high-level resistance. New mutations acquired by the low-level mutants may explain this evolutionary trend.

appear, by the acquisition of secondary mutations in the genes encoding topoisomerases II or IV (Hooper, 1999). Over the past twelve years, the introduction of new drugs of this family (associated with marketing-induced high levels of use) has increased again the rate of fluoroquinolone resistance, now selecting for both types of mutations. The second example (Figure 5) concerns the emergence of penicillin-resistance among *Streptococcus pneumoniae* isolates in the same country (Fenoll *et al.* 1998). Again, low-level resistance populations appeared long before those exhibiting high-level resistance. In this case, it is known that the progressive acquisition of mutations in genes involved in PBP synthesis lead to increased penicillin resistance (Hakenbeck, 1999a; Hakenbeck *et al.*, 1999b). Both examples may be considered as cases of directional selection of resistance, with the emergence of new mutations due to changes in antibiotic consumption. Such observations are difficult to make when high-level resistance is already prevalent in some places in the global village. In countries lacking antibiotic resistant strains the import of a high-level resistant strain will produce a sudden increase in high-level resistance without a prior period of low-level resistance, such as for instance in the case of *S. pneumoniae* penicillin resistance in Iceland.

Both cases, the development of fluoroquinolone-resistance in *E. coli*, and penicillin-resistance in *S. pneumoniae*, illustrate the sequential shift from low-level to high-level resistance, when essentially the same mechanism is involved in both types of resistance. As was said before, in the case that two different mechanisms occur, the low-level mechanism may facilitate the acquisition of the high-level mechanism.

In conclusion, the existence of a selective process leading to the increase in low-level resistant populations can also be detected by comparing the MIC of bacterial isolates of the same species collected over a period of time. A badly defined point in this context is the length of time required by bacterial populations for this directional shift towards high-level resistance. That is expected to be species-, antibiotic-, and consumption-dependent, but probably epidemiological factors are also involved. Low-level resistance may be sufficient to increase the absolute number of a given variant strain in a treated patient. Within-host selection will probably favor between-host bacterial spreading. Spreading by itself may increase variation. In any case, in antibiotic-polluted societies, sequential rounds of selection in differently treated hosts will increase the absolute number of variant cells, thus increasing the probability of emergence of secondary, and more effective, resistance mechanisms. Because of that, a critical point of research is the epidemigenicity and virulence of antibiotic-resistant bacterial populations (Bjorkman *et al.*, 1998).

SELECTION OF LOW-LEVEL RESISTANCE

In organisms with very large effective population numbers, as with microorganisms, it seems highly probable that the usual pattern for natural selection is to act on very small fitness differences. Low-level resistance has been operationally defined in earlier paragraphs as the genetically-based property of a population leading to ''slightly higher'' MIC values than those expected to occur for the entirely susceptible ancestor population. That leads to the important evolutionary problem of the possibility of selection of very small differences. Accordingly, with such modest margins, differences in actual survival and reproductive success may be attributed only to chance. The common intuitive opinion is that any efficient selective process requires significant variations both in the phenotype and in the selective environment. For instance, a widespread assumption is that "if the phenotypic effect of a mutation is low, its contribution to the selective advantage should be similarly low". Clearly there is common sense scepticism to taking into account very small differences in antibiotic susceptibility. In a recent publication, the evolutionary significance of some recently isolated β-lactamase low-level resistant mutants was considered obscure by the authors "since we do not know how small an effect constitutes a selective advantage". There are two main reasons supporting the lack of practical interest in the study of low-level antibiotic resistances. First, that these low-level resistant populations (in particular those with very low level resistance) are rarely reported in clinical practice; second, even if low-level resistant populations can be detected in laboratory experiments, the presence of such populations does not represent any particular risk for the patients. This is based on the conviction that the standard antibiotic dosages assure antibiotic concentrations in the human body well over the MIC of the low-level resistant population, and that the immunological response of the host will drive to extinction these minority populations. In reality, low-level and very-low level resistances are rarely reported because (as pointed out earlier) susceptibility testing procedures rarely include very-low antibiotic concentrations. On the other hand, standard antibiotic dosages assure sufficient antibiotic concentrations in serum, but a wide range of peri-inhibitory or sub-inhibitory concentrations probably occur in many heavily colonized mucosal surfaces, in the inflammatory area, and, in general, in the highly compartmentalized environment of the human body. The importance of melting drug pharmacokinetics, habitat compartmentalization and microbial population dynamics has been recently emphasized (Austin *et al.*, 1998; Baquero and Negri, 1997b; Lipsitch and Levin, 1998). In the open-air environment (soil, surface water, sewage), where antibiotics are present as natural or pollutant molecules, and in raw foodstuffs derived from animals or vegetables previously treated with antibiotics (to prevent infections or to enhance growth), and colonized by microorganisms, a variety of potential low-level antibiotic concentrations

certainly occurs, with potential selective effects for low-level antibiotic resistant populations.

It is clear that very small differences in selection require the existence of slight differences between individuals (with respect to a potential trait, in our case antibiotic resistance) and a weak selective process. We consider such a selective pressure to be one that will be unable to provoke selection among individuals with big differences among them. To a certain extent, selection of very small differences should be able to shape the evolution of a complex biological system when there is a fine-grained variation between individuals, and not just a coarse-grained variation. In the following paragraphs, the conditions and possibilities for low-level antibiotic selection are considered in more detail.

ANTIBIOTIC SELECTIVE CONCENTRATIONS AND SELECTIVE COMPARTMENTS FOR LOW-LEVEL RESISTANT VARIANTS

A classical rhetorical point is how much difference is needed to hook a selective process. The main concept in starting to think about selection of low-level resistance is that the variant selective process is expected to have a sharp maximum at a particular antibiotic concentration or small range of close low-level concentrations. These concentrations should be able to decrease the growth rate or to suppress the original ancestor population without affecting the low-level variant population. Beyond this concentration, antibiotic concentrations may able to slow or suppress in an equivalent way the growth of both susceptible and variant populations, and therefore no selection of the variant is expected to occur. The same applies when the antibiotic concentration is below the MICs of both populations. Therefore, the selection of a particular low-resistant variant may happen only in a narrow range of drug concentrations that can be defined as selective concentrations. The possibility of low-level resistance selection will depend on the availability of a low-level resistant variant population in a particular place where the selective concentration occurs and is maintained during a critical period of time. As mentioned above, and discussed in more detail in the next paragraph, antibiotics are expected to produce gradients of concentrations in the highly compartmentalized human body. The question "how different does be a resistant bacterial variant from the original organism need to be in order to be selected?" is indeed the same as "how different should the concentration between two points in an antibiotic gradient be in order to constitute different selective compartments?" The principle of this type of concentration-specific selection is illustrated in the "balls model" shown in Figure 6. An important point to stress in Figure 6 is that the selection of the low-level resistant population immediately resulting from antibiotic exposure does not imply an increase in the absolute number of the low-level resistant cells, and frequently a certain decrease may occur. The essential point is what happens *after* the elimination of the drug. At the start of the process the ancestor population is frequently predominant, the re-colonization in the formerly non-selective compartments favors such a susceptible population. In contrast, at the formerly selective compartments, the re-colonization process of the habitat increases the frequency of the low-level resistant population, replacing the original susceptible one. Sequential exposure to the antibiotic (dosing) will produce sequential periods of enrichment of the low-level resistant population leading to the predominance of the variant.

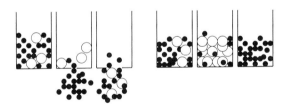

Figure 6. Schematic representation of the principle of concentration-specific selection. Black balls represent a susceptible bacterial population (S); white balls, a low-level resistant sub-population (R). The size of the hole in the bottom of the boxes represents the concentration of the inhibitor. In the left part of the figure, if the hole is very small (first box) no balls are eliminated; if the hole has medium size, only susceptible balls drop; if the size of the hole exceeds a critical diameter, both types of ball are eliminated. In the right part of the figure, the holes are closed after a certain time (antibiotic eliminated), and the "surviving balls" multiply at the same rate. Note that, because the black balls are predominant at the start, there may be a "selection window" for low-level resistant variants.

A HIGH DIVERSITY OF SELECTIVE COMPARTMENTS IN THE HUMAN BODY

Antibiotics used in chemotherapy create, in the human body, a high diversity of spatial and temporal concentration gradients. These gradients are due to pharmacokinetic factors, such as the different diffusion rates into various tissues, metabolism, local inactivation, and variation in the elimination rate from different body sites. The direct effect of microbes of the normal or pathogenic flora, particularly (but not exclusively) if they possess antibiotic-inactivating enzymes, also contributes to the gradient formation. In general, as illustrated in Figure 7, after antibiotic administration, high antibiotic concentrations will be confined to small and ephemeral selective compartments, and low concentrations are expected to be distributed during longer times in large selective compartments (Baquero and Negri, 1997b). Therefore, most bacterial populations in the human microflora probably face a wide range of antibiotic concentrations after each administration of the drug. Moreover, the selective compartments along a natural gradient may change both in space and in time, in such a way altering the potential selectivity for different genetic polymorphisms leading to different low-level resistances. Indeed a given concentration of an antimicrobial agent may have different selective effects on an organism depending upon the time of exposure. After a given period

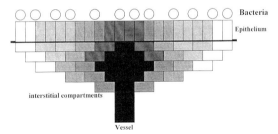

Figure 7. Schematic representation of concentration gradients of a systemic antibiotic in a particular surface of the human body, irrigated by a blood vessel. The bacterial organisms colonizing the area (white circles) are submitted to a variety of antibiotic concentrations. Note that the chance of being in contact with low antibiotic concentrations (medium to light grey) is higher than for high-level concentrations (dark grey to black).

of time, a new set of selective compartments may arise. Obviously, gradients submitted to frequent fluctuations may dissipate the selective power of each compartment, as the process of selection is dependent on a critical period of time, that is variable for each pair of selective agent-selectable population. Further generation of new selective compartments can occur as a result of interference between gradients. Finally, as the bacterial population structure is modified by the selective process, this change may modify the gradient itself, eventually expanding the selective compartment for the low-level resistant population. Considering that the variation from susceptibility to low-level antibiotic resistance is probably much more frequent than to high-level resistance, potentially selective compartments for selection of different types of low-level resistance probably occur during many antibiotic treatments.

PROTEIN EVOLUTION IN REAL TIME: SELECTION OF VARIANT β-LACTAMASES

In vitro experimental systems are able to confirm the concept of concentration-dependent selection. To this end antibiotic-mediated selection of bacteria harboring wild and variant β-lactamases was studied. β-lactamases are enzymes produced by a variety of pathogenic microorganisms, able to inactivate (detoxify) β-lactam antibiotics, such as penicillins. For instance, ampicillin is inactivated by the β-lactamase TEM-1, and the producing organisms are resistant to this drug. To counteract this resistance, new β-lactam antibiotics "resistant" to β-lactamase TEM-1 were developed by the pharmaceutical industry, as the third generation cephalosporins (cefotaxime is an example), that suppress the growth of bacteria carrying TEM-1. In a relatively short period of time, bacteria have counteracted this challenge by the production of variant TEM-1 β-lactamases, now able to hydrolyse the recently introduced cephalosporins, in a typical example of a relationship between enzyme activity and bacterial fitness (Medeiros, 1997; Petrosino *et al.*, 1998). To our surprise, the study of the protein sequence of variant TEM enzymes found in microorganisms involved in infections refractory to cefotaxime therapy, frequently

showed that several amino acids were changed with respect to the original TEM-1 sequence. That suggested that a cryptic evolution may have occurred in hospitals, following a subtle selection of single mutants, and gave rise to populations where a second mutation appeared, was in turn selected, and the repetition of the process produced many of the more efficient antibiotic-inactivating enzymes. Why were medical microbiologists unable to detect in real time the first variant enzymes? The reason was that these variants had only a minimal increase in the ability to inactivate the new antibiotics. For instance, in conventional susceptibility testing essays, bacteria harboring the original TEM-1 enzyme were inhibited by 0.03 μg/ml of cefotaxime; those with a "first" TEM-12 variant, resulting from the single substitution of arginine for serine at the position 164, were inhibited by only 0.08 μg/ml of this drug. This minimal increase was considered meaningless in terms of therapy or evolution of resistance, since the antibiotic concentrations in the patient were expected to be sufficient to suppress the variant. In reality, the organism harboring the TEM-12 variant was indeed selected. Then, the acquisition of a new mutation, now replacing glutamic acid for lysine at the position 240 produced the more efficient enzyme TEM-10, which significantly increased the ability of the host bacteria to survive in relatively high cefotaxime concentrations and created clinical problems for the therapy of human infections.

How could the "first variant" population harboring TEM-12 be selected? According to the hypothesis discussed above, that occurs within a particular selective compartment. The selective compartments can be reproduced using *in vitro* models. Using directed mutagenesis technology we prepared a collection of isogenic *Escherichia coli* strains harboring TEM-1 (wild enzyme), TEM-12 (first variant, with one single mutation) and TEM-10 (two mutations). Double and triple mixtures of these strains were prepared, at the respective 100:10:1 proportions, and then challenged during 4 hours by different cefotaxime concentrations, corresponding to different environmental compartments within a gradient. The antibiotic was enzymatically degraded, and the surviving cells transferred to antibiotic-free medium, and the proportion of cells containing the wild and variant enzymes were analysed in these cultures after overnight growth. In several independent experiments, the same result was obtained: TEM-12 was selected over TEM-1 only at a narrow range of antibiotic concentrations (0.008 to 0.06 μg/ml), which correspond to the TEM-12 selective compartments under the experimental conditions (Baquero *et al.*, 1997c). In triple mixtures, the selective compartments for the TEM-10 variant were identified at higher antibiotic concentrations. Essentially the same type of result was obtained in another model, using this time mixtures of *Streptococcus pneumoniae* populations showing different levels of susceptibility to several beta-lactam antibiotics (Negri *et al.*, 1994). In short, these experiments served to illustrate the basic concept that a particular variant is selected in a particular selective compartment within an environmental (antibiotic) gradient.

Another turn in the complexity of the system results from the interference between gradients, creating new bi-dimensional frames that generate new selective compartments. In the case of antibiotics, that may occur during combined or fluctuating antibiotic therapy schedules. We have shown (Blázquez *et al.*, 1998) that some supposedly ''neutral'' mutations in TEM-1 beta-lactamase (such as the replacement of alanine for threonine in position 237) have probably been selected in this type of compartment, as a result of the optimization of the enzyme to deal with combined antibiotic challenges (ceftazidime + cefotaxime). This type of result was predicted by earlier studies on the evolution of selective neutrality. In conclusion, the significant diversity in β-lactamases may express the diversity of selective antibiotic compartments that occur during therapy (Baquero and Negri, 1997b). New compartments were produced from the introduction of new antibiotics or therapeutic schedules that resulted in the diversification of TEM β-lactamases; this is certainly an elegant testimony to the flexibility of biological systems and the power of natural selection.

SELECTION OF LOW-LEVEL RESISTANCE: PREDICTIONS FROM POPULATION BIOLOGY MODELS

In the study of antibiotic resistance, the design of future experiments should take into consideration the multi-faceted determination of the evolutionary process involved in the emergence and spread of resistant bacteria. Multi-variable experiments are hard to perform and even harder to interpret. Fortunately, mathematical models can provide reference frames for developing hypotheses based on multifactorial components, that can then be contrasted with reality. A number of population and evolutionary biologists have recently entered the antibiotic resistance arena. Some of this work, summarized below, has provided predictions useful for the study of low-level resistance (Levin *et al.*, 1997; Lipsitch and Levin, 1997, 1998; Austin *et al.*, 1998; Levin and Anderson, 1999; Bangham *et al.*, 1999). For example, mathematical modelling of the population biology of microorganisms under antimicrobial therapy suggests that heterogeneities in drug activities, either in the space (different body compartments) or in time (essentially due to reductions in doses or increases in dosing intervals) increases the risk for selection of low-level resistance. Some other models have predicted that the emergence of resistance is also low if the decline of the bacterial population by antibiotics is comparable to the bacterial division rate. Sustained exposure to sub-inhibitory concentrations should select resistance. In general, intermediate antibiotic concentrations tend to be the more selective for resistance. Indeed models predict that heterogeneity in drug concentration or efficacy of the antibiotic in different body compartments may facilitate such evolution of drug resistance, because relatively narrow windows of drug concentrations may allow the local evolution of resistant variants (Kepler and Perelson, 1998; and Lipsitch and Levin, 1998). Complementing this view, other models suggested that the use of large dosages at the start of the treatment may be useful to drive the early extinction of low-level resistant variants, particularly if these are present in low proportions, in such a way suppressing further evolution to high-level resistance. In the case of higher initial proportions, high dosages should be maintained much longer to achieve such a goal. Similarly, combination therapy (two or more antibiotics) in a single patient may delay the emergence of resistance, depending on the activity of each antibiotic against low-level mutants resistant to the other antibiotic. In some cases, there is a "false combination therapy", when the antibiotics are not simultaneously present in the compartments where bacteria are located within the human body, and selection of low-level mutants may occur. Obviously, models predict that non-adherence to the therapy (irregular uptake of drugs by the patient) favors the emergence of low-level resistance, particularly among immunocompromised patients.

LOW-LEVEL RESISTANCE: LIFE AND EVOLUTION ON THE MARGINS

Low-level resistant populations frequently live on the margins. During antibiotic therapy, these populations are frequently confronted with drug concentrations that may exceed for limited periods of time the protective possibilities of the low-level mechanism. Part of the low-level resistant population will be periodically eliminated, but part may "persist under challenge". Low-level populations are in such a way maintained under exposure to challenging concentrations much more frequently than susceptible bacteria (that are rapidly killed) or highly resistant bacteria (as the challenge does not exist any more). Sub-lethal conditions have been revealed as potentially inductive for bacterial variation. The regulation of the emergence of antibiotic resistant mutants by stress is poorly studied, but some observations are already available. Some antimicrobials, such as quinolones, increase mutation rates (Mamber *et al.*, 1993; Phillips *et al.*, 1997); and low-level concentrations of quinolones increase the emergence of antibiotic resistant mutants in *E. coli* (Riesenfeld *et al.*, 1997). Sub-inhibitory concentrations of β-lactam agents may also increase the mutation rates to unrelated antibiotics, such as rifampicin (Blázquez, 1999). An obviously important field of research is the characterization of the signals associated with antibiotic-mediated stress. These signals could be generated by non-integrated metabolites, or by alterations in DNA supercoiling, or may be similar to those of early stationary phase. Indeed stationary-phase signals and regulators influence the rate of mutation in *E. coli* (Maenhaut and Shapiro, 1994; Gómez-Gómez *et al.*, 1997b), and general starvation stress signals (such as ppGpp) may be involved (Aize-

man *et al.*, 1996). The involvement of two-component signal transduction systems in sensing stress is also very likely. In any case, a population under stress appears to increase its rate of mutation, as a result of the emergence in the population of "mutator cells", most frequently resulting from structural defects or altered expression in the methyl-directed mismatch repair (MR) genes (Miller, 1996; Taddei *et al.*, 1997a, 1997b). Recently, the term "mutases" has been coined to describe a number of stress-inducible DNA polymerases that increase mutation rates (Radman, 1999). At "critical antibiotic concentrations", where a part of the population is dying, these cells may generate favorable mutations significantly increasing survival to the antibiotic challenge. Again, low-level resistance is expected to be the most frequent outcome of these mutations. Low-level antibiotic challenge probably produces weaker mutator alleles than high-level antibiotic challenge. Nevertheless, at low antibiotic concentrations, these mutator strains are expected to be selected by the antibiotic (hitch-hiking); in such a way the mutator population should increase its absolute number. In compartments where these organisms are confronted with higher antibiotic levels, there will be a high possibility of generation of high-level mutants. Certainly the mutator phenotype is useful to cross bottle-necks, but, in conditions of no-antibiotic therapy, hyper-mutating strains will be disadvantaged (increase in mutational load). The hyper-mutable phenotype tends to reverse to the normal mutation state; in this process, the increased rate of recombination of mutators may facilitate the recovery of the damaged MR gene. In other cases, the hyper-mutable phenotype could be only transiently expressed, but the conditions of this adaptive burst (under stress?) are not completely understood (Heinemann, 1999: Rosche *et al.*, 1999). The exposure to antibiotics may increase the observed rate of mutation; that has been observed not only for DNA-damaging agents, such as ciprofloxacin (Riesenfeld *et al.*, 1997), and protein-synthesis inhibitors, such as streptomycin (Ren *et al.*, 1999), but also in cell-wall active agents (Blázquez *et al.*, 1999). In some circumstances, a mutator population will be fixed to a certain extent: in highly compartmentalized habitats, this population may be confined in areas where competition with the ancestor strain is low, and similarly reduce the opportunities for reacquisition of the MR wild gene. When these areas are under frequent antibiotic pressure, they may constitute hot-points for the development of antibiotic resistance. That is the case for *Pseudomonas* strains located in the very compartmentalized lung of cystic fibrosis patients, where more of $\frac{1}{3}$ of the recovered colonies may display a mutator phenotype, and these strains have a significantly higher resistance to antibiotics (Oliver *et al.*, 1999). It is also conceivable that the pathogenic process may accelerate the rate of genetic variation in bacteria, with increased *in-vivo* mutation rates (Andersson, personal communication). In any case, in very dense bacterial populations, some fixation of mutator alleles is expected to occur. In summary, low-level antibiotic resistance may enable bacterial populations to survive to a certain extent in critical antibiotic concentrations, which may favor the emergence of variability, and accelerate the evolution towards more effective antibiotic resistance.

ACKNOWLEDGEMENTS

To my friends Jesús Blázquez, Rafael Cantón, Cristina Negri, and Marisa Morosini, from our Department, for fruitful work and discussions on low-level resistance. To Bruce Levin and Marc Lipsitch, respectively from Emory University and Harvard Medical School, for intense interactions around theory and practice of selective events. To Jose-Luis Martínez, from the Institute of Biotechnology in Madrid, for endless exchanges of views and hypothesis; to him, to Patrice Courvalin, from the Pasteur Institute in Paris and the University of California at San Diego, and to Dan Andersson and Anna Smed Sörensen, from the Swedish Institute for Infectious Disease Control in Solna, Sweden, for facilitating access to unpublished scientific information at the time of writing.

REFERENCES

Aguiar, J.M., Chacon, J., Cantón, R. and Baquero, F. (1992) The emergence of highly fluoroquinolone-resistant *Escherichia coli* in community-acquired urinary tract infections. *J. Antimicrob. Chemother.* **29**, 349–350.

Aizenman, E., Engelberg-Kulka, H. and Glaser, G. (1996) An *Escherichia coli* chromosomal "addiction module" regulated by guanosine $3',5'$-bispyrophosphate: a model for programmed bacterial cell death. *Proc. Natl. Acad. Sci. USA.* **93**, 6059–6053.

Alekshun, M.N. and Levy, S.B. (1999) The *mar* regulon: multiple resistance to antibiotics and other toxic chemicals. *Trends Microbiol.* **7**, 410–413.

Andersson, D.I., Slechta, S.E. and Roth, J.R. (1998) Evidence that gene amplification underlies adaptive mutability on the bacterial *lac* operon. *Science* **282**, 1133–1135.

Austin, D.J., White, N.J. and Andersson, R.M. (1998) The dynamics of drug action on the within-host population growth of infectious agents: melting pharmacokinetics with pathogen population dynamics. *J. Theor. Biol.* **194**, 313–339

Arthur, M., Depardieu, F., Reynolds, P. and Courvalin, P. (1999) Moderate-level resistance to glycopeptide LY333328 mediated by genes of the *vanA* and *vanB* clusters in Enterococci. *Antimicrob. Agents Chemother.* **43**, 1875–1880.

Baquero, F., Vicente, M.F. and Pérez-Díaz, J.C. (1985) β-lactam coselection of sensitive and TEM-1 β-lactamase producing subpopulations in heterogeneous *Escherichia coli* colonies. *J. Antimicrob. Chemother.* **15**, 151–157.

Baquero, F. and Blázquez, J. (1997a) Evolution of antibiotic resistance. *Trends Ecol. Evol.* **12**, 482–487.

Baquero, F. and Negri, M.C. (1997b) Selective compartments for resistant microorganisms in antibiotic gradients. *BioEssays* **19**, 731–736.

Baquero, F., Negri, M.C., Morosini, M.I and Blazquez, J. (1997c) The antibiotic selective process: concentration-specific amplification of low-level resistant populations. In: *Antibiotic Resistance: Origin, Evolution, Selection and Spread.* J. Wiley & Sons, Ciba Found Symp. **207**, 93–105.

Baquero, F., Negri, M.C., Morosini, M.I. and Blázquez, J. (1998) Antibiotic-selective environments. *Clin. Infect. Dis.* **27**, S5–S11.

Bangham, C., Anderson, R., Baquero, F., Bax, R., Hastings, I., Koella, J., *et al.* (1999) Evolution of infectious diseases: the impact of vaccines, drugs and social factors. In: *Evolution in Health and Disease* (Stephen C. Stearns, Ed.). Oxford University Press, pp. 152–160.

Bjorkman, J., Hughes, D. and Andersson, D.I. (1998) Virulence of antibiotic-resistant *Salmonella typhimurium*. *Proc Natl Acad Sci USA* **95**, 3949–3953.

Blázquez, J., Negri, M.C., Morosini, M.I., Gómez, J.M. and Baquero, F. (1998) A237T as a modulating mutation in naturally occurring extended-spectrum β-lactamases. *Antimicrob. Agents Chemother.* **42**, 1042–1045.

Blázquez, J., Baquero, M.R., Gómez-Gómez, J.M., Coque, T., Lobo, M. and Baquero, F. (1999) β-lactams increases bacterial mutation rate. *Proc. 39th Interscience Conference on Antimicrobial Agents and Chemotherapy*. San Francisco, September 1999.

Bongaerts, G.P.A. and Kaptijn, G.M.P. (1981) Aminoglycoside-phosphotransferase II-mediated amikacin resistance in *Escherichia coli*. *Antimicrob. Agents Chemother.* **20**, 344–350.

Bottger, E. C., Springer, B., Pletschette, M. and Sander, P. (1998) Fitness of antibiotic-resistant microorganisms and compensatory mutations. *Nature Medicine* **4**, 1343–1344.

Courvalin, P. (1999) Combinatorial approach of bacteria to antibiotic resistance. *Res. Microbiol.* **150**, 1–7.

Charles, I.G., Harford, S., Brookfield, J.F. and Shaw. W.V. (1985) Resistance to chloramphenicol in *Proteus mirabilis* by expression of a chromosomal gene for chloramphenicol acetyltransferase. *J Bacteriol.* **164**, 114–122.

Charvalos, E., Tselentis, Y., Hamzehpour, M.M., Köhler, T. and Pechère, J.C. (1995) Evidence for an efflux pump in multidrug-resistant *Campylobacter jejuni. Antimicrob. Agents Chemother.* **39**, 2019–2022.

Cockerill, F.R. (1999) Genetic methods for assessing antimicrobial resistance. *Antimicrob. Agents Chemother.* **43**, 199–212.

Cohen, S.P., Levy, S.B., Foulds, J. and Rosner, J.L. (1993) Salicylate induction of antibiotic resistance in *Escherichia coli*: activation of the *mar* operon and a *mar* independent pathway. *J. Bacteriol.* **175**, 7856–7862.

Deplano, A., Zekhnini, A., Allali, N., Couturier, M. and Struelens, M.J. (1997) Association of mutations in *grl*A and gyrA topoisomerase genes with resistance to ciprofloxacin in epidemic and sporadic isolates of methicillin-resistant *Staphylococcus aureus. Antimicrob Agents Chemother.* **41**, 2023–2025.

Fenoll, A., Jado, I., Vicioso, D., Pérez, A. and Casal, J. (1998) Evolution of *Streptococcus pneumoniae* resistance in Spain: update (1990–1996). *J. Clin. Microbiol.* **36**, 3447–3454.

Garcia-Arata, M.I., Baquero, F., de Rafael, L., Martin de Argila, C., Gisbert, J.P., Bermejo, F., *et al.* (1999) Mutations in 23S rRNA in *Helicobacter pylori* conferring resistance to erythromycin do not always confer resistance to clarithromycin. *Antimicrob Agents Chemother.* **43**, 374–376.

Gilbert, P., Collier, P.J. and Brown, M.R.W. (1990) Influence of growth rate on susceptibility to antimicrobial agents: biofilms, cell cycle, dormancy and stringent response. *Antimicrob. Agents Chemother.* **34**, 1865–1868.

Gilbert, P. and Brown, M.R.W. (1995) Phenotypic plasticity and mechanisms of protection of bacterial biofilms from antimicrobial agents. In: *Microbial Biofilms* (Lappin-Scott, H.E and Costerton, J.W., eds) pp. 118–132

Goldman, J.D., White, D.G. and Levy, S.B. (1996) Multiple antibiotic resistance (*mar*) locus protects *Escherichia coli* from rapid cell killing by fluoroquinolones. *Antimicrob. Agents Chemother.* **40**, 1266–1269.

Gómez-Gómez, J.M., Blázquez, J., de los Monteros, L.E.E., Baquero, M.R., Baquero, F. and Martínez, J.L. (1997a) *In vitro* plasmid-mediated encoded resistance to quinolones. *FEMS Microbiol. Letters* **154**, 271–276.

Gomez-Gomez, J.M., Blazquez, J., Baquero, F. and J.L. Martinez. (1997b) H-NS and RpoS regulate emergence of Lac Ara(+) mutants of *Escherichia coli* MCS2. *J. Bacteriol.* **179**, 4620–4622.

Gould, I.M. and MacKenzie, F.M. (1997) The response of Enterobacteriaceae to β-lactam antibiotics-"round forms, filaments and the root of all evil". *J. Antimicrobial Chemother.* **40**, 495–499.

Gustafson, J.E., Candelaria, P.V., Fisher, S.A., Goodridge, J.P., Lichocik, T.M., McWilliams T.M., *et al.* (1999). Growth in the presence of salicylate increaees fluoroquinolone resistance in *Staphylococcus aureus. Antimicrob. Agents Chemother.* **43**, 990–992.

Hakenbeck, R. (1999a) β-lactam-resistant *Streptococcus pneumoniae*: epidemiology and evolutionary mechanism. *Chemotherapy* **45**, 83–94.

Hakenbeck, R., Grebe, T., Zähner, D. and Stock, J.B. (1999b) β-lactam resistance in *Streptococcus pneumoniae*: penicillin-binding proteins and non-penicillin-binding proteins. *Mol. Microbiol.* **33**, 673–678.

Heinemann, J.A. (1999) How antibiotics cause antibiotic resistance. *Drug Discov. Today* **4**, 72–79.

Hooper, D.C. (1999) Mechanisms of fluoroquinolone resistance. *Drug Res. Updates* **2**, 38–55.

Jacob, F. (1977) Evolution and tinkering. *Science* **196**, 1161–1163.

Jaffe, A., Chabbert, Y.A. and Semonin, O. (1982) Role of porin proteins OmpF and OmpC in the permeation of beta-lactams. *Antimicrob. Agents Chemother.* **22**, 942–948.

Kauffman, S.A. (1993) The structure of rugged fitness landscapes. In: *The Origins of Order*. Oxford University Press, Inc., Oxford, pp. 33–67.

Kepler, T.B. and Perelson, A.S. (1998) Drug concentration heterogeneity facilitates the evolution of drug resistance. *Proc. Natl. Acad. Sci. USA* **95**, 11514–11519.

Korona, R., Nakatsu, C.H., Forney, L.J. and Lenski, R.E. (1994) Evidence for multiple adaptive peaks from populations of bacteria evolving in a structured habitat. *Proc. Natl. Acad. Sci. USA* **91**, 9037–9041.

Lenski, R. (1997) The cost of antibiotic resistance-from the perspective of a bacterium. In: *Antibiotic Resistance: Origins, Evolution, Selection and Spread*, J. Wiley & Sons, Ciba Found Symp. **207**, 131–151.

Levin, B.R. and Anderson, R,M. (1999a) The population biology of anti-infective chemotherapy and the evolution of drug resistance: more questions than answers. In: *Evolution in Health and Disease* (Stephen C. Stearns, Ed.). Oxford University Press, pp. 125–137.

Levin, B.R., Lipsitch, M., Perrot, V., Schrag, S., Antia, R., Simonsen, L., *et al.* (1997) The population genetics of antibiotic resistance. *Clin. Infect. Dis.* **24**, S9–S16.

Levin, B.R., Lipsitch, M. and Bonhoeffer, S. (1999b) Population biology, evolution, and infectious disease: convergence and synthesis. *Science* **283**, 806–809.

Lipsitch, M. and Levin, B.R. (1997) The population dynamics of antimicrobial chemotherapy. *Antimicrob. Agents Chemother.* **41**, 363–373.

Lipsitch, M. and Levin, B.R. (1998) Population dynamics of tuberculosis treatment: mathematical models of the roles of non-compliance and bacterial heterogeneity in the evolution of drug resistance. *Int. J. Tuberc. Lung Dis.* **2**, 187–199.

Lorian, V. (1986) Effect of low antibiotic concentrations on bacteria: effects on ultrastructure, their virulence, and susceptibility to immunodefences. In: *Antibiotics in Laboratory Medicine*, 2nd edition (Lorian, V. ed.). pp. 596–668. Williams and Wilkins, Baltimore, MD.

Ma, D., Cook, D.N., Hearst, J.E. and Nikaido, H. (1994) Efflux pumps and drug resistance in Gram negative bacteria. *Trends Microbiol.* **2**, 489–493.

Maenhaut Michel, G. and Shapiro, J.A. (1994) The roles of starvation and selective substrates in the emergence of araB-lacZ fusion clones. *EMBO J.* **13**, 5229–5239.

Malathum, K., Coque, T.M., Singh, K.V. and Murray, B.E. (1999) *In vitro* activities of two ketolides, HMR3647 and HMR3004, against Gram positive bacteria. *Antimicrob. Agents Chemother.* **43**, 930–936.

Mamber, S.W., Kolek, B., Brookshire, K., Bonner, W.D., D.P. and Fung-Tomc, J. (1993) Activity of quinolones in the Ames *Salmonella* TA102 mutagenicity test and other bacterial genotoxicity assays. *Antimicrob Agents Chemother.* **37**, 213–217.

Markham, P.N. and Neyfack, A.A. (1996) Inhibition of the multidrug transporter NorA prevents emergence of norfloxacin resistance in *Staphylococcus aureus. Antimicrob. Agents Chemother.* **40**, 2673–2674.

Markham, P.N. (1999) Inhibition of the emergence of ciprofloxacin resistance in *Streptococcus pneumoniae* by the multidrug efflux inhibitor reserpine. *Antimicrob. Agents Chemother.* **43**, 988–989.

Martinez, J.L., Martinez-Suarez, J., Culebras, E., Perez-Diaz, J.C. and Baquero, F. (1989) Antibiotic inactivating enzymes from a clinical isolate of *Agrobacterium radiobacter*. *J Antimicrob Chemother.* **23**, 283–284.

Masuda, N., Gotoh, N., Ishii, C., Sakagawa, E., Ohya, S. and Nishino, T. (1999). Interplay between chromosomal β-lactamase and the MexAB-OprM efflux system in intrinsic resistance to β-lactams in *Pseudomonas aeruginosa. Antimicrob. Agents Chemother.* **43**, 400–402.

Medeiros, A.A. (1997) Evolution and dissemination of β-lactamases accelerated by generations of β-lactam antibiotics. *Clin. Infect. Dis.* **24**, S19–S45.

Menard, R., Molinas, C., Arthur, M., Duval, J., Courvalin, P. and Leclerq, R. (1993) Overproduction of 3'-aminoglycoside phosphotransferase type I confers resistance to tobramycin in *Escherichia coli. Antimicrob. Agents Chemother.* **37**, 78–83.

Miller, J.H. (1996) Spontaneous mutators in bacteria: Insights into pathways of mutagenesis and repair. *Ann. Rev. Microbiol.* **50**, 625–43.

Molinari, G., Guzman, C.A., Pesce, A. and Schito, G.C. (1993) Inhibition of *Pseudomonas aeruginosa* virulence factors by subinhibitory concentrations of azithromycin and other macrolide antibiotics. *J. Antimicrob. Chemother.* **31**, 681–688.

Moreillon, P. and Tomasz, A. (1988) Penicillin resistance and defective lysis in clinical isolates of pneumococci: evidence for two kinds of antibiotic pressure operating in the clinical environment. *J. Infect. Dis.* **157**, 1150–1157.

Moxon, E.R., Rainey, P.B., Nowak, M.A. and Lenski, R.E. (1994) Adaptive evolution of highly mutable loci in pathogenic bacteria. *Curr. Biol.* **4**, 24–33.

Moxon, E.R. and Thaler, D.S. (1997) The tinkerer's evolving tool-box. *Nature* **367**, 659–662.

Muñoz-Bellido, J.L., Alonso Manzanares M.A., Yagüe Fuirao, G.H., Gutierrez Zufiaurre, M.N., Toldos, M.C.M., *et al.* (1999) *In vitro* activities of 13 fluoroquinolones against *Staphylococcus aureus* isolates with characterized mutations in *gyrA, gyrB, grlA,* and *norA* and against wild-type isolates. *Antimicrob. Agents Chemother.* **43**, 966–968.

Nash, K.A. and C.B. Inderlied. (1995) Genetic basis of macrolide resistance in *Mycobacterium avium* isolated from patients with disseminated disease. *Antimicrob Agents Chemother.* **39**, 2625–2630.

NCCLS. (1999) *Performance Standards for Antimicrobial Susceptibility Testing: Ninth Informational Supplement.* M100-S9. NCCLS, Wayne, Penn.

Negri, M.C., Morosini, M.I., Loza, E. and Baquero, F. (1994) *In vitro* selective concentrations of beta-lactams for penicillin-resistant *Streptococcus pneumoniae* populations. *Antimicrob. Agents Chemother.* **38**, 122–125.

Nikaido, H. (1998) Multiple antibiotic resistance and efflux. *Curr. Opin. Microbiol.* **1**, 516–523.

Novak, R., Henriques, B., Charpentier, E., Normark, S. and Tuomanen, E. (1999) Emergence of vancomycin tolerance in *Streptococcus pneumoniae. Nature* **399**, 590–593.

Okusu, H., Ma D. and Nikaido, H. (1996) AcrAB efflux pump plays a major role in the antibiotic resistance phenotype of *Escherichia coli* multiple-antibiotic-resistance (Mar) mutants. *J. Bacteriol.* **178**, 306–308.

Oliver, A., Cantón, R., Campos, P., Baquero, F. and Blázquez, J. (2000) High frequency of hypermutable *Pseudomonas aeruginasa* in cystic fibrosis lung infection. *Science,* **288**: 1251–1253.

Perlin, M.H. and Lerner, S.A. (1979) Amikacin resistance associated with a plasmid-borned aminoglycoside phosphotransferase in *Escherichia coli. Antimicrob. Agents Chemother.* **16**, 598–604.

Perlin, M.H. and Lerner, S.A. (1986) High level amikacin resistance in *Escherichia coli* due to phosphorylation and impaired aminoglycoside uptake. *Antimicrob. Agents Chemother.* **29**, 216–224.

Pestova, E., Beyer, R., Cianciotto, N.P. Noskin, G.A. and Peterson, L.R. (1999) Contribution of topoisomerase IV and DNA Gyrase mutations in *Streptococcus pneumoniae* to resistance to novel fluoroquinolones. *Antimicrob. Agents Chemother.* **43**, 2000–2004.

Petrosino, J., Cantu III, C. and Palzkill, T. (1998) β-lactamases: protein evolution in real time. *Trends Microbiol.* **6**, 323–327.

Phillips, I., Culebras, E., Moreno, F. and Baquero, F. (1987) Induction of the SOS response by new 4-quinolones. *J. Antimicrob. Chemother.* **20**, 631–638.

Podglajen, I., Breuil, J. and Collatz, E. (1994) Insertion of a novel DNA sequence, IS*1186,* upstream the silent carbapenemase gene *cfiA,* promotes expression of carbapenem resistance in clinical isolates of *Bacteroides fragilis. Mol. Microbiol.* **12**, 105–114.

Prammananan, T., Sander, P., Springer, B. and Böttger, E.C. (1999) RecA mediated gene conversion and aminoglycoside resistance in strains heterozygous for rRNA. *Antimicrob. Agents Chemother.* **43**, 447–453.

Radman, M. (1999) Enzymes of evolutionary change. *Nature* **401**, 866–869.

Rainey, P.B. and Travisano, M. (1998) Adaptive radiation in a heterogeneous environment. *Nature* **394**, 69–72.

Reguera, J.A., Baquero, F., Páerez-D'az, J.C. and Martínez. J.L. (1991) Factors determining resistance to β-lactams combined with β-lactamase inhibitors in *Escherichia coli. J. Antimicrob. Chemother.* **27**, 569–575.

Ren, L., Radman, M. and Humayun, M.Z. (1999) *Escherichia coli* cells exposed to streptomycin display a mutator phenotype. *J. Bacteriol.* **181**, 1043–1044.

Riesenfeld, C., Everett, M., Piddock, L.J. and Hall, B.G. (1997) Adaptive mutations produce resistance to ciprofloxacin. *Antimicrob Agents Chemother.* **41**, 2059–2060.

Rosche, W.A., Foster, P.F. and Cairns, J. (1999) The role of transient hypermutators in adaptive mutation in *Escherichia coli. Proc. natl. Acad. Sci. USA* **96**, 6862–6887.

Sánchez, L., Pan, W., Vinas, M. and Nikaido, H. (1997) AcrAB homolog of *Haemophilus influenzae* codes for a functional multidrug efflux pump. *J. Bacteriol.* **179**, 6855–6857.

Sander, P., Prammananan, T., Meier, A., Frischkorn, K. and Bottger E.C. (1997) The role of ribosomal RNAs in macrolide resistance. *Mol Microbiol.* **26**, 469–480.

Schrag, S.J. and Perrot, V. (1997a) Reducing antibiotic resistance. *Nature* **381**, 120–121.

Schrag, S.J., Perrot, V. and Levin, B.R. (1997b) Adaptation to the fitness costs of antibiotic resistance in *Escherichia coli. Proc. Royal Soc. of London Series B* **264**, 1287–1291.

Scott Champney, W. and Tober, C. (1999) Molecular investigation of postantibiotic effects of clarithromycin and erythromycin on *Staphylococcus aureus* cells. *Antimicrob. Agents Chemother.* **43**, 1324–1328.

Shibl, A.M. (1983) Effect of antibiotics on production of enzymes and toxins by microorganisms. *Rev. Infect. Dis.* **5**, 865–875.

Stevens, D.L., Yan, S. and Bryant, A.E. (1993) Penicillin-binding protein expression at different growth stages determines penicillin efficacy *in vitro* and *in vivo:* an explanation for the inoculum effect. *J. Infect. Dis.* **167**, 1401–1405.

Taddei, F., Matic, I., Godelle, B. and Radman, M. (1997a) To be a mutator, or how pathogenic and commensal bacteria can evolve rapidly. *Trends Microbiol.* **5**, 427–429.

Taddei, F., Radman, M., Maynard-Smith, J., Toupance, B., Gouyon, P.H. and Godelle, B. (1997b) Role of mutator alleles in adaptive evolution. *Nature* **387**, 700–702.

Yoshimura, F. and Nikaido, H. (1985) Diffusion of β-lactam antibiotics through the porin channels of *Escherichia coli* K12. *Antimicrob. Agents Chemother.* **27**, 84–92.

Ziha-Zarifi, I., Llanes,C., Kohler, T., PechÈre, J.C. and Plesiat, P. (1999) *In vivo* emergence of multidrug-resistant mutants of *Pseudomonas aeruginosa* overexpressing the active efflux system MexA-MexB-OprM. *Antimicrob. Agents Chemother.* **43**, 287–291.

9. Possible Impact on Antibiotic Resistance in Human Pathogens Due to Agricultural Use of Antibiotics

Abigail A. Salyers[1] and Patty McManus[2]
[1]Department of Microbiology, University of Illinois, Urbana, IL, USA.
[2]Department of Plant Pathology, University of Wisconsin, Madison, WI, USA.

AGRICULTURAL USE OF ANTIBIOTICS: AN AREA OF GROWING CONCERN

Extent of Agricultural Use of Antibiotics

Much has been written about the threat to human health posed by antibiotic-resistant bacterial pathogens (see, for example, Davies, 1996). Most scientists agree that the main factor contributing to the rise of antibiotic resistance is the abuse and overuse of antibiotics by physicians to treat or prevent human disease. But is this the only factor contributing to the rise of resistance? Antibiotic-resistant bacteria can be found in the environment and in the intestines of humans who have not received antibiotics (Salyers and Amabile-Cuevas, 1997). The distribution of resistant strains in nature suggests that use of antibiotics in hospitals and in clinics is not the only factor contributing to the rise in incidence of antibiotic resistant bacteria. Another setting in which large quantities of antibiotics are used is agriculture (McDonald *et al.*, 1997; WHO Report, 1997; IOM report, 1988).

In the U.S., accurate figures for the total amount of antibiotics used in agriculture and aquaculture are not available, although estimates place the amount used in agriculture at about equal the amount used to treat or prevent human disease (IOM report, 1988). Figures are now available for Europe (*http://www.fedesa.be*) and Australia (*http://www.health.gov.au/pubs/jetacar.htm*). The European estimate finds about half of antibiotics going to farm use. In Australia, agricultural use of antibiotics is double that used in human medicine. In any event, the total amount of antibiotics used in agriculture and aquaculture is probably less important than the way these antibiotics are used. The U.S. FDA has recently recognized this by incorporating the type of antibitoic use in its framework document proposed for evaluating agricultural antibiotics for approval (FDA framework document, 1999). Everything we have learned about how bacteria become resistant to antibiotics suggests that long-term, low-dose use of antibiotics is much more likely to select for resistant strains than short-term, high dose use.

Antibiotics are used in three ways in agriculture. Two uses have parallels in human medicine: the use of antibiotics to treat sick animals and the prophylactic use of antibiotics to prevent disease. The third agricultural use of antibiotics, which has no parallel in human medicine is as growth-promoters. That is, the long-term administration of low levels of antibiotics to some animals increases the weight gain per unit of food and allows animals to reach full weight more quickly. The practice of using antibiotics as growth promoters is the one currently under attack, both because it is the type of use most likely to select for resistant bacteria and because it is not as critical to animal health as treatment or prophylaxis. In Australia, most of the antibiotics used in agriculture were used as growth promoters or for prophylaxis with only about 10% being used for therapy, according to the Jetacar figures. The most recent estimate from the European Union (1997, see website) found 66% of the agricultural use going for growth promotion and the rest for therapy/ prophylaxis.

The boundaries between the three categories of use – therapy, prophylaxis and growth promotion – are themselves not well defined. Therapy is the most clearly delineated of these terms, although one could argue whether antibiotics administered to all members of a group of animals after several have become ill and the remainder are likely to contract the disease is therapy or prophylaxis. The line between prophylaxis and growth promotion is even less clearly drawn. If one believes that the reason antibiotics promote growth of animals is because they keep infections in check – and in some cases this is clearly the mechanism – then the feeding of antibiotics to animals for purposes of growth promotion could be regarded as prophylaxis. This distinction has become important politically and economically because the public and consumer groups have become concerned about the use of antibiotics as growth promoters. Thus, farmers' groups and pharmaceutical companies may be tempted to replace the term "growth-promotion" with "prophylaxis" wherever possible. And given the grey area in the definitions of these two terms, they would be justified in doing so. Another issue is the context in which these terms are used. Most people would probably find the use of prophylaxis to prevent sickness in animals acceptable, but might balk at the spraying of antibiotics on fruit trees to prevent bacterially-induced rot (see later section).

What Antibiotics Should Receive Primary Attention?

A subject that has become increasingly contentious is the question of what types of resistance in which bacteria should be the primary focus of concern. Many of the antibiotics used extensively in agriculture, aquaculture and horticulture, such as

β-lactam antibiotics, tetracycline and gentamicin are also used in human medicine, but mainly for less serious illnesses for which alternative therapies are available. Of course, β-lactam antibiotics are also used to treat life-threatening diseases such as pneumonia, but in many cases effective alternatives to this class of drug are available.

In recent years, attention has focused not so much on these drugs but on drugs that select for resistance to antibiotics that are the last ones available for treating serious human infections (FDA framework document). Among these are antibiotics such as tylosin and avoparcin (Table 1). These antibiotics have been used widely in agriculture but not in human medicine. They are of concern, however, because they cross-select for resistance to some of the current front line human use antibiotics such as Synercid, vancomycin, and some of the newer macrolide antibiotics currently in human clinical trials. Examples of infections for which only one or a few antibiotics are still effective are endocarditis and septicemia caused by *Staphylococcus* species and *Enterococcus* species, and pneumonia and septicemia caused by *Streptococcus pneumoniae*. Although multi-drug resistant strains of these bacteria are seen more frequently in hospital intensive care wards than in the community, they are being seen more commonly in patients with community-acquired disease and thus consitute an even greater threat than before (see, for example, Herold *et al.*, 1998;

Table 1. Antibiotics used in agriculture that cross-select for resistance to important human use antibiotics. Avoparcin is now banned in Europe and was never allowed in the U.S., but it is still being used in Southeast Asia.

Agricultural Drug	Class of Antibiotic	Human Use Drug(s)
Avoparcin Enrofloxacin Spiramycin Tylosin	Glycopeptide Fluoroquinolone Macrolide	Vancomycin Ciprofloxacin Erythromycin Azithromycin Synercid
Virginiamycin	Streptogramin	Erythromycin Azithromycin Synercid
Avilomycin	Oligosaccharide	Everinomycin (SCH 27899)

Anon., 1999). The European Union has now banned the use of avoparcin, tylosin, spiramycin, bacitracin and virginiamycin as growth promoters.

Another factor that must be taken into account is changing patterns of antibiotic use and new drugs either in human clinical trials or recently approved. In the past, for example, macrolides and streptogramins were not considered important human use antibiotics. Macrolides had relatively few uses in human medicine and streptogramins had not been used at all in the U.S. and only occasionally in Europe. With the introduction of azithromycin, a powerful new macrolide, and the impending approval of some even newer macrolides, which are candidates for control of otherwise untreatable infections, macrolide antibiotics have taken on a new importance. Similarly, the recent approval of the streptogramin mixture, Synercid, for use in treating infections caused by multi-drug resistant Gram positive cocci has given streptogramins an importance they did not previously have. Clearly changes in human use patterns and the appearance of new antibiotics on the market will continue to occur. This makes it difficult to predict with confidence what antibiotics will be acceptable in the current or future regulatory climate, a circumstance that makes the development of new antibiotics for agricultural use in therapy, prophylaxis and growth promotion much less attractive for the pharmaceutical companies.

Impact of Agricultural Use of Antibiotics

There is no question that antibiotic use in agriculture leads to increases in the number of resistant bacteria in the intestines of animals exposed to antibiotics (Endtz et al., 1991; Helmuth and Protz, 1996; Witte, 1998; WHO Report, 1997). This increase could have serious consequences. Increasing antibiotic-resistance could have a direct impact on farmers in the form of resistant animal pathogens.

A question that is closer to home for most people, however, is whether antibiotic use on the farm is compromising the effectiveness of antibiotics used to treat human infections. At first glance, it might seem unlikely that the use of antibiotics on farms would have any impact on resistance profiles of human pathogens, especially

those that are human-specific, because farms are usually located far from large population centers. Farms and cities may be located far from each other geographically, but they are linked very closely through the food supply and the water supply. They are also linked through people who travel from the farm to towns and cities, especially if they enter hospitals. Since both agriculture and food processing have become highly centralized, food from one site travels to many different locations. The water supply is somewhat more localized but runoff from farms could easily transmit resistant bacteria or antibiotic residues to groundwater or rivers used by nearby towns or cities for drinking water. Water used for irrigation or for washing fruits and vegetables at the harvest site, if contaminated by farm runoffs, could also transmit antibiotic-resistant bacteria to vegetables and fruits. A study of groundwater in a rural U. S. state found a surprisingly high incidence of antibiotic-resistant E. coli (McKeon et al., 1995). The authors did not determine the origin of the antibiotic-resistant bacteria, but since the watershed is in a rural area, runoff from nearby farms is high on the list of possible sources. More recently, Waters and Davies (1997) found a high incidence of fluoroquinolone resistance in soil bacteria. An important question to be answered is whether such findings reflect contamination of apparently pristine areas by runoffs from farm or human wastes or whether they reflect the existence of selective pressures other than antibiotics. This latter possibility will be discussed later in this chapter.

ANTIBIOTIC USE PATTERNS – AN IMPORTANT ECOLOGICAL FACTOR

How antibiotics are used – the amount administered and the time frame of administration – would be expected to have an effect on the frequency with which resistant strains will be selected in a particular setting (Lipsitch and Levin, 1997). When antibiotic use is transient, bacteria resistant to the antibiotic are selected initially, but may then be unable to complete with susceptible bacteria when antibiotic selection is removed. This may be the reason why drastically reducing the use of a previously overused antibiotic in human medicine is usually followed by a decrease in the incidence of resistant strains (see, for example, Westh, 1995; Klare *et al.*, 1999). If antibiotic selection is continuous over long periods of time, however, bacteria have the chance to accumulate mutations that allow them to retain their resistance phenotype without losing fitness in the absence of selection. This has been shown in laboratory experiments using *E. coli* as the test subject (Lenski *et al.*, 1994; Bottger *et al.*, 1998; Schrag *et al.*, 1998). We do not know the extent to which this type of adaptation occurs in nature or how long selection must be maintained for the adaptation to occur. In fact, the answer is likely to be different for different bacterial species and for different animals.

Control of Antibiotic Use

Antibiotic use in agriculture is not as tightly controlled as antibiotic use in humans. In most developed countries, a veterinarian's prescription is needed to purchase antibiotics. But except for treatment of sick animals, and sometimes in that case too, the antibiotic is administered by farm workers who may not have been trained adequately in the proper use of antibiotics. Human patients can misuse prescriptions too, but misuse of a few grams of antibiotic is not a problem of the same magnitude as the misuse of enough antibiotic to medicate a herd of cattle or a flock of turkeys. In the developing world, which produces a quarter of the world's meat supply, the only control on antibiotic use in most cases is the cost of antibiotics and medicated feed. Although the use of antibiotics in agriculture appears to be controlled to some extent, an issue that is never mentioned is black market sales of antibiotics. This question may be particularly pertinent in Europe, where the European Union has already banned the use of avoparcin as a growth promoter and is in the process of banning other growth promoters. Will farmers who have come to rely on these antibiotics accept the ban and change their management practices or will they stockpile antibiotics or acquire them from countries where they are still sold?

Antibiotic Preparations As Carriers of Resistance Genes

Antibiotics used to treat sick animals are generally of the same quality as antibiotics used to treat humans. This is not necessarily true for antibiotics used as growth promoters. Since antibiotics used as growth promoters are used in large quantities, they must be cheap enough to be economical for the farmer. Pharmaceutical companies that sell antibiotics for use on the farm market antibiotic preparations that are much less pure than those intended for human consumption. One of the contaminants of these preparations is DNA from the antibiotic producing microbe. Antibiotic-producing bacteria have resistance genes to protect them from the antibiotic they produce. Webb and Davies (1993) showed that agricultural antibiotic preparations contained levels of resistance genes from the producer that were high enough to be easily detectable by PCR. Thus, resistance genes are being fed along with the antibiotics that select for strains that acquire them. Of course, the resistance gene would have to enter intestinal bacteria, integrate into the chromosome and be expressed in their new host. This series of steps would occur at low frequency because most intestinal bacteria are not naturally transformable and would thus not take up the DNA. Also, DNA would have to integrate by illegitimate recombination, an inefficient process. If the gene was not expressed at first, gene expression would have to be activated by mutations in the promoter or insertion of an insertion sequence in the promoter region to provide a new promoter. Although this series of steps is extremely unlikely, the constant presence of the antibiotic would quickly select for a bacterium that acquired the resistance gene. Recently, the vancomycin resistance genes in the bacterium that produces vancomycin have been cloned and sequenced (Marshall *et al.*, 1998). Their sequences are only about 60% identical to vancomycin resistance genes currently being found in vancomycin-resistant strains. This is strong proof that the vancomycin resistance genes that are turning up in animal isolates in Europe did not arise as a result of uptake of DNA from the antibiotic preparations being fed to animals.

Why Is Antibiotic Use So Widespread In Agriculture?

If antibiotic use has the potential to cause serious health problems for animals and humans, why do farmers use them? It is important to understand why farmers have come to rely so heavily on antibiotics because only realistic solutions to the problem, which take into account the needs of farmers as well as the needs of the public, are going to be successful ones. The increased centralization of agriculture has been widely deplored, but the fact is that this centralization has been driven by strong economic factors. As farms have become larger and animal quarters more crowded, the danger of infectious disease outbreaks within herds or flocks has increased dramatically. Farmers have been accused of using antibiotics as a substitute for good hygienic practices and reports in the literature claim that proper animal housing and care practices more than make up for the advantages conferred by antibiotic use (Wierup, 1998, 1999; Aaerstrup, 1999). Still, it is hard to believe that in the very large-scale animal production operations antibiotic use can be wholly discontinued, especially therapeutic and prophylactic use. It is also important to realize that poor hygienic practices may themselves have been encouraged by the economic climate in which farmers operate. Without relieving pressures that foster bad practices or rewarding good practices, it is unlikely that bad practices will disappear.

The use of antibiotics as growth promoters is another practice that has been adopted by farmers because the meat production industry has become so competitive that small margins are critically important. A growth yield increase of only 4–5% may mean the difference between profit and loss. Clearly, farmers believe in antibiotics as growth promoters because they spend millions of dollars each year to purchase them. Agricultural use of antibiotics has also been defended on the basis that banning their use would drive food prices upward. A recent analysis by the Institute of Medicine has placed the actual cost of eliminating growth promoters at about $10 per person per year (IOM Report, 1998). Such estimates must be viewed with caution, but further evidence that the effect on prices would not be too drastic comes from the Coleman Natural Beef Comany (Colorado, USA), which uses the fact that it does not use antibiotics as growth promoters as a marketing tool for its beef. The Carson company charges about 25% more for its beef than other companies that do use antibiotics. This is more than the $10 per person

per year estimate, but is still a relatively low premium to pay for reduced antibiotic use.

EFFECT OF AGRICULTURAL USE OF ANTIBIOTICS ON ANIMAL COMMENSALS

Incidence of Antibiotic-Resistance

There have been a number of reports of antibiotic-resistance in bacteria isolated from farm animals (DANMAP, 1997; Aaerstrup et al., 1998; Bates et al., 1994; Welton et al., 1998; WHO report, 1998; Witte, 1998; Wray et al., 1993). A few recent examples of such reports will serve to illustrate some important features of resistance in animal isolates. Van den Braak et al. (1998) found that 79% of poultry carcasses taken from supermarkets or local meat markets contained vancomycin-resistant enterococci. Aaerstrup et al. (1998) found that 59% of E. faecium isolated from live chickens were resistant to vancomycin. The drug avoparcin, a vancomycin analog, was used in Europe as a growth promoter until 1997 when approval of its use was suspended. There are two questions that should be raised in connection with reports of high levels of resistance in animal bacteria. First, since most of the surveillance efforts were started within the last several years, there are no baseline data from previous times to provide some indication of how much the incidence of resistance has increased or to show that it is linked with antibiotic use. Resistance levels can rise because the resistance gene is linked genetically (e.g., on a plasmid) with other types of resistance genes, so that one type of selective pressure can select for a resistance to an unrelated drug (Woodford et al., 1995; Salyers and Amabile-Cuevas, 1997; Aaerstrup, 1998). A second question is how the investigators define antibiotic resistance (breakpoint between susceptible and resistant strains). Although breakpoints are now being established for many animal pathogens, the appropriate breakpoints for commensals have not been determined. Certainly, there is nothing comparable to the precisely spelled-out protocols for determining resistance in human pathogens.

Not all investigators find such high levels of resistant strains. Klein et al. (1998) found no vancomycin-resistant enterococci, in fact few enterococci of any kind, in samples of minced beef and pork from supermarkets in Germany. Aaerestrup *et al.* (1998) also found no vancomycin-resistant enterococci in the intestines of cattle in Denmark, but they did find that 29% of enterococci from live pigs were resistant. In general, a Danish surveillance program (DANMAP, 1997) found less antibiotic resistance in bacteria from the intestines of cattle than in those isolated from pigs or chickens. These differences may reflect differences in the way antibiotics are used in cattle farming as opposed to pig farming or chicken farming, a point that has not yet been addressed. Another possible explanation is suggested by a careful perusal of Aarestrup *et al.* (1998), the only paper to provide a detailed breakdown of the data on levels of resistance in the strains they tested. The breakpoint between resistance and susceptibility for vancomycin was 8 ug/ml in this study. Klein *et al.* (1998) did not provide MIC values or indicate the breakpoint they used. Van den Braak *et al.* (1998) used 6 ug/ml as the definition of resistance. In the study of Aarestrup *et al.* (1998), 29% of the strains were listed as resistant to avoparcin, but 55% of the strains had MICs of 2–8 ug/ml – still technically susceptible but getting close to resistant. Similarly, none of the E. faecalis strains from pigs were scored resistant, but over half had an MIC value of 2 or more and 20% had MIC values of 4–8 ug/ml, right below the breakpoint. Thus, a slight adjustment in the breakpoint could have had significant effects on the number of strains classified as resistant. Clinical microbiologists have learned this lesson from seeing ''sensitive'' species become largely ''resistant'', apparently almost overnight. But in reality, the distribution within the ''sensitive'' category had been inching toward the breakpoint for a long time. It is better to think of susceptibility as a population distribution. As the population becomes less susceptible and the mean moves toward the breakpoint, resistance levels can go rather rapidly from low to high, making it seem as if resistance arose within a very short time, when in fact susceptibility had been decreasing for some time.

One of the clearest examples of a cause and effect relationship between antibiotic use in agriculture and increasing resistance in animal commensals is the impact of the use of noursethricin in East Germany (Witte, 1998). Within two years of the initiation of use in 1983, resistance to noursethricin appeared and continued to increase in incidence until 1990 when its use was discontinued. We can be reasonably sure that the resistance arose from agricultural use because noursethricin has not been used in human medicine.

Ecology of Antibiotic Resistance

Antibiotic use patterns can differ from one country to another and such differences will have an effect on the ecology of resistant strains. In Europe, avoparcin was used heavily on farms from 1995 to 1997, when approval was suspended, whereas avoparcin was never approved for use in the United States. Vancomycin has been used much less extensively in European hospitals than in U.S. hospitals. Thus, in Europe, the principal use of glycopeptides is in agriculture whereas in the U.S. the principal use is to treat human infections. Accordingly, patterns of resistance development would be expected to be different in Europe and the U.S. In fact, this proves to be the case. Coque *et al.* (1996) surveyed hospital, community and animal isolates obtained in the U.S. state of Texas. They found vancomycin-resistant enterococci in hospitals but not in the intestines of members of the community, in the intestines of animals or on meat products. Studies carried out in Europe, by contrast, found few if any vancomycin-resistant enterococci in hospitals but Van den Bogaard and Stobberingh (1999) have found vancomycin-resistant *Enterococcus faecium* in the fecal microflora of urban adults and on chicken carcasses offered for sale in meat markets. Defending the use of avoparcin on farms in Europe by arguing that U.S. experience has shown that hospital use not farm use is the important selective pressure is inappropriate because use patterns differ so much.

Unfortunately, most of the scientists currently doing surveillance of antibiotic resistant bacteria were not trained as microbial ecologists. Even in cases where scientists may have such training, funding agencies usually have very specific goals in mind that discourage good ecological design. If a group is funded to monitor resistant bacteria in the intestines of cattle, swine and chickens, the agency supporting their work may not take to kindly to having part of the funds spent on sampling soil, local water and wild animals. Nonetheless, this type of ecological study needs to be done if we are to make sense of the bewildering and sometimes apparently contradictory mass of resistance data that is beginning to become available.

RANK-ORDERING AGRICULTURAL ANTIBIOTICS IN TERMS OF POTENTIAL RISKS

Not only are antibiotic use patterns different in different countries but some antibiotics used in agriculture are a greater threat to human therapy than others. Examples of antibiotics in this class are shown in Table 1. Avoparcin was just mentioned. Other antibiotics of primary concern are the fluoroquinolones, another class of front-line human antibiotics, and virginiamycin, a streptogramin which can elicit cross-resistance to streptogramins and macrolide antibiotics currently used to treat human infections. Avilomycin has been reported to elicit cross-resistance to a new human use antibiotic everninomycin (Aaerstrupm, 1998). Streptogramins have not been used previously the treatment of human infections, but some of the new antibiotics currently being considered for approval for treatment of human infections are streptogramins (e.g. Synercid). Since these antibiotics are needed badly for treatment of patients with multiply antibiotic-resistant infections, it seems prudent to make sure that these new drugs are not compromised before the first vials destined for treatment of human disease are opened.

Focusing on future uses of antibiotics to treat human disease rather than current use has become more important than ever, given new developments in the understanding of human disease. This is well-illustrated by a recent clinical trial, which demonstrated that administration of the macrolide erythromycin to pregnant women significantly reduced the risk of preterm births, presumably because it controlled cases of bacterial vaginosis (Hauth et al., 1995). Currently, another macrolide, azythromycin, is being tested for its ability to prevent heart disease. Mounting evidence supports a role for the bacterium Chlamydia pneumoniae in at least some cases of atherosclerosis. Another macrolide, clarithromycin, is one of the antibiotics

used to treat ulcers caused by Helicobacter pulori. These new uses of macrolides could make the macrolides even more important in human medicine than they are today. Thus, in deciding what antibiotics are to be avoided or more strictly controlled with respect to agricultural use, it is important to consider future use as well as present use.

Another consideration that should be taken into account is genetic linkage of resistance genes. If two types of resistance gene are found together on the same transmissible element, either antibiotic could select for both of them. For example, in Bacteroides species, erythromycin resistance is often accompanied by tetracycline resistance (Salyers et al., 1997). This type of linkage has also been seen, although not universally, in Gram-positive cocci. Examples of more than one type of resistance gene on a plasmid have also been reported in many studies. In fact, there are special elements in the Enterobacteriaciae, called integrons that recruit resistance genes of various sorts to create gene clusters (Hall and Collis, 1995).

There are some agriculturally used antibiotic that can be ranked as least likely to cause problems in human medicine, at least for the present. For example, monensin, an antibiotic that forms channels in bacterial membranes and collapses the proton gradient, is used exclusively in agriculture. It has never been proposed as a human use antibiotic and does not cross-select for resistance to any antibiotic being used to treat humans. Bacitracin might also be included in this category. It is included in some topical ointments designed for human use but it is by no means a frontline human drug. Even here, however, caution is in order because there have been reports that bacitracin can eliminate carriage of vancomycin-resistant enterococci (O'Donnovan et al., 1994). If this is borne out by subsequent studies, the view of bacitracin's importance in human medicine may change dramatically. As the infectious disease picture continues to change, today's irrelevant-to-human-medicine antibiotic could easily become tomorrow's front-line human therapy. This fact of life is going to make it increasingly difficult for scientists in regulatory agencies to make choices about what antibiotics are safe for agricultural use.

Another fact of life is that the timing of resistance development varies widely from species to species and from antibiotic to antibiotic. Some antibiotics seem to generate resistant strains more readily than others. It is important to understand this to avoid a fallacy that is often committed: the fallacy of assuming that since resistance has not arisen after years of use of an antibiotic, resistance will never develop. An instructive example is provided by penicillin resistance in streptococci. *E. coli* and many other Gram-negative species began to develop resistance to ampicillin and penicillin almost immediately after these antibiotics were first used in human medicine. Since this did not happen with *S. pneumoniae*, which remained sensitive to penicillin for decades, physicians lulled themselves into believing that *S. pneumoniae* would never become resistant to penicillin. When penicillin-resistant strains finally arrived, physicians were not prepared, either medically or intellectually to see *S. pneumoniae* go from an easily treated pathogen to a multiply-resistant one. During the long period of penicillin sensitivity, *S. pneumoniae* strains were picking up resistances to other antibiotics such as tetracycline, erythromycin and aminoglycosides, but clinicians did not pay attention to this trend because they assumed that penicillin would always be effective. This example illustrates the fact that there is usually some warning about the imminent emergence of multi-drug resistant strains if the surveillance apparatus is in place and scientists in the area understand how to look for emerging resistance trends.

POSSIBLE IMPACTS OF ANTIBIOTIC-RESISTANT ANIMAL COMMENSALS ON HUMAN MEDICINE

Antibiotic-resistant Zoonotic Pathogens

There are at least three ways in which antibiotic-resistant animal commensals could have an impact on human medicine. The simplest and most familiar effect is the transmission to humans of such multi-drug resistant foodborne pathogens as nontyphoid *Salmonella* serotyes and *Campylobacter* species (Glynn *et al.*, 1998).

Some physicians have downplayed multi-drug resistant *Salmonella* species as a threat to human health because salmonellosis is usually not treated with antibiotics and antibiotic resistance is thus irrelevant. This view overlooks the fact that in any large outbreak, there will be some people who develop a systemic form of the disease, which can be fatal if not treated with antibiotics. So far, the pentaresistant *Salmonella typhimurium* strain DT104 has produced outbreaks of disease but no major public health disasters. This is a good thing, of course, but its down side is that the perceived lack of impact on human health could lull the public into complacency about antibiotic-resistant foodborne bacteria. An ominous indicator that the worst may be beginning to happen is a recent report by Molbak *et al.* (1999) of an outbreak of *Salmonella typhimurium* DT104 that caused two deaths due to treatment failure. This is the first report clearly linking agricultural use of antibiotics with human treatment failures resulting in death.

Antibiotic-Resistant Commensals of Animals and Humans

There are other, potentially as serious, outcomes of the agricultural use of antibiotics than gastroenteritis or septicemia caused by *Salmonella* serotypes and *Campylobacter* species. The pathogens clinicians and the public really worry about are bacteria such as *Streptococcus pneumoniae* and *Staphylococcus aureus* that killed many people in the preantibiotic era and are still taking a heavy toll even when they are still susceptible to some antibiotics. Bacterial pneumonia, of which *S. pneumoniae* is the leading cause, is currently the most common cause of infectious disease death in the U.S. Opportunistic pathogens such as *Enterococcus* species are also cause for concern because they are common causes of life-threatening hospital-acquired infections (Coque *et al.*, 1996). A person who enters a hospital colonized with multi-drug resistant *S. pneumoniae*, *S. aureus* or *Enterococcus* spp. has an increased risk of dying from a post-surgical infection caused by members of his or her own bacterial flora. Such patients also bring into the hospital antibiotic-resistant bacteria that can spread to other patients on the hands of hospital staff members.

S. pneumoniae and *S. aureus* are primarily human pathogens. Enterococci are colonizers of many animals as well as humans.

One way these major human pathogens could be affected by agricultural use of antibiotics is through exposure of farm workers to antibiotics. Many people are transiently colonized with *Streptococcus pneumoniae* or other potentially pathogenic bacterial species at various times in their lives. Continuous exposure to antibiotic dusts could encourage the development of resistant strains of these species, which could later cause serious disease. Resistant strains could also be passed to family members and other contacts. Such events would probably have a relatively localized effect, although the increasing tendency of people to travel widely might increase the impact of multi-drug resistant strains. A second, and potentially more serious, way human-specific pathogens could be affected by antibiotic-resistant animal bacteria is through the transfer of antibiotic-resistance genes from the animal strain to human strains during transient colonization of the human intestine. That is, resistant strains could be transmitted through the food supply to the human intestine where horizontal gene transfer would move the resistance genes into human strains. There have been a number of reports, including some of those already mentioned, documenting the presence of antibiotic-resistant enterococci in human foods. One limitation of these studies is that they have focused exclusively on meat and animal products such as milk and cheese.

Water could also serve as a vehicle for the transfer of antibiotic resistant bacteria to humans. Manure management remains a large and largely unsolved problem in agriculture. Antibiotic-resistant bacteria are being found in water supplies (McKeon *et al.*, 1995). Antibiotics themselves, especially the fluoroquinolones, are also being detected at fairly high levels in wastewater (Raloff, 1998). The unanswered question is where do the antibiotic resistant bacteria and the antibiotic residues come from – agricultural sources or human fecal pollution or a combination of the two? The use of untreated or partially treated water for irrigation or for washing vegetables, or the use of manure as fertilizer for vegetables and fruits could contaminate food plants with antibiotic-resistant bacteria. To what extent this actually occurs is

unknown. Few studies have been done of resistance patterns in the intestinal microflora of vegetarians, but one study by Elder *et al.* (1993) found higher levels of multi-drug resistant bacteria in the intestines of vegetarians than in the bacteria of meat eaters.

Antibiotic-resistant enterococci and other members of the normal flora of animals, which are carried on food, could colonize humans or even cause extraintestinal disease. Kluytmans *et al.* (1995) analyzed a 1994 outbreak of methicillin-resistant *Staphylococcus aureus*, in which the bacteria appear to have been spread on food. The source of this outbreak was almost certainly one of the health care workers who contaminated the food, and not a case involving contamination on the farm or during processing, but it shows that the hospital food supply can serve as a vector for pathogens that cause extraintestinal disease. The patients fed the contaminated food developed sepsis and post-operative infections and a number of hospital staff members were colonized with the strain, making them a possible new source of MRSA for critically ill patients. This is the only food-initiated outbreak of its type to be reported so far, but such events may be more common than we think since food is not commonly considered as a likely vector for hospital acquired infections. Transmission events of this type would be limited to hospitals or nursing homes, where large numbers of people with impaired immune systems are housed, and is less likely to be a problem in other institutions such as schools and military installations.

Assessing Resistance Gene Transfer in Humans

Once an antibiotic-resistant bacterium has been ingested by someone, the bacterium can transfer its resistance genes to another bacterium. At first glance, this might appear to be a very infrequent event, but new studies suggest that this sort of horizontal transfer event is unexpectedly common. So far, the main evidence supporting the hypothesis that horizontal gene transfer events occur in the intestine comes from finding the same resistance genes in very different hosts. If copies of resistance genes in these different hosts have virtually identical DNA sequences, horizontal transfer of some sort is the most likely explanation. An alternative explanation of such events is that selection pressures to keep the amino acid sequence of the protein constant also keep the DNA sequence constant. Thus, two genes could arise independently in two different bacterial species but end up having very similar sequences. This is called convergent evolution. Since, however, the third base in many codons can vary without changing the amino acid, two genes with identical amino acid sequences could differ by as much as 20% at the DNA sequence level. For this reason, copies of the same gene in different bacterial species that are 95–100% identical at the DNA sequence level are virtually guaranteed to have been acquired by horizontal gene transfer. The best examples are those where the DNA sequence identity is 99–100%. Such transfer events are likely to have been recent. An obvious limitation of this type of analysis is that, although one can say with some certainty that the gene has been transferred horizontally, it is usually not possible to deduce the direction of transfer simply from sequence identity, unless the resistance gene has a %G+C or codon usage that differs substantially from that of the host chromosome or unless some sequence signature has been identified. Moreover, it is not possible to deduce whether the transfer occurred directly between the two species or indirectly by means of a third or fourth participant. Nonetheless, this type of analysis gives some indication of the extent to which gene transfer actually occurs in nature.

Over the past few years, a number of examples have been found that support the hypothesis that horizontal gene transfers do occur in nature between very distantly related species and appear to occur rather frequently (Teuber *et al.*, 1996; McDonald *et al.*, 1997; Van den Bogaard *et al.*, 1997; Van den Braak *et al.*, 1998; Teuber *et al.*, 1999). No one has yet measured the frequency of transfer directly in a farm animal or human, but the fact that it was easy to find such examples suggests that they are not rare events. Some examples are provided in Table 2. These show that gene transfer occurs within the human colon and within the intestines of farm animals. Often the genes are still transmissible, but in some cases either the conditions needed to promote transfer in the laboratory were not achieved or the gene had been transferred in but was no longer transferred out. Some surprising conclusions emerge from the evidence summarized in Table 2. First, transfer of antibiotic resistance genes can occur between bacteria normally found in the intestines of animals and bacteria normally found in the intestines of humans. Nikolich *et al.*, 1994) found alleles of *tetQ* in animal species of *Prevotella ruminicola* and in human colonic *Bacteroides* that shared more than 95% DNA sequence identity. By contrast, the chromosomes of *P. ruminicola* and *Bacteroides* spp. share less than 5% DNA-DNA homology. In this case, on the basis of DNA sequences upstream and downstream of *tetQ*, the authors deduced that the direction of transfer had probably been from human to animal bacteria (Nikolich *et al.*, 1994).

Table 2. Some of the cases in which virtually identical resistance genes have been found in distantly related species of bacteria, suggesting that gene transfer has occurred between these species in nature (Salyers and Shoemaker, 1996; Teuber *et al.*, 1996).

Resistance Gene	Bacterial species where found	Site of isolation
tetQ	*Bacteroides* spp.	Human colon
	Porphyromonas spp.	Human mouth
	Prevotella ruminicola	Rumen of cattle
ermG	*Bacteroides* spp.	Human colon
	Bacillus sphaericus	Soil
tetK	*Staphylococcus xylosus*	Cheese
	Staphylococcus aureus	Human body, infections
tetS	*Lactococcus lactis*	Cheese
	Listeria monocytogenes	Human infection
cat	*Enterococcus faecalis*	Sausage
	Staphylococcus aureus	Human infection

In the case of a *vanA* gene found in animal and human strains of *Enterococcus faecium*, the transfer may have been from animal to human bacteria (Van den Braak *et al.*, 1998; Werner *et al.*, 1997). The evidence for this was that the *vanA* gene cluster in animals had one DNA sequence signature, and some of the human strain *vanA* clusters had a different signature, but the human strains also carried *vanA* clusters with the animal signature. Copies of virtually identical alleles of *tetQ* and *tetM* have been found in human oral and intestinal bacteria, another indication that bacteria normally found in different sites can exchange genes. Perhaps the most troubling evidence comes from examples where the same resistance gene was found in bacteria isolated from unpasteurized cheese or uncooked meat and in bacteria isolated from human patients (Teuber *et al.*, 1996; Teuber *et al.*, 1999). Most of the human clinical isolates were species that are normally found in the human colon, such as *Enterococcus* spp. These examples support the hypothesis that bacteria carried on human foods, are exchanging antibiotic resistance genes with bacteria in the human intestine during their brief passage through the human intestine.

Most of the examples given in Table 2 were found by accident, not by design. In such cases, the design of the experiment and control of parameters was not optimal and the results should be interpreted with caution. What emerges from these examples, however, is the growing conviction that well-designed real time experiments should be conducted to determine how rapidly genes from bacteria in food are transferred to human or animal intestinal bacteria. Such experiments are expensive enough that some justification for undertaking them is necessary. Table 2 contains enough evidence to justify moving to the next step. The examples given in Table 2 all came from retrospective studies, where the results of gene transfer events are seen in the present and scientists try to guess how they came to pass. Until prospective experiments are done, where the parameters are more carefully controlled and the direction of transfer is more easily deduced, many people will remain unconvinced that such events as the transfer of genes from foodborne bacteria to human intestinal bacteria occur with appreciable frequency.

Clearly, the ideal way to determine gene transfer rates would be to seed humans or animals with a resistant strain of bacteria and monitor the spread of the resistance gene. This has been done with laboratory rodents (see for example, Doucet-Populaire *et al.*, 1992), and transfer rates were detectable over fairly short periods of time. The East German experience with noursthricin use also supports the possibility of gene transfer: the noursthricin reisistance gene initially found in commensals turned up in *Shigella* strains (Witte, 1998). *Shigella* is a human specific pathogen. An advantage of this prospective type of study is that the direction of transfer is known and rates of transfer can be deduced, at least approx-imately. This approach has some disadvantages, however. First, it is very expensive to do such experiments with farm animals or humans rather than laboratory rodents, and these are the experiments that will be most convincing. Second, gene transfer elements vary considerably in frequency of transfer. Some have regulated transfer and will only transfer if the inducing conditions pertain. The nature of the recipient also affects transfer frequencies. So unless a single species is targeted and only transfers within this species are monitored, interpretation becomes complex and problematic. These are not arguments against this type of study, but rather indicate the need to develop an adequate interpretational framework in which to analyze the outcome.

Antibiotic Treatment of Companion Animals

Perhaps the biggest hole in the antibiotic resistance database is the lack of information about antibiotic resistance in bacteria found in the mouth and intestine of dogs and cats. Antibiotics are used to treat sick companion animals. Since companion animals are in constant propinquity with their owners, there are plenty of opportunities for bacteria to make the animal to human jump. In particular, bacteria such as staphylococci or enterococci that can cause opportunistic infections should be monitored. Companion animals may also acquire resistant bacteria through the food chain. There have been a couple of reports of antibiotic resistance in companion animals (Devriese *et al.*, 1996; Van Belkun *et al.*, 1996), but this is an area scientists have tended to overlook.

ANTIBIOTICS FOR PLANT DISEASE CONTROL

Emergence of Resistance in Plant-Associated Bacteria

The use of antibiotics in horticulture and the emergence of antibiotic-resistant plant pathogens have in many respects paralleled antibiotic use and resistance development in clinical settings. Streptomycin and oxytetracycline were first applied as dusts and solution to the aerial parts of fruit and vegetable crops in the 1950s. By the early 1960s, streptomycin resistance was reported in *Xanthomonas campestris pv. vesicatoria*, the bacterial spot pathogen of tomato (Stall and Thayer, 1962). Pear growers in the western U.S. typically made 10–18 applications of streptomycin (120–240 μg/ml) during the bloom and up to 30 days before harvest in order to control fire blight disease (Moller *et al.*, 1981). By 1973, streptomycin-resistant strains of *Erwinia amylovora*, the cause of fire blight disease, were being isolated from areas encompassing half of California's pear-growing acreage (Moller *et al.*, 1981). Streptomycin-resistant strains of *E. amylovora* were subsequently found in many other states of the U.S. and in New Zealand. Streptomycin resistance has also been reported in plant pathogenic *Pseudomonas* species. Oxytetracycline has been used under emergency registrations on apple in some regions where streptomycin-resistant strains have been documented. This is an example of how growing resistance to one antibiotic creates pressure to admit the use of another class of antibiotic, in this case, one more important in human medicine than streptomycin. Surveys in Oregon and Michigan did not find tetracycline resistant strains of *E. amylovora* but did identify tetracycline-resistance determinants in nonpathogenic orchard bacteria (Palmer and Jones, 1997).

Currently, in the U.S., streptomycin (Agri-mycin 17) is registered for use on at least 12 fruit, vegetable and ornamental plants. Oxytetracycline (Mycoshield) is registered for use on on 3 fruit crops. Gentamicin, sometimes in combination with oxytetracycline is used on certain fruits, vegetables and ornamental plants in Mexico and Central America and registration is pending in several other countries. Tree fruits account for the majority of antibiotic use on plants in the U.S. In 1995,

approximately 25,000 lbs. and 13,700 lbs. of streptomycin and oxytetracycline, respectively, were applied to fruit trees in the major tree-fruit states (National Agricultural Statistics Service, 1996). Antibiotics were applied to less than 20% of apple, 35%–40% of pear, and 4% of peach acreage.

Molecular analyses of streptomycin-resistant *E. amylovora* have revealed that in general resistance in strains collected in the western U.S. and New Zealand is conferred by a point mutation in rpsL that renders the bacterial ribosome insensitive to streptomycin (Chiou and Jones 1995b). This type of resistance is not transmissible, but some species of plant pathogens now carry the streptomycin resistance genes, *strA* and *strB*, which are usually found on conjugative plasmids (Sundin and Bender, 1996a, 1996b). These genes have now been identified in at least 17 species of environmental and clinical bacteria capable of colonizing diverse sites. Thus, antibiotic-resistance genes selected by agricultural use of antibiotics on plants are not necessarily restricted to the orchards where they arose.

In most plant studies, the crop of interest is sampled but not neighboring plants or the microflora of workers who apply the antibiotics. Also, few studies span the full range of niches such as soil, roots and aerial parts and aerosol particles in which bacteria are transported. Recent studies showed that neighboring crops, including cover crops, quantitatively and qualitatively affected the bacteria on trees (Lindow and Andersen, 1996). The movement of bacteria among plants by rain splash or in aerosol particles is well-documented (Hirano and Upper, 1983). Thus, it seems likely that antibiotics being applied in an orchard would impact bacterial populations associated with cover plants, through selection pressure from the antibiotic or by acquisition of antibiotic-resistant bacteria from trees.

Special Aspects of the Plant Protection Story

While the medical, veterinary and crop protection fields share a history of overuse of antibiotics, with the resultant emergence of antibiotic-resistant bacteria, antibiotic use in the agroecosystem presents unique circumstances that could strongly impact the buildup and persistence of resistance genes in the environment. First, streptomycin and oxytetracycline are applied to plants in high concentrations with airblast spray equipment. While growers strive to minimize drift by spraying during calm conditions, and advances in sprayer technology continue to improve the efficiency of pesticide applications, non-target bacteria on plants, in soil and in water are exposed to antibiotics. Antibiotics are frequently applied over physically large expanses. In regions of dense fruit production, antibiotics may be applied to hundreds of hectares of contiguous orchard. Moreover, the past decade has seen a dramatic increase in the planting of fruit trees that are susceptible to fire blight. This has created a situation analogous to clinical settings where immune-compromised patients are housed in crowded conditions – a setting associated with the breeding and spread of antibiotic-resistant bacteria.

The purity of antibiotics used in crop protection is unknown, but it is expected that they would be no purer than antibiotics used in animal production. As mentioned in a previous section, Webb and Davies (1993) found antibiotic-resistance genes from the antibiotic-producing bacterium in veterinary-grade antibiotics. The genes found in these preparations (*otrA* and *aphE*) are different from the resistance genes found so far in plant pathogens, but plant-grade antibiotic preparations must be viewed as a potential source of antibiotic resistance genes in the agroecosystem. It is interesting to note that fertilizers and fungicides applied to fruit trees are rich sources of divalent cations. The concentrations of Mn^{+2} and Ca^{+2} in these preparations are on the order of 10–50 mM, similar to the concentration used for chemical transformation of bacteria in the laboratory. Also, the current formulation of oxytetracycline (Mycoshield) is a calcium complex. Although natural transformation has been documented for only a few species, the concurrent application of antibiotics (with resistance genes) and unnaturally high concentrations of divalent cations could produce chemical transformation *in planta*. Once introduced into the bacteria, the resistance genes would have to be integrated in the bacterial genome by illegitimate recombination, an inefficient process. However, if bacteria acquired resistance genes, even at very low frequency, exposure to antibiotics would provide the

selection pressure needed to convert the initially rare resistant bacteria into a prominent component of the population.

Implications for Resistance in Human Commensals and Human Pathogens

What implications does antibiotic use on plants have for human health? At least one consumer advocacy group has argued that applying antibiotics to crops is an imprudent luxury that could lead to the demise of life-saving drugs (Lieberman and Wooten, 1998). Growers, however, defend their practice as being so limited in scope as to be inconsequential to human and environmental health. Unfortunately, neither side has sound data to uphold its position. For now, this leaves us with a contentious debate based on circumstantial evidence and fueled by passion. On the one hand, fruit and vegetable producers have sizeable economic interests (including their very livelihoods) at stake when managing bacterial diseases. Bacterial diseases of plants are notoriously difficult to manage and in many cases antibiotics are the most effective control option. The diversity and quantity of antibiotics used in plant disease control is meager compared to medical and veterinary use, and no apparent human health issues have arisen after four decades of use on crops. On the other hand, medical experts have witnessed the failure of one antibiotic after another in clinical settings, which, at least superficially, appear to be much more confined and strictly regulated than farm settings.

Antibiotic use on crops and ornamental plants is regulated by the U.S. Environmental Protection Agency. An antibiotic purchased through an agricultural chemical supplier is accompanied by a product label and supplemental literature specifying permitted use and safety precautions. The label clearly spells out what type of personal protective equipment (PPE; i.e., special clothing, boots, gloves, respirators) must be worn by mixers, applicators and persons entering a treated area after and application has been made. These documents are legally binding and it is a violation of federal law to use a pesticide in a manner inconsistent with its labeling. Also, the federal Worker Protection Standard for Agricultural Pesticides requires employers to provide farm workers with appropriate PPE and education related to pesticide use. In addition to federal laws, states have pesticide safety laws and help enforce the federal mandates. The bottom line is, although the application of antibiotics to plants is markedly different from clinical uses and may appear to be done with reckless abandon, it is a highly regulated activity and farmers are bound by stringent measures to protect the health of workers and the environment.

If farm workers diligently wear PPE, their exposure to antibiotics is minimal. However, the efficacy of PPE and its level of comfort seem to be directly correlated. Also, workers may grow complacent toward antibiotics, which are not acutely toxic as are many insecticides or as carcinogenic as many fungicides. Thus, despite education and regulation, some farm workers are undoubtedly exposed to high levels of antibiotics. The effects, if any, on the resistance profiles of their epidermal and intestinal bacteria are not known. The greatest effect on the human population at large could be mediated by ingestion of uncooked fruits and vegetables. Most consumers conscientiously wash produce before serving it, but washing may not remove all of the bacteria adhering to the vegetable surface. Thus, antibiotic-resistant plant pathogenic and saprophytic bacteria, many of which are enterics, might be introduced into the human intestine, where they would have ample opportunity to transfer resistance genes to intestinal bacteria.

Acquisition of streptomycin-resistant intestinal bacteria as part of the human intestinal microflora is not of great concern because streptomycin is rarely used to treat human diseases. Moreover, *strA* and *strB*, the resistance genes commonly found in plant-associated bacteria, do not confer resistance to most other aminoglycoside antibiotics. Similarly, tetracycline resistance is already widespread among intestinal bacteria. Even so, this route of introduction of antibiotic resistance genes into the intestinal microflora should not be dismissed. First, resistance to front-line drugs could be genetically linked to genes conferring resistance to streptomycin or tetracycline. In fact, the conjugative transposons found in *Bacteroides* spp. and the Gram-positive cocci, which carry tetracycline resistance genes, often carry resistance genes that confer resistance to macrolides and streptogramins. Furthermore, in some cases exposure to tetracycline triggers the

transfer of conjugative transposons to new recipients. Thus, the use of tetracycline could stimulate the transfer of and select for maintenance of macrolide- and streptogramin-resistance genes. A second concern is that overuse of antibiotics could render plant pathogens resistant to streptomycin and tetracycline, giving rise to demands for approval of more advanced antibiotics for use on plants. This has already happened in veterinary medicine and animal production.

BIOCONTROL AGENTS AND BIOPESTICIDES

Nonpathogenic bacteria are being exploited as biological controls to prevent crop losses from fire blight caused by *E. amylovora* (Johnson and Stockwell, 1988). *Pantoea agglomerans* (formerly *Erwinia herbicola*) occupies some of the same niches on apple as the plant pathogen *Erwinia amylovora*. The presence of an identical plasmid (pEa34) carrying the streptomycin-resistance transposon in natural isolates of both *E. amylovora* and *P. agglomerans*, plus the demonstration of conjugal transfer between these two species in the laboratory, shows that horizontal transfer betwen pathogen and biocontrol agent could occur. Just as introduction of plant pathogens into the intestinal tract on raw fruits and vegetables might lead to gene transfer events involving human intestinal species – some of which are closely related to *E. amylovora* – similar transfer events involving antibiotic-resistant biocontrol agents could also occur. Determination of the antibiotic-resistance profile of biocontrol agents should become a part of the safety assessment procedure.

Concern about resistance genes and their transfer to human intestinal bacteria should also arise in connection with biopesticides. The biopesticide *Bacillus popilliae* has been used for years to prevent infestations of Japanese beetles. *B. popilliae* has never been known to cause human disease, but it could create an antibiotic-resistance problem. Recently Rippere *et al.* (1998) found vancomycin-resistance genes in *B. popilliae* (both *vanA* and *vanB*). Whether these genes can be transferred by conjugation is unknown, but since vancomycin is such an important human antibiotic, this possibility should be assessed.

CONCLUSIONS

The oft-expressed view that the main resistance problem arising from agricultural use of antibiotics is transmission of antibiotic-resistant zoonotic pathogens to humans far underestimates the complexity, magnitude and potential impacts of antibiotic use in agriculture. A potentially greater danger is that antibiotic resistant bacteria from animals, whether pathogen or commensal, will be transmitted to humans through the food supply and will transfer resistance genes to human intestinal bacteria and through them to serious human pathogens. That is, the spread of genes could be a problem, not just the spread of bacteria. Evidence is mounting that transfer of antibiotic resistance genes between bacteria normally found in the animal intestinal tract and bacteria found in the human intestine occurs far more frequently than would have been expected from laboratory experiments. The direction of transfer is uncertain in most cases, but the fact that a genetic conduit of some sort is open between animal and human bacteria increases the possibility of resistance gene flow from animal bacteria to human bacteria via the food supply. The possibility of gene transfer should be considered seriously in any deliberations over safety issues. Another factor that has received insufficient attention is the stability of resistance genes in most hosts. Contrary to earlier beliefs, antibiotic resistance genes do not take a fitness toll in most cases. This is probably the case because under selective pressure, bacteria make genetic changes that improve the fit between a newly acquired resistance gene and its bacterial host. Resistance genes are not only easy to get, but also hard to lose (Salyers and Amabile-Cuevas, 1997). Thus, the spread of a particular type of resistance gene may be difficult or impossible to reverse. It is of primary importance to prevent the spread of resistance genes in the first place, by prudent use of antibiotics.

Debates on antibiotic resistance have centered on livestock animals such as pigs, chickens and cattle. Yet, antibiotics are also used by vegetable and fruit growers and in aquaculture. Use in these areas may have as great an impact on human health and the environment as antibiotic use in animal production. Another area deserving more attention is the antibiotic resistant bacteria in the intestines of companion animals. Antibiotic-resistant bacteria in dog or cat food could be spread to companion animals and through them to humans. No one disputes the important contribution to the spread of antibiotic-rersistant bacteria made by the abuse and overuse of antibiotics in human medicine. The fact that this is probably the main cause of the rise in antibiotic resistance, however, is not an argument for ignoring antibiotic overuse and abuse in agriculture. The public views the use of antibiotics to treat human infections as necessary and desirable. By contrast, most agricultural uses other than treatment of sick animals, appears to those who know little about it to be optional and in some cases frivolous. Thus, the public may support an end to use of antibiotics as growth promoters or as plant protection agents despite the fact that these uses probably do not make as great a contribution to resistance in human pathogens as human clinical use. Moreover, it is important to realize that the extent of the contribution agricultural use of antibiotics makes to resistance in human pathogens has not been established. It may be greater than we think.

Perhaps the most serious difficulty in trying to define safe and prudent use of antibiotics in agriculture is that antibiotics may not be the only selective pressure maintaining antibiotic-resistance genes in the environment. Antibiotic resistance genes can be genetically linked to other genes such as metal resistance genes. In such cases, metal pollution could select for resistance to antibiotics as well as resistance to the metal in question. Studies of the effects of antibiotic use in human medicine or agriculture seldom establish a resistance baseline in the surrounding area of the site being tested. Nor do they establish whether the resistances they detect are linked to some other trait that could be the actual target for local selective pressures. Langlois *et al.* (1988) reported that age and housing location affect the number of antibiotic resistant bacteria in the intestines of pigs in a herd of pigs that had not been exposed to antibiotics. What is the selective pressure in cases such as this? There are many such observations that cannot be easily explained simply by invoking antibiotic exposure. The debate over agricultural use of antibiotics is never going to be settled scientifically without more information on how likely resistance genes move through the food chain to human bacteria and what are the most important selective pressures that maintain resistant bacteria in the environment. In the past, the agriculture industry has been able to counter challenges to its use of antibiotics by exerting political and economic pressure to avoid unwelcome regulations. So far, the public has accepted this or has ignored the issue completely. If, however, there are cases of human treatment failures that can be traced to agricultural use of antibiotics, public opinion could change swiftly and dramatically. It is time for the industry and the public to take a careful and critical look at the way it uses antibiotics and how this use might influence human health.

ACKNOWLEDGMENTS

Some of the work mentioned in this chapter was supported by grant AI/GM 22383 from the U.S. National Institutes of Health. A website that is posting information on antibiotic resistance in commensals (supported by NIH grant AI/GM 22383) can be reached at http://www.antibiotic.org (click on roar) or http://www.roar.antibiotic.org.

REFERENCES

Aarestrup, F.M. (1998) Association between decreased suscept-ibility to a new antibiotic for treatment of human diseases, everinomicin (SCH 27899), and resistance to an antibiotic used for growth promotion in animals, avilamycin. *Microbial Drug Resistance* **4**, 137–141.

Aarestrup, F.M. (1999) The European perspective on anti-microbial related regulations and trade. Agriculture's Role in Managing Antimicrobial Resistance. Proceedings from October 24–26 meeting sponsored by the Ontario Ministry of Agriculture, Food and Rural Affairs. pp. 131–135.

Aarestrup, F.M., Bager, F., Jensen, N.E., Madsen, M., Meyling A., and Wegener, H.C. (1998) Surveillance of antimicrobial resistance in bacteria isolated from food animals to anti-microbial growth promoters and related therapeutic agents in Denmark. *APMIS* **106**, 606–622.

Anonymous. (1999) Four pediatric deaths from community-acquired methicillin-resistant *Staphylococcus aureus* – Minne-sota and North Dakota, 1997–1999, *Morbidity and Mortality Weekly Reports* **48**, 707–710.

Bates, J., Jordens, J.Z., and Griffiths, D.T. (1994) Farm animals as a putative reservoir for vancomycin-resistant enterococcal infection in man. *J. Antimicrob. Chemother.* **34**, 507–514.

Bottger, E.C., Springer, B., Pletschette, M., and Sander, P. (1998). Fitness of antibiotic-resisttant microorganisms and compensatory mutations. *Nature Med.* **12**, 1343–1344.

Chiou, C.S., and Jones, A.L. (1995) Molecular analysis of high-level streptomycin resistance in *Erwinia amylovora*. *Phyto-pathology* **85**, 324–328.

Coque, T.M., Tomayko, J.F., Ricke, S.C., Okhyusen, P.C., and Murray, B.E. (1996) Vancomycin-resistant enterococci from nosocomial, community and animal sources in the United States. *Antimicrob. Agents Chemother* **40**, 2605–2609.

DANMAP. (1997) Consumption of antimicrobial agents and occurrence of antimicrobial resistance in bacteria from food animals, food and humans in Denmark. No. 1, 1997, Danish Zoonosis Centre, Danish Veterinary Laboratory, Bulowsvej 27. D-1790 Copenhagen V.

Davies, J. (1996) Bacteria on the rampage. *Nature* **383**, 219–220.

Devriese, L.A., Ieven, M., Goossens, H., Vandamme, P., Pot, B., and Hommez, J., *et al.* (1996) Presence of vancomycin-resistant enterococci in farm and pet animals. *Antimicrob. Agents Chemother* **40**, 2285–2287.

Doucet-Populaire, F., Trieu-Cuot, P., Andremont, A., and Courvalin, P. (1992) Conjugal transfer of plasmid DNA from Enterococcus faecails to Escherichia coli in the digestive tracts of gnotobiotic mice. *Antimicrob. Agents Chemother.* **36**, 502–504.

Elder, H.A., Roy, I., Lehman, S., Phillips, I.A., and Kass, E.H. (1993) Human studies to measure the effect of antibiotic residues. *Vet. Human Toxicol.* **35**(Suppl 1), 31–36.

Endtz, H.P., Ruijs, G.J., van Klingeren, B., Jansen, W.H., van der Reyden, T., and Mouton, R.P. (1991) Quinolone resistance in campylobacter isolated from man and poultry following the introduction of fluoroquinolones in veterinary medicine. *J. Antimocrob. Chemother.* **27**, 199–208.

U.S. Food and Drug Administration framework document. (1999) A proposed framework for evaluating and assuring the human safety of the microbial effects of antimicrobial new animal drugs intended for use in food-producing animals. Meeting of the Veterinary Medicine Advisory Committee, Jan. 25–26, 1999. Gaithersburg, Md.

Glynn, M.K., Bopp, C., Dewitt, W., Dabney, P., Mokhtar, M., and Angulo, F.J. (1998) Emergence of multidrug-resistant Salmonella enterica serotype typhimurium DT104 infections in the United States. *New Engl. J. Med.* **338**, 1333–1338.

Hall, R.M., and Collis, C.M. (1995) Mobile gene cassettes and integrons: capture and spread of genes by site-specific recombination. *Mol. Microbiol.* **15**, 593–600.

Hauth, J.C., Goldenberg, R.L., Andrews, W.M., DuBard, M.B., and Copper, E.L. (1995) Reduced incidence of preterm delivery with metronidazole and erythromycin in women with bacterial vaginosis. *New Engl. J. Med.* **333**, 1732–1736.

Helmuth, R., and Protz, D. (1997) How to modify conditions limiting resistance in bacteria in animals and other reservoirs. *Clin. Infect. Dis.* **24**(Suppl.), S136–139.

Herold, B.D., *et al.* (1998) Community-acquired methicillin-resistant *Staphylococcus aureus* in children with no identified predisposing risk. *JAMA* **279**, 593–598.

Hirano, S.S., and Upper, C.D. (1983) Ecology and epidemiology of foliar bacteria plant pathogens. *Annu. Rev. Phytopathol.* **21**, 243–269.

Institute of Medicine (IOM) and National Research Council. (1998). *The use of drugs in food animals: benefits and risks.* National Academy Press.

Johnson, K.B., and Stockwell, V.O. (1998). Management of fire blight: a case study of microbial ecology. *Annu. Rev. Phytopathol.* **36**, 227–248.

Klare, I., Badstubner, D., Konstabel, C., Bohme, G., Clause, H., and Witte, W. (1999) Decreased incidence of VanA-type vancomycin-resistant enterococci isolated from poultry meat and from fecal samples of humans in the community after discontinuation of avoparcin usage in animal husbandry. *Microb. Drug Resist.* **5**, 45–52.

Klein, G., Pack, A., and Reuter, G. (1998) Antibiotic resistance patterns of enterococci and occurrence of vancomycin-resistant enterococci in raw minced beef and pork in Germany. *Appl. Environ. Microbiol.* **64**, 1825–1830.

Kluytmans, J., *et al.* (more than 10 authors). (1995) Food initiated outbreak of methicillin-resistant Staphylococcus aureus analyzed by pheno- and genotyping. *J. Clin. Microbiol.* **33**, 1121–1128.

Langlois, B.E., Dawson, K.A., Leak, I., and Aaron, D.K. (1988) Effect of age and housing location on antibiotic resistance of fecal coliforms from pigs in a non-antibiotic-exposed herd. *Appl. Environ. Microbiol.* **54**, 1341–1344.

Lenski, R.E., Simpson, S.C., and Nguyen, T.T. (1994) Genetic analysis of a plasmid-encoded, host genotype-specific enhancement of bacterial fitness. *J. Bacteriol.* **176**, 3140–3147.

Lieberman, P.B., and Wootan, M.G. (1998) Protecting the crown jewels of medicine. Center for Science in the Public Interest Newsletter.

Lindow, S.E., and Andersen, G.L. (1996) Influence of immigration on epiphytic bacterial populations on navel orange leaves. *Appl. Environ. Microbiol.* **62**, 2978–2987.

Lipsitch, M., and Levin, B.R. (1997) The population dynamics of antibiotic therapy. *Antimicrob. Agents Chemother.* **41**, 363–373.

Loper, J.E., Henkels, M.D., Roberts, R.G., Grove, G.G., Willett, M.J., and Smith, T.J. (1991) Evaluation of streptomycin, oxytetracycline, and copper resistance in *Erwinia amylovora* isolated from pear orchards in Washington state. *Plant Dis.* **75**, 287–290.

Manulis, S.F., Kleitman, O., Dror, I., Davif and Zutra, D. (1998) Characterization of the *Erwinia amylovora* population in Israel. *Phytoparasitica* **26**, 39–46.

Marshall, C.G., Lessard, I.A., Park, I.S., and Wright, G.D. (1998) Glycopeptide antibiotic resistance genes in glycopeptide-producing organisms. *Antimicrob. Agents Chemother.* **42**, 2215–2220.

McDonald, L.C., Kuehnert, M.J., Tenover, F.C., and Jarvis, W.R. (1997) Vancomycin-resistant enterococci outside the health care setting: Prevalence, sources and public health implications. *Emerging Inf. Dis.* **3**, 311–317.

McKeon, D.M., Calabrese, J.P., and Bissonnette, G.K. (1995) Antibiotic resistant gram-negative bacteria in rural groundwater supplies. *Water Res.* **29**, 1902–1908.

McManus, P.S., and Jones, A.L. (1994) Epidemiology and genetic analysis of streptomycin-resistant Erwinia amylovora from Michigan and evaluation of oxytetracycline for control. *Phytopathology* **84**, 627–633.

Molbak, K., *et al.* (more than 8 authors). (1999) An outbreak of multidrug-resistant quinolone-resistant *Salmonella enterica* serotype *typhimurium* DT104.

Moller, W.J., Schroth, M.N., and Thomson, S.V. (1981) The scenario of fire blight and streptomycin resistance. *Plant Disease* **65**, 563–568.

National Agricultural Statistics Service. (1996) Agricultural Chemical Usage, 1995 Fruits Summary. No. 96172. U.S. Dept. of Agriculture.

Nikolich, M., Hong, G., Shoemaker, N.B., and Salyers, A.A. (1994) Evidence that conjugal transfer of a tetracycline resistance gene (*tetQ*) has occurred very recently between the normal microflora of animals and the normal microflora of animals. *Appl. Environ. Microbiol.* **60**, 3255–3260.

Norelli, J.L., *et al.* (1991) Homologous streptomycin resistance gene present among diverse Gram-negative bacteria in New York state apple orchards. *Appl. Environ. Microbiol.* **57**, 486–491.

O'Donovan, C.A., Fan-Harvard, P. Tecson-Tumang, F.T., Smith. S.M., and Eng, R.H. (1994) Enteric eradication of

vancomycin-resistant Enterococcus faecium with oral bacitracin. *Diagn. Microbiol. Infect. Dis.* **18**, 105–109.

Palmer, E.L., and Jones, A.L. (1997) Tetracycline-resistant determinants present in Michigan apple orchards. *Phytopathology* **87**, S74.

Raloff, J. (1998) Drugged waters. *Science News* **153**, 187–189.

Rippere, K., *et al.* (1998) DNA sequence resembles vanA and vanB in the vancomycin-resistant biopesticide Bacillus popilliae. *J. Infect. Dis.* **178**, 584–588.

Salyers, A.A., and Amabile-Cuevas, C. (1997) Why are antibiotic resistance gene so resistant to elimination? *Antimicrob. Agents Chemother.* **41**, 2321–25.

Salyers, A.A., and Shoemaker, N.B. (1996) Resistance gene transfer in anaerobes: New insights, new problems. *Clin. Infect. Dis.* **23** (Suppl.), S36–43.

Scheck, H.J., Pscheidt, F.W., and Moore, L.D. (1996) Copper and streptomycin resistance in strains of Psuedomonas syringae from Pacific Northwest nurseries. *Plant Dis.* **80**, 1034–1038.

Schrag, S.J., Perrot, V., and Levin, B.R. (1997) Adaptation to the fitness costs of antibiotic resistance in *Escherichia coli.* *Proc. R. Soc. Lond. B. Biol. Sci.* **264**, 1287–1291.

Sobiczewski, P., Chiou, C.S., and Jones A.L. (1991) Streptomycin-resistant epiphytic bacteria with homologous DNA for streptomycin resistance in Michigan apple orchards. *Plant Disease* **75**, 1110–1113.

Stall, R.E., and Thayer, P.L. (1962) Streptomycin resistance of the bacterial spot pathogen and control with streptomycin. *Plant Dis. Rep.* **46**, 389–392.

Sundin, G.W., and Bender, C.L. (1996). Dissemination of the strA-strB streptomycin-resistance genes among commensal and pathogenic bacteria from humans, animals and plants. *Molec. Ecology* **5**, 133–143.

Sundin, G.W., Monks, D.E., and Bender, C.L. (1996) Molecular genetics and ecology of transposon-encoded streptomycin resistance in plant pathogenic bacteria. In: *Molecular genetics and evolution of pesticide resistance* (T.M. Brown, ed.). American Chemical Society, Washington, D.C.

Teuber, M., Meeile, L., and Schwarz, F. (1999) Acquired antibiotic resistance in lactic acid bacteria from food. *Antonie van Leeuwenhoek.* **76**, 115–137.

Teuber, M., Perreten, V., and Wirsching, F. (1996) Antibiotikumresistente bakterien: eine neue dimension in der lebensmittel-mikrobiologie. *Lebensmittle-technologie* **29**, 182–199.

Van Belkun, A., van den Braak, N., Thomassen, R., Verbrugh, H., and Endtz, H. (1996) Vancomycin-resistant enterococci in cats and dogs [letter]. *Lancet* **348**, 1038–1039.

Van den Bogaard, A.E., van den Jensen, L.B., and Stobberingh, E.E. (1997) An identical VRE isolated from a turkey and a farmer. *New Engl. J. Med.*, in press.

Van den Braak, N., Van Belkum, A., Van Keulen, M., Vliegenthart, J., Verbrugh, H.S., and Endtz, H.P. (1998) Molecular characterization of vancomycin-resistant enterococci from hospitalized patients and poultry products in the Netherlands. *J. Clin. Microbiol.* **36**, 1927–1932.

Vanneste, J.L. (1995) *Erwinia amylovora*. In: *Pathogenesis and host specificity in plant diseases: Histopathological, biochemical, genetic and molecular bases* (U.S. Singh, R.P. Singh and K. Kohmoto, eds). Pergamon Press, Oxford, pp. 21–41.

Waters, B., and Davies, J. (1997) Amino acid variation in the GyrA subunit of bacteria potentially associated with natural resistance to fluoroquinolone antibiotics. *Antimicrob. Agents Chemother.*

Webb, V., and Davies, J. (1993). Antibiotic preparations contain DNA: a source of drug resistance genes? *Antimicrob. Agents Chemother.* **37**, 2379–2384.

Welton, L.A., *et al.* (1998) Antimicrobial resistance in enterococci isolated from turkey flocks fed virginiamycin. *Antimicrob. Agents Chemother.* **42**, 705–708.

Werner, G., Klare, I., and Witte, W. (1997) Arrangement of the vanA gene cluster in enterococci of different ecological origin. *FEMS Microbiol. Lett.*, **155**, 55–61.

Westh, H. Influence of erythromycin consumption on erythromycin resistance in Staphylococcus aureus in Denmark. *APUA Newsletter* **13**(1), 1–4.

WHO Report. (1997) The medical impact of the use of antimicrobials in food animals. WHO/EMC/ZOO/97.4

Wierup, M. (1998) Preventive methods replace antibiotic growth promoters: Ten years experience from Sweden. *Aliance for Prudent Antibiotic Use Newsletter* **16**, 1–4.

Wierup, M. (1999). The Swedish experience with limiting antimicrobial use. Agriculture's Role in Managing Antimicrobial Resistance. Proceedings from October 24–26 meeting sponsored by the Ontario Ministry of Agriculture, Food and Rural Affairs. pp. 131–135.

Witte, W. 1998. Medical consequences of antibiotic use in agriculture. *Science* **279**, 996–997.

Woodford, N., Adebiyi, A.-M., Palepou, M.F., and Cookson, B.D. (1998) Diversity of VanA glycopeptide resistance elements in enterococci from humans and nonhuman sources. *Antimicrob. Agents Chemother.* **42**, 502–508.

Woodford, N., Jones, B.L., Baccus, Z., Ludlam, H.A., and Brown, D.F. (1995) Linkage of vancomycin and high-level gentamicin resistance genes on the same plasmid in a clincal isolate of *Enterococcus faecalis. J. Antimicrob. Chemother.* **35**, 179–184.

Wray, C., McLaren, I.M., and Carroll, P.J. (1993) *Escherichia coli* isolated from farm animals in England and Wales between 1986 and 1991. *Vet. Rec.* **133**, 439–442.

10. Fitness and Virulence of Antibiotic Resistant Bacteria

Dan I. Andersson[§], Johanna Björkman[§#] and Diarmaid Hughes[#]

[§]Swedish Institute for Infectious Disease Control, Department of Bacteriology, S-171 82 Solna, Sweden
[#] Uppsala University, Department of Cell and Molecular Biology, BMC, Box 596, S-751 24 Uppsala, Sweden

INTRODUCTION

The use, overuse and misuse of antibiotics, since their introduction about 50 years ago, has resulted in the evolution and spread of antibiotic resistance with a concomitant increase in the frequency of resistant bacteria (Cohen, 1992, Davies, 1994; Levy, 1992; McCormick, 1998). A consequence of this is that infections previously treatable with antibiotics might require treatment with more expensive or less efficient drugs. Clearly, antibiotic resistance represents a growing threat to public health and is a cause of major medical and economic concern. Whether this unwanted trend can be slowed down or reversed is unclear. That depends on some factors which we can in principle control, such as the volume of antibiotic use, but also on factors outside our control, such as the fitness costs resistance places on bacteria and the rate and extent to which natural selection reduces or eliminates these costs.

A problem in the above context is that, at present, we lack the experimental knowledge required to propose and implement rational strategies on how to reduce the frequency of resistance or to evalute the success of any such strategy. This is mainly due to a lack of data on the factors affecting the rate and stability of antibiotic resistance combined with a lack of knowledge on how to incorporate these data into quantitative models which describe and predict resistance development. Thus, there is a need for experimental, epidemiological and theoretical studies where we obtain: (i) the experimental and conceptual background knowledge required to rationally predict and assess the risks of antibiotic resistance development in society, (ii) the data required to allow us to develop policies for the use of antibiotics in clinical settings so as to minimize resistance development without compromising treatment efficacy, and (iii) the data required to provide criteria for pharmaceutical companies, drug-licensing agencies and physicians on how to evaluate the risks of resistance development towards both new compounds and already marketed antibiotics.

This chapter deals mainly with resistances conferred by target alterations due to chromosomal mutations. The reasons for this bias are two-fold. Firstly, few detailed studies have been done on the fitness costs of plasmid-/transposon-borne resistances and secondly, among the ones done the general conclusions appear similar to what is observed from more thorough studies of chromosomal resistances (see Andersson and Levin, 1999 and references therein).

WHAT QUESTIONS AND WHAT DATA?

How Rapidly does Antibiotic Resistance Develop?

This is a multifaceted problem which is influenced by a number of factors, most of which remain to be experimentally defined. **First**, the volume of antibiotic usage will set the selection pressure for resistance development. How this volume relates to the frequency of resistance has been described in theoretical models in which the population genetics and transmission dynamics of the resistant and sensitive bacteria are incorporated (Austin et al., 1998; Levin et al., 1997). These models make clear predictions and provide a framework within which to incorporate experimental and epidemiological data. However, in practice, the relationship between the volume of antibiotic usage and the frequency of resistance might be obscured by other factors, such as genetic drift, the random spread of certain resistant clones, and the co-selection of resistance genes. **Second**, bacterial mutation rates will have an impact on the rate of resistance development by affecting the rate of the initial appearance of resistant mutants. In this context stable mutator bacteria are of great interest since they are expected to be enriched in high-use antibiotic environments (such as intensive-care units) and increase the rate of resistance development (LeClerc et al., 1996; Mao et al., 1997; Miller, 1996; see also chapters by Baquero and Cebula et al.). During the past decade it has also been shown that apparent mutation rates in bacteria may vary significantly in response to various physiological and selective conditions (Andersson et al., 1998; Hall, 1998; Foster, 1998; Rosenberg et al., 1998). Most of these studies have been made under laboratory conditions, and the influence of the eukaryotic host on rates and mechanisms of resistance development in pathogenic bacteria is essentially unknown. Recent experimental data suggest however that growth in hosts can increase mutation rates to resistance (Björkman et al., unpubl., see below). **Third**, genetic factors in bacteria may predispose certain populations to develop resistance. Thus, there are mutations which by themselves do not confer resistance but might facilitate the formation or selection of resistant mutants by, for example, prolonging survival in the presence of antibiotics.

Of special interest is the tolerance phenomenon and other cell wall alterations that might facilitate β-lactam resistance (Novak et al., 1999; Tuomanen et al., 1991). **Finally**, low antibiotic concentrations caused by spatial or temporal variation in antibiotic concentrations may have a critical influence on how rapidly resistance develops, by allowing for example the stepwise selection of resistant mutants (Austin et al., 1998; Baquero et al., 1993; Baquero et al., 1997, Kepler and Perelson, 1998; see also chapter by Baquero).

How Stable is Antibiotic Resistance?

The frequency, stability and reversibility of resistance in a bacterial population at a given antibiotic pressure is mainly determined by the fitness and transmission costs of resistance combined with the ability of the resistant bacteria to compensate for these costs by new mutations (Andersson and Levin, 1999). The fitness of an antibiotic resistant pathogen is affected by the relative rates with which sensitive and resistant bacteria (i) grow and die within and outside hosts, (ii) are transmitted between hosts, and (iii) are cleared from infected hosts. In experimental studies resistance has been found to usually confer some cost, although a few exceptions are known (see below). For the combinations of antibiotics and bacteria where the fitness costs of resistance are the highest and where the ability to compensate is the lowest it is most likely that resistance is reversible. Such experimental knowledge can be used to decide which combinations of antibiotics and bacteria maximize the chances of reversibility of resistance in clinical situations. In addition, the biological cost and genetic compensation will be important parameters to consider in the development of new antibiotics.

What Effect does Resistance have on Virulence and Disease Pathology?

This is a question which deserves further study but which to date has been largely neglected. Thus, resistant mutants could conceivably have an altered virulence as well as a changed disease pathology compared with the sensitive strains. This could be especially relevant for resistances caused by target alterations (e.g. RNA polymerase, ribosome mutations) which could have pleiotropic effects on cell growth and physiology. If so, antibiotic pressure and the associated increase in the frequency of resistant mutants could in the long term result in an altered outcome of the disease caused by affected bacteria.

HOW TO MEASURE THESE VARIOUS COMPONENTS?

The Biological Cost

The biological cost of resistance can be measured in three ways, (i) retrospectively, by fitting quantitative models to the changes in frequencies of humans infected with sensitive and resistant bacteria following known changes in the volume of antibiotic use in human populations, (ii) prospectively, by measuring the rates with which individuals become infected with and cleared of sensitive and resistant bacteria, and (iii) experimentally, by estimating the rates of growth, transmission and clearence of sensitive and resistant bacteria in laboratory media and experimental animals.

Experimental studies of the costs of resistance are usually aimed at determining the rates of growth, survival and competitive performance of sensitive and resistant bacteria, usually by pairwise competition experiments (see Andersson and Levin, 1999 and references therein). Mixtures of isogenic sensitive and resistant strains are grown in chemostats, batch cultures or laboratory animals and changes in their relative frequencies are followed by selective plating, using either a neutral genetic tag (e.g. a transposon) or the resistance itself as the selective marker. These pairwise competition experiments estimate several components of the competitive fitness of sensitive and resistant bacteria: their lag periods, rates of exponential growth, resource utilization efficiencies, and their mortality in the presence or absence of host defenses. By repeated measurements, these pairwise competition experiments allow detection of fitness differences that are <1% (Lenski, 1991).

One problem in determining the cost of resistance in laboratory experiments is how to interpret negative results; i.e. when the resistant mutant appears to be as fit as the wild type. If resistance confers no cost in competition experiments performed under many different *in vitro* and *in vivo* conditions, it is likely that the costs are low or non-existent. However, it remains possible that there are natural conditions where these costs are considerable. In contrast, if costs of resistance are observed in laboratory experiments, it is reasonable to assume that there are also some natural conditions where that resistance would impose a fitness burden. One limitation of the experimental studies of the cost of resistance performed to date is the focus on growth and competitive performance, whereas experiments to determine transmission and clearence costs are much rarer. The technical difficulties in determining transmission rates have clearly obstructed these types of studies.

Estimating the Amelioration of the Costs of Resistance

The rate, degree and nature of evolution of the costs of resistance can be followed by growing the resistant bacteria in chemostats, or by serial passage in batch cultures or in experimental animals. By pairwise competition with the original strains (or from the decrease in the frequency of a selectively neutral indicator strain present at low frequency) it is possible to determine if mutants with enhanced fitness have appeared and increased in frequency. In these compensated mutants one can determine whether the reduction in the cost of resistance is conferred by (i) reversion to the wild-type parent or alternatively by (ii) compensation via intragenic or extragenic suppressor mutations.

Determining Effects on Virulence and Disease Pathology

The effect of antibiotic resistance on the virulence and disease pathology of pathogenic bacteria can be assessed by measuring LD_{50} values in experimental animals or by examining the specific pathology of the disease (e.g. inflammatory responses, histopathology etc.). Few studies have been directed towards these questions but resistant mutants can clearly show altered virulence (Wilson *et al.*, 1995; Zhongming *et al.*, 1997). Whether disease pathology may be altered in resistant mutants is still an open question (see below).

WHAT ARE THE COSTS OF RESISTANCE?

Laboratory Studies

In most studies, resistance caused by target alterations has been found to confer some cost (Table 1). As measured by pairwise competition between sensitive and resistant bacteria in experimental animals or laboratory media these costs can be considerable with selection coefficients ranging from 0.01 to 1–2. Also apparent from these studies is that the costs are not easily predictable, neither from the type of target molecule (e.g. ribosome, RNA polymerase, DNA gyrase etc.) nor from the nature of the specific mutation. Mutants with no measurable cost have also been observed. One example of a ''no cost'' resistance mutation is the 42nd codon AAA (Lys) → AGA (Arg) substitution of the *rpsL* gene, responsible for resistance to high concentrations of streptomycin in *S. typhimurium* and other enteric bacteria (Kurland *et al.*, 1996). While other substitutions at the same position cause severe reductions in fitness both *in vitro* and in mice, the *rpsL* AGA mutation appears to be selectively neutral and may even confer a slight advantage over wild type, at least in mice (Björkman *et al.*, 1998).

From these studies it is clear that the fitness cost associated with resistance strongly depends on the growth conditions. For example, resistant mutants that show no cost in laboratory medium may have large costs in laboratory mice, and conversely, mutants that show no cost in mice may have substantial costs *in vitro* (Björkman *et al.*, 2000). The reasons for this are at present unclear but the conditionality supports the view that experimental estimates of the cost of resistance should be done under a variety of environmental conditions to obtain relevant data.

Estimating Costs from Clinical Studies and Epidemiology

Experimental studies of the costs of resistance and adaptation to those costs make a number of predictions that can be tested by examining the resistance genes found in bacteria isolated from humans to see if the same mutations are found. The results of such tests suggest that chromosomal mutations responsible for resistance in clinically isolated bacteria are likely to be the same as those observed to have low cost experimentally. For example, Böttger *et al.* (1998) found that the *rpsL* mutations responsible for resistance to streptomy-

Table 1. The biological cost of resistance conferred by target-altering chromosomal mutations.

Bacteria	Resistance	Mutation	Cost	Assay system	Reference
S. typhimurium	Streptomycin	*rpsL*	yes/no	mice, *in vitro*	Björkman *et al.*, 1998, 1999
	Rifampicin	*rpoB*	yes	mice, *in vitro*	Björkman *et al.*, 1998
	Nalidixic acid	*gyrA*	yes	mice, *in vitro*	Björkman *et al.*, 1998
	Fusidic acid	*fusA*	yes/no	mice, *in vitro*	Johansson *et al.*, 1996, Björkman *et al.*, 2000
E. coli	Streptomycin	*rpsL*	yes/no	*in vitro*	Schrag *et al.*, 1997 Schrag and Perrot, 1996
	Rifampicin	*rpoB*	yes/no	*in vitro*	Reynolds, pers. comm.
M. tuberculosis and *M. bovis*	Isoniazid	*katG*	yes	mice	Heym *et al.*, 1997 Wilson *et al.*, 1995 Zhongming *et al.*, 1997
S. aureus	Fusidic acid	*fus*	yes/no	rats, *in vitro*	Nagaev *et al.*, 1999 unpubl.
H. pylori	Clarithromycin	*rrn*	yes	mice	Björkholm *et al.*, 1999 unpubl.
P. aeruginosa	Ciprofloxacin	?	yes	*in vitro*	Björkman, unpubl.
	Gentamicin	?	yes	*in vitro*	Björkman, unpubl.
	Rifampicin	?	yes	*in vitro*	Björkman, unpubl.
P. fluorescens and *Putida*	Rifampicin	?	yes	soil	Compeau *et al.*, 1988
L. monocytogenes	class IIa bacteriocin	?	yes	*in vitro*	Dykes and Hastings, 1998
N. meningitidis	Sulfonamide	*sul2*	yes	*in vitro*	Fermer and Swedberg, 1997

cin in clinical isolates of *M. tuberculosis*, were those which showed no cost in *S. typhimurium* and *E. coli* under experimental conditions. These results were interpreted to suggest that no cost mutants are preferentially selected in humans. However, it is also possible that they were the result of evolution during growth in the host to compensate for a more costly original mutation, as has also been observed experimentally in *S. typhimurium* (Björkman *et al.*, 1998). To test for compensatory evolution, it would be necessary to perform a prospective study and follow the progression from the first appearance of primary resistance in a patient and determine whether this is compensated by second-site suppressor mutations.

EFFECTS OF RESISTANCE ON VIRULENCE AND DISEASE PATHOLOGY

How resistance affects virulence as assessed by LD_{50} tests (or similar assays) is in most cases unclear. In cases where both fitness (as measured by growth rates in host and environment) and virulence have been measured it is generally observed that resistant mutants show both decreased fitness and virulence, even though there is no ''a priori'' reason to assume that they should be postively correlated. Studies of isoniazid resistant *M. bovis/tuberculosis*, which are resistant due to *katG* mutations, indicate that virulence is severely reduced by this particular resistance mutation as measured either by killing

or histopathology (i.e. granuloma formation) (Wilson *et al.*, 1995; Zhongming *et al.*, 1997).

One example in which disease pathology might be different in the resistant mutant compared to the wild-type is in cephalosporin resistant opportunistic gram negative bacteria such as, for example, *Citrobacter freundii* and *Enterobacter cloacae* which can express a chromosomal β-lactamase that is inducible by β-lactam antibiotics. These organisms remain sensitive to third generation cephalosporins. However, mutations arise at a high frequency leading to constitutive β-lactamase production, and as a consequence, resistance also to third generation cephalosporins. These mutations occur in one single gene, *ampD*, encoding a cytosolic anhydro-muramyl-peptide amidase required for efficient peptidoglycan recycling. Since resistance is due to a loss of function of AmpD, the mutation frequency *in vitro* is very high and *ampD* mutants may also arise during ongoing therapy resulting in clinical failure (Jacobs *et al.*, 1994; Jacobs *et al.*, 1997; Tuomanen *et al.*, 1991). β-lactam resistant *ampD* mutants are unable to remove the peptide portion from recycling anhydro-muramyl peptides and as a result accumulate large amounts of anhydro-N-acetyl-muramyl-tripeptide in the cytoplasm. The accumulation of anhydro-muramyl-peptides in an *ampD* mutant has been shown to result in induction of NO in both epithelial and phagocytic cells infected with *E. chloace* (S. Normark, pers. comm.) and this might alter, for example, the inflammatory response.

Table 2.   Amelioration of fitness losses caused by chromosomal mutations.

Bacteria	Resistance mutation (resistance)	Compensatory mutation (resistance in compensated mutant)	Selection for compensation in	Reference
S. typhimurium	*rpsL* (streptomycin)	intragenic, *rpsL* (maintained)	mice	Björkman *et al.*, 1998
	rpsL (streptomycin)	extragenic, *rpsD/E*, (maintained)	laboratory medium	Björkman *et al.*, 1998, 1999
	gyrA (nalidixic acid)	intragenic, *gyrA*, (maintained)	mice	Björkman *et al.*, 1998
	rpoB (rifampicin)	intragenic, *rpoB*, (maintained)	mice	Björkman *et al.*, 1998
	fusA (fusidic acid)	true reversion, *fusA*, (lost)	mice	Björkman *et al.*, 2000
	fusA (fusidic acid)	intragenic, *fusA*, (often maintained)	laboratory medium	Johansson *et al.*, 1996
S. aureus	*fus* (fusidic acid)	intragenic, *fus*, (maintained/lost)	laboratory medium	Nagaev *et al.*, unpubl.
E. coli	*rpsL* (streptomcyin)	extragenic, *rpsD/E*, (maintained)	laboratory medium	Schrag *et al.*, 1997, Schrag and Perrot, 1996
	rpoB (rifampicin)	intragenic, *rpoB* (maintained)	laboratory medium	Levin *et al.*, unpubl.
M. tuberculosis	*katG* (isoniazid)	extragenic, *ahpC*, (maintained)	humans	Sherman *et al.*, 1996
N. meningitidis	*sul2* (sulfonamide)	intragenic (maintained)	humans	Fermer and Swedberg, 1997

REVERSION, COMPENSATORY EVOLUTION AND AMELIORATION OF FITNESS COSTS

In the absence of antibiotics, evolution to reduce the cost of resistance by second-site mutations is more common than reversion to drug sensitivity (Table 2, and Andersson and Levin, 1999). This is mainly caused by two processes: (i) compensatory mutations are more common than true reversion, which requires a specific single nucleotide substitution, and (ii) serial passages are sometimes associated with severe population bottlenecks resulting in the selection of the most frequently occuring clones of mutants rather than the best (Levin *et al.*, 2000). Thus, low frequency revertants, which are likely to appear late during growth are rarely sufficiently common to be transferred in a serial passage experiment. In contrast, compensated mutants, which may be generally less fit than the true revertants, have a higher probability of being transferred in each passage simply because they are more common. Furthermore, if compensated mutants become fixed in the population, true revertants are unlikely to ascend in frequency because in the genetic background of the compensated mutant they may be at a substantial disadvantage because the compensatory mutation by itself often reduces fitness (Björkman *et al.*, 1999; Schrag and Perrot, 1996).

The physiological mechanisms by which compensatory mutations restore fitness have been determined for a few cases. For example, compensation for the cost of streptomycin resistance in *E. coli* and *S. typhimurium rpsL* mutants can be achieved by extragenic mutations that restore the efficiency and rate of translation to wild-type or nearly wild-type levels (Björkman *et al.*, 1999; Bohman *et al.*, 1984; Ruusala *et al.*, 1984). Similarly, compensation for the cost of rifampicin resistance in *rpoB* mutants has been through intragenic compensatory mutations which restore the rate of transcription to wild type levels (Reynolds, pers. comm.).

Recent data suggest that mutation rates for bacteria grown inside a host might be higher than that seen in laboratory medium (Björkman *et al.*, 2000). Thus, it is conceivable that the rates of both resistance development and genetic compensation might be faster in hosts than

anticipated from laboratory determined mutation rates. It is therefore important to understand this putative environmental influence on mutation rates if we are to correctly assess the speed and trajectory of molecular evolution that occurs in resistant bacteria.

The degree of restoration of fitness by the compensatory mutations varies greatly, and in some cases restoration appears complete whereas in others it is only partial. An important question in understanding resistance development is what determines the type of mutations found. From our studies it appears as if the selection conditions (laboratory media vs. hosts) affect the spectrum of mutations. One example is the previously discussed streptomycin resistant *rpsL* mutants in *S. typhimurium*. When selected in mice, compensatory mutations are only intragenic. In contrast, when selection for restored fitness is done in laboratory medium compensation occurs by extragenic suppressor mutations (Björkman *et al.*, 1998; Björkman *et al.*, 1999; Björkman *et al.*, unpubl.). The most likely explanation for the difference in compensatory mutations for the *rpsL* mutants is that the mutational spectrum in bacteria differs between laboratory media and mice. Also for fusidic acid resistant *fusA* mutants compensation occurs by different mutations in laboratory medium and experimental animals. Thus, in laboratory medium compensation is almost exclusively by intragenic compensatory mutations whereas in mice almost only true revertants are found. For the *fusA* mutants, selection could account for the difference. In these mutants, compensation of the growth defect in laboratory medium can be conferred either by true reversion (which is rare since it occurs only by one specific amino acid substitution) or by intragenic second-site compensatory mutations (which are common because they can occur by at least 16 different substitutions). When the relative fitness of the bacteria was determined by pairwise competition assays in mice, only the true revertants were fully compensated whereas the intragenic second-site compensated mutants showed only partial compensation. In laboratory medium, both the intragenically compensated mutants and the true revertants grew as well as the wild type (Johanson *et al.*, 1996). Thus, the revertant mutants, even though they were expected to be rarer than the second-site suppressors because of their smaller genetic target size, were predominantly selected in mice because they were the only mutants that had a fully restored fitness.

Why can some specific compensatory mutations ameliorate fitness losses for bacteria grown in laboratory medium but not in mice? I.e., why are the defects in the mutant EF-G's only seen during growth in mice? The fusidic acid resistant mutants have altered levels of the nucleotides ppGpp and pppGpp (Macvanin *et al.*, 2000). Because these nucleotides are pleiotropic regulators of gene expression (Cashel *et al.*, 1996), it is conceivable that changes in their concentrations affect expression of virulence functions or other functions needed for rapid growth in mice. This could alter fitness in mice, without any effect on growth in laboratory medium.

The only evidence for compensatory evolution occurring in bacteria isolated from patients comes from a study of isoniazid resistant *M. tuberculosis* (Sherman *et al.*, 1996). These bacteria become resistant by virtue of knock-out mutations in the *katG* gene, which cause a loss of catalase activity. This results in slower growth in experimental animals and avirulence as measured by competition experiments, LD_{50} tests and histopathology (Wilson *et al.*, 1995; Zhongming *et al.*, 1997). The majority of clinical isolates with the *katG* mutation also contain a promoter-up mutation in the *ahpC* gene which causes an increase in the level of alkyl hydroxyperoxidase reductase (AhpC). Even though no direct causality has been established, it is likely that the overproduction of AhpC due to promoter-up mutations compensates for the lack of catalase in the isoniazid resistant *katG* mutants and restores virulence (Heym *et al.*, 1997; Sherman *et al.*, 1996).

CONCLUSION

The limited amount of data currently available on the biological costs of antibiotic resistance and its genetic compensation suggest that antibiotic resistance might be less easily reversed than previously anticipated. The commonly observed decrease in fitness caused by resistance mutations may be rapidly and efficiently restored by compensatory mutations. Hence, compensatory evolution may stabilize antibiotic resistant bacterial populations and make them as fit as the sensitive clones even in the absence of antibiotics. These conclusions further emphasize the importance of using antibiotics prudently in order to prolong their useful lifespans.

ACKNOWLEDGEMENTS

This work was supported by the Swedish Natural Science Research Council, the Swedish Medical Research Council, the Swedish Institute for Infectious Disease Control, the Leo Research Foundation, and the EU Biotechnology Research Programme.

REFERENCES

Andersson, D.I., Slechta, S.E. and Roth, J.R. (1998) Evidence that gene amplification underlies adaptive mutability of the bacterial *lac* operon. *Science* **282**, 1133–1135.

Andersson, D.I. and Levin, B.R. (1999). The biological cost of resistance. *Curr. Op. in Microbiol.* **2**, 487–491.

Austin, D.J., White, N.J. and Anderson, R.M. (1998) The dynamics of drug action on the within-host population growth of infectious agents: melding pharmacokinetics with pathogen population dynamics. *J. Theor. Biol.* **194**, 313–339.

Austin, D.J., Kristinsson, K.G. and Anderson, R.M. (1999) The relationship between the volume of antimicrobial consumption in human communities and the frequency of resistance. *Proc. Natl. Acad. Sci. USA* **96**, 1152–1156.

Baquero, F., Negri, M.C., Morosini, M.I. and Blazquez, J. (1997) The antibiotic selective process: concentration-specific amplification of low-level resistant populations. In: Antibiotic resistance: origins, evolution, selection and spread (Ciba Foundation Symposium 207) pp. 93–111.

Baquero, F., Negri, M.C., Morosini, M.I. and Blazquez, J. (1993) Effect of selective antibiotic concentrations on the evolution of antimicrobial resistance. *APUA Newsletter* **11**, 4–5.

Björkman, J., Hughes, D. and Andersson, D.I. (1998) Virulence of antibiotic-resistant *Salmonella typhimuirium*. *Proc. Natl. Acad. Sci. USA* **95**, 3949–3953.

Björkman J., Samuelsson, P., Andersson, D.I. and Hughes, D. (1999) Novel ribosomal mutations affecting translational accuracy, antibiotic resistance and virulence of *Salmonella typhimurium*. *Mol. Microbiol.* **31**, 53–58.

Björkman, J., Nagaeu, I., Berg, O.G., Hughes, D. and Anderson, D.I. (2000) Effects of environment on compensatory mutations to ameliorate costs of antibiotic resistance. *Science*, **287**, 1479–1482.

Bohman, K., Ruusala, T., Jelenc, P.J. and Kurland, C.G. (1984) Kinetic impairment of restrictive steptomyicn-resistant ribosomes. *Mol. Gen. Genet.* **198**, 90–99.

Böttger, E.C., Springer, B., Pletschette, M. and Sander, P. (1998) Fitness of antibiotic-resistant microorganisms and compensatory mutations. *Nature Med.* **4**, 1343–1344.

Cashel, M., Gentry, D.R., Hernandez, V.J. and Vinella, D. (1996) The stringent response. in *Escherichia coli* and *Salmonella*: Cellular and Molecular Biology, (F.C. Neidhardt *et al.*, Eds.). ASM Press, Washington, DC, pp. 1458–1496.

Cohen, M.L. (1992) Epidemiology of drug resistance: implications for a post-antimicrobial era. *Science* **257**, 1050–1055.

Compeau, G., Al-Achi, B.J., Platsouka, E. and Levy, S.B. (1988) Survival of rifampicin resistant mutants of Pseudomonas fluorescens and Pseudomonas putida in soil systems. *Appl. Environ. Microbiol.* **54**, 2432–2438.

Davies, J. (1994) Inactivation of antibiotics and the dissemination of resistance genes. *Science* **264**, 375–382.

Dykes, G.A. and Hastings, J.W. (1998) Fitness costs associated with class IIa bacteriocin resistance in Listeria monocytogenes B73. *Lett. Appl. Microbiol.* **26**, 5–8.

Fermer, C. and Swedberg, G. (1997) Adaptation to sulfonamide resistance in *Neisseria meningitidis* have required compensatory changes to retain enzyme function: kinetic analysis of dihydropteroate synthases from *N. meningitidis* expressed in a knockout mutant of *Escherichia coli. J. of Bacteriol.* **179**, 831–837.

Foster, P.L. (1998) Adaptive mutation: has the unicorn landed? *Genetics* **148**, 1453–1459.

Hall, B.G. (1998) Adaptive mutagenesis: a process that generates almost exclusively benficial mutations. *Genetica* **102–3**, 109–125.

Heym, B., Stavropoulus, E., Honore, N., Domenech, P., Saint-Joanis, B., Wilson, T., *et al.* (1997) Effects of overexpression of the alkyl hydroxyperoxidase reductase *AphC* on the virulence and isoniazid resistance of *Mycobacterium tuberculosis. Infect. Immun.* **65**, 1395– 1401.

Jacobs, C., Park, J., Huang, L., Bartowsky, E. and Normark, S. (1994). Bacterial cell wall recycling provides cytosolic muropeptides as effectors for β-lactamase induction. *EMBO J.* **13**, 4684–4694.

Jacobs, C., Frere, J.-M. and Normark, S. (1997) Cytosolic intermediates for cell wall biosynthesis and degradation control β-lactam resistance in Gram-negative bacteria. *Cell* **88**, 823–832.

Johanson, U., Ævarsson, A., Liljas, A. and Hughes, D. (1996) The dynamic structure of EF-G studied by fusidic acid resistance and internal revertants. *J. Mol. Biol.* **258**, 420–432.

Kepler, T.B. and Perelson, A.S. (1998) Drug concentration heterogeneity facilitates the evolution of drug resistance. *Proc. Natl. Acad. Sci. USA* **95**, 11514–11519.

Kurland, C.G., Hughes, D. and Ehrenberg, M. (1996) The limitations of translational accuracy. in *Escherichia coli* and *Salmonella*: Cellular and Molecular Biology. (F.C. Neidhardt *et. al.*, Eds.). ASM Press, Washington, DC, pp. 979–1004.

LeClerc, J.E., Li, B., Payne, W.L. and Cebula, T.A. (1996) High mutation frequencies among *Escherichia coli* and *Salmonella typhimurium. Science* **274**, 1208–1211.

Lenski, R.E. (1991) Quantifying fitness and gene stability in microorganisms. *Biotechnology* **15**, 173–192.

Levin, B.R., Lipsitch, M., Perrot, V., Schrag, S., Antia, R., Simonsen, L., *et al.* (1997) The population genetics of antibiotic resistance. *Clin. Inf. Dis.* **24**, S9–16.

Levin, B.R., Perrot, V. and Walker, N.M. (2000) Compensatory mutations, antibiotic resistance and the population genetics of adaptive evolution in bacteria. *Genetics* **154**, 985–997.

Levy, S.B. (1992) *The antibiotic paradox: how miracle drugs are destroying the miracle.* Plenium Press, New York.

Macvanin, M., Johanson, U., Ehrenberg, M. and Hughes, D. (2000) Fusidic acid resistant EF-G perturbs the accumulation of ppGpp. *Mol. Microbiol.* **37**, 98–107.

Mao, E.F., Lane, L., Lee, J. and Miller, J.H. (1997) Proliferation of mutators in a cell population. *J. Bacteriol.* **179**, 417–422.

McCormick, J.B. (1998) Epidemiology of emerging/re-emerging antimicrobial-resistant pathogens. *Curr. Op. Microbiol.* **1**, 125–129.

Miller, J.H. (1996) Spontaneous mutators in bacteria: insights into pathways of mutagenesis and repair. *Ann. Rev. Microbiol.* **50**, 625–643.

Novak, R., Henriques, B., Charpentier, E., Normark, S. and Tuomanen, E.S. (1999) Emergence of vancomycin tolerance in *Streptococcus pneumoniae. Nature* **399**, 590–593.

Rosenberg, S.M., Thulin, C. and Harris, R.S. (1998) Transient and heritable mutators in adaptive evolution in the lab and nature. *Genetics* **148**, 1559–1566.

Ruusala, T., Andersson, D.I., Ehrenberg, M. and Kurland, C.G. (1984) Hyper-accurate ribosomes inhibit growth. *EMBO J.* **3**, 2575–2580.

Schrag, S., Perrot, V. and Levin, B.R. (1997) Adaptation to the fitness cost of antibiotic resistance in *Escherichia coli. Proc. R. Soc. London* **264**, 1287–1291.

Schrag, S. and Perrot, V. (1996) Reducing antibiotic resistance. *Nature* **381**, 120–121.

Sherman, D.R., Mdluli, K., Hickey, M.J., Arain, T.M., Morris, S.L., Barry III, C.E., *et al.* (1996) Compensatory *ahpC* gene expression in isoniazid-resistant *Mycobacterium tuberculosis. Science* **272**, 1641–1643.

Tuomanen, E.S., Lindquist, S., Sande, S., Galleni, M., Light, K., Gage, D., *et al.* (1991). Coordinate regulation of β-lactamase induction and peptidoglycan composition by *ampD. Science* **251**, 201–204.

Wilson, T.M., de Lisle, G.W. and Collins, D.M. (1995) Effect of *inhA* and *katG* on isoniazid resistance and virulence of *Mycobacterium bovis. Mol. Microbiol.* **15**, 1009–1015.

Zhongming, L., Kelley, C., Collins, F., Rouse, D. and Morris, S. (1997) Expression of *katG* in *Mycobacterium tuberculosis* is associated with it growth and persistence in mice and guinea pigs. *J. Infect. Dis.* **177**, 1030–1035.

11. Evolutionary Consequences and Costs of Plasmid-Borne Resistance to Antibiotics

Santiago F. Elena
Institut *Cavanilles* de Biodiversitat i Biologia Evolutiva and Departament de Genètica, Universitat de València, Apartado 2085. 46071 València, Spain

Most clinically important resistance to antibiotics is the result of plasmid-encoded genes (Falkow, 1975; Levy and Novick, 1986; Cohen, 1992; Neu, 1992a; Gómez-Lus, 1998). Many bacterial species have become resistant to antibiotics as a result of plasmid exchange by transformation or conjugation. Conjugation with plasmid transfer is a very common phenomenon among the *Enterobacteriaceae*, *Pseudomonas*, and some anaerobic species, such as *Staphylococcus aureus* or *Serratia marcescens* (Bryan, 1988; Finland, 1979; Neu, 1984; Schaberg *et al.*, 1981). Some examples are the emergence of plasmid-encoded penicillinase-producing *Neisseria gonorrhoeae*, the plasmid mediated resistance to chloramphenicol of *Haemophilus*, the plasmid NTP2 of *Salmonella typhimurium* that encodes for resistance to streptomycin and sulfonamide, and the plasmid mediated resistance to penicillins of *Pseudomonas* among many other examples.

Plasmids have the advantage of being easily exchangeable between different bacterial species. This extremely promiscuous behavior of plasmids helps bacteria to gain resistance to antibiotics against which they have never been challenged. For example, hospital outbreaks of *S. aureus* strains bearing R-plasmids were preceded by outbreaks of *Staphylococcus epidermidis* showing exactly the same pattern of plasmid mediated drug-resistance (Cohen, 1992). Thus, the existence of a *S. epidermidis*

reservoir, in which both species coexisted, was important for the emergence of resistant strains of *S. aureus*.

Plasmids are transferred not only within bacterial genera. Plasmid transfer between different genera has been also described. For example, an outbreak of multidrug-resistant *Salmonella flexneri* reported on a Hopi Indian reservation was traced back to a single patient who had recurrent urinary tract infections of *Escherichia coli* and was treated with long-term prophylactic trimethoprim sulfamethoxale (Cohen, 1992; Neu, 1992a). It was later demonstrated that *S. flexneri* became drug-resistant by acquiring an R-plasmid from *E. coli* (Tauxe *et al.*, 1989). It has been also proposed that *Haemophilus influenza* acquired its ability to produce (β-lactamases from *E. coli* (Neu, 1992a). Plasmid mediated glycopeptide resistance in *Enterococcus faecium* is transferable by conjugation to different gram-positive bacteria, such as *Streptococcus spp.* and *Listeria monocytogenes* (Wallace *et al.*, 1989, 1990). Some enteric pathogens, such as *Campylobacter* have become tetracycline-resistant as a result of acquiring the *tet*M gene from enterococci (Neu, 1992b). *Kleibsiella* carries a plasmid encoding genes for both, (β-lactamase production and adhesion to intestinal epithelium cells. It is similar to the plasmids found in the pathogenic *Shigella dysenteriae* responsible for outbreaks in the 1970s (Neu, 1992a). In general, gram-positive species transfer resistance to gram-negative species but the reverse is uncommon (Neu, 1992a).

Genes responsible for plasmid-borne antibiotic resistance are usually embedded in transposon-like structures (Cohen, 1976). It is long known that these transposon-like structures allow the transfer of antibiotic resistance genes to the chromosome (Iyobe et al., 1969, 1970), eliminating the necessity of carrying extra unnecessary DNA. This should save the energy and chemicals required for proper plasmid replication. In this case however, the advantages of having plasmid-borne resistance (such as ease of spread or high copy number) are lost. In this sense, plasmids can be seen either as the resistance factor itself or simply the carrier of resistance.

Bacteria that were isolated from the wild, years before of the widespread use of antibiotics, already contained plasmids similar to the ones isolated today (Hughes and Datta, 1983), although these plasmids did not encode for antibiotic resistance. Instead, the antibiotic resistance coding regions observed at present must have evolved in minority strains whose populations increased due to anti-biotic selection. Once these primitive resistant strains arose, chances for the resistance gene to spread increased with its association with transferable genetic elements. In fact, the discovery of resistance genes on plasmids was considered the first clue of the importance of gene transfer in resistance evolution (Watanabe, 1963).

The organization of this review will be as follows: first, I will provide a simple mathematical model that exhibits all the principal features relating to R-plasmid spread and maintenance in a bacterial population. Second, I will review experimental evidence concerning the magnitude and importance of each parameter in the model. Third, I will present a few alterations in the bacterial host that change cell fitness, as a consequence of the new selective forces imposed by the presence of R-plasmids with special attention to the regulation of resistance gene expression. Next, I will give a quick review of the different ways of generating new R-plasmids. Finally, I will discuss one point that has received much attention in recent years: coevolution between R-plasmids and host cells and its implication for the control of resistance in natural bacterial populations.

A SIMPLE MODEL CAN HELP TO UNDERSTAND THE FATE OF R-PLASMID BEARING CELLS IN BACTERIAL POPULATIONS

A mathematical model is a simplistic and formal description of what happens in nature. In theory, we can construct a complicated model with many parameters and so get an almost perfect description of empirical observations. However, the complexity of a model can then make it useless for non-expert readers and generate negative reactions. Instead, the presentation of a simple model can help us to analyze which parameters are most important in understanding the population dynamics and evolutionary implications of R-plasmids. Several models have been proposed to illustrate how bacteria carrying R-plasmids behave both in laboratory and in wild populations (Stewart and Levin, 1977; Freter et al., 1983; Esty et al., 1986; Lenski and Bouma, 1987; Cooper et al., 1987; Lenski and Nguyen, 1988; Sýkora et al., 1989; Simonsen, 1990; Summers, 1991; Proctor, 1994). The model I describe here is close to the one proposed by Sýkora et al. (1989).

Let us imagine a population of bacteria living in the gut that consists of two different types of cells: R_t represents the number of R-plasmid carrying cells and S_t the number of plasmid-free antibiotic-sensitive cells at a given moment t. The model takes into consideration three different factors. (1) Cells carrying the R-plasmid and plasmid-free cells replicate at different rates, μ_R and μ_S, respectively. In absence of antibiotic in the environment, plasmid-free cells replicate faster than resistant cells ($\mu_R < \mu_S$), while in the presence of sub-inhibitory concentrations of antibiotic, resistant cells grow faster than sensitive cells ($\mu_R > \mu_S$). In other words, we assume a cost associated with bearing an R-plasmid in the absence of antibiotics in the environment. When the concentration of antibiotic is larger than the minimum inhibitory concentration (MIC), sensitive cells do not grow at all, while resistant cells grow without problem ($\mu_R > \mu_S = 0$). (2) Due to imperfect vertical plasmid transmission during cell division, a fraction of cells loses the plasmid by segregation. Let us call ν the rate of spontaneous plasmid segregation ($\nu \geq 0$). (3) The third factor incorporated into the model is the conjugation rate, γ. If the plasmid is conjugative ($\gamma > 0$), a fraction of sensitive cells can be converted to resistant cells by plasmid transfer. (Non-conjugative plasmids are only a particular case in which $\gamma = 0$.) For convenience, let us assume that bacteria feed on some carbon source that has a finite concentration in the environment. Finally, as a consequence of the limitation of carbon resources, to keep population size constant we will assume density-dependent growth with a maximum carrying capacity of K cells. With these assumptions, the change in R_t and S_t with time can be described by the following set of non-lineal differential equations:

$$\frac{dR_t}{dt} = (\mu_R R_t - \nu R_t + \gamma R_t S_t)\left(1 - \frac{R_t + S_t}{K}\right)$$

$$\frac{dS_t}{dt} = (\mu_s S_t + \nu R_t - \gamma R_t S_t)\left(1 - \frac{R_t + S_t}{K}\right)$$

Figure 1 shows numerical solutions for the above system with several different sets of parameters that illustrate the temporal dynamics of plasmid-bearing cells with no selective pressure for plasmid maintenance (i.e. no antibiotic added to the environment). Note that the solution when the antibiotic is present is trivial: plasmid-free cells are immediately eliminated. Instead of plotting the absolute densities, Figure 1 shows the frequency of plasmid-bearing cells in the population [$p_t = R_t/(R_t + S_t)$]. Notice that despite various combinations of ν and γ, plasmid-free cells always outcompete R-plasmid-bearing cells. The only exception to this was the trivial case of both cells growing at the same rate (i.e. the plasmid does not have any substantial effect on fitness). The rate at which plasmid-bearing cells are eliminated from the population depends on their effect on fitness, and for small effects the time that plasmid-bearing cells are maintained at low frequency in the population can be very long. It is also interesting to note that when the conjugation rate was large enough, the frequency of plasmid-bearing cells had a sudden initial increase, and after a short period of time this frequency starts declining if $\mu_R < 1$. Similar effect, but of a smaller magnitude can be also observed with intermediate conjugation rates and small segregation rates. The lower the conjugation rate, the faster the elimination of the resistant cells. Comparing the different panels in Figure 1, one can confirm that segregation rate has less effect on the frequency of plasmid-bearing cells than conjuga-

tion rate. This is due to the second-order nature of the conjugation term in the above equations.

It is interesting to see whether both types of cells can coexist in a stable equilibrium under certain conditions. For this, we must equate the above equations to zero and solve for the equilibrium values, R^* and S^*. In doing so, it is easy to show that a stable equilibrium is possible only when the following condition is met:

$$\gamma K = \gamma(R^* + S^*) > \left(1 - \frac{\mu_S}{\mu_R}\right)(\nu - \mu_R)$$

This equation is easily interpreted: in the absence of antibiotic in the environment, resistant cells can be maintained in the population only if the transfer of plasmids by conjugation (first term in the equation) is large enough to compensate for both, an excessive loss by segregation, ($\nu - \mu_R > 0$) and their selective disadvantage in replication ($1 - \mu_S/\mu_R < 0$). In other words, the frequency of resistant cells in the population increases with the rate of conjugation and decreases with the rate of segregation, and the final density of each cell type will be determined by these two factors as well as the difference in growth rates. In the presence of an antibiotic concentration higher than the MIC, it is expected sensitive cells that will be completely eliminated from the population despite the fact that segregation is always generating plasmid-free cells. Once antibiotic concentration decreases below the MIC, new sensitive bacteria will appear by segregation and their frequency will increase in the population due to their larger growth rate.

Later in this chapter I will review experimental evidence concerning the importance of several factors in the stability and persistence of plasmid encoded resistance: differences in growth rate between plasmid-bearing and plasmid-free cells, the effect of segregation, and the effect of conjugation.

PLASMID CARRIAGE REDUCES HOST GROWTH RATE

Definition of the Cost of Plasmid Carriage in Terms of Host Fitness

It is possible to define the cost of carrying a plasmid as the magnitude of reduction in host cell

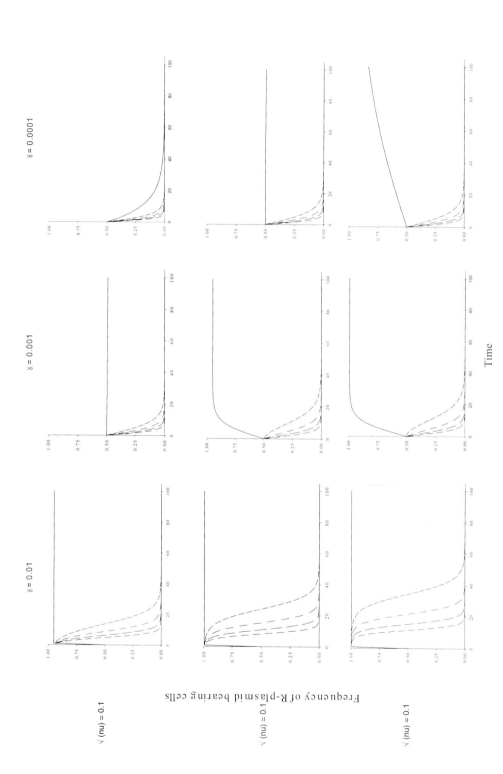

Figure 1. Trajectories of the R-plasmid bearing cells under different conditions. Values for the growth rate of resistant cells (μ_R) have been changed from 1 (solid line) to 0.5 with 0.1 increments. Growth rate for the plasmid-free cells was always set to 1. Segregation (ν) and conjugation rates (γ) were changed as indicated in the figure. The simulations were initiated with 100 resistant cells and 100 sensitive cells. Population grew to a total density of $K = 10^6$ cells. Plots were generated from the set of differential equations given in the main text.

fitness due to any activity related to replication and maintenance of the plasmid, as well as with the expression of any genes present on the plasmid. In addition, the presence of the plasmid may interfere with some basic cellular structures and processes. For example, regulatory proteins coded on a plasmid-borne gene can inhibit chromosomal genes. In the presence of a selective factor in the environment, for instance antibiotics, this cost can be compensated for by the ability to quickly and efficiently respond to the selective factor. In the absence of antibiotics, the expression of the resistance genes does not provide any advantage but instead wastes energy and resources through the production of an excess of unneeded proteins.

All the studies analyzing the effect of R-plasmid carriage, in both presence and absence of antibiotic, involve competition between sensitive and resistant strains that otherwise must be isogenic (Levin, 1980). During head-to-head competition for a limiting-resource, it is possible to measure how the relative densities of each competitor (R_t/S_t) change with time. If the presence of a plasmid has an effect on fitness, the carrying cells will replicate slower and their frequency will decrease in the population. In contrast, the frequency of the plasmid-free cells will increase. However, if the plasmid does not have an effect on fitness, the frequency of both genotypes will not change with time. The slope of the linear regression of $\ln(R_t/S_t)$ over time gives an estimate of the difference in growth rates, $\mu_R - \mu_S$. By definition, this difference indicates how much faster the R-plasmid bearing cells replicate when compared with the plasmid-free cells, i.e. their fitness (Lenski, 1997).

Evidence for Effects of Plasmid Carriage on Fitness

Godwin and Slater (1979) studied the effect of different growth environments on the stability of R-plasmids. They found that for TP120, a multidrug resistance plasmid, plasmid-bearing cells were always less fit than plasmid-free cells. They also found that growth under carbon- and phosphorus-limited conditions always induced the loss of one or more of the drug resistances. Under carbon-limitation, resistance to tetracycline (Tc^R) was lost

but the other resistances remained. However, under phosphorus-limited conditions, all the resistances were lost. Strains that had lost resistance to one or more drugs grew at a higher rate than their resistant parentals. This easy loss of resistances suggested that TP120 could partially fragment, producing smaller plasmids, and hence gain an advantage in growth. Large multiple resistance plasmids are the result of the linkage of separately-derived resistance genes to a basic replicon and this is a reversible process (Cohen, 1976). The fact that Tc^R was regularly lost indicates that the module containing it was easier to remove from the plasmid than the other modules.

Lee and Edlin (1985) found that cells carrying plasmid pBR322 or any of its derivatives had a reduced fitness when compared with the plasmid-free cells in a glucose-limited environment. They showed that this fitness effect was not only due to carrying the plasmid itself, but also to the expression of the Tc^R gene in the absence of the drug. The pBR322-derivative plasmids usually express Tc^R constitutively. In fact, they demonstrated that when the Tc^R genes were inactivated by fragment deletion, plasmid-bearing cells grew at the same rate as plasmid-free cells. This last experiment examined the effect of constitutive expression of Tc^R but did not examined the possible effect of repressed expression. Nguyen et al. (1989) made a more in depth study exploring the effect of alternative modes of regulation of the resistance in a Tc-free environment. They constructed three different plasmids with different modes of regulation and intensity of expression. The first one, pBT107, was characterized by a strong *tet*A promoter and a functional *tet*R repressor (inducible gene expression) (hereafter pInd-Hi). The second plasmid (pBT1071S) had a weak promoter and its expression was also inducible (pInd-Lo). The last plasmid (pBT1071) had a weak promoter and a non-functional repressor (constitutive gene expression) (pCon-Lo). Their first observation indicated that the fitness of plasmid-bearing cells decreased with increased induction of the Tc^R operon. They also found that in the absence of antibiotic selection cells carrying the constitutive pCon-Lo were approximately 3% less fit than cells carrying the inducible pInd-Lo, despite the fact that the former plasmid had a weak *tet*A promoter. It has been previously observed by Moyed and

Bertrand (1983) that constitutive expression of the *tet*A gene carried on multicopy plasmids was lethal for cells unless there was an associated reduction in the strength of the *tet*A promoter. However, when the expression was artificially induced in both cells by the addition of an appropriate concentration of the inductor 5a,6-anhydrotetracycline (At), both plasmids had similar effects on fitness. In a complementary set of experiments, Nguyen *et al.* (1989) found other interesting results of the effect of resistance expression on fitness: (i) the repression of the Tc^R operon must be extremely tight and the residual expression of the *tet*A gene could be at most reduce fitness by 0.2%; (ii) the carriage of the resistance determinant by itself had at most 0.3% effect on fitness; (iii) the carriage of the plasmid by itself, excluded the full Tc^R operon, was small (about 1.5%) but still significant, and (iv) the hyper-expression of the repressor *tet*R did not have detectable effect on fitness.

Valenzuela *et al.* (1996a) observed that pKH47, a high copy number derivative of pBR322, has a strong negative fitness effect on bearing cells due to a gene dosage-dependent expression of the *tet*A gene. They also demonstrated that the effect was due to the entire *tet*A gene product rather than to any part of the protein. Previously, it was proposed that the N-terminal membrane binding domain of the TetA had pleiotropic effects such as an increase in the susceptibility to amino glycosides, an increase in the plasmid's supercoiling (Lodge *et al.*, 1989) or complementing mutations affecting potassium uptake (Griffith *et al.*, 1988). In a latter paper, Valenzuela *et al.* (1996b) demonstrated that the reduction in fitness was due to a reduction in the activity of some ATPase enzymes bounded to membrane.

In general, all of the above evidences supports the idea that a fitness cost for the bacteria is associated with the presence of the plasmid. This cost is the result of two different components. The first cost is associated with the carriage of the plasmid itself [around 1.5% in the experiments of Nguyen *et al.* (1989)]. The second cost, which seems to be larger (about 3% in the same experiments), is due to the expression of the resistance genes in the absence of the antibiotic in the environment. This second cost can be mini-mized by a tight repression of the resistance expression in absence of antibiotics. Most of the

evidence above dealt with the Tc^R operon. Whether or not these findings are general for any other antibiotic is an open question at this point.

Despite this apparent general cost of R-plasmid carriage, some examples indicated that this cost is context-dependent. For instance, in a classic set of experiments, Adams *et al.* (1979) and Helling *et al.* (1981) found than on rich medium, without limited resources, no differences in growth rate for both colicin-producing plasmid-containing cells and plasmid-free cells were observed. However, when the carbon source becomes scarce, plasmid-free cells have a selective advantage when their density is higher than that of plasmid-bearing cells. In contrast, when the frequency of plasmid-bearing cells in the mixture is higher, colicin-producing cells become fixed in the population due to their ability to eliminate plasmid-free colicin-sensitive cells. Furthermore, Simonsen (1990) found that whereas in liquid cultures the presence of plasmids R1 or R1*drd*19 imposes a cost to *E. coli* K12 cells in terms of fitness, in surface-growing colonies no significant effect was detected, and both plasmid-bearing and plasmid-free cells grew at the same rate.

PLASMID SEGREGATION IS LESS IMPORTANT THAN DIFFERENCES IN GROWTH RATE

There is a common observation that although spontaneous plasmid segregation happens at a measurable rate, its magnitude, and therefore its influence on the frequency of plasmid-bearing cells is far less than the influence due to differences induced by differential growth rates. However, it will be useful to review a few experiments in which the rate of spontaneous segregation was measured and compared with differences in growth rate.

In the same experiment reviewed in the previous section, Adams *et al.* (1979) made more specific experiments to test whether the observed loss of the plasmid-bearing cells was actually due to differ-ences in growth rates or due to an increased segregation rate. They constructed isogenic strains with an easy-to-score phenotypic marker in the chromosome that allowed them to unambiguously identify segregant cells from ancestral plasmid-free cells. When plasmid-bearing cells (with the marker) and plasmid-free cells were mixed at a 1:1 proportion, invariably the plasmid-free cells increased their frequency without any increase in the number of cells with the chromosomal marker but without the plasmid. This demonstrated that, although segregational loss of the plasmid happens, plasmid-bearing cells are selected against on the basis of their reduced growth rate. This differences in growth rate, rather than segregation, is the main factor for eliminating plasmid-bearing cells in the absence of antibiotic.

Jones *et al.* (1980) studied the segregational stability of several different plasmids. In chemostat cultures without antibiotic selection, they found that after 120 generations Col-derived plasmid-bearing cells were still the only class in the population and that no segregant had arisen. However, in a separate experiment, pBR322-related plasmid-bearing cells were replaced, although never completely eliminated, by their plasmid-free partners irrespectively of the host cell genetic background. Even though the Col-related plasmids were maintained in the population, some physiological change happened. Their copy-number fell five-fold during the experiment, which suggests that loss of pBR322-derived plasmids occurred by a similar mechanism. Initially copy number decreases as a response to the nutrient limitation (rather than selection of mutants with a lower copy-number) followed by defective segregation to daughter cells. I will consider this experiment in greater depth in a later section.

Similar results were also obtained by Lenski and Bouma (1987). They set up two populations, one consisting of only pACYC184 plasmid-bearing cells and another containing equal numbers of plasmid-bearing and isogenic plasmid-free cells. They tracked the frequency of plasmid-bearing cells through time in an antibiotic-free environment (i.e. no selective pressure for maintaining the plasmid). Independently of the population, the result was always the same: plasmid-free cells increase in the population by eliminating the plasmid-bearing cells. Using a simple mathematical model, they were able to determine both the segregation rate and the selection coefficient (1 − fitness). From their data, it was straightforward to conclude that intense selection (~30% fitness effect) against plasmid carriage was the reason for the rapid loss of pACYC184. They were unable to

estimate the segregation rate and concluded that it was too low as to be measured.

In conclusion, segregation must be important in generating the first plasmid-free cell, but not for increasing its frequency in the population. The frequency will be increased by the fitness differences induced by plasmid carriage.

HORIZONTAL PLASMID TRANSFER CAN ALSO PLAY AN IMPORTANT ROLE IN PLASMID STABILITY

Horizontal gene transfer in laboratory conditions is a common and easy-to-induce phenomenon (either by conjugation, transformation or transduction). Unfortunately, we know too little about the rate at which horizontal gene transfer occurs in natural environments such as soil and the gastrointestinal tract, whose inhabitants comprise a large number of different bacterial species and genera. Regardless, conjugation should be the most effective way to spread an R-plasmid within and between species of bacteria.

Simonsen (1990) made a set of interesting experiments comparing the rate and the importance of conjugation, both in liquid and surface bacterial cultures. This study was of great importance because in nature bacteria usually grow attached to surfaces as colonies, rather than in suspension. The dynamics of plasmid transfer on surfaces are expected to be much more complex than in liquid, because cells are growing in clusters, and the opportunity for a given cell to participate in mating is not the same but depends on its location in the colony. Simonsen employed two different multi-drug resistance conjugative plasmids, R1 and its derepressed mutant R1*drd*19, to analyze the effect of colony density on a surface in the rate of conjugation. For high cell densities, where the cells grew as a lawn, the kinetics of plasmid transfer was similar to that observed in liquid culture for both plasmids. However, as the colony density decreased, the frequency of conjugation for R1*drd*19 was strongly reduced whereas for R1 it was less affected.

The above evidence for plasmid transfer were obtained under *in vitro* conditions. However, plasmid transfer in natural environments, such as the gut of animals, is difficult to demonstrate

(Jaromel and Kemp, 1969; Smith, 1970; Farrar *et al.*, 1972; Anderson *et al.*, 1973; Smith, 1976). In a rigorous analysis, Freter *et al.* (1983) found that the conjugation rate was around 10^{-13}–10^{-9} mL/cell/hour both in chemostat cultures and in mice. More evidence in the same vein comes from a recent study done by Licht *et al.* (1999). They measured the rate of R1*drd*19 plasmid transfer in batch cultures, biofilms and mice intestinal extracts. For homogeneous batch cultures they estimated the rate of transfer to be ~3 × 10^{-11} mL/cell/hour, with a constant transfer through experimental time. In contrast, in the biofilm experiments the transfer happens at a high rate at the beginning of the experiment, but further transfer was not observed. The kinetics of plasmid transfer in the intestine shows similarities with the biofilm: a brief initial period of plasmid transfer followed by a long period without noticeable transfers. The rate of transfer was estimated to be ~7 × 10^{-12} for caecal mucus and ~5 × 10^{-13} for caecal contents. From their data, the authors suggested that the viscous mucus layer on the apical surface of intestinal epithelial cells is where the majority of plasmid transfer occurs.

The small figures for transfer rate reported above are consistent with the previous lack of empirical observations of transconjugants *in vivo*. They also indicate that the observed low rates can be explained by differences in growth rate rather than the existence of different kinds of conjugation inhibitors (bile salts, acid metabolic products, pH, fatty acids, anaerobic conditions, etc) (Anderson, 1975; Burman, 1977). In addition, it has been demonstrated that a number of different antibiotics increase the rate of plasmid transfer between different bacteria (Mazodier and Davies, 1991). The implications of this finding are tremendous: not only does the expression of resistance depend on the drug's presence in the environment, but also the antibiotic itself may induce the transfer of the resistance genes (Torres *et al.*, 1991).

Despite the importance of conjugation in the spread of R-plasmids, the transfer of some plasmids among different bacterial species is not as general as expected. The existence of incompatibility systems that reduce the fitness of the recipient cells have been known for a long time. For example Holĉik and Iyer (1996) found that the transfer of the pCU1 plasmid from its natural host *E. coli* to

Kleibsiella oxytoca induced the death of more than 90% of the recipient cells. These authors characterized the basis of this killer phenotype and found the responsible region *kik*A, composed by two ORFs. The *kik*A region is conserved among all the plasmids belonging to the IncN incompatibility group. However, for *K. oxytoca* cells growing in an environment with antibiotic this is not a bad deal at all, despite this great death induced by the plasmid, since a small fraction of the recipients survived (less than 10%) and therefore these survivors will be positively selected and a new population of resistant cells will be generated. This affirmation implies that unless the incompatibility systems guarantees 100% death of the recipients, survivors will constitute a reservoir of resistants. These resistants will play a key role in case of presence of antibiotic in the environment. Otherwise, they will be eliminated due to reduced replication ability.

A simple message can be taken from all these experiments. The rate of horizontal transfer strongly depends on the cell density. If bacteria are growing at low densities, for instance, as isolated colonies on surfaces, the likelihood of plasmid transfer is low. As density increases, either by growing in homogeneous liquid environments or as biofilms, the rate of transfer increases as well. However, even the larger estimates of plasmid transfer under optimal conditions, are small. These small values imply that to be an important factor, population size has to be enormous, at least inversely proportional to the rate of plasmid transfer. Therefore, plasmid transfer is not going to play an important role in the spread of antibiotic resistances compared with the role played by the differences in growth rate of plasmid-free and plasmid-bearing cells. However, the role of transfer must be extremely important in the initial spread of the plasmid, but once it has been transferred to a few new host cells in the presence of sub-inhibitory concentrations of antibiotic where its presence becomes beneficial, their frequency in the population will increase mostly due to positive fitness effects.

Another consideration that needs to be made regarding the transfer of plasmids is the mode of transmission and the effect that it has on the relationship between plasmid and host cell. Recently, Turner *et al.* (1998) made a detailed analysis of the different modes of plasmid transfer. Theoretically, a plasmid can be considered an intracellular parasite, and as with many parasites, there should be a trade-off between vertical transmission (from mother to daughter cells) and horizontal transmission (conjugation between donor and recipient cells). For a mostly-vertically transmitted plasmid, it is beneficial for the plasmid to reduce its deleterious fitness effects because it must rely on its present host for transmission. On the other hand, a mostly-horizontally transmitted plasmid does not need to worry about its harmful fitness effects because its strategy is to exploit the host and move to another as soon as possible. Hence, a trade-off between these two opposing plasmid strategies is expected. Turner *et al.* (1998) found a significant, positive correlation between conjugation rate and fitness cost, supporting the above hypothesis of trade-off between transmission rates and cost: plasmids that evolved higher conjugation rates were more deleterious to their hosts, whereas those plasmids that evolved lower conjugation rates were less harmful. Interestingly, they found that the Tc^R encoded by these plasmids was lost by the plasmids with low conjugation rates. As pointed out by the authors, it is difficult to explain the reason for the association between Tc^R and conjugation rate. One possibility is that the reduction in fitness cost of carrying and transmitting the plasmid vertically was due to the knock out by point mutation or by deletion of the *tet*A gene. Alternatively, the *tet*R could have become permanently repressed, reducing the deleterious effect of the constitutive expression of the *tet*A gene and allowing then for less costly vertical transmission.

SOME PHYSIOLOGICAL IMPLICATIONS OF PLASMID-MEDIATED RESISTANCE

A detailed look into the metabolic and physiological changes induced in a cell bearing an R-plasmid is not the intent of this review. However, it is interesting to examine a few examples that illustrate the parameters relevant to the maintenance of R-plasmids. These parameters are: (i) increase in generation time, which relates with the replicative fitness of the R-plasmid bearing cells (ii) decrease in the copy number, possibly due to segregation, and (iii) negative dosage effect as

responsible of modulating the expression of resistance genes.

Lengthening of Generation Time

Zünd and Lebek (1980) observed that the transformation of *E. coli* K12 with R-plasmids isolated from different wild hosts (i.e. *Salmonella, Enterobacter, Klebsiella*) had the general effect of lengthening their host's generation time. Out of 101 plasmids tested, one-fourth increased the generation time by more than 15%, with an extreme case, pGL207, that doubled it. When the authors measured the size of these large-effect plasmids, they found that all were larger than 80 Kb. Those smaller than 80 Kb did not increase generation time. Also, there was no clear relationship between plasmid size and generation time. Zünd and Lebek (1980) proposed two explanations for this generation time-effect: (i) a result of utilization of cell facilities for synthesis of plasmid products, and (ii) the genes carried by the plasmid had some effect on membrane permeability to nutrients.

Clearly, this increase in generation time has the negative effect of reducing the fitness of R-plasmid bearing cells. During competition with plasmid-free cells in the absence of selective pressure imposed by antibiotics, a longer generation time implies a disadvantage. For a given time unit, a plasmid-free cell will be able to produce more offspring that a plasmid-bearing one.

Decrease in the Copy Number

Many different mechanisms have been proposed to explain the maintenance of a certain copy number, with each family of plasmids maintaining their own level. Copy numbers vary from the small high copy number ColE1-like plasmids that do not require any plasmid-encoded protein for *in vivo* replication to the large high copy number plasmids from the FII incompatibility group that contains *tra* genes which encode functions involved in the conjugative transfer of the plasmids. A detailed review on this topic can be found in Scott (1984).

As previously mentioned, one possible method to reduce the deleterious effect of plasmid carriage in the absence of a selective pressure is through a decrease in plasmid copy number. Jones *et al.* (1980) observed that in *E. coli* chemostat cultures RP1, pDS4101 (a ColEK derivative) and, pDS1109 (a ColE1 derivative) copy number fell five-fold, but they failed to detect segregants. As a consequence of the reduction in copy number, they observed a four-fold reduction in the MIC for ampicillin (Ap). In contrast, plasmid-free segregants of pBR322 and pMB9 were easily obtained, irrespective of the host background and the chemostat conditions. The authors suggested that the mechanism for losing pBR322 and pMB9 followed a similar copy-number decrease and, once the copy number was reduced, the likelihood of segregation increases as an immediate consequence: fewer plasmids are more difficult to equally distribute among dividing cells. These authors also suggested the interesting idea that decrease in copy number was a phenotypic response to nutrient limitation rather than to positive selection for mutants with lower copy numbers. This is supported by the observation that it was enough with a single cycle of growth in complex media to recover the MIC for Ap characteristic of the ancestral strains. Therefore, they might recover the original plasmid copy number.

Evidence in the same direction is abundant. Modi and Adams (1991) found that segregation rate increased as a way to reduce the deleterious effect of plasmid carriage during 800 generations of coevolution between *E. coli* and pBR322Δ5. In their experiments, they found that the segregational loss increased up to two orders of magnitude in the evolved strains ($\sim 10^{-2}$ per cell) when compared with the ancestral ones ($\sim 10^{-4}$ per cell). Furthermore, the strain exhibiting the highest fitness also possessed the higher segregation rate. They failed, however, to find a significant correlation between the increase in segregation rate and the reduction in copy number. It has also been demonstrated that copy number is negatively correlated with growth rate in the absence of selective pressure. Lin-Chao and Bremer (1986) observed that in continuous chemostat cultures, as growth rates increased from 0.6 and 2.5 doublings/hour, the number of pBR322 plasmids per copy of chromosomal DNA (the copy number) was decreased by half. They explained their observations in terms of a model of negative control and suggested that the activities of the plasmid's replication inhibitor (RNAI), and the preprimer (RNAII) and the efficiency with which

RNAI binds the plasmid, inhibiting its replication, are controlled by the growth rate.

All of these observations taken together suggest that decreasing the number of plasmids per host cell will also decrease the deleterious effect of R-plasmid carriage in the absence of antibiotic, and therefore increase the fitness of host cells. However, evidence suggests that this reduction in copy number is not a consequence of genetic changes in the plasmid but a phenotypic result of limitation in the amount of resources available for cell replication (Jones *et al.* 1980). In the long run, a decrease in copy number induced by a limitation in resources can generate plasmid-free cells as a consequence of imperfect distribution of plasmids among daughter cells. Once plasmid-free cells arise in the population, their frequency will logistically increase as a consequence of their selective advantage in the absence of antibiotic. However, plasmid-bearing cells will still remain in the population at frequencies determined by the join action of segregation, horizontal transfer and differential growth.

Plasmid-Encoded Resistance Gene Expression and Regulation

Chopra *et al.* (1981), Coleman and Foster (1981), Moyed *et al.* (1983) and Moyed and Bertrand (1983) observed that *tet* has a negative dosage effect. Moyed *et al.* (1983) made *E. coli* strains carrying the Tn*10 tet* gene inserted in pACYC117, a multicopy plasmid. These constructs showed between 4- and 12-fold less resistance to Tc than strains having a single insertion of Tn*10* into the chromosome. Furthermore, these constructs were hypersensitive to the *tet* artificial inductor At at concentrations that are harmless to plasmid free cells (see above for more details). These chemical is 50- to 100-fold more effective than Tc as inductor of the resistance, but in contrast, it is 2- to 4-fold less effective than Tc in the inhibition of protein synthesis. Therefore, the hypersensitivity of strains carrying multiple copies of *tet* is due to the over expression of the gene rather than to side effects associated with the At. The observed effects were also strongly dependent on the chromosomal genetic background, which indicates some epistatic interactions between *tet* and chromosomally encoded proteins. Intuitively, this suggests that an increase in copy number will cause an excess production of the repressor protein, and therefore, more repression of the *tet*A gene. However, Coleman and Foster (1981) showed that increased synthesis of the TetR repressor protein is not the basis of the negative gene dosage effect. Instead, they suggest that feedback inhibition at the level of translation, results in reduction of the TetA protein and therefore, reduced resistance to the drug.

Moyed and Bertrand (1983), however, proposed that high level expression of the *tet*A gene inhibits the growth of *E. coli* by interaction of the TetA protein with some membrane protein. Unfortunately, it was impossible for the authors to obtain a plasmid that constitutively expressed the *tet*A gene. Recently, Valenzuela *et al.* (1996b) observed that the reduction on fitness associated with the expression of Tc^R is due to a decrease in membrane-bound ATPase activity; an overproduction of TetA causes a collapse of the membrane potential.

As shown above, Nguyen *et al.* (1989) made a more complete study of the effect of different Tc^R modes of regulation on the fitness of *E. coli*. They demonstrated that either inducible or constitutive expression of the *tet*A gene could be responsible for the large fitness effects associated with its expression. Nevertheless, an inducible system does not imply a significant burden in the absence of the antibiotic. Lenski *et al.* (1994b) explored the effects of constitutive *versus* inducible Tc^R expression and found two significant disadvantages for inducible resistance expression: (i) in a fluctuating environment in which the antibiotic concentration changes dramatically between zero and higher than MIC concentrations, the inducible genotype has a longer lag phase, and (ii) the full induction of the inducible genotype is never reached due to the continued action of the repressor. Lenski *et al.* employed constructs combining different promoters and repressors: pInd-Lo, which has a weak *tet*A and is inducible (functional *tet*R); pCon-Lo, with a weak *tet*A and constitutive (non functional *tet*R); pInd-Md, with a intermediate promoter and inducible; pCon-Md, which has an intermediate promoter and constitutive expression, and pInd-Hi, with a strong promoter and inducible. As expected, in the absence of Tc they found that, for a weak promoter, the effect of a constitutive expression is about 3% (comparing pCon-Lo *versus* pInd-Lo). However, if drug is added

to the medium, then pCon-Lo confers a large fitness advantage (20%) over its inducible counterpart. This enormous advantage of the constitutive construct, when the antibiotic is present in the medium, was demonstrated to be partly due to a longer lag-phase prior to initiation of exponential growth in the inducible strains. If cells were acclimated to Tc previous to the competition experiments, the 20% fitness disadvantage of pInd-Lo is reduced to a 15%, but not completely eliminated. This finding indicated that differences are also due to a submaximal expression of the *tet*A gene in the inducible genotype, because of residual repression by the TetR protein. To ascertain the likelihood of this possibility, in a second set of experiments Lenski *et al.* (1994b) compared plasmids with inducible but strong expression to plasmids with constitutive but weak expression. It was expected that an inducible genotype with a strong *tet*A promoter (pInd-Hi) would show a higher concentration of TetA protein than its counterpart with low expression (pInd-Lo). Therefore, it was expected that the fitness of the constitutive constructs would be reduced. The experiments showed that pInd-Hi was only approximately 6% less fit than pCon-Lo, the constitutive construct with the weak promoter. Again, although reduced, the difference between constitutive and inducible expression was not eliminated. In a new set of competition experiments Lenski *et al.* (1994b) demonstrated that this effect was caused by the overproduction of TetA by pInd-Hi. Finally, they also found that constitutive genotypes did only a 4% better with an intermediate promoter (pCon-Md) than with the weak promoter (pCon-Lo). To clarify all these findings, Figure 2 shows the fitness of each of the constructs employed by Lenski *et al.* (1994b) relative to pCon-Lo. A side conclusion of the Lenski *et al.* (1994b) experiments is that there are strong epistatic interactions between the *tet*R and *tet*A loci. This means that the selective value of a particular allele depends not only on the environmental conditions but also upon the genetic context.

MECHANISMS FOR GENERATING NEW R-PLASMIDS

The role of transposable elements and more concretely, the role of transposases have been demonstrated in promoting the fusion of two R-plasmids by interaction between the recombination

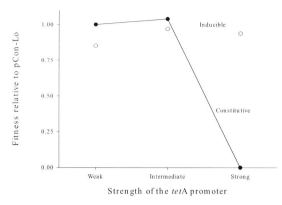

Figure 2. Fitness of each plasmid studied by Lenski *et al.* (1994b) relative to pBT1071 (weak promoter and constitutive expression). Plasmids characteristics are: pBT107-39L-6A, intermediate promoter and constitutive expression; pBT1071S, weak promoter and inducible expression; pBT107-6A, intermediate promoter and inducible expression; pBT107 strong and inducible promoter. The construct with strong and constitutive promoter was lethal. [From Lenski *et al.* (1994b)].

hotspot of one plasmid and a secondary integrase target site on the second plasmid (Francia *et al.*, 1993). Recently, several papers have characterized and presented new R-plasmids that combined several resistances, along with proposed mechanisms for the origin of these novel plasmids.

Leelaporn *et al.* (1996) characterized a family of plasmids (pSK639, pSK697 and pSK818) from *S. epidermidis* which carried resistance to the antibiotic trimethoprim (Tp) and studied the role of the insertion sequence IS257 in the evolution of these plasmids. They found that the three plasmids share sequence homology, and also have a transposon-like structure Tn*4003* bounded by IS257. This structure encodes the enzyme S1 dihydrofolate reductase responsible for the Tp^R phenotype. Moreover, pSK697 and pSK818 contain a third copy of the IS257 and an additional DNA fragment that encodes for a functional *smr* gene (multidrug resistance to antiseptics and disinfectants). This new integrated fragment was found to be nearly identical to the *smr* plasmid pSK108 previously identified in *S. epidermis*. This example clearly illustrates that cointegration of plasmids mediated by IS, resulting from non-resolved replicative transposition, is a potentially important way to generate new plasmids carrying

more complex resistances to antibiotics. Recently, Berg *et al.* (1998) found a new multidrug-resistance plasmid of the pSK family obtained by IS257-mediated cointegration of up to four smaller plasmids, each one coding for a different resistance.

Guessouss *et al.* (1996) worked with two conjugative plasmids of *E. coli* 2418: R2418, which encodes resistance to β-lactam antibiotics, and R2418S, which also encodes resistance to streptomycin (Sm). Restriction analysis showed that the latter evolved from the former by means of two different mutations, the insertion of a 2.5 Kb fragment responsible for the Sm^R phenotype, and the deletion of a 0.5 Kb fragment. These changes also implied a reduction in the ability for conjugational transfer of the plasmid R2418S. In order to properly transfer this plasmid, it becomes necessary for the presence of both plasmids in the donor cell. The 2.5 Kb fragment encoding Sm^R was probably present in a transposon, as demonstrated by Southern blot analysis showing multiple hybridization signals scattered all over the genome of the bacterium and by the fact that it was possible to obtain Sm^R transconjugants sensitive to β-lactam antibiotics.

These examples, among others, show how easily new combinations of resistances arise by transposon-mediated recombination. Multidrug-resistance plasmids can be seen as the result of natural selection playing puzzles. Each piece in the puzzle represents a cassette containing the resistance to a single antibiotic embedded within a transposon structure. Once a new combination arises by chance, if it provides a certain advantage to the host cell in the presence of different antibiotics in the environment, this combination will be quickly selected. Prior to generating any of these new combination it is required that the single units coexist in the same host cell. And this is not a hard requirement to meet due to the promiscuous behavior of bacteria, as discussed above.

COEVOLUTION BETWEEN HOST CELL AND R-PLASMIDS

Up to this point, I have presented evidence of R-plasmid costs in a novel host. However, coevolution between host cell and the plasmid will result in a reduction of the harmful effects of the carriage. Many empirical examples support this particular point. The implications of this fitness compensation are tremendous from a clinical standpoint, since the elimination of any burden associated with the presence of a plasmid eliminates the possibility of eliminating the R-plasmids from a bacterial population by suspending the administration of antibiotics to susceptible patients. I will give a brief review of evidences for coevolution.

In the seventies, Smith (1975) already observed that despite the prohibition of Tc for promoting the growth of farm pigs, the incidence of pigs excreting Tc^R bacteria was still 100% four years later. He argued that during the long time in which this practice was common, resistant bacteria arose and that these bacteria were as good as sensitive strains to replicate in the absence of Tc. In some way, resistant bacteria were able to eliminate the harmful effect on fitness of carrying and expressing the resistance plasmids.

Compensatory Mutations can take Place on the Bacterial Chromosome...

In their experiments Helling *et al.* (1981) also found that, despite the cost associated with the ColE1 plasmid carriage, in some lines the plasmid-bearing cells outgrew plasmid-free cells. They argued that the fitness increase of the plasmid-bearing cells was due to the appearance of some new compensatory mutation. They tested whether this compensatory mutation occurred in the plasmid or on the chromosome. For this, they transformed isogenic plasmid-free cells with the "compensated" plasmid. If the mutation was in the plasmid, the transformed cell would not suffer the fitness decline associated with the plasmid carriage. Instead, they found that transformed cells always lost fitness and were rapidly outcompeted by their plasmid-free counterparts. These results indicated that the compensatory mutation was chromosomal rather than on the plasmid. Similar results were obtained by Blot *et al.* (1994). In absence of bleomycin (Bm), *E. coli* cells with a Bm^R gene on a plasmid had improved survival and growth compared with the plasmid-free cells. Also, these cells were more resistant to unrelated antibiotics such us amikacin or Sm (Blázquez *et al.*, 1993).

Bouma and Lenski (1988) designed the first experiment specifically to test for coevolution

between non-conjugative plasmid, pACYC184, and its bacterial host. These authors demonstrated that after 500 generations of evolution in medium containing chloramphenicol (Cm) to eliminate plasmid-free segregants, the 6% deleterious effect of plasmid carriage was not only eliminated but a ~10% beneficial effect appeared in the absence of antibiotic. The relationship between plasmid and host changed from parasitic to mutualistic in the absence of Cm. Experiments in which the authors moved plasmids from their coevolved cells to all the other possible hosts, showed a significant chromosomal effect but no significant effects associated with either the plasmid or the plasmid-chromosome association. These results, along with those from Helling *et al.* (1981), showed that compensatory changes occurred in the bacterial chromosome, not in the plasmid.

However, Bouma and Lenski (1988) did not show whether the beneficial effect was directly associated with the Cm^R-plasmid itself or with another function encoded by the pACYC184. In a later paper, Lenski *et al.* (1994a) created new plasmids by deletion of fragments from pACYC184. They found that the beneficial effect observed was not from the Cm^R. Moreover, Cm^R function was demonstrated to have a deleterious effect both in the ancestral and coevolved hosts. What could be the origin of the beneficial effect? pACYC184 also encoded for Tc^R. Lenski *et al.* (1994a) found that the presence of the *tet* coding region was not responsible for the fitness cost for the ancestral genotype, and in fact, the presence of the Tc^R function was necessary for the beneficial effect of pACYC184 in the coevolved background. The expression of Tc^R function was responsible for ~70% of the fitness increase in the coevolved host. These results clearly illustrate two main features: (i) the host acquired some new benefit from a plasmid-encoded function different from the selectively important antibiotic resistance, and (ii) genes linked to antibiotic resistance can be selected that stabilize the presence of the plasmid in the host cell (Levin *et al.*, 1997). The beneficial effect of Tc^R on fitness was surprising, because as we have seen above, many examples have shown the deleterious effect of *tet* gene overexpression. Lenski *et al.* (1994a) argued that the reason for this difference was that all these experiments were done under different conditions and with different tetracycline resistance determinants. They also proposed two hypothetical mechanisms to explain the beneficial effect associated with Tc^R function: (i) the coevolved host may use the Tc efflux mechanism for some other purpose (e.g., eliminating some toxic metabolite), or (ii) the coevolved host might have adapted to the environment by means of some interaction between membrane-bound proteins and the membrane Tet protein. This interaction can, for example, speed up the uptake of glucose or some other metabolite.

At this point, it is interesting from an evolutionary point of view to test the effect of the chromosomal mutations in the absence of the coevolved plasmid. This is an important question, since although it is generally accepted that beneficial effects associated with new mutations are context dependent, in the sense that they are beneficial in the specific situation where they arose and got fixed by the action of natural selection but they could be neutral or even detrimental if the conditions change, some mutations affecting basic functions such us DNA replication or metabolite uptake can be seen as unconditionally beneficial. This question has also important implications in designing strategies for controlling plasmid-mediated resistances. Let's get some insight into this question.

...But Also in the Plasmid

In contrast with the above observation that fitness compensation occurs usually by changes in the bacterial chromosome, Modi and Adams (1991) found that after 800 generation of coevolution between *E. coli* and plasmid pBR322Δ5 under chemostat conditions, changes associated with fitness increases were located both in the chromosome and in the plasmid. Furthermore, a significant interaction between both types of changes was also detected. The authors cured the evolved strains of their plasmids and competed these strains with their ancestral counterparts. Their results showed increases in fitness due to chromosomal changes that ranged between 18% and >461%. In another set of experiments, they found that the increase in fitness due to changes in the plasmids ranged between 4% and 12%. Finally,

they also found that the beneficial fitness effect of a plasmid was larger when assayed in the genomic background in which it had evolved than in any other.

Continuing with this last experiment, Modi *et al.* (1991) analyzed which types of changes in the plasmid were responsible for the reduction of deleterious effect on bacterial. Under growth conditions where plasmid-carriage was deleterious (i.e. absence of antibiotic), they founded chemostat populations with cells bearing pBR322Δ5. After 100 generations of evolution, plasmid-free cells were detectable and increasing their frequency in the population. After 300 generations, plasmid-free cells constituted the majority of the population, as expected. However, plasmid-bearing cells still remained in the population. When the plasmid was purified and characterized, Modi *et al.* (1991) observed a spontaneous 2.25 Kb deletion encompassing most of the *tet* operon and a region upstream of it sized approximately 1.5 Kb. These deletions resulted in a 20% fitness increase for both the coevolved cell and also for the ancestral cell genotype, although the effect was greater for the coevolved chromosomal genotype than for the ancestral. In further experiments, Modi *et al.* (1991) analyzed whether the elimination of deleterious effects on fitness were due to the deletion itself or, by contrast, to structural changes within the plasmid, independent of the deletion. To determine this, the authors constructed artificial plasmids by deleting different fragments of pBR322Δ5. All the constructs contained portions of the *tet* operon. Their results clearly showed that deletions always produced significant (between 10% and 25%) fitness increases. Deletion and inactivation of a portion of the *tet* coding locus resulted in fitness increases equivalent to those obtained with larger deletions. Therefore, it was possible to conclude that the deletion encompassing the *tet* operon was responsible for the major proportion of the observed decrease in deleterious fitness effect.

This latter results contrast with the ones reported by Lenski *et al.* (1994a). While Modi *et al.* (1991) found a beneficial effect associated with the elimination of the *tet* coding locus, Lenski *et al.* (1994a) found that the beneficial effect was associated with the expression of the Tc^R function see above).

From a Harmful Parasite to a Good Buddy

Coevolution between a plasmid and its host cell means a fine tuning of the common metabolism of both entities, eliminating any possible interference and promoting any possible benefit. Therefore, it will be expected that if after coevolution one of the coevolving units is eliminated, the fitness of the other will be negatively affected. In fact after 500 generations of coevolution between *E. coli* and the plasmid pACYC184, Lenski *et al.* (1994a) observed that segregation of the plasmid had a negative fitness effect for the coevolved bacterial genotype. Not only has evidence about coevolution between R-plasmids and host cells been reported. There is also plenty of experimental evidence showing reduction of the antibiotic resistance-induced deleterious fitness effects by coevolution and the appearance of compensatory mutations for chromosomally encoded factors (Schrag and Perrot, 1996; Schrag *et al.*, 1997; Björkman *et al.*, 1998). For example, similar results of those of Lenski *et al.* (1994a) were observed by Schrag *et al.* (1997). Deleterious fitness effect of a chromosomally encoded Sm^R was compensated, after 180 generations of coevolution, by mutation. Interestingly, if the *rps*L locus, responsible for the Sm^R phenotype, was eliminated, the compensatory mutations were deleterious by themselves. Due to this background-dependent deleterious effect of compensatory mutations, the elimination of the antibiotic-selective pressure is not necessarily going to imply a reduction in the frequency of resistant bacteria, due to the reduced fitness of plasmid-free cells when compared with the coevolved resistant cells.

These last results obtained for both, plasmid- and chromosome-encoded resistances, indicate that mutations that compensate for the deleterious effect of carrying a resistance, or even increase the fitness of resistant cells in the absence of antibiotic are conditionally beneficial, as suggested above. If the resistance is removed from the cell, then the mutations that fixed in the chromosome as a consequence of the coevolutionary process, are not beneficial anymore, but they become deleterious for the cell.

CONCLUSIONS: HOW MANY CHANCES DO WE HAVE TO BEAT (PLASMID-MEDIATED) RESISTANT BACTERIA?

The evidence reported above about the ability of bacteria to adapt genetically to the cost of plasmid carriage has strong implications for the persistence of resistant bacteria in the wild and, indeed, stronger clinical implications. As I have reviewed in the last section, the coevolution of plasmid-host cell interactions can result in a reduced cost for the cell and, in some cases, even a beneficial relationship. This increased R-plasmid stability implies stronger efforts are needed to control the existence and spread of antibiotic resistant cells in human or animal populations.

Bacteria with plasmid-encoded antibiotic resistance will spread into a population as the result of positive selection, imposed by the use of antimicrobial agents. If drug treatment is long enough, coevolution between hosts and plasmids will take place, and compensatory mutations will quickly arise by chance and fix in the population, enhancing the fitness of the plasmid-host association. Once the use of antibiotic is eliminated or reduced, these coevolved cells were already well adapted to the antibiotic-free environment, and any possible plasmid-segregant cell will suffer from the deleterious side effect of compensatory mutations in the absence of the plasmid, which indeed, can not be considered as compensatory anymore, but as putative deleterious mutations. In conclusion, plasmid-free cells will be outcompeted by the fitter plasmid-containing parental cells. Therefore, it is common sense that if the chances for host-plasmid coevolution are reduced by therapies based on short periods with strong antibiotic load, compensatory mutations will arise less frequently. Consequently, once the antibiotic is eliminated from the environment, it is more likely that plasmid-bearing cells are eliminated by plasmid-free ones due to fitness disadvantage of the former.

With these consequences of evolution acting on bacterial population, it can be proposed as a guideline for drug administration, that long-term use of antibiotics may pose a great risk for antimicrobial resistance, whereas short-term therapeutic or prophylactic use will not. Unfortunately, the elimination of therapeutic dosages to patients is not correlated with the real elimination of antibiotic from the environment. For example, tetracycline is quite stable and typically is never broken down by resistant strains (Johnson and Adams, 1992). This long-life of antibiotics in the environment can help to promote coevolution in plasmid/bacteria exposed to it.

The coevolution phenomenon, and the fitness advantage obtained when both host cell and plasmid evolve together, imposes a serious restriction to the approach proposed by Levy (1994) to solve the antibiotic resistance problem. He proposed a proper use of *current* antibiotics in ways that encourage the re-establishment of susceptible bacteria. As we have seen above, after coevolution, the resistant bacteria should be the winners in a competition with susceptible cells even in the absence of drugs in the media. As an alternative, I will propose an approach based on the proper use of *new* drugs. Ideally, treatment should consist of a short prophylactic use, instead of long-term treatments, combined if necessary, with alternations of different drugs with different mechanisms of action. Of course, it would be desirable that, after the eradication of the infection, the mean life of the drug be as short as possible, in order to avoid the development of low-level resistants.

The easy-to-spread plasmid resistances make it conceivable that within a few years we will see bacteria resistant to new antibiotics to which they have never before been exposed. For example, Neu (1992a) predicted that, due to the easy transfer of genetic material between *Bacteroides* and *E. coli*, within this decade *Enterobacteriaceae* resistant to carbapenems will emerge. Moreover, in *Enterobacteriaceae* isolated from faeces of wild rodents that have never been exposed to antibiotics, resistances to common-use antibiotics, such us amoxycillin, cefuroxime, Cm, Tc or Tp were found (Gilliver *et al.*, 1999). Tolerance to vancomycin, the first step to a complete resistance, has been recently described in *Streptococcus pneumoniae* (Novak *et al.*, 1999). These are just a few examples of what is becoming a dangerous problem.

Epilogue. Evolution has been finding the way to "optimal solutions" for many problems during the history of life on Earth. It is just a matter of time and chance to find the correct mutation and then, natural selection will do the rest. From the point of view of bacteria, becoming resistant to an antibiotic is just an evolutionary problem. And bacterial

populations are large enough to have a good chance of always finding the solution. Then, the remaining is easy . . .

ACKNOWLEDGEMENTS

I wish to thank Drs. Hughes and Andersson for inviting me to write this chapter. I also want to thank Drs. F. González-Candelas, A. Latorre, R.E. Lenski and P.E. Turner, Ms. B. Sabater and an anonymous reviewer for critical reading of the manuscript and constructive suggestions. My research is being supported by grant PM97-0060-C02-02 from Spanish D.G.E.S.

REFERENCES

Adams, J., Kinney, T., Thompson, S., Rubin, L. and Helling, R.B. (1979) Frequency-dependent selection for plasmid containing cells of *Escherichia coli*. *Genetics* **91**, 627–637.
Anderson, J.D. (1975) Factors that may prevent transfer of antibiotic resistance between gram-negative bacteria in the gut. *J. Med. Microbiol.* **8**, 83–88.
Anderson, J.D., Gillespie, W.A. and Richmond, M.H. (1973) Chemotherapy and antibiotic-resistance transfer between enterobacteria in the human gastro-intestinal tract. *J. Med. Microbiol.* **6**, 461–473.
Berg, T., Firth, N., Apisiridej, S., Hettiaratchi, A., Leelaporn, A. and Skurray, R.A. (1998) Complete nucleotide sequence of pSK41: evolution of staphylococcal conjugative multiresistance plasmids. *J. Bacteriol.* **180**, 4350–4359.
Björkman, J., Hughes, D. and Anderson, D.I. (1998) Virulence of antibiotic-resistant *Salmonella typhimurium*. *Proc. Natl. Acad. Sci. USA* **95**, 3949–3953.
Blázquez, J., Martinez, J.L. and Baquero, F. (1993) Bleomycin increases amikacin and streptomycin resistance in *Escherichia coli* harboring transposon Tn5. *Antimicrob. Agents Chemother.* **37**, 1982–1985.
Blot, M., Hauer, B. and Monnet, G. (1994) The Tn5 bleomycin resistance gene confers improved survival and growth advantage on *Escherichia coli*. *Mol. Gen. Genet.* **242**, 595–601.
Bouma, J.E. and Lenski, R.E. (1988) Evolution of a bacteria/plasmid association. *Nature* **335**, 351–352.
Burman, L.G. (1977) Expression of R-plasmid functions during anaerobic growth of an *Escherichia coli* K-12 host. *J. Bacteriol.* **131**, 69–75.
Bryan, L.E. (1988) General mechanisms of resistance to antibiotics. *J. Antimicrob. Chemother.* **22**, 1–15.
Chopra, I., Shales, S.W., Ward, J.M. and Wallace, L.J. (1981) Reduced expression of Tn10-mediated tetracycline resistance in *Escherichia coli* containing more than one copy of the transposon. *J. Gen. Microbiol.* **126**, 45–54.
Cohen, S.L. (1976) Transposable genetic elements and plasmid evolution. *Nature* **236**, 731–738.
Cohen, M.L. (1992) Epidemiology of drug resistance: implications for a post-antimicrobial era. *Science* **257**, 1050–1055.
Coleman, D.C. and Foster, T.J. (1981) Analysis of the reduction in expression of tetracycline resistance determined by

transposon Tn*10* in the multicopy state. *Mol. Gen. Genet.* **182**, 171–177.

Cooper, N.S., Brown, M.E. and Caulcott, C.A. (1987) A mathematical method for analyzing plasmid stability in microorganisms. *J. Gen. Microbiol.* **133**, 1871–1880.

Esty, W.W., Miller, R.V. and Sands, D.C. (1986) A model for fluctuations in the fraction of a bacterial population harboring plasmids. *Theor. Pop. Biol.* **30**, 111–124.

Falkow, S. (1975) *Infectious Multiple Drug Resistance*, Pion, London.

Farrar, W.E., Eidson, M., Guerry, P., Falkow, S., Drusin, L.M. and Roberts, R.B. (1972) Interbacterial tranfer of and R-factor in the human intestive: *in vivo* acquisition of R-factor-mediated kanamycin resistance by a multi-resistant strain of *Shigella sonnei. J. Infect. Dis.* **126**, 27–33.

Finland, M. (1979) Emergence of antibiotic resistance in hospitals, 1935–1975. *Rev. Infect. Dis.* **1**, 4–22.

Francia, M.V., de la Cruz, F. and García Lobo, J.M. (1993) Secondary-sites for integration mediated by the Tn*21* integrase. *Mol. Microbiol.* **10**, 823–828.

Freter, R., Freter, R.R. and Brickner, H. (1983) Experimental and mathematical models of *Escherichia coli* plasmid transfer *in vitro* and *in vivo. Infect. Immun.* **39**, 60–84.

Gilliver, M.A., Bennett, M., Begon, M., Hazel, S.M. and Hart, C.A. (1999) Antibiotic resistance found in wild rodents. *Nature* **401**, 233–234.

Godwin, D. and Slater, J.H. (1979) The influence of the growth environment on the stability of a drug resistance plasmid in *Escherichia coli* K12. *J. Gen. Microbiol.* **111**, 201–210.

Gómez-Lus, R. (1998) Evolution of bacterial resistance to antibiotics during the last three decades. *Internatl. Microbiol.* **1**, 279–284.

Griffith, J.K., Kogoma, T., Corvo, D.L., Anderson, W.L. and Kazim, A.L. (1988) An N-terminal domain of the tetracycline resistance protein increases susceptibility to aminoglycosides and complements potassium uptake defects in *Escherichia coli. J. Bacteriol.* **170**, 598–604.

Guessouss, M., Ben-Mahrez, K., Belhadj, C. and Belhadj, O. (1996) Characterization of the drug resistance plasmid R2418: restriction map and the role of insertion and deletion in its evolution. *Can. J. Microbiol.* **42**, 12–18.

Helling, R.B., Kinney, T. and Adams, J. (1981) The maintenance of plasmid-containing organisms in populations of *Escherichia coli. J. Gen. Microbiol.* **123**, 129–141.

Holčík, M. and Iyer, V.N. (1996) Structure and mode of action of *kikA*, a gene region lethal to *Kleibsiella oxytoca* and associated with conjugative antibiotic-resistance plasmids of the IncN group. *Plasmid* **35**, 189–203.

Hughes, V.M. and Datta, N. (1983) Conjugative plasmids in bacteria of the "pre-antibiotic" era. *Nature* **302**, 725–726.

Iyobe, S., Hashimoto, H. and Mitsuhashi, S. (1969) Integration of chloramphenicol resistance gene of an R factor on *Escherichia coli* chromosome. *Jap. J. Microbiol.* **13**, 225–232.

Iyobe, S., Hashimoto, H. and Mitsuhashi, S. (1970) Integration of chloramphenicol-resistance genes of an R factor into various sites of an *Escherichia coli* chromosome. *Jpn. J. Microbiol.* **14**, 463–471.

Jaromel, H. and Kemp, G. (1969) R-factor transmission *in vivo. J. Bacteriol.* **99**, 487–490.

Johnson, R. and Adams, J. (1992) The ecology and evolution of tetracycline resistance. *Trends Ecol. Evol.* **7**, 295–299.

Jones, M., Primrose, S.B., Robinson, A. and Ellwood, D.C. (1980) Maintenance of some ColE1-type plasmids in chemostat culture. *Mol. Gen. Genet.* **180**, 579–584.

Lee, S.W. and Edlin, G. (1985) Expression of tetracycline resistance

in pBR322 derivatives reduces the reproductive fitness of plasmid-containing *Escherichia coli. Gene* **39**, 173–180.

Leelaport, A., Firth, N., Paulsen, I.T. and Skurry, R.A. (1996) IS*257*-mediated cointegration in the evolution of a family of staphylococcal trimethoprim resistance plasmid. *J. Bacteriol.* **178**, 6070–6073.

Lenski, R.E. (1997) The cost of antibiotic resistance – from the perspective of a bacterium. In Ciba Foundation Symposium 207, *Antibiotic resistance: origins, evolution, selection and spread*, John Wiley & Sons, U.S.A., pp. 131–151.

Lenski, R.E. and Bouma, J.E. (1987) Effects of segregation and selection on instability of plasmid pACYC184 in *Escherichia coli* B. *J. Bacteriol.* **169**, 5314–5316.

Lenski, R.E. and Nguyen, T.T. (1988) Stability of recombinant DNA and its effects on fitness. *Trends Ecol. Evol.* **3**, S18–S20.

Lenski, R.E., Simpson, S.C. and Nguyen, T.T. (1994a) Genetic analysis of a plasmid-encoded, host genotype-specific enhancement of bacterial fitness. *J. Bacteriol.* **176**, 3140–3147.

Lenski, R.E., Souza, V., Duong, L.P., Phan, Q.G., Nguyen, T.N.M. and Bertrand, K.P. (1994b) Epistatic effects of promoter and repressor functions of the Tn*10* tetracycline-resistance operon on the fitness of *Escherichia coli. Mol. Ecol.* **3**, 127–135.

Levin, B.R. (1980) Conditions for the existence of R-plasmids in bacterial population. In S. Mitsuhashi, L. Rosival, and V. Krcmery, (eds.), *Antibiotic resistance: transposition and other mechanisms*, Springer, Germany, pp. 197–202.

Levin, B.R., Lipsitch, M., Perrot, V., Schrag, S., Antia, R., Simonsen, L., *et al.* (1997) The population genetics of antibiotic resistance. *Clin. Infect. Dis.* **24**, S9–S16.

Levy, S.G. (1994) Balancing the drug-resistance equation. *Trends Microbiol.* **2**, 341–343.

Levy, S.B. and Novick, R.P. (1986) *Antibiotic Resistance Genes: Ecology, Transfer and Expression*, CSHL Press, Cold Spring Harbor (N.Y.).

Licht, T.R., Christensen, B.B., Krogfelt, K.A. and Molin, S. (1999) Plasmid transfer in the animal intestine and other dynamic bacterial populations: the role of community structure and environment. *Microbiology* **145**, 2615–2622.

Lin-Chao, S. and Bremer, H. (1986) Effect of the bacterial growth rate on replication control of plasmid pBR322 in *Escherichia coli. Mol. Gen. Genet.* **203**, 143–149.

Lodge, J.K., Kazic, T. and Berg, D.E. (1989) Formation of supercoiling domains in plasmid pBR322. *J. Bacteriol.* **171**, 2181–2187.

Mazodier, P., And Davies, J. (1991) Gene transfer between distantly related bacteria. *Annu. Rev. Genet.* **25**, 147–171.

Modi, R.I. and Adams, J. (1991) Coevolution in bacterial-plasmid populations. *Evolution* **45**, 656–667.

Modi, R.I., Wilke, C.M., Rosenzweig, R.F. and Adams, J. (1991) Plasmid macro-evolution: selection of deletions during adaptation in a nutrient-limited environment. *Genetica* **84**, 195–202.

Moyed, H.S. and Bertrand, K.P. (1983) Mutations in multicopy Tn*10 tet* plasmids confer resistance to inhibitory effects of inducers of *tet* gene expression. *J. Bacteriol.* **155**, 557–564.

Moyed, H.S., Nguyen, T.T. and Bertrand, K.P. (1983) Multicopy Tn*10 tet* plasmids confer sensitivity to induction of *tet* expression. *J. Bacteriol.* **155**, 549–556.

Neu, H.C. (1984) Current mechanisms of resistance to antimicrobial agents in microorganisms causing infections in the patient at risk for infection. *Am. J. Med.* **76**, 11–27.

Neu, H.C. (1992a) The crisis of antibiotic resistance. *Science* **257**, 1064–1073.

Neu, H.C. (1992b) Quinolone antimicrobial agents. *Annu. Rev. Med.* **43**, 465–486.

Nguyen, T.N.M., Phan, Q.G., Duong, L.P., Bertrand, K.P. and Lenski, R.E. (1989) Effects of carriage and expression of the Tn*10* tetracycline-resistance operon on the fitness of *Escherichia coli* K12. *Mol. Biol. Evol.* **6**, 213–225.

Novak, R., Henriques, B., Charpentier, E., Normark, S. and Tuomanen, E. (1999) Emergence of vancomycin tolerance in *Streptococcus pneumoniae*. *Nature* **399**, 590–593.

Proctor, G.N. (1994) Mathematics of microbial plasmid instability and subsequent differenctial growth of plasmid-free and plasmid-containing cells, relevant to the analysis of experimental colony number data. *Plasmid* **32**, 101–130.

Schaberg, D.R., Rubens, C.E., Alford, R.H., Farrar, W.E., Schaffner, W. and McGee, Z.A. (1981) Evolution of antimicrobial resistance and nosocomial infections. Lessons from the Vanderbilt experience. *Am. J. Med.* **70**, 445–448.

Schrag, S.J. and Perrot, V. (1996) Reducing antibiotic resistance. *Nature* **381**, 120–121.

Schrag, S.J., Perrot, V. and Levin, B.R. (1997) Adaptation to the fitness costs of antibiotic resistance in *Escherichia coli*. *Proc. R. Soc. Lond. B* **264**, 1287–1291.

Scott, J.R. (1984) Regulation of plasmid replication. *Microb. Rev.* **48**, 1–23.

Simonsen, L. (1990) Dynamics of plasmid transfer on surfaces. *J. Gen. Microbiol.* **136**, 1001–1007.

Smith, H.W. (1970) The transfer of antibiotic resistance between strains of enterobacteria in chicken, calves and pigs. *J. Med. Microbiol.* **3**, 165–180.

Smith, H.W. (1975) Persistence of tetracycline resistance in pig *E. coli*. *Nature* **258**, 628–630.

Smith, H.W. (1976) R factor transfer in vivo in sheep with *E. coli* K12. *Nature* **261**, 348.

Stewart, F.M. and Levin, B.R. (1977) The population biology of bacterial plasmids: *a priori* conditions for the existence of conjugationally transmitted factors. *Genetics* **87**, 209–228.

Summers, D.K. (1991) The kinetics of plasmid loss. *TibTech.* **9**, 273–278.

Sýkora, P., Foltýnova, Z. and Smitálova, K. (1989) A kinetic model for plasmid curing. *Plasmid* **21**, 85–98.

Tauxe, R.V., Cavanagh, T.R. and Cohen, M.L. (1989) Interspecies gene transfer *in vivo* producing an outbreak of multiply resistant shigellosis. *J. Infect. Dis.* **160**, 1067–1070.

Torres, O.R., Korman, R.Z., Zahler, S.A. and Dunny, G.M. (1991) The conjugative transposon Tn*925*: enhancement of conjugal transfer by tetracycline in *Enterococcus faecalis* and mobilization of chromosomal genes in *Bacillus subtilis* and *E. faecalis*. *Mol. Gen. Genet.* **225**, 395–400.

Turner, P.E., Cooper, V.S. and Lenski, R.E. (1998) Tradeoff between horizontal and vertical modes of transmission in bacterial plasmids. *Evolution* **52**, 315–329.

Valenzuela, M.S., Ikpeazu, E.V. and Siddiqui, K.A.I. (1996a) *E. coli* growth inhibition by a high copy number derivative of plasmid pBR322. *Biochem. Biophys. Res. Commun.* **219**, 879–883.

Valenzuela, M.S., Siddiqui, K.A.I. and Sarkar, B.L. (1996b) High expression of plasmid-encoded tetracycline resistance gene in *E. coli* causes a decrease in membrane-bound ATPase activity. *Plasmid* **36**, 19–25.

Wallace, R.J., Jr., Steingrube, V.A., Nash, D.R., Hollis, D.G., Flanagan, C., Brown, B.A., *et al.* (1989) BRO (β-lactamases of *Branhamella catarrhalis* and *Moraxella* subgenus *Moraxella*, including evidence for chromosomal (β-lactamase transfer by conjugation in *B. catarrhalis*, *M. nonliquefaciens*, and *M. lacunata*. *Antimicrob. Agents Chemother.* **33**, 1845–1854.

Wallace, R.J., Jr., Nash, D.R. and Steingrube, V.A. (1990) Antibiotic susceptibilities and drug resistance in *Moraxella (Branhamella) catarrhalis*. *Am. J. Med.* **14**, 46–50.

Watanabe, T. (1963) Infective heredity of multiple drug resistance in bacteria. *Bacteriol. Rev.* **27**, 87–115.

Zünd, P. and Lebek, G. (1980) Generation time-prolonging R plasmids: correlation between increases in the generation time of Escherichia coli caused by R plasmids and their molecular size. *Plasmid* **3**, 65–69.

12. Structural Insights into Antibiotic-Target Interactions

Elspeth Gordon[¥], Nicolas Mouz[*¥], Thierry Vernet[*] and Otto Dideberg[¥]
* Institut de Biologie Structurale *Jean-Pierre Ebel* (CEA-CNRS-UJF),
Laboratoire d'Ingénierie des Macromolécules, 41, rue Jules Horowitz, F-38027
Grenoble Cedex 1, France
[¥] Institut de Biologie Structurale *Jean-Pierre Ebel* (CEA-CNRS-UJF),
Laboratoire de Cristallographie Macromoléculaire, 41, rue Jules Horowitz,
F-38027 Grenoble Cedex 1, France

INTRODUCTION

Since the discovery of penicillin in the 1940s and its first clinical use in the 1950s, bacterial infections have posed little threat. The beauty of penicillin lies in its target, a set of enzymes unique to bacteria, namely the enzymes involved in the construction and maintenance of the bacterial cell wall; these are known as the penicillin-binding proteins (PBPs) (Ghuysen, 1994). Rather conveniently, these enzymes are found in the periplasmic space and are thus readily accessible to the drug. However, in recent years, antibiotic-resistant strains have emerged and resistance is rapidly transmitted to other bacterial strains and species.

Two approaches can be used to design drugs active against pathogenic bacteria. One is to search for drugs with novel modes of action, as a result of the identification of a new target; at the moment, there is a dearth of new antibiotics of this type. A second is to ''rejuvenate'' existing drugs and, by constantly modifying existing antibiotics, try to keep one step ahead of the bacteria; this approach is that used for many years by the pharmaceutical industry. After the 1960s, certain new drugs, the cephalosporins and, more recently, the monobactams, appeared on the market; they have with the penicillin one common chemical feature, the presence of the four-membered β-lactam ring, and are thus known as the β-lactam antibiotic family. Both approaches have been helped considerably by information on the structure of the target enzymes.

Here, we review some of the structural information available for the family of enzymes that interact with β-lactam antibiotics, and, by analyzing the structures of the native and complexed enzyme, we try to correlate these structural observations with the functional characteristics of the proteins.

PROTEINS INTERACTING WITH β-LACTAM ANTIBIOTICS

Since the discovery of penicillin's powerful action against bacterial infections, much effort has been expended on identifying proteins that interact with β-lactam antibiotics. After more than fifty years of research, a large number of such proteins have been identified by molecular biological techniques; some of these have also been well characterized at the protein level (Ghuysen, 1991; Ghuysen, 1994; Knox, 1995; Jamin *et al.*, 1995; Massova and Mobashery, 1998; Matagne *et al.*, 1998) and, in a few cases, the detailed three-dimensional structure is known.

Substrates and Inhibitors

The targets of the β-lactam antibiotics are D-alanyl-D-alanine carboxypeptidases/transpeptidases (DD-peptidases) which catalyze the final step in bacterial cell wall synthesis. DD-peptidases crosslink two pentapeptide chains:

$$R_1\text{-D-Ala-}\textbf{D-Ala} + R_2\text{-L-Lys-D-Ala-D-Ala} \rightarrow$$
$$|$$
$$\textbf{NH}_2$$
$$R_1\text{-D-Ala-CO-}\textbf{NH}\text{-L-Lys-}R_2 + \textbf{D-Ala} + H_2O$$
$$|\_\text{D-Ala-D-Ala}$$

β-lactam antibiotics (Figure 1a) are structurally similar to the cell wall peptide, L-Lys-D-Ala-D-Ala (Figure 1b) (Tipper and Strominger, 1965). All DD-peptidases so far characterized have an active site serine and form a covalent complex with the antibiotic. Since the complex is quite stable (several hours), the DD-peptidases can be labeled with radioactive β-lactam, which classifies these proteins as penicillin-binding proteins (PBPs). Often, the DD-peptidase domain comprises only one part of a multi-functional protein. In general, DD-peptidases are membrane-bound enzymes involved in the synthesis and remodeling of peptidoglycan. Due to the complexity of the processes involved

(e.g. elongation and division), a given bacterium contains many DD-peptidases (4–10). The best known PBPs are those from *E. coli*. Ten proteins have been identified and are numbered by order of decreasing molecular mass from PBP1a (93,636 daltons) to PBP7 (34,245 daltons); a few have well defined functions, e.g. PBP1a, 1b and 2 are involved in cell elongation and PBP3 in cell division (septation).

Some bacteria which are protected from β-lactam antibiotics because of their natural niche have far fewer PBPs, e.g. *Helicobacter pylori* and *Borrelia burgdorferi* each have 3 PBPs, which were identified in the complete genome.

Protein Family

Preliminary X-ray crystallographic results (Kelly *et al.*, 1986; Samraoui *et al.*, 1986) revealed a possible link between DD-peptidases and the β-lactam-hydrolysing enzymes ("β-lactamases") having an active-site serine residue. Other β-lactamases with an active site containing Zn ions (Class B or Zn^{2+}-β-lactamases) and a different fold (Carfi *et al.*, 1995) are not discussed here. The serine β-lactamases catalyze the cleavage of the cyclic amide bond of penicillin (Figure 2) by a method involving the covalent binding of the substrate to a serine residue in the protein. Over the years, much new structural information has strengthened this association and has clearly shown that both families, the DD-peptidase domain of PBPs and the serine β-lactamases, form a large protein superfamily. Based on X-ray crystallographic results, three sequence motifs have been identified as markers for **A**ctive-site **S**erine **P**enicillin **R**ecognizing **E**nzymes (**AS-PRE**) (Joris *et al.*, 1988):

Motif 1: a **S*-X-X-K** tetrad, in which S* is the active serine located at the N-terminus of a long

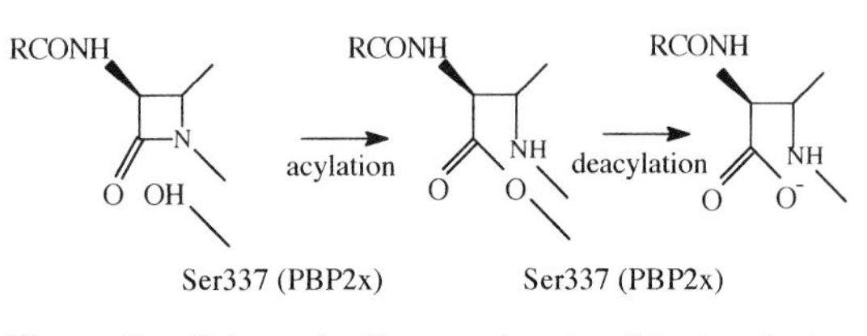

(a)   (b)

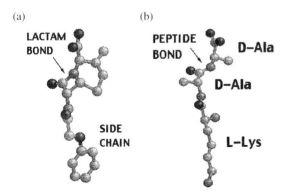

Figure 1.   Benzylpenicillin (a) and L-Lys-D-Ala-D-Ala (b).

Figure 2.   Schematic diagram showing β-lactam hydrolysis.

helix. The side chains of both the S and K residues are located at the bottom of the active site cleft;

Motif 2: a **S/Y**-X-**N** triad situated on one side of the active site;

Motif 3: a **K/H-S/T-G** triad on a β-strand at the opposite side of the active site.

These three motifs are not selective enough to accurately identify a protein as being a member of the **ASPRE** family; this requires biochemical proof of its interaction with β-lactam antibiotics. On the basis of their sequences, serine β-lactamases and PBPs can be subdivided. Serine β-lactamases are monofunctional and usually soluble proteins, which fall into three classes (A, C and D). PBPs are monofunctional or bifunctional proteins and are usually membrane-bound; only two DD-peptidases from *Streptomyces* R61 and *Actiromadura* R39 are water-soluble (Ghuysen, 1991). The PBPs have been divided into four classes:

1. The high molecular mass class A (HMM-class A) PBPs, which consists of two-module proteins with glycosyltransferase and transpeptidase activity.
2. The high molecular mass class B (HMM-class B) PBPs, which consists of two-module proteins with only transpeptidase activity.
   Both these classes have a N-terminal region of about 30 amino acids, responsible for anchoring the protein in the membrane.
3. The high molecular mass class C (HMM-class C) PBPs, formed from two penicillin sensory-transducers with many membrane anchor helical regions. These proteins are not involved in peptidoglycan synthesis. The C-terminal domain is related to the class D β-lactamases.
4. The low molecular mass class A (LMM-class A) PBPs consist of DD-carboxypeptidase and/ or transpeptidase enzymes, either bound to the membrane via a C-terminal amphipathic helix or loosely membrane-bound.

Figure 3 represents the molecular organization of the proteins described in this paper.

In *Streptococcus pneumoniae*, or the related species, *Streptococcus pyogenes*, 7 genes encoding proteins containing the three classical motifs of **ASPRE** enzymes have been identified; only 6 (PBP1a, 1b, 2a, 2b, 2x, and 3) have been characterized at the protein level. PBP1a,1b and

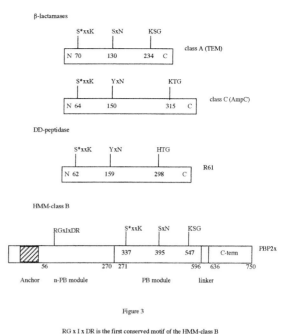

Figure 3

RG x I x DR is the first conserved motif of the HMM-class B

Figure 3. Schematic diagram showing the conserved fingerprint markers for members of the **ASPRE** family, a class A β-lactamase (TEM), a class C β-lactamase (AmpC), DD-peptidase R61 from *Streptomyces* R61 and an HMM-class B PBP from *S. pneumoniae* (PBP2x). PBP2x is made of six sequence regions: a short cytoplasmic domain, an anchor embedded in the membrane, a non penicillin-binding (n-PB), a penicillin-binding (PB or transpeptidase domain), a linker region and a C-terminal domain.

2a belong to HMM-class A, PBP2b and 2x to HMM-class B and PBP3 to LMM-class A.

RESISTANCE

β-lactam antibiotics exert their lethal effect by inhibiting the final stages of peptidoglycan synthesis. Resistance to β-lactams has developed in three ways:

1. Secretion of β-lactamases which hydrolyze the β-lactam
2. Modification of the target PBPs
3. In Gram-negative bacteria, changes in outer membrane permeability and appearance of efflux pumps

The third method does not fall within the scope of this review; interested readers are directed to a recent review (Hancock, 1997).

β-lactamase-mediated resistance can involve one of two mechanisms, either the production of β-lactamases with extended substrate specificity, e.g. due to the incorporation of point mutations in the plasmid encoded β-lactamase TEM (Figure 3), or the increased production of existing β-lactamases, often accompanied by an outer membrane permeability change.

Target modification involves either the modification of an existing PBP, as in *Streptococcus pneumoniae* in which all five PBPs are subject to mutations (Spratt, 1994; Hakenbeck *et al.*, 1998), or the expression of a novel PBP induced by antibiotic stress, the mechanism employed by Methicillin-resistant *Staphylococcus aureus* (MRSA), in which large amounts of a low affinity PBP2a are produced in methicillin-resistant strains (Hakenbeck and Coyette, 1998).

DIFFERENCE BETWEEN β-LACTAMASES AND PBPs

The interaction of **ASPRE**'s with β-lactam antibiotics is a three-step process, involving two anionic intermediates (Figure 2). The first stage, the reversible formation of a Michaelis complex, is followed by acylation of the enzyme by the β-lactam, which involves the opening of the β-lactam ring and the formation of a covalent bond between the antibiotic and the active site serine, while the final step is deacylation, during which the covalent bond is broken and the modified β-lactam released. It is this final deacylation step that defines the difference between a PBP and a β-lactamase; in PBPs, deacylation is very slow and the half-life of the acyl-enzyme complex of the order of 15–20 hours, while, in β-lactamases, deacylation is very fast, with a half-life of the order of 10^{-4}–10^{-2}sec (Guillaume *et al.*, 1997).

STRUCTURE OF PBP2x

Despite much academic and pharmaceutical interest, the elucidation of the structure of a PBP from a pathogenic organism has proved difficult. To date, only two structures for LMM-PBPs are available. The structure of R61 from *Streptomyces* R61 (Kelly *et al.*, 1985; Kelly and Kuzin, 1995) has been used

as a model for other PBPs. More recently, the structure of another LMM-PBP, *Streptomyces* strain K15, has been determined (Fonzé, 1999), but, as yet, the coordinates were not available when we prepared this review. Both are atypical PBPs in that they are soluble proteins lacking membrane anchors, and *in vitro* show carboxypeptidase and/or transpeptidase activity. They both originate from bacteria of little clinical importance.

PBP2x from *S. pneumoniae* is more typical of the HMM-PBPs: it has a membrane anchor and contains three domains (Parès *et al.*, 1996). *S. pneumoniae* is the main causative agent of pneumonia and the emergence of drug-resistant strains of this bacterium is a real clinical threat. *S. pneumoniae* has six PBPs, subdivided into three classes. PBP2x belongs to the HMM-class B PBPs. A functional PBP2x is essential for bacterial viability (Kell *et al.*, 1993), and modifications in PBP2x are implicated in the development of drug-resistant strains (Grebe and Hakenbeck, 1996). PBP2x is also believed to be involved in cell septation (see below).

GENERAL TOPOLOGY

The crystallization of a modified soluble PBP2x in which the small cytoplasmic domain and the membrane anchor region were removed (Charlier *et al.*, 1993), allowed its structural determination to a resolution of 3.5Å (Parès *et al.*, 1996). The structure consists of three domains, with the central domain structurally resembling members of the **ASPRE** family (Figure 3 and 4). More recently, a different crystal form allowed the determination of the structure of the transpeptidase and C-terminal domains to 2.5Å resolution (Gordon *et al.*, 2000). The following structural description is based on data from these two studies.

N-terminal Domain

The N-terminal domain has a sugar-tong topology (Parès *et al.*, 1996) (Figure 4). It predominantly consists of long β-strands, with a more globular subdomain, remote from the central/transpeptidase domain. The function of the N-terminal domain is unknown, although it possesses several sequence motifs that are conserved in HMM-class B PBPs;

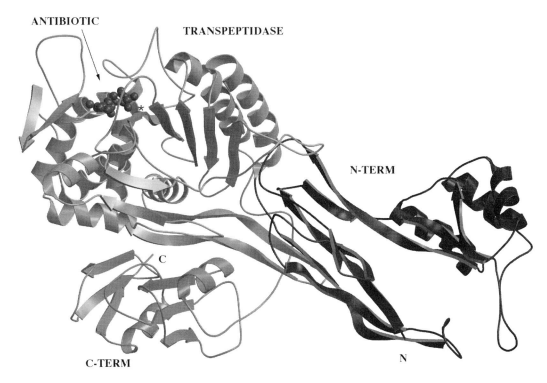

Figure 4. Ribbon diagram showing the overall folding and arrangement of the three domains of PBP2x. An antibiotic is shown in the active site. The asterisk labels the active site serine residue.

for clarity, only the localization of the first motif is shown in Figure 3. In the case of PBP3 from *E. coli*, an analogous PBP, the N-terminal domain has been suggested to be responsible for intermodular contacts (Nguyen-Disteche *et al.*, 1998). More specifically, it is possible that this domain plays a role in the formation of the divisome complex, the multi-enzymatic complex involved in division-specific peptidoglycan synthesis (Bramhill, 1997; Holtje, 1998) Interactions of this type have not yet been demonstrated for *S. pneumoniae* PBP2x (see below). A second hypothesis, also based on *E. coli* studies, is that a complex consisting of PBP2x and another molecule, FtsW, has transglycosylase activity (Matsuhashi *et al.*, 1990).

Transpeptidase Domain

The transpeptidase domain of PBP2x is involved in peptidoglycan metabolism and assembly, and is the target of penicillin; it belongs to the **ASPRE** family (Figure 3 and 4). It has an all α subdomain and an

α/β subdomain, which consists of a core 5-stranded β-sheet, protected by helices on both faces. The active site cleft lies at the interface of these two subdomains. The active site serine contained in the SXXK motif lies deep in the cavity, at the amino end of helix $\alpha 2$, the **SXN** motif in the all α subdomain lies opposite the active site serine and the conserved **KSG** motif is found on the last β-strand of the sheet, at the extreme edge of the α/β domain, $\beta 3$. In terms of the active site, this transpeptidase domain more closely resembles the class A, rather than the class C β-lactamases (Parès *et al.*, 1996).

C-terminal Domain

The C-terminal domain, which is only rarely found in HMM-class B PBPs, contains a repeated $\alpha\beta\beta\beta$ motif, probably arising from gene duplication. Sequence alignment with the C-terminal domains of other HMM-class B PBPs (*Streptococcus mitis* PBP2x, *Streptococcus oralis* PBP2x, *Streptococcus pyogenes* PBP2x, *Enterococcus hirae* PBP3s, *Enterococcus faecalis* PBP2, *Staphylococcus aureus* PBP1, *Bacillus subtilis* stage V sporulation protein D and *Clostridium acetobutylicum* stage V sporulation protein D) shows that glycine and proline residues are conserved, implying they all share a similar 3-dimensional structure. Interestingly, a PBP (BB0136) from *Borrelia burgdorferi*, a Gram-negative bacteria, is predicted as having a single $\alpha\beta\beta\beta$ motif. The function of the C-terminal domain is unclear at this time.

THE TRANSPEPTIDASE ACTIVE SITE

The catalytically important residues (**SXXK, SXN** and **KSG**) (Figure 5) lie at the center of the cleft, between the α and the α/β subdomains, and are centered on the amino terminus of helix $\alpha2$ (Parès *et al.*, 1996). Remarkably, the same spatial arrangement of these residues is seen in all **ASPRE** enzymes (see below). In the native enzyme, a sulfate ion is hydrogen bonded to Ser337 Oγ, Ser395 Oγ, Thr550 N and a water molecule. A large number of water molecules fill this cavity, and there are two strongly bound structural water molecules. The structure of the acyl-enzyme, consisting of an antibiotic covalently complexed with PBP2x, shows a more disordered structure,

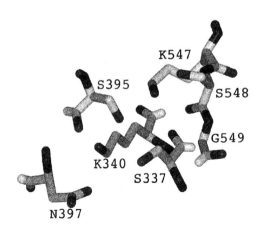

Figure 5. Active site cleft of PBP2x. Conserved residues: the serine and lysine of the **SXXK** motif lie at the amino terminus of helix $\alpha2$, the serine and asparagine of the **SXN** motif are shown on the left side of the figure and the lysine of the **KSG** motif, which lies on $\beta3$, on the right.

especially in the α subdomain, suggesting that this subdomain has moved to accommodate the antibiotic; in addition, residues Gln452 and Gln552 have had to change rotamers so that the antibiotic can lie close to, and in line with, $\beta3$.

PBP2x AND RESISTANCE

Penicillin-resistant strains of *S. pneumoniae* contain mosaic genes encoding low affinity PBPs (Smith *et al.*, 1991), and five of the six PBPs exist as low affinity forms (Hakenbeck *et al.*, 1998). Remarkably, sensitive or clinical isolate PBP contains the same number of codons, an obvious advantage when exchanging all, or part, of the gene between species. The ease with which Streptococci transfer genes between species is a frightening prospect for therapy. PBP2x is a primary resistance determinant (Grebe and Hakenbeck, 1996; Krauss *et al.*, 1996; Hakenbeck and Coyette, 1998), i.e. a modified PBP2x is an essential step in the development of high level resistance, which results from the mutation of several PBPs to low affinity forms. Clinically resistant isolates with modified PBP2xs can contain as many as 40 mutations in the transpeptidase domain alone. The analysis of sequences from clinically resistant strains has revealed 14 families. The alignment of these sequences with those for three sensitive PBP2xs shows that the mutation most frequently associated with a resistant phenotype is at Thr338, while the second most common is at Gln552. The 2.5Å resolution structure shows that Thr338 binds to a water molecule which lies in a buried cavity behind the active site serine (Figure 6). Site-directed mutagenesis of Thr338 and the other water ligands, Ser571 and Tyr586, produced proteins with a reduced affinity for the tested antibiotics (Figure 7) and a reduced hydrolytic activity for selected substrate analogues (Mouz *et al.*, 1998). The Thr338/Ser571 double mutants (for example, T338A/S571P) have reaction profiles similar to that of a resistant clinical isolate, Sp328. We propose that these mutations result in the displacement or loss of the water molecule normally in contact with Thr338, which, in turn, affects the reactivity of the active site serine, Ser337. When tested *in vivo*, the point mutation T338A results in a transformant with an increased MIC (minimum

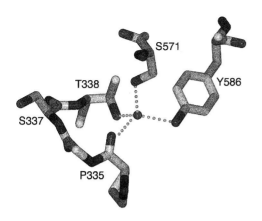

Figure 6. Water molecule bound in the buried cavity behind the active site serine 337 with the protein ligands shown as a ball-and-stick representation. The water molecule ligands are conserved in all PBPs found in the DCW of Gram-positive bacteria.

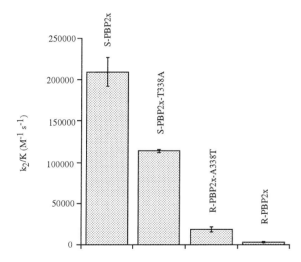

Figure 7. Histogram showing the k2/K ratio for cefotaxime: the proteins used are a sensitive PBP2x (S-PBP2x), a mutant T338A, a clinically resistant PBP2x (R-PBP2x) and the reverse mutant A338T.

inhibitory concentration) value (Hakenbeck, personal communication).

PBP2x AND THE CELL DIVISION CLUSTER

The water molecule ligands, Thr338, Ser571 and Tyr586, are strictly conserved in a subclass of HMM-class B PBPs, which are all found in the cell division cluster (DCW) of Gram-positive bacteria. A search of current data bases, often originating from the entire genomic sequence of the bacterium, shows that *Streptococcus mitis* PBP2x, *Streptococcus oralis* PBP2x, *Streptococcus pyogenes* PBP2x, *Enterococcus hirae* PBP3s, *Enterococcus faecalis* PBP2, *Staphylococcus aureus* PBP1, *Bacillus subtilis* stage V sporulation protein D, and *Clostridium acetobutylicum* stage V sporulation protein D all contain these water molecule ligands. The location of these genes in the DCW suggests that the corresponding proteins are involved in cell division. Interestingly, all these HMM-class B PBPs possess a C-terminal domain.

COMPARISON OF NATIVE STRUCTURES

In all structures so far determined for members of the **ASPRE** family, the active site is highly conserved. Fingerprint motifs are linked by a network of hydrogen bonds. The study of the hydrogen bonding patterns in the native enzyme and acyl-enzyme complexes provides some insight into possible enzymatic mechanisms. The aim of the following comparisons is to highlight similarities and differences between the structures of the two PBPs available and those of two representative β-lactamase structures.

Comparison of PBP2x and TEM, a Class A β-lactamase

Comparison of the transpeptidase domain of PBP2x with that of TEM β-lactamase (PDB code: 1btl) (Jelsch *et al.*, 1993) shows that the overall topology of the molecule is conserved, with changes occurring at a distance from the active site. The structures have been determined at a resolution of 2.5Å and 1.8Å, respectively. Since these enzymes are functionally different, we would expect subtle differences in the active site region. They share certain substrate specificities, e.g. their recognition of β-lactam antibiotics, but differ in their rate of hydrolysis of β-lactams. Structurally, the first difference is the size and shape of the active site. The active site of PBP2x consists of a long cleft that is 25Å long, 10Å wide and 14–16Å deep, while TEM has an active site pocket that is 16Å long, 10Å wide and 8Å deep on the α domain side and 16Å deep on the α/β domain side. In TEM, β3 is found on one side of the pocket (Figure 8), and the base of the pocket is formed by the Ω loop, containing Glu166; this Ω loop is not found in PBP2x. In PBP2x, the upper part of the active site, above Ser337, is filled with water molecules, but few well defined water molecules are found below Ser337. In TEM, the entire pocket contains ordered water molecules, including one bound in the oxyanion hole. Superposition of the active sites shows a remarkable spatial conservation of the fingerprint markers, **SXXK**, **SXN** and **K**(S/T)**G**. The hydrogen bonding network between these residues is mainly conserved (Figure 8). In PBP2x, an additional hydrogen bond exists between Ser395 Oγ and Lys340 Nζ. The upper part of the active site appears to be more highly conserved between these two enzymes. The area around the active site serine bears a strong positive charge, and both structures contain a sulfate ion bound, in PBP2x, to Ser337 Oγ, Thr550 Oγ, Thr550 N and Ser395 Oγ and, in β-lactamase, to Ser130 Oγ, Ser235 Oγ and Arg244 Nζ. Arg244, which is believed to act as an electrostatic catalyst (Maveyraud *et al.*, 1996), lies on β4, and its side chain reaches over to lie above the active site serine; no equivalent to Arg244 is found in the PBP, in which this position is occupied by Phe570. In the β-lactamase, a water molecule is bound in the oxyanion hole, whereas the sulfate ion in the PBP hinders such an interaction. In TEM, Glu166 is hydrogen bonded to a water molecule, and it is now widely accepted

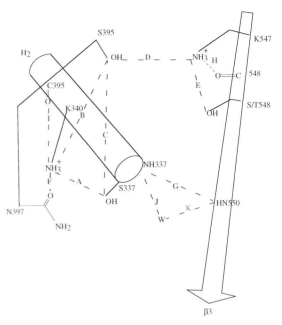

Figure 8. Schematic representation showing distances between active site residues in the **ASPRE** family. See the accompanying table (Table 1) for the actual values. The PBP2x numbering is used.

that Glu166 activates this water molecule, which then acts as the nucleophile involved in deacylation. As expected, Glu166 is not found in PBP2x, in which it is replaced by Phe450, which is not a suitable activator; in addition, no likely candidate for a deacylating water molecule is present, explaining why deacylation is very slow in PBP2x. Despite the changes in size and shape of the active site pockets, the regions close to the "fingerprint" motifs are conserved in the TEM β-lactamase and PBP2x (Table 1). Since the Ω loop of TEM

Table 1.   Distances (Å) between heavy atoms of the active site

	PBP2X	Class A (ave)	Class C (ave)	R61
A($337O^\gamma$–$340N^\zeta$)	3.14	2.70	3.93	4.15
B($340N^\zeta$–$395O^\gamma$)	2.81	3.59	3.21	2.75
C($337O^\gamma$–$395O^\gamma$)	3.29	3.25	3.20	3.75
D($395O^\gamma$–$547N^\zeta$)	3.24	2.90	2.74	4.70
E($547N^\zeta$–$548O^\gamma$)	4.27	4.11	4.21	4.30
F($340N^\zeta$–$397O^\delta$)	2.95	2.70	2.51	2.85
G(NH-NH)	4.64	4.70	3.75	3.70
H($547N^\zeta$–$548O$)	2.90	2.99	2.84	2.84
I($340N^\zeta$–$395O$)	2.60	3.34	7.62	8.20
J	–	3.25	3.60	–
K	–	2.85	2.85	–

(involved in deacylation in class A β-lactamase) is not present in PBP2x, these enzymes must have different mechanics of deacylation.

Comparison of R61 and AmpC, a Class C β-lactamase

As in the above case, the two structures compared have different functions, with R61 being responsible *in vitro* for both transpeptidase and carboxypeptidase reactions, while the β-lactamase has hydrolase activity. The structure of R61 (PDB code: 3pte) (Kelly and Kuzin, 1995) was compared to that of the AmpC β-lactamase from *E. coli.* (PDB code: 2bls) (Usher *et al.*, 1998). Both structures have been determined at high resolutions, 1.6Å and 2.0Å, respectively. In both cases, the active sites are small pocket-like cavities, that in AmpC being more open than that in R61. In R61, the active site pocket is 20Å long, 12Å wide and 16Å deep, while, in AmpC, it is more than 20Å long, 12Å wide and 8Å deep; in addition, it is open at the bottom and narrower at the top than that in R61. The catalytically important residues in these two cases differ in motif 3, the KTG in the class C β-lactamase being replaced by HTG in R61, but the positions of the active atoms are nevertheless superimposable. Atoms Ser62 Oγ, Lys65 Nζ, Tyr159 OH, His298 Nϵ2, and Asn161 Nδ2 in R61 are equivalent to Ser64 Oγ, Lys67 Nζ, Tyr150 OH, Lys315 Nζ, and Asn152 Nδ2 in the class C β-lactamase. The most notable difference is that, in R61, the distance (D, Table 1) between Tyr159 OH and H298 Nϵ2 is 4.70Å, while the equivalent Tyr150 OH and Lys315 Nζ in the β-lactamase are separated by 2.74Å. Although Class C β-lactamases have a fast deacylation rate, they contain no catalytic water and no amino acid equivalent to Glu166. In R61, there are two possible catalytic water molecules, W507 and W644, which lie below the oxyanion hole; however, there are no neighboring residues that could activate these water molecules, making them nucleophilic. It is clear from the active sites of R61 and AmpC that the proteins are closely related. Again, the β-lactamase has a water molecule in the oxyanion hole, while R61 does not. The difference in distance between Tyr159/Tyr150 and His298/Lys315 is notable, but at this point, of unknown significance.

Comparison of PBP2x and R61

Superficially, the active site of PBP2x is much bigger than that of R61. In R61, the cleft seen in the PBP is blocked by a cluster of aromatic residues. The functions and specificities of these two enzymes are probably very similar. *In vitro*, R61 has both transpeptidase and carboxypeptidase activity, whereas PBP2x has never been shown to have *in vitro* transpeptidase activity, its activity being measured by reaction with a thiol-substrate or acylation by β-lactam antibiotics. Although presumed to have a common function, they contain different ''conserved'' catalytic residues, with the fingerprints for PBP2x being **SXXK**, **SXN** and **KSG** and those for R61 **SXXK**, **YXN** and **HTG**. However, as mentioned above, the important catalytic residues are more or less superimposable, the differences being in the distances between Ser337/62 and Lys340/65 (distance A in Table 1: 3.14Å in PBP2x and 4.15Å in R61) and between Ser395 Oγ/Tyr159 OH and Lys547 Nζ/His298 Nϵ2 (distance D in Table 1:3.24Å and 4.7Å, respectively). Neither R61 nor PBP2x contains a water molecule in the oxyanion hole. Interestingly, both R61 and PBP2x have Thr 550/301 and Gln552/303 on β3, both of which have been implicated in resistant forms of PBP2x (Laible and Hakenbeck, 1991; Spratt, 1994). The mutation T550A in PBP2x is found in low-level resistant laboratory mutants, resistant to cefotaxime; however, this mutation is hypersensitive to penicillin (Sifaoui *et al.*, 1996; Mouz *et al.*, 1999). The equivalent position in TEM has a direct consequence for the specificity: the mutation A237T has altered enzymatic parameters for all β-lactams (Matagne *et al.*, 1998).

COMPARISON OF THE STRUCTURES OF ENZYME-ANTIBIOTIC COMPLEXES

The structures of the complexes show the enzyme in a stable transition state in which the active site functional groups are disposed around the reaction center, analogous to the arrangement expected in the transition state during normal catalysis. The design of new drugs requires an understanding of the recognition processes involved in PBPs and β-lactamases. In addition, the study of existing enzyme/drug complexes may suggest ways of optimizing drug-enzyme interactions. In terms of β-lactamases, blocking the deacylation reaction is essential for their effective inhibition.

a) Comparison of R61 Acyl-enzyme and PBP2x Acyl-enzyme

Two structures of R61-antibiotic complexes have been published (Kuzin *et al.*, 1995). The clinically important antibiotics, cephalothin and cefotaxime, form covalent bonds with Ser62. The structure of the complexes (PDB codes; 1ceg and 1cef) have been refined to 1.8Å and 2.0Å, respectively. The structure of an acyl-enzyme complex of PBP2x with the therapeutically important antibiotic, cefuroxime, has been determined at 2.8Å (Gordon *et al.*, 2000). Schematic representations of the drug-enzyme complex interactions are shown in

Figure 9a–c. On acylation of R61, the only apparent changes are rotation of the side chain of Thr301 and the displacement of some water molecules. The structure of the complexed enzyme is very similar to the native structure. Two water molecules possibly involved in deacylation, W507 and W644, are shifted 0.5–0.8Å relative to their positions in the native structures. In contrast, acylation of PBP2x involves the displacement of a sulfate ion and water molecules, as well as side chain rearrangements of Gln552 and Gln452, which accommodate the cefuroxime molecule. Notably, in both cases, there is no displacement of water from the oxyanion hole, as none is present in either native structure. The most striking difference between these three enzyme-antibiotic complexes appears to be the different conformations adopted by the antibiotic in the bound state: in the cephalothin structure, the hydrophobic thiophene group is unrestrained and points out of the active site, in the cefotaxime structure, the methoxy

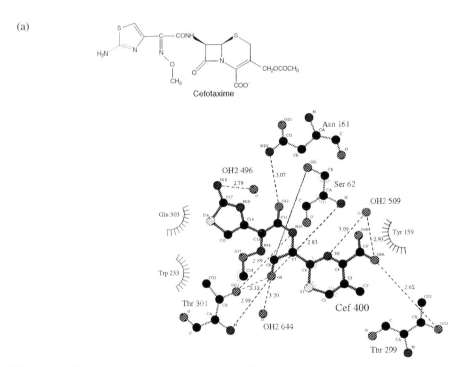

(a)

Cefotaxime

Figure 9. Schematic views showing protein/antibiotic interactions. (the views are prepared with the LIGPLOT program)

a) Cefotaxime bound to the DD-peptidase R61 from *Streptomyces* R61

b) Cephalothin bound to R61

c) Cefuroxime bound to PBP2x from *S. pneumoniae*

d) 6α-(hydroxymethyl)penicillinate bound to TEM1 from *E. coli*

All the spatially conserved amino acids represented in Figure 9a–d for the three proteins are listed in Table 2.

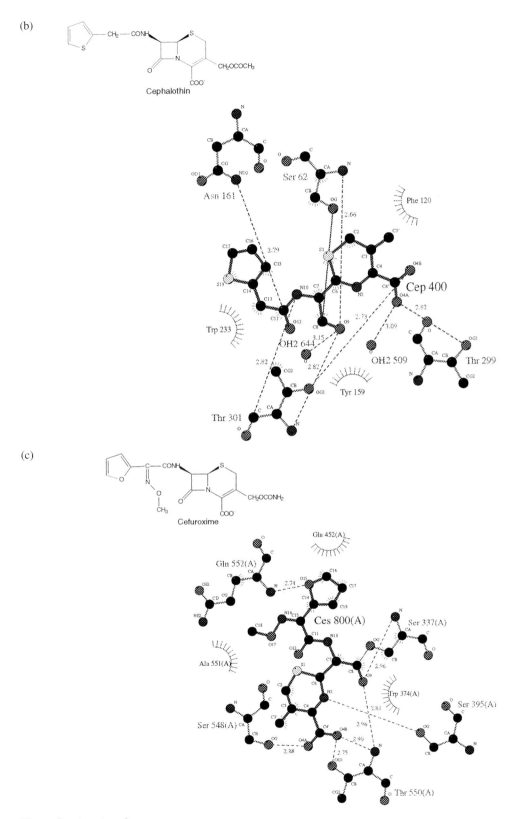

Figure 9. (*continued*)

(d)

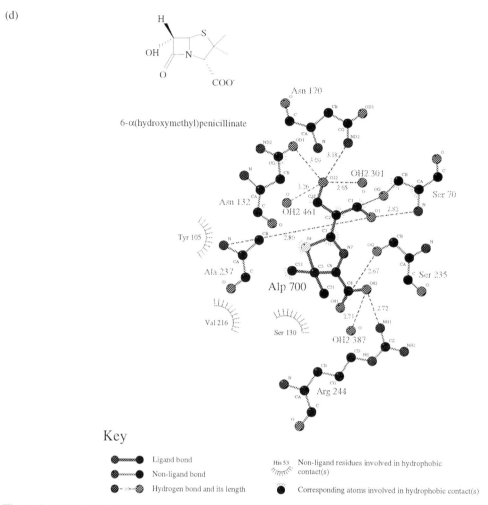

6-α(hydroxymethyl)penicillinate

Key

● Ligand bond

● Non-ligand bond

● Hydrogen bond and its length

His 53 Non-ligand residues involved in hydrophobic contact(s)

● Corresponding atoms involved in hydrophobic contact(s)

Figure 9. (*continued*)

Table 2. Relationship between structurally conserved amino acids displayed in figures 9(a-d)

Function	PBP2x	Class A	R61
Active-site serine	337S	70S	62S
	374W	105Y	116T
(S/Y)XN loop	395S	130S	159Y
	397N	132N	161N
	452Q	170N	233W
Amino acids in the			
β₃ strand	548S	235S	299T
	550T	237A	301T
	551A	238G	302V
	552Q	240E	303Q
	570F	244R	306Y

group is buried and the aminothiazole ring exposed, while, in the cefuroxime/PBP2x structure, the opposite is found, with the furan ring being buried close to β3, forcing Gln552 and Gln452 to adopt alternative rotamers. In both the cephalothin and cefotaxime complexes, the carboxylate substituent (O4B) at the C4 position of the 6-membered dihydrothiazine ring hydrogen bonds to Thr299 Oγ and to a water molecule (W509), which is, in turn, bound to Arg285 NH2. In the PBP2x/cefuroxime structure, only the interaction with Ser548 Oγ is conserved, but, in addition, the O4A substituent hydrogen bonds to Thr550 Oγ. An analogous interaction is seen for the R61/cephalothin complex in which O4A hydrogen bonds to Thr301 Oγ, thus positioning the cephalothin so that a hydrogen bond is formed between the 7-amido linkage and the main chain carbonyl group of Thr301. In the R61/cefotaxime complex, there is a hydrogen bond between the 7-amido linkage and the side chain Thr301 Oγ. In the cefuroxime/PBP2x structure, there is no hydrogen bond to the 7-amido group, but this group is much closer to the main chain carbonyl group than to the side chain, the distances being 3.2Å and 4.2Å, respectively. The limited resolution of this structure may hinder the definition of the exact hydrogen bonding pattern. In R61, the carbonyl moiety of the ester group of cefotaxime and cephalothin is tightly bonded in the oxyanion hole and water molecules W507 and W644 lie 3.98/3.80Å and 3.20/3.15Å, respectively, from the ester carbonyl carbon. In PBP2x, the carbonyl moiety of the ester group of cefuroxime also lies in the oxyanion hole, 2.95Å and 2.96Å from Ser337 N and Thr550 N, respectively; no equivalent water molecules are found in the PBP2x acyl-enzyme structure. The nearest contacts between Oγ of the active site serine and potential general bases are about 3Å in all cases, with the Tyr159/Ser395 distance being the shortest. In R61, it has been proposed that the serine-bound carbonyl oxygen is strongly polarized because of its proximity to Ser62 N and Thr301 N, which would lead to increased nucleophilicity of the C8 atom, making it more susceptible to attack by a water molecule during deacylation. It is probably the lack of an activator residue analogous to Glu166 in Class-A β-lactamases that makes deacylation less efficient in the PBPs.

Despite the differences in the resolution of the complexes, similar, but not identical, patterns of binding have been observed. Although all the complexes studied involved cephalosporin antibiotics, these belong to different generations and the data suggest that the first and second generation cephalosporins, cephalothin and cefuroxime, bind in a way different from that of the third generation cephalosporins.

Comparison of TEM Acyl-enzyme with PBP2x Acyl-enzyme

The structure of TEM β-lactamase complexed with 6α-(hydroxymethyl)penicillanate (PDB code: 1tem) has been determined at 1.95Å (Maveyraud et al.,

1996). It is the only known structure for an acyl-enzyme intermediate involving a ''real'' substrate for a native class A β-lactamase (the acyl-enzyme intermediate is stable in this case because the inhibitor blocks deacylation); all other TEM complexes studied contain either mutant enzymes or substrate analogues. In the TEM/6α-(hydroxymethyl)penicillanate complex, the carboxylate oxygen substituent (O82) on the C3 position of the 5-membered ring forms hydrogen bonds with Arg244 NH1 and with a water molecule, which is, in turn, bound to the backbone carbonyl of Val216; this water molecule plays an important role in the mechanism of inhibition of class A enzymes (Maveyraud et al., 1996). The other carboxylate oxygen, O81, hydrogen bonds to Ser235 Oγ. In the acyl-enzyme structure of PBP2x, only one hydrogen bond is formed, that between O4A (equivalent to O82) and Oγ Ser548 on β3. These interactions through the carboxylate substituent strongly influence the tilt of the antibiotic in the active site. In the TEM complex, the carbonyl oxygen of the ester group lies in the oxyanion hole and hydrogen bonds to Ser70 N and Ala237 N; these hydrogen bonds are in the plane of the peptide bonds, so there is a strong nucleophilicity-inducing force on C8 of the antibiotic. In the PBP2x complex, the interactions appear to be weaker, but this may be a result of the lower resolution of the structure. 6α-(hydroxymethyl)penicillanate contains no equivalent to the cefuroxime aminothiozole ring. The positioning of 6α-(hydroxymethyl)penicillanate in the active site is determined by three factors, covalent bond formation, the carboxylate substituent on the 6-membered dihydrothiazine ring and the hydroxyl group of the 6α-(hydroxymethyl). At the base of the binding site, Glu166 Oϵ2 hydrogen bonds to Asn170 Nδ2, while Glu166 Oϵ1 and Asn170 Oδ1 hydrogen bond to the proposed hydrolytic water, which is 3.0Å from the ester group of the antibiotic. The hydroxyl group of the 6α-(hydroxymethyl) moiety also hydrogen bonds to the hydrolytic water, and it is this interaction that arrests the enzyme complex in the acyl-enzyme state. In the PBP2x structure, the entry of the hydrolytic water, found in TEM, is prevented by the presence of the hydrophobic residue, Phe450. In contrast to the other complexes, the relative distances between the active serine and the potential bases in the TEM complex is reversed, the distance Ser70 Oγ/Lys73 Nζ being shorter than Ser70 Oγ/Ser130 Oγ. This novel inhibitor of TEM, 6α-(hydroxymethyl)penicillanate, was designed so that interactions between it and the hydrolytic water are optimized. In the PBP2x complex, extensive interactions occur between cefuroxime and Thr550, highlighting the importance of this residue.

CONCLUSIONS

β-lactam antibiotics are recognized by, and bind covalently to, the target PBPs, forming a inter-mediate acyl-enzyme complex which is stable for several hours, while β-lactamases show similar recognition of antibiotics, but form a covalent complex which dissociates within 10^{-2} seconds. Structural similarities between PBP and β-lacta-mase suggest a common mechanism of recognition, but different deacylation mechanisms. Despite a number of high resolution studies of β-lactamase, the exact enzymatic mechanism remains contro-versial. Mutant enzymes have demonstrated very subtle changes in substrate specificity and selectiv-ity (Guillaume et al., 1997). Systematic mutation of β-lactamases has highlighted residues involved in recognition and function, but we are a long way from a full understanding of the role played by each amino acid in enzymatic function (Huang et al., 1996; Matagne et al., 1998).

PBP2x is a modular enzyme and, although the functions of some of the domains are not under-stood, their elucidation will be dependent on the determined structure. In the case of the transpepti-dase domain, naturally occurring mutants have allowed us to suggest the structural basis for the most frequent resistance mechanism seen in clinical resistant PBP2x, i.e. the loss of a water molecule from behind the active site serine. Other mechan-isms are certainly involved in producing multi-resistant clinical isolates, and the determination of the structure will allow the molecular mechanisms involved to be explored. As with the β-lactamases, very small changes in structure can result in very large changes in substrate specificity. However, the PBPs are much more restricted, since they must continue to carry out an enzymatic function and recognize their natural substrates, the muropep-tides. Comparison of the structures of the class A β-lactamase and PBP2x shows a remarkable con-

servation of the fingerprint motifs, while notable differences include the lack – in the PBP – of a water molecule in the oxyanion hole and of the Ω loop bearing a glutamic acid.

REFERENCES

Bramhill, D. (1997) Bacterial cell division. *Ann. Rev. Cell Dev. Biol.* **13**, 395–424.

Carfi, A., Pares, S., Duée, E., Galleni, M., Duez, C., Frère, J.-M., et al. (1995) The 3-D structure of a zinc metallo-β-lactamase from *Bacillus cereus* reveals a new type of protein fold. *EMBO J.* **14**, 4914–4921.

Charlier, P., Buisson, G., Dideberg, O., Wierenga, J., Keck, W., Laible, G., et al. (1993) Crystallization of a genetically engineered water-soluble primary penicillin target enzyme. The high molecular mass PBP2x of *Streptococcus pneumo-niae*. *J. Mol. Biol.* **232**, 1007–1009.

Fonzé, E., Vermeire M., Nguyen-Disteche, M., Brasseur, R., and Charlier, P. (1999) The crystal structure of a penicilloyl-serine transferase of intermediate penicillin sensitivity. The DD-transpeptidase of *Streptomyces* K15. *J. Biol. Chem.*, **274**, 21853–21860.

Ghuysen, J.M. (1991) Serine β-lactamases and penicillin-binding proteins. *Annu. Rev. Microbiol.* **45**, 37–67.

Ghuysen, J.M. (1994) Molecular structures of penicillin-binding proteins and β-lactamases. *Trends Microbiol.* **2**, 372–380.

Gordon, E., Mouz, N., Duée, E., and Dideberg, O. (2000) The crystal structure of the penicillin-binding protein 2x from *Streptococcus pneumoniae* and its acyl-enzyme form: im-plication in drug resistance. *J. Mol. Biol.*, **299**, 501–509.

Grebe, T., and Hakenbeck, R. (1996) Penicillin-binding proteins 2b and 2x of *Streptococcus pneumoniae* are primary resistance determinants for different classes of β-lactam antibiotics. *Antimicrob. Agents Chemother.* **40**, 829–834.

Guillaume, G., Vanhove, M., Lamotte-Brasseur, J., Ledent, P., Jamin, M., Joris, B., et al. (1997) Site-directed mutagenesis of glutamate 166 in two beta-lactamases. Kinetic and molecular modeling studies. *J. Biol. Chem.* **272**, 5438–5444.

Hakenbeck, R., and Coyette, J. (1998) Resistant penicillin-binding proteins. *Cell Mol. Life. Sci.* **54**, 332–340.

Hakenbeck, R., Konig, A., Kern, I., van der Linden, M., Keck, W., Billot-Klein, D., et al. (1998) Acquisition of five high-M_r penicillin-binding protein variants during transfer of high-level beta-lactam resistance from *Streptococcus mitis* to *Streptococcus pneumoniae*. *J. Bacteriol.* **180**, 1831–1840.

Hancock, R.E. (1997) The bacterial outer membrane as a drug barrier. *Trends Microbiol.* **5**, 37–42.

Holtje, J.V. (1998) Growth of the stress-bearing and shape-maintaining murein sacculus of *Escherichia coli*. *Microbiol. Mol. Biol. Rev.* **62**, 181–203.

Huang, W., Petrosino, J., Hirsch, M., Shenkin, P.S., and Palzkill, T. (1996) Amino acid sequence determinants of beta-lactamase structure and activity. *J. Mol. Biol.* **258**, 688–703.

Jamin, M., Wilkin, J.M., and Frere, J.M. (1995) Bacterial DD-transpeptidases and penicillin. *Essays in Biochemistry* **29**, 1–24.

Jelsch, C., Mourey, L., Masson, J.M., and Samama, J.P. (1993) Crystal structure of *Escherichia coli* TEM1 beta-lactamase at 1.8 A resolution. *Proteins* **16**, 364–383.

Joris, B., Ghuysen, J.M., Dive, G., Renard, A., Dideberg, O., Charlier, P., et al. (1988) The active-site-serine penicillin-recognizing enzymes as members of the *Streptomyces* R61 DD-peptidase family. *Biochem. J.* **250**, 313–324.

Kell, C.M., Sharma, U.K., Dowson, C.G., Town, C., Balganesh, T.S., and Spratt, B.G. (1993) Deletion analysis of the essentiality of penicillin-binding proteins 1A, 2B and 2X of *Streptococcus pneumoniae. FEMS Microbiol. Lett.*, **106**, 171–175.

Kelly, J.A., Knox, J.R., Moews, P.C., Hite, G.J., Bartolone, J.B., Zhao, H., *et al.* (1985) 2.8-Å Structure of penicillin-sensitive D-alanyl carboxypeptidase-transpeptidase from *Streptomyces* R61 and complexes with beta-lactams. *J. Biol. Chem.* **260**, 6449–6458.

Kelly, J.A., Dideberg, O., Charlier, P., Wery, J.P., Libert, M., Moews, P.C., *et al.* (1986) On the origin of bacterial resistance to penicillin: comparison of a beta-lactamase and a penicillin target. *Science* **231**, 1429–1431.

Kelly, J.A., and Kuzin, A.P. (1995) The refined crystallographic structure of a DD-peptidase penicillin-target enzyme at 1.6 Å resolution. *J. Mol. Biol.* **254**, 223–236.

Knox, J.R. (1995) Extended-spectrum and inhibitor-resistant TEM-type beta-lactamases: mutations, specificity, and three-dimensional structure. *Antimicrob. Agents Chemother.* **39**, 2593–2601.

Krauss, J., van der Linden, M., Grebe, T., and Hakenbeck, R. (1996) Penicillin-binding proteins 2x and 2b as primary PBP targets in *Streptococcus pneumoniae. Microb. Drug Resist.* **2**, 183–186.

Kuzin, A.P., Liu, H., Kelly, J.A., and Knox, J.R. (1995) Binding of cephalothin and cefotaxime to D-ala-D-ala-peptidase reveals a functional basis of a natural mutation in a low-affinity penicillin-binding protein and in extended-spectrum beta-lactamases. *Biochemistry* **34**, 9532–9540.

Laible, G., and Hakenbeck, R. (1991) Five independent combinations of mutations can result in low-affinity penicillin-binding protein 2x of *Streptococcus pneumoniae. J. Bacteriol.* **173**, 6986–6990.

Massova, I., and Mobashery, S. (1998) Kinship and diversification of bacterial penicillin-binding proteins and beta-lactamases. *Antimicrob. Agents Chemother.* **42**, 1–17.

Matagne, A., Lamotte-Brasseur, J., and Frere, J.M. (1998) Catalytic properties of class A beta-lactamases: efficiency and diversity. *Biochem. J.* **330**, 581–598.

Matsuhashi, M., Wachi, M., and Ishino, F. (1990) Machinery for cell growth and division: penicillin-binding proteins and other proteins. *Res. Microbiol.* **141**, 89–103.

Maveyraud, L., Massova, I., Birck, C., Miyashita, K., Samama, J.P., and Mobashery, S. (1996) Crystal structure of 6 α-hydroxymethylpenicillanate complexed to the Tem-1 β-lactamase from *Escherichia coli:* evidence on the mechanism of of action of a novel inhibitor designed by a computer-aided process. *J. Am. Chem. Soc.* **118**, 7435–7440.

Mouz, N., Gordon, E., Di Guilmi, A., Petit, I., Petillot, Y., Dupont, Y., *et al.* (1998) Identification of a structural determinant for resistance to beta-lactam antibiotics in gram-positive bacteria. *Proc. Natl. Acad. Sci. USA* **95**, 13403–13406.

Mouz, N., Di Guilmi, A.M., Gordon, E., Hakenbeck, R., Didelberg, O., and Vernet, T. (1999) Mutations in the active site of penicillin-binding protein PBP2x from *Streptococcus pneumoniae*. Role in the specificity for beta-lactam antibiotics. *J. Biol. Chem.*, **274**, 19175–19180.

Nguyen-Disteche, M., Fraipont, C., Buddelmeijer, N., and Nanninga, N. (1998) The structure and function of *Escherichia coli* penicillin-binding protein 3. *Cell Mol. Life. Sci.* **54**, 309–316.

Parès, S., Mouz, N., Petillot, Y., Hakenbeck, R., and Dideberg, O. (1996) X-ray structure of *Streptococcus pneumoniae* PBP2x, a primary penicillin target enzyme. *Nat. Struct. Biol.* **3**, 284–289.

Samraoui, B., Sutton, B.J., Todd, R.J., Artymiuk, P.J., Waley, S.G., and Phillips, D.C. (1986) Tertiary structural similarity between a class A beta-lactamase and a penicillin-sensitive D-alanyl carboxypeptidase-transpeptidase. *Nature* **320**, 378–380.

Sifaoui, F., Kitzis, M.-D., and Gutmann, L. (1996) In vitro selection of one-step mutants of *Streptococcus pneumoniae* resistant to different oral β-lactam antibiotics is associated with alterations of PBP2x. *Antimicrobial Agents and Chemotherapy* **40**, 152–156.

Smith, J.M., Dowson, C.G., and Spratt, B.G. (1991) Localized sex in bacteria. *Nature* **349**, 29–31.

Spratt, B.G. (1994) Resistance to antibiotics mediated by target alterations. *Science* **264**, 388–393.

Tipper, D.J., and Strominger, J.L. (1965) Mechanism of action of penicillians: a proposal based on their structural similarity to acyl-D-alanyl-D-alanine. *Proc. Natl. Acad. Sci. USA* **54**, 1133–1141.

Usher, K.C., Blaszczak, L.C., Weston, G.S., Shoichet, B.K., and Remington, S.J. (1998) Three-dimensional structure of AmpC beta-lactamase from *Escherichia coli* bound to a transition-state analogue: possible implications for the oxyanion hypothesis and for inhibitor design [In Process Citation]. *Biochemistry* **37**, 16082–16092.

13. New Targets and Strategies for Identification of Novel Classes of Antibiotics

Molly B. Schmid*

Microcide Pharmaceuticals, Inc. 850 Maude Avenue Mountain View, CA 94043, USA.

INTRODUCTION

With the completion of numerous bacterial genome sequences, antibacterial drug discovery has fully entered the genomic era. The combination of genetic, genomic and biochemical methods that take advantage of improvements and innovations in genomics and high throughput screening capacity are changing the drug discovery process. Both classical and novel approaches have been established to expand the number of genes that are recognized as antimicrobial targets. Several novel approaches allow rapid, target-based screen development even when the biochemical functions of the genes are not known. Increasingly, new methods that exploit genomic information and technologies are being developed to determine mechanisms of action for novel antimicrobials and to predict any potential toxicities. The availability of a large number of antimicrobial targets and screens creates new challenges, decision points and opportunities in the search for novel antimicrobial compounds. Strategies that emerge may have general applicability for use of genomic information in other therapeutic areas.

MICROBIAL GENOMICS

The development of technologies and strategies for efficient genomic sequencing and contig assembly has resulted in the completion of numerous microbial genome sequences within the past five years (Fleischmann *et al.*, 1995; Fraser *et al.*, 1995; Blattner *et al.*, 1997; Fraser *et al.*, 1997; Kunst *et al.*, 1997). With numerous academic and commercial organizations engaged in sequencing microbial genomes, the molecular understanding of genomic information content is advancing rapidly. In publicly available databases, there are currently 15 completed sequences for eubacterial genomes and at least 58 in-progress. Additional microbial sequences are available in commercial sequence databases, such as those offered by Incyte Pharmaceuticals, Inc. (*www.incyte.com*) and Genome Therapeutics Corp. (*www.genomecorp.com*). Coupled with a widely recognized need for new antimicrobial agents due to the rapid emergence of drug resistance (Levy, 1998), microbial genomics has been embraced by the pharmaceutical industry. Underlying the pharmaceutical interest in microbial genome sequences is the belief that target-specific screening and new targets will allow identification of new structural classes of antimicrobial agents.

* Current address: Genencor International, Inc. 925 Page Mill Road, Palo Alto, CA 94304. Phone: (650) 846-4011; Fax: (650) 621-7901; Email: mschmid@genencor.com

Several key features of microbial genomes have emerged from the sequences that have been obtained thus far. The smallest known bacterial genomes are those of the mycoplasmas. The genome of *M. genitalium* is 580 Kbp, and is predicted to encode 470 proteins (Fraser *et al.*, 1995). The largest bacterial genome fully sequenced to date is *E. coli* K-12 (Blattner *et al.*, 1997), which at 4.6 Mbp is predicted to encode 4288 proteins. Bacterial genomes are information-rich: over 85% of most sequenced bacterial genomes encode protein or RNAs, with only a small fraction of the genome dedicated to control regions and non-coding features. The *E. coli* K-12 genome is relatively typical: 87.8% of the genome is predicted to encode protein, 0.8% stable RNA, 0.7% noncoding repeat sequences, leaving 11% as regulatory, undetermined, and non-functional DNA sequence (Blattner *et al.*, 1997).

Bacterial genes are relatively small (due to the general lack of introns, an average bacterial gene is roughly 1 Kbp), which allows the majority of information to be obtained without assembly of the genome sequence into a single contig. Thus, for most pharmaceutical applications, "nearly complete" bacterial genome sequences provide adequate information for target identification and pathway assessment, and "sample sequencing" provides significant information on a pathogen (Baltz *et al.*, 1998).

The conservation of protein sequence information among eubacteria is sufficient to recognize proteins with similar structure and anticipated function ("orthologs") among all groups of eubacteria (Fraser *et al.*, 1995). Many genes are shared among the bacterial species that have been sequenced to date, and currently about 75% of genes in a eubacterial genome can be found in at least one other sequenced microbial genome.

Genome rearrangements, such as gene duplications, fusions, insertions and deletions, are common differences between closely related microorganisms (Bergthorsson *et al.*, 1998; Karaolis *et al.*, 1998). Acquired segments of DNA are frequently associated with virulence, leading to the term "pathogenicity islands" to describe the acquired segments that encode virulence determinants (Lee, 1996; Posfai *et al.*, 1997; Bergthorsson *et al.*, 1998; Karaolis *et al.*, 1998). In microbial genomes, gene duplications and fusions sometimes mask partially orthologous genes, and make comparing microbial genome information more difficult. For example, in *E. coli*, the DnaE protein is the major subunit of DNA polymerase III, and is encoded by a single gene, *dnaE* (Tomasiewicz *et al.*, 1987). The DnaQ accessory protein possesses a 3′–5′ exonuclease, which associates with DnaE in the *E. coli* holoenzyme. However, in *B. subtilis* and *S. aureus*, two genes encode structurally related DNA polymerase III subunits (Hammond *et al.*, 1991; Pacitti *et al.*, 1995). The *S. aureus dnaE* gene encodes a protein with homology to the *E. coli* DnaE protein, while the *S. aureus polC* gene encodes a DnaE-DnaQ fusion protein. The *E. coli* genome does not encode this "PolC-like" protein (Blattner *et al.*, 1997). In *S. aureus*, both the *dnaE* gene and the *polC* gene are essential, so these two gene products are not redundant – they have specialized functions (Boyer *et al.*, 1998). There are numerous additional examples of both gene fusion and gene duplication that are apparent from analysis of bacterial genome sequences.

Two surprises have emerged from the global analysis of microbial genome information. First, "orphan" genes, which lack an ortholog in another sequenced genome, continue to appear in microbial genomes, despite the sequencing of an evolutionarily diverse set of microbial genomes (Clayton *et al.*, 1998). In addition, a surprisingly small number of genes are found in all sequenced microorganisms. These two findings undoubtedly reflect the extensive and unappreciated microbial species diversity, as well as the large size of the composite microbial gene pool. In addition, it is becoming clear that non-orthologous proteins can perform similar functions. Comparative genomics has shown several examples of structurally distinct proteins that serve functionally analogous roles in different microorganisms (Galperin *et al.*, 1998; Forterre, 1999). These examples of convergent evolution provide an unanticipated source of functional redundancy in microbes, as well as an explanation for the apparent dissimilarity in microbial genome information content. Such examples begin to demonstrate evolution's capacity for creating functional redundancy by methods other than gene duplication.

Targets of Current Antimicrobial Agents

The current commercially available antimicrobial agents inhibit the function of a small number of molecular targets. The primary molecular targets of

Table 1.   Commercially available antibiotics.

Antibiotic class	Molecular target	Examples
Beta-lactam	Penicillin binding proteins	Penicillins (benzylpenicillin, ampicillin, amoxycillin); cephalosporins (cefotaxime, cephalexin, ceftazidime); carbapenems (imipenem, meropenem) monobactams (aztreonam)
Aminoglycoside	30S, 50S Ribosome subunits	Gentamicin, neomycin, tobramycin, amikacin, streptomycin
Trimethoprim, sulfonamides	Folate synthesis	Trimethoprim, sulphadiazine
Quinolone	Gyrase, topoisomerase IV	Nalidixic acid, fluoroquinolones (ciprofloxacin, ofloxacin, norfloxacin)
Macrolide	50S ribosome subunit	Erythromycin, lincomycin, azithromycin, clarithromycin
Tetracycline	30S ribosome subunit	Tetracycline, doxycycline, minocycline
Glycopeptide	Cell wall peptidoglycan	Vancomycin, teicoplanin
Chloramphenicol	50S subunit	chloramphenicol
Rifamycins	RNA polymerase	Rifamycin B
Polymyxin	Membranes	Polymyxin B, colistin

these agents are the components of the translation process, cell wall biosynthetic enzymes, RNA polymerase, and type II topoisomerases (see Table 1). Numerous studies have shown that many additional genes involved in the tasks of microbial DNA replication, DNA repair, transcription, cell division, protein localization, nucleoid segregation, and biosynthesis of cell wall, lipid and anionic surface features (teichoic acids, lipopolysaccharides) are necessary for microbial growth and should provide effective targets for new antibacterial agents.

Among the hypotheses to explain the limited classes of inhibitors are the bias of previous screening strategies toward targets that had been validated by successful inhibitors, the difficulty of identifying the molecular mechanism of action for a novel inhibitor (Silverman et al., 1993; Urban et al., 1996; McMurray et al., 1998; Matassova et al., 1999; Stewart et al., 1999), the severe hurdles that a compound class must overcome before it has a chance to become a drug (Gootz, 1990), the limited chemical diversity represented within current chemical libraries (Ecker et al., 1995; Bures et al., 1998), and the inability of chemicals to access the molecular target due to the barriers of compound entry and efflux (Hsieh et al., 1998; Lomovskaya et al., 1999). The key problems

whose solutions will allow breakthroughs in identifying novel classes of antimicrobials remain to be established.

The pharmaceutical and medical desire for broad spectrum compounds probably also contributes to the difficulty of identifying new antimicrobial agents. Until recently, it was believed that the desire for broad spectrum agents might limit the number of targets that were shared among the clinically relevant species. However, genomic analysis shows that this is unlikely. For example, of 22 proteins known to be involved in E. coli DNA replication (Marians, 1996), eight are found in six fully sequenced eubacterial genomes, and eleven are found in five of the six genomes.

Identification of Pharmaceutically-Relevant Targets

Genomic information provides a wealth of new potential targets for antimicrobial chemotherapy. However, not all genes in a clinically relevant microorganism are good targets for antimicrobial drug discovery. Several methods are now used to identify the pharmaceutically-relevant genes in microbial genomes.

Initially, it was thought that *in silico* methods could predict the pharmaceutically-relevant genes (Maniloff, 1996; Mushegian *et al.*, 1996; Arigoni *et al.*, 1998). Comparison of completed genomes of *H. influenza* and *M. genitalium* showed that approximately 250 genes were found in both of the bacterial genomes and these were hypothesized to comprise a "minimal genome set" necessary to support bacterial viability (Mushegian *et al.*, 1996). The computational strategy assumed that bacteria accomplish essential functions through common mechanisms and that the genes encoding these functions would be evolutionarily conserved. However, as additional genomes were sequenced, the set of completely conserved genes became very small (Koonin, 1997). In addition, experiments showed that many genes that were highly conserved among bacteria, could be disrupted without loss of viability, and thus are not genetically essential. For genes that were conserved between *E. coli* and the small *M. genitalium* genome, 20 of 26 conserved genes could be disrupted without loss of viability (Arigoni *et al.* 1998). Thus, highly conserved genes are not necessarily essential for viability under typical laboratory growth conditions. Collectively, these results suggest that different species may perform certain essential functions using structurally different proteins (Mushegian *et al.*, 1996; Arigoni *et al.*, 1998).

There are several experimental strategies to identify the pharmaceutically relevant genes within a microbial genome (Isaacson, 1997). Current antimicrobial chemotherapy relies on compounds that prevent microbial growth (bacteriostatic compounds) or kill the microbe (bactericidal compounds). Pharmaceutically relevant genes are those that cause these antimicrobial effects when the gene product is inactivated. Genetic methods can be used to identify these genes and to predict the effects of chemical inhibition of these genes (Bostian *et al.*, 1997; Hartwell *et al.*, 1997).

Experimentally, genetic inactivation of a potential target can be accomplished by gene disruption, conditional expression or conditional-function mutations. Gene disruption has been applied on a case-by-case basis, to test whether an individual gene is essential (Martin *et al.*, 1999), or in a high throughput mode to provide a genome-wide assessment of the essential genes in an organism (Akerley *et al.*, 1998). Most high throughput methods use an efficient transposon system to generate a population of random insertions, then provide a molecular display of the insertion sites that remain in the population after a period of growth (Shoemaker *et al.*, 1996; Smith *et al.*, 1996). Strains that carry a disruption of a gene that causes inviability or slow growth are lost from the population, resulting in a gap in the molecular display of chromosomal insertion sites.

Disruption strategies have disadvantages and pitfalls that should be recognized when interpreting data and making predictions of the essentiality of bacterial genes. The methods use negative evidence – the inability to isolate a viable strain – as evidence of essentiality. In addition, most bacteria exhibit the phenomenon of transcriptional polarity, in which blocking the normal translation of an upstream gene decreases transcription of a downstream gene. Thus, disruption of an upstream gene can decrease downstream transcription, and can cause erroneous conclusions about the essentiality of a gene. Disruption strategies designed to circumvent the problems of bacterial polarity have been devised (Link *et al.*, 1997).

Spontaneous tandem duplications in bacteria can also confound disruption analysis. When a gene is duplicated, one copy of the gene can be disrupted, while the other can provide gene function (Anderson *et al.*, 1977), causing an essential gene to appear non-essential. Experimental estimates of the frequency of gene duplication showed that 0.1–2% of cells in a population of *S. typhimurium* carried a duplication of any gene (Anderson *et al.*, 1979). Thus, when using certain methods for assessing gene essentiality there is a need to remain cognizant of the relatively frequent occurrence of tandem duplication.

All of the methods to define essential genes require a definition of the growth environment. In general, enriched medium in standard laboratory conditions has been used for identifying the pharmaceutically relevant essential genes, though there has been a growing recognition that there are additional genes that provide functions necessary for growth in the host environment (Bowe *et al.*, 1998), and methods have been developed to identify these genes (see below).

The number of essential genes in bacterial genomes is not well defined, but current estimates are surprisingly low. The small *M. genitalium* genome has only 470 genes that encode all of the genetic information necessary for growth of this wall-less bacterium. Results from another mycoplasma species suggest that not all genes in these small genomes are essential for viability in laboratory conditions. *M. pneumoniae*, with a genome size of 816Kb, contains a number of paralogs (Himmelreich *et al.*, 1996) and disruption of several genes can occur without loss of viability (Dhandayuthapani *et al.*, 1999). Statistical analyses, based on the frequency of genes causing lethality when mutated, have been performed in *B. subtilis* and *S. typhimurium*. In *B. subtilis*, disruption in only 6 of 79 genes caused inability to grow on an enriched medium, yielding an estimate that between 318–562 Kbp of the *B. subtilis* genome (7.6–13.4%) encoded indispensable genes (Itaya, 1995). A statistical estimate made from genetic saturation within a conditional lethal collection suggested that 100–200 non-redundant genes are required for growth on enriched medium in *S. typhimurium* (Schmid *et al.*, 1989).

Functional redundancy of gene products may decrease the number of genetically essential genes. If two gene products perform a similar essential function, neither gene will appear essential by the genetic tests that are currently applied. Functional redundancy may come from either structurally-related paralogs or from non-paralogous gene products that provide functional redundancy. The possession of back-up systems and failsafe mechanisms is apparently an evolutionary success strategy for microbes, and may help to explain the apparently low number of genetically essential genes.

Genomic Characteristics of Bacterial Essential Genes

The essential genes of *S. aureus* have been identified through a conditional lethal mutant approach by researchers at Microcide Pharmaceuticals (unpublished data). Comparison of the DNA sequence information of the *S. aureus* essential genes with the publicly available sequence information of *E. coli*, *H. influenzae*, *H. pylori*, *M. genitalium*, *B. subtilis* and *Synechocystis* showed the general distribution of these genes in bacterial genomes (D. Biek, personal communication). The majority of these genes exist in another sequenced organism (see Figure 1). Only 7% of the 96 genes used in this

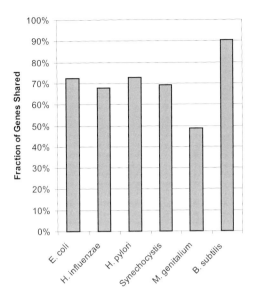

Figure 1. The percent of *S. aureus* essential genes found in other genomes was determined by BLAST search analysis. The number of orthologous genes identified and the number tested for each species were: *E. coli* 69 orthologs found among 95 genes examined; *H. influenzae* 64 orthologs found among 94 genes examined; *H. pylori* 62 orthologs found among 85 genes examined; *Synechocystis* 52 orthologs found among 75 genes examined; *M. genitalium* 46 orthologs found among 94 genes examined; *B. subtilis* 86 orthologs found among 95 genes examined.

analysis were unique to staphylococci, while an additional 16% of the essential genes existed only in another gram-positive species (*B. subtilis*). Some of the genes that are unique to *S. aureus* encode proteins that produce unique features of *S. aureus*, such as the pentaglycine bridge of the cell wall. Although most of these essential genes exist in other species, much remains to be learned about them: over one-quarter of the essential genes were orthologs of predicted genes with no known function at the time of the analysis (D. Biek, personal communication, 1998).

Non-Traditional Targets

Most of the current commercially available antibacterial agents cause loss of microbial viability (bactericidal agents), or cessation of microbial growth (bacteriostatic agents). Recently, proposals have suggested developing agents that target drug resistance mechanisms or virulence factors (Coleman et al., 1994; Chopra et al., 1997; Ehlert, 1999).

Inhibitors of drug-resistance mechanisms would revitalize existing agents by inactivating a resistance mechanism. These new types of compounds would be used in combination with existing antimicrobial agents to provide therapeutic relief. Identification of genes that encode drug-resistance is relatively straightforward, and a large body of information has been gathered on the molecular mechanisms underlying drug-resistance (Davies, 1997). Thus, numerous drug-resistance targets are known (Brakstad et al., 1997; Van Veen et al., 1997; Chambers, 1999; Ehlert, 1999; Lomovskaya et al., 1999), many of which have accessible biochemical assays. This approach will have highest likelihood of commercial success when the resistance mechanism has transferred horizontally, or has a common genetic origin in different bacterial species, allowing a single inhibitor to have broad spectrum applicability.

Inhibitors of drug resistance targets have been described by several investigators. Inhibitors of the beta-lactamase enzymes, such as sulbactam and clavulanic acid were identified and successfully developed as fixed combinations with beta-lactams. The huge commercial success of Augmentin™ (1998 worldwide sales of $1.6B), a fixed combination of amoxicillin and clavulanic acid provides a successful precedent for the development of combination agents. Inhibitors of other drug resistance targets have been described (Chamberland et al., 1995; Blais et al., 1999), providing potential for the development of inhibitors of additional resistance targets.

Additional targets for novel therapeutics may be found among genes necessary for microbial virulence (Isaacson, 1997). There are several strategies for identifying these genes. Certain approaches identify genes that are expressed in vivo, but not in vitro. These methods, including differential display, IVET and microarray technology (Chee et al., 1996; Heithoff et al., 1997; Isaacson, 1997) rely on the close correspondence of genes that are differentially expressed under in vivo conditions, and those that cause virulence. After identifying genes that are differentially expressed, gene disruptions are created to monitor the loss of virulence caused by loss of the gene product. Unfortunately, in many species, it has

been found that many more genes are differentially expressed than have a significant role in effecting virulence (Heithoff et al., 1997), making this a somewhat inefficient process.

Other approaches identify genes that are required for growth in an animal host, but not under laboratory conditions. Genes that are necessary for growth in the unique in vivo environment can be efficiently identified by screening pools of mutant strains, in which each member of the pool is uniquely identified by a molecular tag (Shoemaker et al., 1996; Mei et al., 1997). Such strategies allow methods to perform genome-wide assessment of the genes required for microbial survival in the mammalian host environment (Chiang et al., 1998; Coulter et al., 1998). Unique understanding of the in vivo nutritional environment, host defenses, and interactions between mammalian and microbial cells is resulting from assessment of the identified genes.

Agents that inhibit virulence targets should behave as standard antibiotics that prevent growth or kill microbes; creative manipulation of in vitro conditions should sufficiently recreate the mammalian environment and allow assessment of inhibitors, as well as whole cell high throughput screening strategies.

Prioritization of Targets

With several hundred potential pharmaceutically relevant genes in clinically important pathogens, prioritization of targets must occur unless the high throughput screening resources allow consideration of screening against all available targets (Dove, 1999). Good antimicrobial agents must possess a number of different properties: strong inhibition of microbial growth; little interaction with mammalian targets; low resistance development; broad antimicrobial activity spectrum; and desirable physical, chemical, and pharmacological properties.

Certain features of the antibiotic may be influenced by the choice of target, and may be predicted from genetic and genomic information, while other features of an antibiotic are relatively independent of the target choice. In random screening approaches, it is difficult to predict which targets will yield leads with desirable physical, chemical, and pharmacological properties. Reasonable generalizations can be made – membrane protein targets may be more likely to be inhibited by highly lipophilic compounds – but the evidence supporting even these general assumptions is weak. For genes with known substrates, rational drug design may provide a route to identify inhibitors with more desirable physical properties.

Low resistance emergence has been associated with antibiotics that have more than one genetic target, since it is expected that each gene will require mutation for the cell to become resistant. An excellent example supporting this contention is the highly successful beta-lactam antibiotic class, which inhibits penicillin binding proteins (PBP's). In *E. coli*, several of these PBP's, are known to be genetically essential, and most other eubacterial genomes also possess multiple PBP's. Resistance to beta-lactams due to minor modifications of the target does not arise frequently, and the only clinically important resistance mechanisms result from beta-lactamases, efflux mechanisms and acquisition of beta-lactam resistant PBPs by horizontal transfer. The fluoroquinolones are also highly successful and inhibit the two bacterial Type II topoisomerases, DNA gyrase and Topoisomerase IV (Drlica *et al.*, 1997). These two topoisomerases are structurally related, genetically essential, and found in most eubacterial genomes (Huang, 1996). However, while commercially successful, the fluoroquinolones are subject to both target-based and efflux-mediated resistance development. Thus, choosing multi-gene targets does not guarantee low resistance frequencies. Nonetheless, there are several unexploited gene families that may provide additional multi-gene targets for inhibitors, such as the two component signal transduction systems (Barrett *et al.*, 1998; Martin *et al.*, 1999), tRNA synthetases (Schimmel *et al.*, 1998), RNA polymerase sigma subunits, helicases, amino acyl ligases of peptidoglycan biosynthesis, and several enzymatic steps of fatty acid biosynthesis.

Genomic information is being used to predict the potential of a target for broad antimicrobial activity spectrum and high selectivity. Within the bacterial kingdom, orthologs can be determined with relative certainty. In addition, mammalian, yeast, nematode and fruitfly DNA sequence information shows significant conservation of cellular processes, pathways and proteins across all eukaryotes (Miklos *et al.*, 1996). This creates

the hope of identifying antimicrobial targets that are unique to prokaryotes, and that will more readily yield inhibitors with high selectivity. However, the amount of sequence dissimilarity that is necessary to assure selectivity is unknown; similar 3-dimensional protein structures can be determined by significantly different primary sequences (Orengo *et al.*, 1994), and the success of the fluoroquinolones shows that selectivity can be achieved, even when the microbial target has a mammalian homolog. These factors introduce uncertainty about the wisdom of hoping for selectivity based on simple sequence analysis, and of using sequence-based selectivity arguments to eliminate otherwise good antibacterial targets.

Many other factors can enter the process of target prioritization. The bacterial two-component regulators (2CR's) provide an interesting example of attractive, but so far commercially unexploited targets. The 2CR's are generally composed of a histidine kinase sensor protein and a response regulator that controls transcription of a regulon. The development of biochemical assays measuring both kinase and phosphorylase activity allowed development of a high throughput screen to identify inhibitors (Barrett *et al.*, 1998). Because this mechanism of regulation is widespread in eubacteria, but not eukaryotes, these targets may be more likely to yield broad spectrum and selective antibiotic agents. Most eubacterial genomes possess a large number of genes involved in 2CR's (*E. coli* has at least 37 genes putatively involved in 2CR's), suggesting that a single inhibitor might affect several structurally related targets, potentially reducing the frequency of resistance development. Members of the 2CR family that are necessary for virulence (Cirillo *et al.*, 1998) and viability (Martin *et al.*, 1999), have been described in *S. aureus*, providing confidence that inhibition of the target would provide a traditional antibiotic capable of inhibiting bacterial growth and diminishing virulence. Furthermore, a conditional lethal mutant of the *S. aureus* HPK gene, *yycF*, was reported to lose viability rapidly after a shift to non-permissive conditions and also to have increased susceptibility to macrolide antibiotics. These characteristics of the *yycF* conditional mutant suggest that that chemical inhibition of this target might cause bactericidal effects and show synergy with macrolides (Martin *et al.*, 1999).

Genomics and Screen Development

For genomics to increase its value in the pharmaceutical setting, methods for using the information and technology must be developed to improve and accelerate steps downstream of target identification. Several methods are emerging to improve and expand the possibilities for target-directed screen development.

Traditional target-directed screening methods employed a biochemical screen against a particular target to assure high sensitivity and specificity for the target (Umezawa, 1982). DNA sequence information enables simpler methods for obtaining large quantities of a target protein from a clinically relevant organism. However, this strategy limits the targets to those that have well-developed biochemical assays, which is a relatively small subset of the pharmaceutically-relevant genes. Over 25% of the essential genes identified in *S. aureus* have no functional characterization, identified only as orthologs of an ORF in another genome; many of the remaining genes do not have biochemical assays that could be readily implemented in high throughput.

Biochemical approaches for high throughput screening have traditionally required extensive knowledge of the biochemical function of a protein. Recently, methods have been developed and implemented in high throughput that can differentiate ligand-bound and unbound protein, generally based on the increased conformational stability of a protein when it is bound to a ligand (Bowie *et al.*, 1996). These methods enable generic high throughput biochemical screens that rely only on over-expression and purification of large quantities of folded protein, and can be used with proteins having little functional information. These biochemical approaches enjoy the advantage of all biochemical assays, in their ability to provide quantitative and sensitive measures of the inhibitor's ability to bind the target protein and do not require the inhibitor to surmount permeability barriers in order to be detected.

New genome-based whole cell screening approaches have used genetically altered strains to create novel, target-specific phenotypes (Bostian *et al.*, 1997; Hartwell *et al.*, 1997; Giaever *et al.*, 1999). Several of the whole cell screening approaches use hypomorphic phenotypes created

by mutation or underexpression of an essential gene. Cells carrying such mutations are hypersensitive to inhibitors of the target and functionally related targets and can detect even weak inhibitors. These screens are the chemical equivalent of a genetic search for synthetic lethal mutations in which synthetically lethal pairs of mutations arise in functionally redundant genes, and when gene products physically or functionally interact with one another (Davies *et al.*, 1999). By mutation, one cellular component is weakened; chemical inhibitors of that target, or functionally related targets, are identified through the screening process.

Genomics and Mechanism of Action

New approaches evolving from genomics may aid the process of determining mechanism of action for new classes of inhibitors. Historically, mechanism of action studies have been initiated after selection of a lead series. Rapid and high throughput mechanism of action capability would allow examination of the biological effects of compounds earlier in the discovery process, perhaps allowing this information to aid in the selection of lead compounds as well as in the functional characterization of diversity libraries.

Manipulation of the gene dosage of an inhibitor's target was found to cause hypersensitivity or resistance to the inhibitor (Giaever *et al.*, 1999). Similarly, mutants with weakened target function showed hypersensitivity to an inhibitor (Bostian *et al.*, 1997). Using these findings, a set of strains carrying alterations in possible target genes becomes a tool for characterizing the mechanism of action of new inhibitors (Bostian *et al.*, 1997; Hartwell *et al.*, 1997; O'Connor *et al.*, 1997; Weinstein *et al.*, 1997; Shi *et al.*, 1998). Since the cellular machinery is composed of functionally interacting macromolecules, it was confirmed that strains carrying functionally related alterations were hypersensitive to an inhibitor. Similarly, compounds that affect functionally related targets will display their relatedness through their similar pattern of mutant inhibition.

Microarray and proteomic technologies can provide information about the transcriptional and translational responses of cells to inhibitors. The cellular response may be sufficiently unique to provide a signature determining a compound's mechanism of action. Microarray gene expression profiles of strains carrying genetic defects in the target gene can provide a "key" to decipher the microarray patterns of chemical inhibitors (Hartwell *et al.*, 1997; Gray *et al.*, 1998). In all of these approaches, the need for sophisticated statistical analysis of large data sets is necessary to see patterns that may underlie the data (Eisen *et al.*, 1998).

Bacteria pose some technical challenges for transcriptional microarrays due to the low abundance of mRNA (~2% of total RNA), and the inability to simply separate mRNA from rRNA and tRNA in bacteria because of the general lack of poly-A tailing of bacterial mRNA. Nonetheless, successful assessment of bacterial gene expression has been reported using two different methods, oligonucleotide arrays and larger (9–21 kb) DNA segments (Chuang *et al.*, 1993; de Saizieu *et al.*, 1998). Neither of these efforts attempted mRNA isolation or enrichment. Despite the presence of the excess rRNA in the labeled probe, the background was acceptable, and the sensitivity remained high. Using the oligonucleotide array, it was estimated that mRNAs with an abundance of two transcripts per cell were detectable (de Saizieu *et al.*, 1998).

Genomics and Pharmacology

A major barrier in the progression of an antimicrobial compound frequently comes during the testing of a compound for its *in vivo* activity and safety. Historically, such testing is done at relatively late stages in the drug discovery process, because of the need for relatively large amounts of compound and relatively labor-intensive, long-term assays. Microarray technologies and increasing amounts of genomic information for mammalian systems may have the potential to revolutionize drug discovery even further downstream by providing rapid, compound-sparing methods to gain molecular understanding of the *in vivo* effects of compounds at much earlier points in the drug discovery process (Braxton *et al.*, 1998; Debouck *et al.*, 1999; Nuwaysir *et al.*, 1999).

The early prediction and understanding of a compound's potential toxicities are challenges for the future drug discovery process. Several methods

may provide sufficient data to predict biological activities of novel compounds, based on the similarity of their biological profiles to the biological profiles of characterized compounds. For example, the metabolic effects of a compound in mice or rats may be predicted by comparing the expression profile of animals after treatment with a new compound, with a database of information on characterized compounds. Increased sensitivity and molecular resolution of effects should emerge from the use of microarray technology for this purpose. Rapid protein separation and identification provides the ability to look at a large number of the expressed proteins (Blackstock *et al.*, 1999), which may also provide a biological profile for characterization of novel compounds. Pattern matching between the biological profiles of novel compounds and a database of pharmacologically characterized compounds will be needed to provide the link for predicting molecular toxicities.

Other methods have been explored that use surrogate information to provide a biological profile. A well-chosen set of proteins may provide the ability to characterize a compound's binding potential (Kauvar *et al.*, 1998). Alternatively, the expression patterns of both prokaryotic and eukaryotic cell lines, measured in a variety of ways, have been suggested for characterizing a compound's potential toxicities (Todd *et al.*, 1995).

It is increasingly clear that the new genomic technologies, coupled with improvements in computer and database technologies provide the ability to generate large sets of biological data. The ability of the data to provide unique biological characterizations of chemical compounds remains to be determined, but has the potential to revolutionize drug discovery.

Conclusions

Significant challenges remain in the effort to make drug discovery more reliable, more successful and more efficient. Genomics clearly provides a powerful tool for identifying new targets for antimicrobial agents. Methods for rapidly exploiting these targets for high throughput screening are currently in place, though room for improvement still exists. The ability to readily monitor transcription and protein expression patterns is provoking novel uses of genomic-related technologies for improving the difficult, time-consuming and error-prone steps of identifying attractive lead compounds and following an efficient path from lead compound to development candidate. Exploring the range of options that genomic information can provide will undoubtedly lead to improvements in protein structure prediction and in the ability to undertake rational drug design, but also in the ability to find compounds with the appropriate pharmaceutical parameters that allow development, in addition to those that make a good molecular inhibitor.

The recent rate of acquiring new microbial genome information and extending the use of the technology to improve the process of drug discovery has been exceptional. Hopefully, our ability to identify new classes of antimicrobial drugs will exceed the rate at which microbes are becoming resistant.

REFERENCES

Akerley, B.J., Rubin, E.J., Camilli, A., Lampe, D.J., Robertson, H.M. and Mekalanos, J.J. (1998) Systematic identification of essential genes by *in vitro mariner* mutagenesis. *Proc. Natl. Acad. Sci.* **95**, 8927–8932.

Anderson, R.P. and Roth, J.R. (1977) Tandem genetic duplications in phage and bacteria. *Annu. Rev. Microbiol.* **31**, 473–505.

Anderson, R.P. and Roth, J.R. (1979) Gene duplication in bacteria: alteration of gene dosage by sister-chromosome exchanges. *Cold Spring Harb. Symp. Quant. Biol.* **43**, 1083–1087.

Arigoni, F., Talabot, F., Peitsch, M., Edgerton, M.D., Meldrum, E., Allet, E., *et al.* (1998) A genome-based approach for the identification of essential bacterial genes. *Nat. Biotech.* **16**, 851–856.

Baltz, R.H., Norris, F.H., Matsushima, P., DeHoff, B.S., Rockey, P., Porter, G., *et al.* (1998) DNA sequence sampling of the *Streptococcus pneumoniae* genome to identify novel targets for antibiotic development. *Microbial Drug Resistance* **4**, 1–9.

Barrett, J.F., Goldschmidt, R.M., Lawrence, L.E., Foleno, B., Chen, R., Demers, J.P., *et al.* (1998) Antibacterial agents that inhibit two-component signal transduction systems. *Proc. Natl. Acad. Sci.* **95**, 5317–5322.

Bergthorsson, U. and Ochman, H. (1998) Distribution of chromosome length variation in natural isolates of *Escherichia coli. Mol. Biol. Evol.* **1**, 6–16.

Blackstock, W.P. and Weir, M.P. (1999) Proteomics: quantitative and physical mapping of cellular proteins. *Trends Biotechnol.* **17**, 121–7.

Blais, J., Cho, D., Tangen, K., Ford, C., Lee, A., Lomovskaya, O., *et al.* (1999). Efflux pump inhibitors enhance the activity of antimicrobial agents against a broad selection of bacteria. *39th Interscience Conference for Antimicrobial Agents and Chemotherapy*, San Francisco.

Blattner, F.R., Plunkett, G., III, Bloch, C.A., Perna, N.T., Burland, V., Riley, M., *et al.* (1997) The complete genome sequence of *Escherichia coli* K-12. *Science* **277**, 1453–1474.

Bostian, K.A. and Schmid, M.B. (1997) New antibacterial targets and new approaches for drug discovery. In W.-D. Busse, H.-J. Zeiler, H. Labischinski, (eds.), *Antibacterial Therapy: Achievements, Problems and Future Perspectives* Springer, Berlin, pp. 61–68.

Bowe, F., Lipps, C.J., Tsolis, R.M., Groisman, E., Heffron, F. and Kusters, J.G. (1998) At least four percent of the *Salmonella typhimurium* genome is required for fatal infection of mice. *Infect. Immun.* **66**, 3372–3377.

Bowie, J.U. and Oakula, A.A. (1996) Screening method for identifying ligands for target proteins, Scriptgen Pharmaceuticals, Inc., United States Patent, **5**, 585,277.

Boyer, E., Martin, P.K. and Sun, D. (1998). Two separate loci in *Staphylococcus aureus* encode homologues of the alpha subunit of DNA polymerase III. *ASM 98th General Meeting*, Atlanta GA.

Brakstad, O.G. and Maeland, J.A. (1997) Mechanisms of methicillin resistance in staphylococci. *APMIS* **105**, 264–276.

Braxton, S. and Bedilion, T. (1998) The integration of microarray information in the drug development process. *Curr. Opin. Biotechnol.* **9**, 643–649.

Bures, M.G. and Martin, Y.C. (1998) Computational methods in molecular diversity and combinatorial chemistry. *Curr. Opin. Chem. Biol.* **2**, 376–380.

Chamberland, S., Blais, J., Boggs, A.F., Bao, Y., Malouin, F., Hecker, S.J., *et al.* (1995). MC-207,252 abolishes PBP 2a mediated beta-lactam resistance in staphylococci. *35th Annual Interscience Conference of Antimicrobial Agents and Chemotherapy*, San Francisco, CA.

Chambers, H.F. (1999) Penicillin-binding protein-mediated resistance in pneumococci and staphylococci. *J. Infect. Dis.* **179**, S353–S359.

Chee, M., Yang, R., Hubbell, E., Berno, A., Huang, X.C., Stern, D., *et al.* (1996) Accessing genetic information with high-density DNA arrays. *Science* **274**, 610–614.

Chiang, S.L. and Mekalanos, J.J. (1998) Use of signature-tagged transposon mutagenesis to identify *Vibrio cholerae* genes critical for colonization. *Mol. Microbiol.* **27**, 797–805.

Chopra, I., Hodgson, J., Metcalf, B. and Poste, G. (1997) The search for antimicrobial agents effective against bacteria resistant to multiple antibiotics. *Antimicrobial Agents and Chemotherapy* **41**, 497–503.

Chuang, S.E., Daniels, D.L. and Blattner, F.R. (1993) Global regulation of gene expression in *Escherichia coli. J. Bacteriol.* **175**, 2026–2036.

Cirillo, D.M., Valdivia, R.H., Monack, D.M. and Falkow, S. (1998) Macrophage-dependent induction of the Salmonella pathogenicity island 2 type III secretion system and its role in intracellular survival. *Mol. Microbiol.* **30**, 175–88.

Clayton, R.A., White, O. and Fraser, C.M. (1998) Findings emerging from complete microbial genome sequences. *Current Opinion in Microbiology* **1**, 562–566.

Coleman, K., Athalye, M., Clancey, A., Davison, M., Payne, D.J., Perry, C.R., *et al.* (1994) Bacterial resistance mechanisms as therapeutic targets. *J. Antimicrob. Chemother.* **33**, 1091–116.

Coulter, S.N., Schwan, W.R., Ng, E.Y., Langhorne, M.H., Ritchie, H.D., Westbrock-Wadman, S., *et al.* (1998) *Staphylococcus aureus* genetic loci impacting growth and survival in multiple infection environments. *Mol. Microbiol.* **30**, 393–404.

Davies, A.G., Spike, C.A., Shaw, J.E. and Herman, R.K. (1999) Functional overlap between the *mec-8* gene and five *sym* genes in *Caenorhabditis elegans. Genetics* **153**, 117–134.

Davies, J.E. (1997). Origins, acquisition and dissemination of antibiotic resistance determinants. *CIBA Found Symp.* **207**, 1–27.

de Saizieu, A., Certa, U., Warrington, J., Gray, C., Keck, W. and Mous, J. (1998) Bacterial transcript imaging by hybridization of total RNA to oligonucleotide arrays. *Nature Biotech.* **16**, 45–48.

Debouck, C. and Goodfellow, P.N. (1999) DNA microarrays in drug discovery and development. *Nat. Genet.* **21**, 48–50.

Dhandayuthapani, S., Rasmussen, W.G. and Baseman, J.B. (1999) Disruption of gene mg218 of *Mycoplasma genitalium* through homologous recombination leads to an adherence-deficient phenotype. *Proc. Natl. Acad. Sci.* **96**, 5227–5232.

Dove, A. (1999) Drug screening – beyond the bottleneck. *Nat. Biotech.* **17**, 859–863.

Drlica, K. and Zhao, X. (1997) DNA gyrase, topoisomerase IV, and the 4-quinolones. *Microbiol. Mol. Biol. Rev.* **61**, 377–392.

Ecker, D.J. and Crooke, S.T. (1995) Combinatorial drug discovery: which methods will produce the greatest value? *Biotechnology (N Y)* **13**, 351–360.

Ehlert, K. (1999) Methicillin-resistance in *Staphylococcus aureus* – molecular basis, novel targets and antibiotic therapy. *Curr. Pharm. Design* **5**, 45–55.

Eisen, M.B., Spellman, P.T., Brown, P.O. and Botstein, D. (1998) Cluster analysis and display of genome-wide expression patterns. *Proc. Nat. Acad. Sci.* **95**, 14863–14868.

Fleischmann, R.D., Adams, M.D., White, O., Clayton, R.A., Kirkness, E.F., Kerlavage, A.R., *et al.* (1995) Whole-genome random sequencing and assembly of *Haemophilus influenzae* Rd. *Science* **269**, 496–512.

Forterre, P. (1999) Displacement of cellular proteins by functional analogues from plasmids or viruses could explain puzzling phylogenies of many DNA informational proteins. *Mol. Microbiol.* **33**, 457–465.

Fraser, C.M., *et al.* (1997) Genomic sequence of a Lyme disease spirochaete, *Borrelia Burgdorferi. Nature* **390**, 580–586.

Fraser, C.M., Gocayne, J.D., White, O., Adams, M.D., Clayton, R.A., Fleishmann, R.D., *et al.* (1995) The minimal gene complement of *Mycoplasma genitalium. Science* **270**, 397–403.

Galperin, M.Y., Walker, D.R. and Koonin, E.V. (1998) Analogous enzymes: independent inventions in enzyme evolution. *Genome Research* **8**, 779–790.

Giaever, G., Shoemaker, D., Jones, T.W., Liang, H., Winzler, E.A., Astromoff, A., *et al.* (1999) Genomic profiling of drug sensitivities via induced haploinsufficiency. *Nature Genetics* **21**, 278–283.

Gootz, T.D. (1990) Discovery and development of new antimicrobial agents. *Clin. Microbiol. Rev.* **3**, 13–31.

Gray, N.S., Wodicka, L., Thunnissen, A.M., Norman, T.C., Kwon, S., Espinoza, F.H., *et al.* (1998) Exploiting chemical libraries, structure, and genomics in the search for kinase inhibitors. *Science* **281**, 533–8.

Hammond, R.A., Barnes, M.H., Mack, S.L., Mitchener, J.A. and Brown, N.C. (1991) *Bacillus subtilis* DNA polymerase III: complete sequence, overexpression, and characterization of the *polC* gene. *Gene* **98**, 29–36.

Hartwell, L.H., Szankasi, P., Roberts, C.J., Murray, A.W. and Friend, S.H. (1997) Integrating genetic approaches into the discovery of anticancer drugs. *Science* **278**, 1064–1068.

Heithoff, D.M., Conner, C.P. and Mahan, M.J. (1997) Dissecting the biology of a pathogen during infection. *Trends Microbiol.* **5**, 509–513.

Himmelreich, R., Hilbert, H., Plagens, H., Pirkl, E., Li, B.-C. and Herrmann, R. (1996) Complete sequence analysis of the genome of the bacterium *Mycoplasma pneumoniae. Nucleic Acids Research* **24**, 4420–4449.

Hsieh, P.C., Siegel, S.A., Rogers, B., Davis, D. and Lewis, K. (1998) Bacteria lacking a multidrug pump: a sensitive tool for drug discovery. *Proc. Natl. Acad. Sci.* **95**, 6602–6606.

Huang, W.M. (1996) Bacterial diversity based on type II DNA topoisomerase genes. *Ann. Rev. Genet.* **30**, 79–107.

Isaacson, R.E. (1997) Novel targets for antibiotics – an update. *Exp. Opin. Invest. Drugs* **6**, 1009–1017.

Itaya, M. (1995) An estimation of minimal genome size required for life. *FEBS Lett.* **362**, 257–260.

Karaolis, D., Reeves, P., Kaper, J., Boedeker, E., Bailey, C and Johnson, J. (1998) A *Vibrio cholerae* pathogenicity island associated with epidemic and pandemic strains. *Proc. Natl. Acad. Sci. USA* **95**, 3134–3139.

Kauvar, L.M., Villar, H.O., Sportsman, J.R., Higgins, D.L. and Schmidt, D.E., Jr. (1998) Protein affinity map of chemical space. *J. Chromatogr. B Biomed. Sci. Appl.* **715**, 93–102.

Koonin, E.V. (1997) Genome sequences: genome sequence of a model prokaryote. *Current Biol.* **7**, R656–R659.

Kunst, F., *et al.* (1997) The complete genome sequence of the gram-positive bacterium *Bacillus subtilis. Nature* **390**, 249–256.

Lee, C. (1996) Pathogenicity islands and the evolution of bacterial pathogens. *Infect. Agents Dis.* **1**, 1–7.

Levy, S.B. (1998) The challenge of antibiotic resistance. *Scientific American* **278**, 46–53.

Link, A., Phillips, D. and Church, G. (1997) Methods for generating precise deletions and insertions in the genome of wild-type *Escherichia coli*: application to open reading frame characterization. *J. Bacteriol.* **179**, 6228–6237.

Lomovskaya, O., Lee, A., Hoshino, K., Ishida, H., Mistry, A., Warren, M., *et al.* (1999) Use of a genetic approach to evaluate the consequences of inhibition of efflux pumps in *Pseudomonas aeruginosa. Antimicrobial Agents and Chemotherapy* **43**, 1340–1346.

Maniloff, J. (1996) The minimal cell genome: on being the right size. *Proc. Natl. Acad. Sci.* **93**, 10004–10006.

Marians, K.J. (1996). Replication fork propagation. In F.C. Neidhardt, (ed.) *Escherichia coli and Salmonella typhimurium*, ASM Press, pp. 749–763.

Martin, P., Li, T., Sun, D., Biek, D. and Schmid, M. (1999) Role in cell permeability of an essential two-component system in *Staphylococcus aureus. J. Bacteriol.* **181**, 3666–3673.

Matassova, N.B., Rodnina, M.V., Endermann, R., Kroll, H.P., Pleiss, U., Wild, H., *et al.* (1999) Ribosomal RNA is the target for oxazolidinones, a novel class of translational inhibitors. *RNA* **5**, 939–46.

McMurray, L.M., Oethinger, M. and Levy, S.B. (1998) Triclosan targets lipid synthesis. *Nature* **394**, 531–532.

Mei, J.M., Nourbakhsh, F., Ford, C.W. and Holden, D.W. (1997) Identification of *Staphylococcus aureus* virulence genes in a murine model of bacteraemia using signature-tagged mutagenesis. *Mol. Microbiol.* **26**, 399–407.

Miklos, G.L.G. and Rubin, G.M. (1996) The role of the genome project in determining gene function: insights from model organisms. *Cell* **86**, 521–529.

Mushegian, A. and Koonin, E. (1996) A minimal gene set for cellular life derived by comparison of complete bacterial genomes. *Proc. Natl. Acad. Sci.* **93**, 10268–10273.

Nuwaysir, E.F., Bittner, M., Trent, J., Barrett, J.C. and Afshari, C.A. (1999) Microarrays and toxicology: the advent of toxicogenomics. *Mol. Carcinog.* **24**, 153–159.

O'Connor, P.M., Jackman, J., Bae, I., Myers, T.G., Fan, S., Mutoh, M., *et al.* (1997) Characterization of the p53 tumor suppressor pathway in cell lines of the National Cancer Institute anticancer drug screen and correlations with the growth-inhibitory potency of 123 anticancer agents. *Cancer Res.* **57**, 4285–300.

Orengo, C.A., Jones, D.T. and Thornton, J.A. (1994) Protein superfamilies and domain superfolds. *Nature* **372**, 631–634.

Pacitti, D.F., Barnes, M.H., Li, D.H. and Brown, N.C. (1995) Characterization and overexpression of the gene encoding *Staphylococcus aureus* DNA polymerase III. *Gene* **165**, 51–56.

Posfai, G., Blattner, F., Kirkpatrick, H. and Koob, M. (1997) Versatile insertion plasmids for targeted genome manipulations in bacteria: isolation, depletion, and rescue of the pathogenicity island LEE of the *Escherichia coli* O157:H7 Genome. *J. Bacteriol.* **179**, 4426–4428.

Schimmel, P., Tao, J. and Hill, J. (1998) Aminoacyl tRNA synthetases as targets for new anti-infectives. *The FASEB Journal* **12**, 1599–1609.

Schmid, M.B., Kapur, N., Isaacson, D.R., Lindroos, P. and Sharpe, C. (1989) Genetic analysis of temperature-sensitive lethal mutants of *Salmonella typhimurium*. *Genetics* **123**, 625–633.

Shi, L.M., Fan, Y., Myers, T.G., O'Connor, P.M., Paull, K.D., Friend, S.H., *et al.* (1998) Mining the NCI anticancer drug discovery databases: genetic function approximation for the QSAR study of anticancer ellipticine analogues. *J. Chem. Inf. Comput. Sci.* **38**, 189–99.

Shoemaker, D.D., Lashkari, D.A., Morris, D., Mittmann, M. and Davis, R.W. (1996) Quantitative phenotypic analysis of yeast deletion mutants using a highly parallel molecular bar-coding strategy [see comments]. *Nat. Genet.* **14**, 450–6.

Silverman, R.B., Tran, N., Sinha, R.K. and Neuhaus, F.C. (1993) The oxazolidinone antibacterial agent DuP 105 does not act on cell wall biosynthesis or on a beta-lactamase. *Biochem. Biophys. Res. Commun.* **195**, 1077–80.

Smith, V., Chou, K., Lashkari, D., Botstein, D. and Brown, P.O. (1996) Functional analysis of the genes of yeast chromosome V by genetic footprinting. *Science* **274** 2069–2074.

Stewart, M.J., Parikh, S., Xiao, G., Tonge, P.J. and Kisker, C. (1999) Structural basis and mechanism of enoyl reductase inhibiton by triclosan. *J. Mol. Biol.* **290**, 859–865.

Todd, M.D., Lee, M.J., Williams, J.L., Nalezny, J.M., Gee, P., Benjamin, M.B., *et al.* (1995) The CAT-Tox (L) Assay: A Sensitive and Specific Measure of Stress-Induced Transcription in Transformed Human Liver Cells. *Fundamental and Applied Toxicology* **28**, 118–128.

Tomasiewicz, H.G. and McHenry, C.S. (1987) Sequence analysis of the *Escherichia coli dnaE* gene. *J. Bacteriol.* **169**, 5735–5744.

Umezawa, H. (1982) Low-molecular-weight enzyme inhibitors of microbial origin. *Ann. Rev. Microbiol.* **36**, 75–99.

Urban, C., Mariano, N., Mosinka-Snipas, K., Wadee, C., Chahrour, T. and Rahal, J.J. (1996) Comparative in-vitro activity of SCH 27899, a novel everninomicin, and vancomycin. *J. Antimicrob. Chemother.* **37**, 361–364.

Van Veen, H.W. and Konings, W.N. (1997) Drug efflux proteins in multidrug resistant bacteria. *Biol. Chem.* **378**, 769–777.

Weinstein, J.N., Myers, T.G., O'Connor, P.M., Friend, S.H., Fornace, A.J., Jr., Kohn, K.W., *et al.* (1997) An information-intensive approach to the molecular pharmacology of cancer. *Science* **275**, 343–349.

14. Peptide Antibiotics

Lijuan Zhang and Robert E.W. Hancock

Department of Microbiology and Immunology, University of British Columbia, #300-6174, University Boulevard, Vancouver, BC, V6T 1Z3, Canada

INTRODUCTION

Many peptide antibiotics have been described in the past 55 years (Hancock *et al.*, 1995; Kleinkauf and von Dohren, 1988; Perlman and Bodansky, 1971). These can be fit into two broad classes, non-ribosomally synthesized peptides, such as the gramicidins, polymyxins, bacitracins, glycopeptides etc. and ribosomally synthesized (natural) peptides. The former are synthesized, usually in bacteria, by large complexes of modular enzymes termed peptide synthetases, and are often substantially modified. In contrast, the latter are produced as pro-peptides, using the normal process of protein synthesis. They are found in all species of life (including bacteria) as a major component of the natural host defense molecules of these species.

NON-RIBOSOMALLY SYNTHESIZED PEPTIDES

The non-ribosomally synthesized peptides have been well described to date (Kleinkauf and von Dohren, 1988; Perlman and Bodansky, 1971) and will be briefly summarized here with emphasis on their clinical importance, similarities in function to the natural peptides, and future prospects.

Non ribosomally synthesized peptides can be defined as peptides, derived from bacteria, fungi or streptomycetes, which contain two or more moieties derived from amino acids (Kleinkauf and von Dohren, 1988; Perlman and Bodansky, 1971).

Rather than being synthesized on ribosomes in the normal method of proteins, all of these peptidic molecules are made on multi-enzyme complexes. Interestingly, many of the antibiotics used in our society fall into this class. For example, the natural penicillins comprise structures based on monosubstituted acetic acid, L-cysteine and D-valine residues, while cephalosporin C, the basic building block of many semi-synthetic cephalosporins contains substituted D-α-aminoadipic acid, L-cysteine, α,β-dehydrovaline and acetic acid units. In addition, the glycopeptide antibiotics, including vancomycin and teichoplanin, have sugar-substituted peptide backbones.

However, given the enormous volume of literature on the above two classes, as well as the large number of peptides in this class that have never found their way into the clinic, we are restricting ourselves here to the high molecular weight peptide antibiotics which have been used clinically. These include bactracin, the cationic lipopeptide polymyxin B, and another cationic cyclic decapeptide, gramicidin S. All three are effective antimicrobials and are very highly utilized in topical preparations, often in combination. Colymycin, the methosulphate derivative of the cationic lipopeptide colistin (also called polymyxin E), has been utilized quite successfully in aerosol formulation against *Pseudomonas aeruginosa* lung infections (Jensen *et al.*, 1987; Diot *et al.*, 1997) and has found its way into other clinical uses. Because of its chemical modification as a pro-drug, it is considerably less toxic than its parent drug.

Pathogenesis Inc. is currently planning clinical trials for this agent in patients with cystic fibrosis.

Biosynthesis

There is a strong consensus that non-ribosomal peptide synthesis is performed according to the multiple carrier thiotemplate mechanism (Stein *et al.*, 1996). According to this mechanism, a series of very large multi-functional, modular peptide synthetases perform the assembly and modification of amino acid residues in an ordered fashion. A single peptide synthetase gene can be as large as 13 kB (4,300 amino acids), and contain 4–6 modules [resulting in the addition of 4–6 residues, e.g. the gramicidin S biosynthetic gene *grsB* (Schneider *et al.*, 1998)]. The minimal module is capable of recognizing and activating one amino acid or hydroxy acid residue, stabilizing the activated residue as a thioester, and polymerizing it in its correct sequence to the previously added residue (with the aid of a covalently attached co-factor, 4′-phospho-pantotheine). These versatile enzymes can utilize hydroxy-, L-, D-, or unusual-amino acids. Additional chemical diversity of peptide products can be engendered by further modification, by these modular enzymes, involving heterocyclic ring formation, N-methylation, acylation, or glycosylation. More than 300 different residues have been demonstrated to result from these processes.

Activities and Mechanisms of Action

Three of the peptides currently used clinically are cationic in nature with the polymyxins having a net charge of +5 and gramicidin S having a charge of +2. Polymyxins tend to demonstrate selectivity for Gram-negative bacteria. In contrast, gramicidin S has traditionally been considered selective for Gram-positives. However, if MIC measurements are done in the correct fashion, gramicidin S has excellent activity against Gram negative bacteria and the fungus *Candida albicans*, although it has a very poor therapeutic index (Kondejewski *et al.*, 1996ab). Both classes of cationic peptidic antibiotics seem to act in the same way on cells as the cationic antimicrobial peptides described below

(i.e. self-promoted uptake across the cytoplasmic membrane followed by depolarization of the cytoplasmic membrane).

In contrast, the Gram-positive specific antibiotic bacitracin works by inhibiting the transfer of cytoplasmically synthesized peptidoglycan precursors to bactoprenol pyrophosphate. Another antibiotic peptide of non-ribosomal origin, streptogramin, is a protein synthesis inhibitor, whereas β-lactams act on cell wall synthesis and glycopeptides by binding to cell walls. Thus it seems that peptidic drugs can, in principle, act in any process that involves proteins or the binding of amino acids.

Future Prospects

Although most of the non-ribosomal antimicrobial peptides described here have been known for decades, many others with antibiotic activity have been described in the literature, and these peptides offer a potentially rich source of novel antimicrobials. Indeed these may be one of the major untapped sources of novel drugs, given their enormous potential variety of mechanisms of action. To exploit these, three approaches are being used.

The first approach involves the modification of existing peptides (and presumably also isolation of novel peptides from nature and modification of these). For example, the streptogramins were discovered in the 1950s as a family of cyclic peptides. They tended to demonstrate reasonably high antibacterial activities against Gram-positive bacteria, but their utility was limited by their low solubility in aqueous solvents. This problem has now been solved by developing two water-soluble, semisynthetic streptogramins, dalfopristin and quinpristin. These drugs were combined as a combination parenteral agent (Synercid) and are currently undergoing Phase III clinical trials against antibiotic-resistant Gram-positive bacteria. A second, rather exciting approach, involves exploiting the modular nature of synthesis of these non-ribosomal peptide antibiotics. Schneider *et al.* have demonstrated that one can assemble a novel combination of peptide synthesis modules and obtain a new structure (Schneider *et al.*, 1998). This indicates that there is great potential for obtaining significant chemical diversity in the backbone amino acids or their modifications, and a combinatorial approach to generating diversity (i.e. mixing and matching modules) is possible.

The third approach is to use these structures as templates for chemical synthesis and diversity as done with the gramicidins. Variants of Gramicidin S with altered ring size, charge, amino acid sequences, hydrophobicity, etc., have been constructed and demonstrated to have an improved selectivity for bacteria over mammalian cells (Kondejewski et al., 1996).

RIBOSOMALLY SYNTHESIZED CATIONIC ANTIMICROBIAL PEPTIDES

The ribosomally-synthesized peptides, which are largely basic in nature, represent a new opportunity for the biologically oriented pharmaceutical chemist and will be described in more detail with emphasis on the role in natural host defenses (as nature's antibiotics) and the clinical potential of peptides derived from these natural peptides. Most cationic peptides have two unique features in that they are poly-cationic with a net positive charge of more than +2 and they tend to fold into structures that are amphipathic with both a hydrophobic face and a hydrophilic face (Nicolas and Mor, 1995; Hancock, 1997, 1998)

Structures and Diversity

The naturally occurring peptides are generally from 12 to 50 amino acids long and most of them fit into four major groups, namely, α-helices, extended helices, β-sheets and loop structures (Hancock, 1997; Hancock and Lehrer, 1998). However, one has to keep in mind that there are peptides that do not seem to fall easily into any of these groups and there are many peptides that have not been classified structurally. So far more than 500 cationic peptides have been isolated from bacteria, fungi, plants, insects, birds, crustaceans, amphibians, mammals and even man. We will only describe here certain illustrative examples of peptides from each structural group.

α-Helical peptides

The α-helical peptides are short linear polypeptides of less than 40 amino acids and are devoid of disulfide bridges (Boman, 1995). These polypeptides vary considerably in chain length, hydrophobicity, and overall distribution of charges, but share a common amphipathic α-helical structure that is induced upon association with lipid bilayers (Segrest et al., 1990). In free solution, they are, generally speaking, unstructured and pre-folding them into an α-helix actually decreases activity (Houston et al., 1998).

The amphibian dermal glands secrete several families of microbicidal peptides (Bevins and Zasloff, 1990) and most of them can fold into an amphipathic α-helical structure. Magainins were isolated from the skin of the African clawed frog (Xenopus laevis) in 1987 and turned out to be an exciting group of peptides (Zasloff, 1987). They are α-helical peptides of 23 residues and have a broad spectrum of killing activity against Gram-positive and Gram-negative bacteria, fungi, and protozoa. In addition they are non-hemolytic and specifically kill tumor cells (Cruciani et al., 1991). These peptides have been well characterized structurally, and chemical synthesis of magainin analogs has helped our understanding of structure-function relationship of cationic peptides, which in turn has led to the design of more potent magainin variants with expanded antimicrobial spectra (Dathe et al., 1997; Wieprecht et al., 1997). Also isolated from the skin of a South American frog, Phyllmoedusa sauvagii, was dermaseptin, a fungicidal peptide of 34 residues (Mor et al., 1991a, 1991b). A synthetic peptide with dermaseptin residues 1–18 and with a C-terminal amide was found to form an α-helix in membrane-mimicking solvents and to be a more potent antibiotic than the natural peptide (Mor and Nicolas, 1994). In addition to their fungicidal activity, dermaseptins are active against a wide spectrum of microorganisms including bacteria, protozoa and yeast, but killing efficiency varies among the different dermaseptin variants.

Insects are also known to produce a large array of biologically active peptides. Cecropins A and B were initially found in the silk moth Hyalophora cecropia (Steiner et al., 1981), but have also been isolated from the fresh fly (sarcotoxin I), Drosophila, porcine intestine (Lee *et al.*, 1989) and from the blood cells of a marine *protochordate* (Zhao *et al.*, 1997). Although the pig cecropin P1 differs from the insect forms by not having an amidated C-terminus, and is also shorter than the others, all cecropins have a strongly amphiphathic N-terminal region and a long hydrophobic stretch at the C-terminus, joined by a hinge region with proline or glycine residues or both. Cecropins are only toxic to bacteria and thus are particularly of pharmaceutical interest. Melittin is a basic peptide with 26 residues in an amphipathic sequence with most charges at the C-terminus. It is the major component of the European honeybee *Apis mellifera* venom (Habermann and Jentsch, 1967). The peptide is well known for its strong hemolytic potency, but it also kills other eukaryotic and prokaryotic cells.

Hybrids of two different natural peptides provide advantages in the search for potential therapeutic agents. Cecropin and melittin hybrids, comprising the cationic N-terminal sequence of cecropin A followed by the hydrophobic N-terminal sequence of melittin, showed much higher antibiotic activity than cecropin, while being devoid of the hemolytic activity associated with the C-terminal cationic sequence in melittin (Boman *et al.*, 1989; Andreu *et al.*, 1992).

Lactic acid bacteria are known to secrete bacteriocins that may be cationic and normally contain between 30 and 60 amino acid residues (Sahl *et al.*, 1995). The bacteriocins have received considerable attention because one of them, nisin was approved as a natural food preservative decades ago. Like many bacteriocins, nisin is a lantibiotic containing many post-translationally modified amino acids, some of which (e.g., lanthionine) are cyclized by the formation of a thioether bond between an alanine and a cysteine residue 3 to 4 amino acids apart. Despite this extensive modification, nisin forms an α-helical core structure in free solution with additional β-structure induction upon lipid insertion (van der Hooven *et al.*, 1993). One subgroup of bacteriocins includes lactococcin G consisting of two peptides, α- and β-lactococcin G, both of which can form an α-helical structure upon interaction with negatively charged membrane lipids (Hauge *et al.*, 1998).

Extended helical peptides

Various proline-rich antimicrobial peptides have been recovered from the leukocytes of different mammals, from marine invertebrates and from the haemolymph of many insects. The proline-rich apidaecins, originally isolated from honeybees and wasps, have unusually high contents of proline (29%) and argine (17%) (Casteel *et al.*, 1989). These peptides have a circular dichroism spectra in lipid-like environments (liposomes or SDS), which are similar to that of a model poly-L-proline type II helix structure, basically an extended helix. Apidaecins are highly active against Gram-negative bacteria but show hardly any activity against Gram-positive bacteria. Indolicidin is a peptide of 13 amino acids isolated from bovine neutrophils (Selsted *et al.*, 1992). It has five tryptophan and three proline residues. Like other proline-rich peptides, it has been proposed to form an extended helix structure due to its circular dichroism spectrum, however the high trptophan content makes interpretation of these spectra somewhat complex (Falla *et al.*, 1996). Indolicidin is equally active against Gram-negative and Gram-positive bacteria. Two bovine neutrophil peptides, Bac5 and Bac7, and one porcine peptide PR39, were found to have even higher contents of proline (45–49%) and arginine (24–29%) (Frank *et al.*, 1990; Agerberth *et al.*, 1991). These peptides show good activities against both Gram-negative and -positive bacteria.

β-sheet peptides

Defensins are cationic, cysteine-rich peptides that display broad-spectrum antimicrobial activity. They were first found in the granules of phagocytic cells, where they play a role in nonoxidative microbial killing during inflammation. Defensins have a structure characterized by a conserved cysteine motif that forms three-disulfide linkages imposing a characteristic triple stranded β-sheet structure (Kagan *et al.*, 1994; Ganz and Lehrer, 1995). The vertebrate defensin family contains two branches, designated α- and β-defensins. Humans produce six α-defensins and two β-defensins. The α-defensins (HNP1-4) are stored in the cytoplamic granules of neutrophils and make up over 5% of the total cellular protein (Selsted *et al.*, 1985). Because neutrophils are attracted to and accumulate in infected tissue sites, they provide an extremely appropriate defensin delivery system. The α-defensins have also been isolated from the neutrophils of rabbits, rats, and guinea pigs; and the pulmonary macrophages of rabbits. Two other α-defensins HNP-5 and HNP-6 are expressed by small intestinal paneth cells of humans and mice (Jones and Bevins, 1992, 1993). A high-resolution crystal structure for human neutrophil peptide α-defensin-3 revealed that the peptide adopted a triple-stranded β-sheet configuration stabilized by three intramolecular disulfide bonds, and that it formed basket-shaped dimers with a hydrophobic base and a polar top (Hill et al., 1991).

The β-defensins are slightly larger and differ in the placement and connectivity of their six conserved cysteine residues (Selsted et al., 1993). The structure of the β-defensins closely resembles that established for the α-defensins, and these peptides have been isolated from humans, birds and cows. The gene encoding for human β-defensin-1, HBD-1, was cloned and its product was demonstrated to be present in the lung of humans (Goldman et al., 1997). This peptide is salt-sensitive and seems to be inactive at the surface of the epithelium of cystic fibrosis patients due to excessive salt secretion in these patients (Goldman et al., 1997). Data obtained to date indicate that HBD-1 is mainly produced in the kidney and the female reproductive tract. But in situ hybridization also showed that the HBD-1 mRNA is present in the epithelial layers of the loops of Henle, the vagina, ectocervix, endocervix, uterus, and fallopian tubes in the female reproductive tract, and in the distal tubules and collecting ducts of the kidney (Valore et al., 1998). This broad distribution is consistant with the suggestion the epithelial defensins are an important component of the antimicrobial barrier of mucosal surfaces. Another human β-defensin, HBD-2, is made by inflamed skin (Harder et al., 1997). Defensins are also expressed in the trachea, tongue, intestine and respiratory epithelium. Tracheal antimicrobial peptide (TAP), a β-defensin, was isolated from bovine trachea and shown to have bactericidal activity against a broad array of microorganisms (Diamond et al., 1991). A similar but distinct β-defensin called lingual antimicrobial peptide (LAP) was isolated from cow tongue (Schonwetter et al.,

1995). Genes encoding these peptides are up regulated at the transcriptional level in response to bacterial LPS and inflammatory cytokines although other defensins appear to be made constitutively (Diamond et al., 1996; Russelll et al., 1996).

Both plants and insects also produce defensins whose structures are similar to each other but distinct from their mammalian counterparts. The plant defensins are known as thionins which are basic and cysteine-rich low-molecular-mass polypeptides of about 5 kDa found in wheat and barley as well as a variety of plant species (Garcia-Olmedo et al., 1989; Bloch and Richardson, 1991). The thionins are toxic to animals, some yeast strains, fungi, bacteria, plant cells and insect larvae, although most thionins are preferentially fungicidal. Both α- and β-thionins have been known for decades and γ-thionins were isolated in early 90's followed by the discovery of ω-thionin in 1996 (Méndez et al., 1996). The NMR structure of γ-thionins are highly similar to insect defensins such as drosomycin, a 44-residue peptide produced by the larvae and adult stage of drosophila in response to septic injury (Fehlbaum et al., 1994; Terras et al., 1992). Both thionins and drosomycin have eight cysteines that form four intramolecular disulfide bridges that stabilize a triple-stranded antiparallel β-sheet structure. This strongly suggests that plants and insects may have evolved similar defense mechanisms in their innate immunity.

A unique family of cationic peptides comprising tachyplesins (I–III) and polyphemusins (I, II) was isolated from the hemocyte of Japanese horseshoe crabs (Tachypleus tridentatus, Tachypleus gigas and Carcinoscorpius rotundicauda) and American horseshoe crabs (Limulus polyphemus) (Miyata et al., 1989; Muta et al., 1990; Nakamura et al., 1988). All peptides in this family contain four cysteine residues and form a rigid structure containing anti-parallel β-sheets connected by a β-turn and two disulfide bridges resulting in a stable structure resistant to low pH and high temperature. These peptides have attracted special attention due to their anti-viral and anti-endotoxin activities. The protegrins are five analogous antimicrobial peptides of 16–18 residues isolated from porcine leukocytes (Kokryakov et al., 1993). They also have two disulfide bridges and share structural homology and some sequence similarities with the tachyplesins. Protegrin-1 is able to form anion selective channels like human defensin-induced channels (Mangoni *et al.*, 1996).

Looped peptides

Some peptides have a single intramolecular disulfide bond and thus form a loop structure. Depending on the positions of the two cysteines, the size of the loop and tail can vary. Thus bactenecin has a very short tail (Frank *et al.*, 1990), while brevinins and esculentins have relatively long tails (Morikawa *et al.*, 1992; Simmaco *et al.*, 1993). Bactenecin was the first peptide in this group, and has been isolated from bovine and ovine neutrophils. It contains only 12 amino acids, 4 of which are arginines. It has a central loop of 9 residues held together by a disulfide bond, with only one and two residues on either side. Circular dichroism spectroscopy suggests a β-turn structure (Romeo *et al.*, 1988). Bactenecin showed activity against bacteria but also selective toxicity to neuronal and glial cells (Radermacher *et al.*, 1993).

The brevinins and esculentins are characterized by the presence of two cysteine residues at positions 1 and 7 residues from the carboxyl terminus (Morikawa *et al.*, 1992). These peptides were isolated from the skin of *Rana brevipoda* and *Rana esculenta*, frogs of the family Ranidae, and they are highly active against *E. coli, Bacillus megaterium* and *S. aureus* (Simmaco *et al.*, 1994).

Other peptides

Tailor and coworkers recently isolated a group of novel antimicrobial peptides from plant seed (Tailor *et al.*, 1997). These novel peptides, called Ib-AMP1-4, are 20 amino acids long and are the smallest plant-derived antimicrobial peptides isolated to date. They have been shown to inhibit the growth of a range of fungi and bacteria, but are not cytotoxic to cultured human cells. The Ib-AMPs are highly basic and contain four cysteine residues that form two intramolecular disulfide bonds. Interestingly, these peptides do not have an amphipathic nature. The structure of these

peptides has been studied using circular dichroism and two-dimensional NMR. Results reveal that the peptide may include a β-turn, but show no evidence for either α-helical or β-sheet structure over a range of temperatures and pH (Patel *et al.*, 1998). A sequence search of the protein sequence data-base for Ib-AMPs fails to identify any protein/peptide with statistically significant homology. However, the structure of Ib-AMPs, especially the cysteine arrangements, is very close to that of the α-conotoxins isolated from marine snails (Gray *et al.*, 1984; Martinez *et al.*, 1995). The α-conotoxins are known to cause muscular paralysis by antagonising acetylcholine binding to the postsynaptic acetycholine receptor at neuromuscular junction (Hucho *et al.*, 1996; Martinez *et al.*, 1995). Thus the mechanism of action of Ib-AMPs is thought to be different from other amphipathic peptides and being unlikely to function by nonspecific membrane disruption, but are postulated to interact with a membrane bound receptor.

As mentioned above, bacteria also produce cationic peptides that have been tranditionally called bacteriocins. Lantibiotics, a subgroup of these bacteriocins, constitute an unusual class of biologically active peptides that contain both dehydrated amino acids and lanthionine residues, forming intramolecular thioether rings (Jung, 1991). These monosulphide bridges determine the characteristic polycyclic structures of lantibiotics and their intrachain position has been used to group these peptides into linear and circular lantibiotics (Jung, 1991). It is known that a variety of Gram-postive bacteria produce lantibiotics, including *Bacillus, Lactococcus, Lactobacillus, Staphylococcus, Streptococcus, and Streptomyces* sp. (Marugg *et al.*, 1992; Fremaux *et al.*, 1993; Holo *et al.*, 1991). Most, if not all, bacteriocins from both Gram-negative and Gram-positive bacteria are ribosomally synthesized with a short N-terminal leader sequence that does not always function as a signal peptide. Instead several amino acid residues of the precursor lantibiotic are enzymatically modified, whereafter secretion and processing of the leader peptide takes place, yielding the active antimicrobial peptide, and it seems that peptide export involves ABC transporters (Fath and Kolter, 1993, de Vos *et al.*, 1995).

Mechanisms of Action of Cationic Peptides

The most effective cationic peptides have affinities for lipolysaccharide (LPS) that are at least three orders of magnitude higher than those for the native divalent cations Ca^{2+} or Mg^{2+}. Thus they competitively displace these ions and, being relatively bulky, disrupt the normal barrier property of the outer membrane. The affected membrane is thought to develop transient ''cracks'' which permit passage of a variety of molecules, including hydrophobic compounds and small proteins and/or antimicrobial compounds, and, more importantly, promote the uptake of the perturbing peptide itself (hence the term ''self-promoted uptake'') (Hancock, 1997). This mechanism explains both how cationic peptides bind to and inhibit endotoxin and how they act in synergy with conventional antibiotics.

Exactly how cationic peptides cause bacterial cell death is as yet fully understood. The bacterial cytoplasmic membrane has been proposed by many researchers to be the target. Interaction of cationic peptides with bacterial membranes can lead to subsequent membrane depolarization and dissipation of the membrane potential, which may invariably have secondary effects on bacterial metabolism leading to simultaneous cessation of DNA, RNA and protein synthesis and respiration eventually to bacterial cell death (Lehrer et al., 1989). However, some peptides do not dissipate membrane potential at their MICs and it must be stressed that dissipation of membrane potential is not, per se, a lethal event (e.g., the uncoupler dinitrophenol is bacterostatic not bactericidal). Thus we have recently proposed that for some peptides, their effects on the membrane potential is a manifestation of their passage across the membrane and that the actual target may be something else, such as binding to and precipitation of cellular nucleic acids (Wu et al., 1999ab).

Nisin has also been proposed to act by inserting into the cytoplasmic membrane and triggering the activity of bacterial murein hydrolases, resulting in damage or degradation of the peptidoglycan and lysis of cells (Jack et al., 1995). Mammalian defensins have been shown to permeabilize the outer membrane of several Gram-negative species, however, bacterial death only occurred after permeabilization of the inner membrane of E. coli

(Lehrer et al., 1989). Support for this was provided by the observation that S. aureus exposed to defensins developed characteristic mesosome-like (cytoplasmic membrane invaginations) structures but did not show remarkable changes in the cell wall (Shimoda et al., 1995).

The mechanism of peptide-membrane interactions has been well studied. Generally electrostatic interactions between the cationic peptide and the negatively charged membrane allows the peptide to fold into an amphipathic secondary structure with the polar face associating with the lipid head groups and the hydrophobic face somewhat embedded into the lipid acyl chain region of the membrane. The formation of channels has been documented in vitro for many peptides, although in relatively few cases has solid evidence been presented which links channel formation to bactericidal activity. The pores appear to be composed of a dynamic, peptide-lipid supramolecular complex, allowing the mutually coupled transbilayer transport of ions, lipids and peptides. At very high peptide:lipid ratios, it was demonstrated that a substantial fraction of magainin re-orients perpendicularly to the membrane, and pores could be detected by neutron in-plane scattering. The pores were formed by a toroidal mechanism, in which the membrane phosphoilipids bend backwards upon themselves after magainin monomers have expanded the bilayer's head-group region (Ludtke et al., 1996). In contrast Mangavel et al. proposed that the peptides were bound as single-stranded α-helices orientated parallel to the bilayer surface (Mangavel et al., 1998). In the anchoring of phospholipid head groups around the peptides, the lipid molecules were organized around the hydrophilic face of the α-helices like ''wheat grains around an ear'' and protrude outside the bilayer towards the solvent. The axis of the peptide helices remained parallel to the membrane surface even when an electrical potential was applied and did not reorient to give rise to a bundle of helix monomers. Such a lipid arrangement was able to generate transient structural defects responsible for the membrane permeability enhancement (Mangavel et al., 1998).

Studies on melittin demonstrated pores formation with variable half lives (from msec to secs) and conductance in the membrane and the pore size increases with the peptide-to-lipid ratio (Pawlak et al., 1991; Matsuzaki et al., 1997; Rex and

Schwarz, 1998, Wu *et al.*, 1999b). Upon the disintegration of the pore, a fraction of the peptides become translocated across the bilayer. This suggests that the pore may be an intermediate structure during the transbilayer transport of the peptide (Matsuzaki *et al.*, 1997). Some authors have interpreted the conductance events observed in model membranes as true pores, and have invoked the barrel-stave model (in which the peptide forms the stave of a barrel whose size is determined by the number of peptides involved in the formation of a barrel-like pore). However the transient nature of the channels, the rather small amount of evidence for peptides aligning perpendicular to the membrane, and the observation that similar channels form with peptides of different structures and length, e.g., linear or cyclic, and even 12 mers that are too small to span the membrane, do not favor this interpretation. Our own data with many different types of peptides indicated that peptides may not form true pores but rather form micellar aggregates that contain ion pathways across the membrane (Wu *et al.*, 1999b). The peptide-induced channel formation has also been demonstrated using artificial membranes with cecropins (Christensen *et al.*, 1988), defensins (Kagan *et al.*, 1990) and indolicidin (Falla *et al.*, 1996). Thus α-helical, β-sheet and extended-helix peptides are all able to form voltage-gated or induced, cation-selective ion channels in planar lipid bilayer or liposomal membranes.

In contrast to pore formation, an alternative hypothesis is that the peptides cluster at membrane surface and cause a cooperative permeabilization of the cytoplasmic membrane, a phenomenon called the carpet effect or detergent-like activity (Pouny *et al.*, 1992; Shai *et al.*, 1995). This hypothesis suggests a catastrophic, sudden permeabilization of the cytoplasmic membrane. However, this does not explain a gradual permeabilization over a 10-fold range of concentrations straddling the MIC for certain peptides (Wu *et al.*, 1999ab). In order to explain the graded effects of peptides some researchers proposed an all-or-none mechanism based on the interaction of peptides with individual liposomes which either retained all their marker content or lost it totally (Benachir and Lafleur, 1995; Mancheño *et al.*, 1996). However, given that of 14 peptides examined for ability to depolarize the cytoplasmic membrane of *E. coli*, only 4 caused

maximal depolarization at the MIC, we would argue against this explanation for the lethal event in peptide action on bacteria.

Some, but not all, cationic peptides also promote channel formation in eukaryotic cells (Cruciani *et al.*, 1991). Since the eukaryotic cell membranes are composed predominantly of zwitterionic phospholipids, hydrophobic interactions between the peptides and the membrane are likely to account for such channel formation. The reason that most of peptides show selective killing activity against prokaryotes may be partially due to the high acidic phospholipid content in the outer leaflet of the cytoplasmic membrane of bacteria and to the absence of cholesterol. The presence of cholesterol in the liposomes reduces the rate of insertion of cationic peptide into membranes (Saberwal and Nagaraj, 1994). Another factor that influences selectivity for most peptides is the requirement for the high membrane potential (~−140 mV oriented interior negative) in bacteria (Falla *et al.*, 1996; Kordel *et al.*, 1988; Wu *et al.*, 1998ab). Introduction of proline residue into an α-helical peptide changes dramatically the ability to permeabilize bacterial membrane (Zhang *et al.*, 1999). Indeed the removal of a proline-based kink in an α-helical peptide, resulting in a perfect α-helical structure (Figure 1), caused this peptide to become toxic and to have a much lower induction voltage for channel formation, while single proline peptides (Figure 1) retained good antimicrobial activity and less toxicity with decreased channel forming ability. The double proline peptides (Figure 1), however, showed excellent synergy in combination with conventional antibiotics, although having little antimicrobial activity (Zhang *et al.*, 1999).

Biosynthesis of Ribosomally Synthesized Antimicrobial Peptides

Most ribosomally synthesized peptides are derived from precursors. The DNA sequence of a variety of defensin precursors has been determined and the defensin gene(s) were mapped to human chromosome 8, band p23. Human defensins are initially synthesized as 94–100 amino acid preprodefensins that share a highly conserved 19 amino acid N-terminal signal sequence, followed by an anionic propiece, a short cationic region and the 29–34

(a)

(b)

(c)

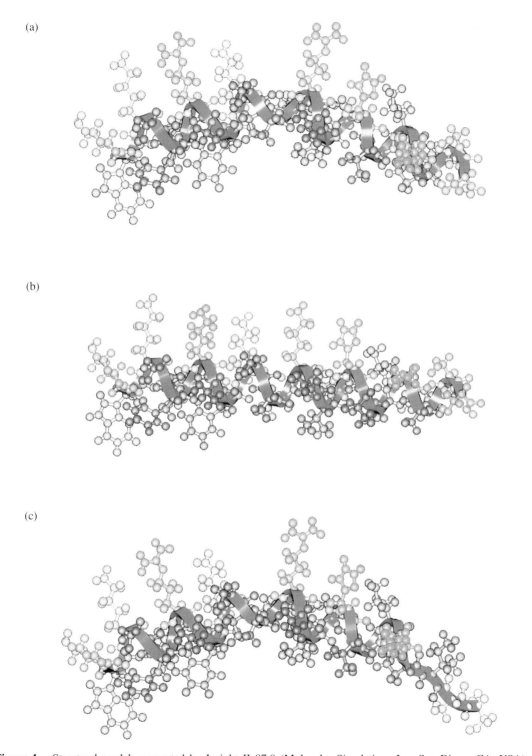

Figure 1. Structural models generated by Insight II 97.0 (Molecular Simulations Inc, San Diego, CA, USA) of α-helical peptides, containing or lacking proline residues. A, Peptide V25$_p$ with single proline residue present in the middle of (α-helical peptide, B, peptide V681$_n$ without proline, C, Peptide V8$_{pp}$ with double proline residues present at both middle and C-terminal of (α-helical peptide. These peptides are described in Zhang *et al.* (1999). *This figure is reproduced in colour at the back of the book.*

amino acid mature peptide at the C-terminus (Daher *et al.*, 1988). Despite the variation in the charge of mature defensins from +8 to +2, the prodefensins have near neutral charge at pH 7, suggesting the importance of ionic interactions between the anionic prepro piece and the cationic defensin.

The family of mammalian neutrophil antimicrobial peptides such as bovine Bac-5, indolicidin, porcine PR-39 and protegrins, and rabbit CAP18, are produced by genes encoding a highly conserved propiece that is homologous to that of the cysteine protease inhibitor, cathelin. The cathelin related peptides are thus termed cathelicidins, and have been isolated from the leukocytes of humans, cattle, pigs, mice, rabbits and sheep (Storici and Zanetti, 1993; Zanetti *et al.*, 1993; 1994). Bac-5 is initially synthesized as a prepropeptide but is processed by removal of the signal sequence to probactenecin before storage in granules. Like prodefensins, probactenecins lack microbicidal activity and exhibit charge neutralization of the mature peptide by the propiece.

The insect peptide cecropins are also synthesized as procecropins although they have relatively shorter propiece. Magainins are synthesized as polyproteins containing six copies of the mature peptide interspaced by segments of acidic hexapeptide (Bevins and Zasloff, 1990). For preprotachyplesins, the anionic propiece is C-terminal to the cationic mature peptide. The prepro region seems to not only be for neutralizing the positively charged peptide but also carrying information for targeting the peptides, such as defensins, to the granules of the phagocytic cells (Valore and Ganz, 1992, Ganz *et al.*, 1993). Similarly, most bacteriocins are produced on the ribosome as a precursor peptide being later processed to release the active moiety. The leader peptide probably keeps the bacteriocin inactive inside the cell and helps to direct its transport to the outside. Recently Cintas *et al.* reported a novel group of bacteriocins, named enterocins L50A and L50B, isolated from *Enterococcus faecium* L50 (Cintas *et al.*, 1998). Unlike lantibiotics or other bacteriocins, the two enterocins are produced without an N-terminal leader sequence. Expression *in vivo* and *in vitro* in transcription/translation experiments confirmed that the peptide gene is the only gene required for expression, indicating that they are not posttranslationally modified (Cintas *et al.*, 1998).

Recombinant Production of Cationic Peptides

Knowing that charge neutralization of cationic mature peptides by anionic propieces is utilized in the biosynthetic pathway of antimicrobial peptides, helps in the design of fusion proteins to permit high level production of cationic peptides via recombinant DNA technology. One potential limitation in the useful application of antimicrobial peptides is the high cost of production. For experimental purposes, peptide antibiotics are usually made by the solid-phase method, and depending on the solid-phase resin used, one can obtain peptides with either a free carboxyl terminus or an amidated one (Merrifield, 1986). However, even using the related but more cost effective method of solution phase chemistry, the cost is still too high (estimated $100 per-gram) to make a reasonably priced therapeutic.

Various recombinant procedures have been developed, but the most broadly effective way appears to be the production of peptides as fusion proteins in bacteria (Piers *et al.*, 1993; Zhang *et al.*, 1998). The basic procedure is to construct a fusion protein partner which mimics the *in vivo* situation in being able to neutralize the positively charged peptide of interest. An anionic segment that can be derived from the prepro sequence of human defensins, is indispensable to prevent both antibiotic activity of the cationic peptide segment against the host bacterium and proteolysis of this segment during recombinant production. A cleavage region containing a single methionine residue may be provided to permit a cyanogen-bromide cleavage procedure to recover the cationic peptide. Affinity tags may also be incorporated into the fusion partner for simplification of downstream purification.

Other recombinant procedures include production in insect cells using baculovirus vector (Andersons *et al.*, 1991, Hellers *et al.*, 1991), in yeast expression system (Reichhart *et al.*, 1992), in tobacco using tobacco-mosaic-virus vectors and in the milk of transgenic mice (Yarus *et al.*, 1996). In addition, transgenic plants and animals carrying peptide genes may be usable. However, no high-level transgenic production precedure has been published to date.

ASSETS OF CATIONIC PEPTIDE ANTIBIOTICS BASED ON *IN VITRO* STUDIES

Antibacterial Activity

To combat super resistant clinical isolates, cationic peptides have two major advantages over conventional antibiotics. Most cationic peptides have MICs in the range of 1–8 μg/ml, that are competitive with those found for even the most potent antibiotics against resistant organisms and these peptides, at concentration around MICs, kill bacteria much more quickly than do conventional antibiotics (Hancock, 1997). More importantly, the MICs against clinically antibiotic resistant and clinically susceptible strains of a given species do not vary greatly, and cationic peptides can include both Gram-positive and negative bacteria as well as fungi in their spectrum of activity. For example, the lantibiotic nisin showed good bactericidal activity *in vitro* against a selected panel of the most important contemporary mutidrug-resistant Gram-positive pathogens such as *S. pneumoniae, S. aureus*, VREF and *Entercoccus faecalis*, which speaks well for nisin as a potential antibacterial agent (Severina *et al.*, 1998).

Protegrins demonstrated efficacy against methicillin-resistant *S. aureus*, vancomycin–resistant *Enterococcus faecalis* (VREF) and *P. aeruginosa*, and administration of protegrin-1 intravenously protected mice from a lethal challenge of *S. aureus* or VREF (Steinberg *et al.*, 1997). When given intraperitoneally it can treat peritoneal infections caused by either *S. aureus* or *P. aeruginosa* (Steinberg *et al.*, 1997). Protegrins are also highly active against mucosal pathogens like *Chlamydia trachomatis* and *Nesseria gonorrhea* (Yasin *et al.*, 1996; Qu *et al.*, 1996). The finding that protegrin-mediated killing is serum independent is more encouraging in using these peptides to protect cervical surfaces by topical therapy. Similarly, both protegrins and human defensins are active against *M. tuberculosis*. Such activity is not inhibited by calcium (1.0mM), magnesium (1.0mM) or sodium chloride (100mM), and is relatively independent of pH from pH5.0 to 8.0 (Miyakawa *et al.*, 1996).

Both protegrins and defensins are effective against periodontopathic bacteria associated with early onset periodontitis, including *Porphyromonas, Prevotella, Treponema, Capnocytophaga* and *Actinobacillus*, suggesting that these peptides have considerable potential for treatment of dental diseases (Miyasaki and Lehrer, 1998).

The other advantage of cationic peptides is synergistic killing in combination with other peptides or conventional antibiotics (Piers *et al.*, 1994). Indeed it has been shown that combination of two frog α-helical peptides, magainin-2 and PGLa, showed much stronger killing activity than either peptide alone against bacteria (Westerhoff *et al.*, 1995). A synergism in membrane depolarization was observed and such functional synergism also extended to cytotoxicity for tumor cells (Westerhoff *et al.*, 1995). The synergistic effect with conventional antibiotics is believed to be due to the cationic peptide disrupting the bacterial outer membrane barrier or possibly perturbing the cytoplasmic membrane or cell wall in the case of Gram-positive bacteria, thus improving the partitioning of antibiotics into bacteria. Magainins demonstrated a significant synergistic effect with the β-lactam cefpirome against *P. aeruginosa* infection in a mouse model (Darveau *et al.*, 1991). A combination of colistin and rifampicin showed much enhanced activity against multi-resistant strains of *Acinetobacter baumannii in vitro* (Hogg *et al.*, 1998). Thus searching for peptides with good synergistic activities will be another alternative approach to overcoming resistance problems.

Antifungal Activity

In addition to bactericidal activity, protegrins are highly active against fungi. Four out of five protegrins displayed potent fungicidal activity against *Candida albicans* and the intramolecular bonds stabilizing the β-sheet structure seemed to be very important for the fungicidal activity of protegrins (Cho *et al.*, 1998). Defensins generally kill Gram-positive bacteria and fungi more effectively than Gram-negative bacteria (Kagan *et al.*, 1994). Human defensin HNP-1 was found to be fungicidal against *Cryptococcus neoformans*, a fungal pathogen frequently causing meningitis and other deep-seated infections in AIDS patients, with a 2 log reduction in fungi after 2 h of exposure to the peptides (Alcouloumre *et al.*, 1993). Rabbit defensins killed *C. albicans* in vitro within minutes (Selsted *et al.*, 1985). The rabbit defensin NP-1, NP-3 and NP-4 were active against arthroconidia of *Coccidioides immitis* (Segal *et al.*, 1985), hyphae and germinating spores of *Rhizopus oryzae* and *Aspergillus fumigatus* (Levitz *et al.*, 1986) and *C. neoformans* (Levitz, 1991). Like human defensins, rabbit defensin-mediated killing is also time and concentration dependent. Interestingly, the rabbit defensin NP-5 lacks significant killing activity against fungi but it enhanced the candidacidal effect of NP-1, NP-2 or NP-3a when added at submicromolar concentrations (Lehrer *et al.*, 1986). In addition, NP-1 acted synergistically with fluconazole in inhibiting the growth of *C. neoformans*, which is encouraging since lower and less toxic doses of antifungal agents may be used to improve overall therapy in cryptococcal disease (Alcouloumre *et al.*, 1993). Some α-helical peptides are fungicidal as well. Structural variants of cecropin-melittin hybrids showed potent antifungal activities against the yeast *Trichosporon begelii* with low cytotoxic activity under physiological conditions, and they seem to act by pore formation in the cell membrane (Lee *et al.*, 1997). Histatins are cationic histidine-rich α-helical peptides secreted by the parotid salivary gland, and they possess a number of antifungal and antibacterial properties (Helmerhorst *et al.*, 1997). Indolicidin when liposomally formulated was showed to have antifungal activity in systemic fungal infections of mice (Ahmad *et al.*, 1995).

Both insect and plant defensins, such as drosomycin and thionins, showed high activity against fungi. Two plant peptides, Pn-AMP-1 and -2 from the seeds of *Pharbitis nil L.* are 40–41 amino acids in size and are fungicidal (Koo *et al.*, 1998). The action is specific, in so far as the peptides have no inhibitory effect on the growth of Gram-negative bacteria and cultured mammalian or insect cells (Koo *et al.*, 1998).

Anti-endotoxin Activity

Lipopolysaccharide (LPS) is the major component of the outer membrane of Gram-negative bacteria. The lipid A portion of LPS is also called endotoxin and activates macrophages and endothelial cells, stimulating the release of potent inflammatory mediators such as tumor necrosis factor (TNF) and free radicals (Beutler and Cerami, 1988; Parrillo, 1993). The release of LPS into systemic circulation caused by Gram-negative bacterial infections leads to a wide variety of toxic effects such as endotoxic shock, and no efficient therapeutic agent is available at present. Mortality due to this syndrome has remained essentially unchanged over the past decade (Parrillo, 1993). LPS molecules contain several phosphate moieties that are strongly negatively charged in the KDO-lipid A part. These anionic residues are the sites of the initial electrostatic interaction of cationic peptides with bacterial outer membranes. Binding of cationic peptides to free LPS prevents it from interacting with macrophages to elicit a TNF response both *in vitro* and *in vivo* (Gough *et al.*, 1996). Therefore the significance is that, unlike most conventional antibiotics which promote endotoxin release and consequent endotoxin shock (Goto and Nakamura, 1980), cationic peptides neutralize endotoxin and prevent endotoxic shock.

The first cationic peptide used for anti-endotoxin treatment was Polymyxin B, a cyclic decapeptide antibiotic obtained from bacterium *Bacillus polymyxa* (Morrison and Jacobs, 1976). It binds to the lipid A moiety with an apparent dissociation constant of 0.4 μM and the resultant complex is virtually devoid of toxicity (Morrison and Jacob, 1976). Polymxin B also kills bacteria by disrupting cell membranes, presumably due to its ionic detergent action. However, the nephrotoxicity of polymyxin B has limited its utility as a therapeutic anti-endotoxin reagent and it is currently used primarily as a topical antibiotic. To overcome the cytotoxicity of polymxin, attempts to develop polymxin-IgG conjugates showed partial success in terms of retaining endotoxin-neutralizing activity with significant protection of mice challenged with LPS (Drabick *et al.*, 1998). The method of creating a pro-drug, in the case of colymycin (polymyxin E methosulphate), while serving to detoxicity polymyxin E, also blocks LPS binding (Hancock, unpublished).

Attempts to search for more potent anti-endotoxin peptides for therapeutics have been made and some have already shown good potential. Two cecropin-melittin hybrid peptides of 26- to 28-

amino acid in size, MBI-27 and MBI-28, are good anti-endotoxin peptides as demonstrated by using both a cultured cell line and an animal model (Gough et al., 1996). Both peptides bind to LPS with an affinity equivalent to that of polymxin B and prevent its ability to induce a TNF response in both a macrophage cell line and galactosamine-sensitized mice. The MBI-28 is particularly interesting, since it blocks LPS induced TNF induction by the macrophage cell line even when added 60 min after the LPS and it also reduces LPS-induced circulating TNF by nearly 90% in the above murine model. As expected both peptides showed significant protection against lethal endotoxemia in mice (Gough et al., 1996).

A human polymorphonuclear leukocyte granule-derived protein CAP37 with a molecular mass of 37 kDa possesses lipid A-binding and week antimicrobial activities. The bactericidal and lipid A binding domains were located to the same region from residues 20 to 44 by epitope mapping technique (Brackett et al., 1997). The synthetic peptide, $CAP37P_{20-44}$, binds to lipid A and attenuates in vivo responses induced during circumstances of endotoxemia such as hypodynamic circulatory shock, hyperlactacidemia, and leukopenia in a dose-related fashion in rats (Brackett et al., 1997). A synthetic peptide of 27 amino acids ($CAP18_{108-135}$) from another granulocyte-derived protein CAP18 is capable of preventing antibiotic-induced endotoxin shock in mice with septicemia (Kirikae et al., 1998). Several cationic proteins, such as bactericidal/permeability increasing protein, LPS binding protein, lysozyme, and lactoferrin etc., also bind to LPS and inhibit endotoxic activity (Appelmelk et al., 1994; Battafarano et al., 1995; Gazzano-Santoro et al., 1995; Ohno and Morrison, 1989; Schumann et al., 1990; Tobias et al., 1986). Peptides derived from above proteins showing better antimicrobial and anti-endotoxin activities are also under development.

Anti-viral Activity

Many defensins of human, rat, rabbit and guinea pig origin neutralized herpes simplex virus, vesicullar stomatitis virus and influenza virus in tissue culture medium (Lehrer et al., 1985; Daher et al., 1986). Tachyplesins (I-III) and polyphemusins (I, II) also displayed anti-viral activity against vesicular stomatitis virus, influenza A virus and human immunodeficiency virus (HIV)-I (Morimoto et al., 1991; Murakami et al., 1991; Masuda et al., 1992). More than 100 peptide analogs of tachyplesins and polyphemusins have been made in order to search for the potential anti-HIV determinants from the natural peptides. One novel peptide, T22 ([$Tyr^{5,12}$, Lys^7]-polyphemusin-II), showed 200 times stronger anti-viral activity than polyphemusin-II with relatively low cytotoxicity in vitro, although the structure of T22, as determined by NMR, is similar to the parent peptide (Tamamura et al., 1993). The same group further demonstrated that the T22 exerts its effect by blocking virus-cell fusion at an early stage of HIV infection. It seems that T22 binds specifically to both gp120 (an envelope protein of HIV) and CD4 (a T-cell surface protein), and both binding events can be inhibited by an anti-T22 antibody (Tamamura et al., 1996). These data may indicate that T22 is an attractive candidate as a new lead compound for anti-AIDS drug development.

Melittins and cecropins have anti-HIV activity as well. Analysis of the effect of melittin on cell-associated virus production revealed decreased levels of gag antigen and HIV-1 mRNAs (Wachinger et al., 1998). Transient transfection assays with HIV long terminal repeat (LTR)-driven reporter gene plasmids indicated that melittin has a direct suppressive effect on activity of the HIV LTR. The HIV LTR activity was also reduced in human cells stably transfected with retroviral expression plasmids for the melittin or cecropin genes (Wachinger et al., 1998).

Indolicidin is another anti-viral peptide. It was reproducibly virucidal against HIV-1 and killing was rapid at 37°C with 50% of killing occurring within 5 min, and nearly 100% viral inactivation achieved by 60 min (Robinson et al., 1998). The anti-viral activity of indolicidin is temperature-sensitive, which is consistent with a membrane-mediated antiviral mechanism (Robinson et al., 1998).

Anti-parasite and Anti-amoeba Activities

Both cecropins and magainins have antiparasitic activities in addition to their antibacterial activities. In vitro studies with the erythrocytic and sexual stages *of Plasmodium falciparum, P. Gallinna-ceum,* and *P. knowlesi* implied that magainin peptides could disrupt the extracellular stages of these parasites (Gwadz *et al.*, 1989). Cecropins also inhibit *Trypanosoma cruzi* (Jaynes *et al.*, 1988). When injected into mosquitoes previously infected with a variety of *Plasmodium* species, both magainins and cecropins disrupt sporogonic development by aborting the normal development of oocysts; sporozoites are not formed and the vector is not able to transmit the parasite to another host (Gwadz *et al.*, 1989). Cecropin-melittin hybrids are antimalarial but non-hemolytic (Boman *et al.*, 1989; Diaz-Achirica *et al.*, 1998). Magainins have been shown to be active *in vitro* against a clinical isolate of *Acanthamoeba polyphaga*, which can cause amoebic keratitis in extended-use contact lens wearers (Schuster and Jacob, 1992). When combined with silver nitrate, magainins demonstrated enhanced amoebastatic and amoebicidal activities, and thus may be useful in treating corneal infections caused by amoebae (Schuster and Jacob, 1992).

The frog-skin peptide dermaseptin is also active against parasites. The killing activity of dermaseptin on parasites has been thoroughly studied in *Leishmania mexicana*, a unicellular protozoa, and these studies indicated that upon treatment of promastigotes with micromolar concentrations of dermaseptin for less than 2 min, the parasites ceased to move and began to swell. After 10 min, large transparent areas were visible in the cytoplasm (Hernandez *et al.*, 1992). The effect of dermaseptin was irreversible since treated promastigotes did not recover motility, nor did they proliferate after a thorough wash, and reincubation in fresh medium over 3 days. The mechanism was shown to be due to pore formation in the membranes (Hernandez *et al.*, 1992).

Most parasites require an insect host in their life cycle. *Trypanosoma cruzi* which causes Chagas disease in humans, uses the reduviid bug *Rhodnius prolixus* as a vector. This bug also carries in its gut lumen an extracellular symbiotic bacterium, *Rhodococcus rhodnii*. Genetic modulation of *R. rhodnii* to express cecropin A and introduction of such altered bacteria into the insect *R. prolixus* resulted in elimination of *T. cruzi* in host insects. Constitutive expression of recombinant peptide led to lysis of developing trypanosomes and the trans-formed bacteria appeared to maintain a stable relationship with the insect host (Durvasula *et al.*, 1997). Toxicity of the recombinant peptide towards gut flora or insect tissues did not seem to be a problem. Thus expression of cationic peptide with anti-parasite activity by genetically transformed symbiotic bacteria of disease-transmitting insects may serve as a powerful approach to control certain arthropod-borne diseases (Durvasula *et al.*, 1997). This approach may help to eliminate other human pathogens, such as spirochetes which are carried by insects, since it has been shown that melittin kills *Borrelia burgdorferi* within minutes (Lubke and Garon, 1997).

Anti-tumor Activity

Defensins killed human and murine tumor cells in a concentration- and time-dependent fashion (Lichtenstein *et al.*, 1986). Synergistic cytolysis ensued when defensins were combined with sublytic concentrations of hydrogen peroxide (Lichtenstein *et al.*, 1986). Magainin and its synthetic analogs are tumoricidal, rapidly and irreversibly lysing hematopoietic and solid tumor cells at concentrations having little effect on differentiated PBLs or PMNs (Cruciani *et al.*, 1991). A combination of the peptides magainin-2 and PGLa showed stronger activity than the individual compounds against the proliferative capacity of human melanoma cells (Westerhoff *et al.*, 1995). Cecropins are toxic to a number of tumor derived cell lines as well. Two analogs of cecropin-like peptides, SB-37 and Shiva-1 preferentially lysed a number of lymphoma and leukemia derived cell lines (Jaynes *et al.*, 1989). Disruption of cytoskeleton was a prerequisite for the preferential killing of tumor cells and both SB-37 and Shiva-1 showed synergistic killing with microtubule and microfilament depolymerization drugs (Jaynes *et al.*, 1989). On the other hand less selective pore or channel formation in tumor cell membranes has also been observed (Cruciani *et al.*, 1991).

Melittin is highly toxic to tumor cells. But its high hemolytic activity prevents it from being used for therapeutic purposes. To overcome this problem, Dunn *et al.* (1996) fused the melittin gene to a gene encoding for an antibody that recognized a specific epitope expressed on the surface of human kappa myeloma and lymphoma cells. This protein-melittin hybrid was expressed in *E. coli*, and purified hybrid protein was found to exhibit specific toxicity towards antigen-bearing target cells. Therefore, linking of cationic peptides to targeting proteins is one way to develop effective anti-tumor agents. Vector-mediated delivery of peptide genes to tumor cells is another potential method for cancer gene therapy. Expression constructs carrying cecropin or melittin have been introduced into a human bladder carcinoma derived cell line. Expression of cecropin resulted in either a complete loss of tumorigenicity or reduced tumorigenicity in mice (Winder *et al.*, 1998).

POTENTIAL RESISTANCE PROBLEMS

Most pathogenic bacteria of medical importance are sensitive to peptide antibiotics but naturally occurring peptide-resistant bacteria also exist, such as *Burkholderia cepacia* and *Serratia marcescens*. Due to their unique killing mechanism, most cationic peptides do not induce resistance in bacteria in the same way as conventional antibiotics. Passage of *E. coli* or *P. aeruginosa* for 14–20 generations in the presence of a half MIC of peptides did not result in the development of resistance to the test peptides (Hancock, 1997; Steinberg *et al.*, 1997). However, a recent study showed that repeated exposure of a nisin-susceptible pneumococcal strain to nisin resulted in the rapid selection of what appears to be a nisin-resistant mutant, with an approximately 15-fold increased MIC (Severina *et al.*, 1998).

The initial interaction of cationic peptides with Gram-negative bacterium involves the LPS. Thus modification of LPS by environmental conditions and genetic regulatory machinery can lead to modulation of peptide susceptibility in the bacterium. The *pmrA-pmrB* regulon, two-component regulatory system, seems to control polymyxin resistance in *S. typhimurium* (McLeod and Spector, 1996). The genetic loci that are regulated by PmrA-PmrB were found to be *pmrE and pmrF*. The *pmrE* contains a single gene encoding for a UDP-glucose dehydrogenase, while the *pmrF* locus seems to encode seven proteins, some with similarity to glycosyltransferases and other complex carbohydrate biosynthetic enzymes. Both *pmrE* and *pmrF*

gene-products modify the LPS core and lipid A regions with ethanolamine and add aminoarabinose to the 4′ phosphate of lipid A resulting in resistance to polymyxin B (Gunn *et al.*, 1998). The same gene *pmrE* (also called ugd) together with two other genes *pbgE* and *pbgP*, is also required for growth on low-Mg^{2+} solid media (Groisman *et al.*, 1997). All these genes are regulated by the *pmrAB* system. Deletion of the *Salmonella pmrA* gene resulted in a mutant that was super susceptible to polymyxin B (Groisman *et al.*, 1997). In addition, the PhoP-PhoQ two component regulatory system has been known for a long time to be associated with resistance to several amphipathic antimicrobial peptides, including defensin, magainins, melittin and cecropins (Baker *et al.*, 1997; Guo *et al.*, 1997; Groisman *et al.*, 1992a, 1992b). Nevertheless the level of resistance achieved through the operation of the PhoP-PhoQ system appears modest.

Recently the endogenous multiple-drug efflux pumps (MDRs) have been implicated to play a role in susceptibility to cationic peptide (Shafer *et al.*, 1998). *Neisseria gonorrhoeae* possesses an energy-dependent efflux pump termed *mtr* (multiple transferable resistance) composed of three genes *mtrCDE*. Studies by Shafer and co-workers demonstrated that independent mutations in the *mtrC, mtrD* or *mtrE* genes significantly increase the sensitivity of gonococci to cationic peptide protegrins (PG-1). Deletion of the *mtrCDE*-encoded efflux pump enhanced gonococcal susceptibility to additional antibacterial peptides such as the linearized PG-1 variant PC8, the cathelicidin peptide LL-37 and the tachyplesin-1 (Shafer *et al.*, 1998). The question as to what extent or how generally efflux pumps contribute to peptide susceptibility among other bacteria remains to be elucidated. In our own studies we have seen no evidence for a role of the homologous *Pseudomonas aeruginosa* efflux pumps in peptide resistance. Another study trying to search for mode of action of mesentericin Y105, a bacteriocin against *Listeria monocytogenes*, identified a sigma 54 (δ^{54}) factor, an *rpoN* homolog showed 38% identity to that of *B. subtilis*, is likely involved in resistance to this particular peptide (Robichon *et al.*, 1997). When this transcriptional factor is not expressed in *L. monocytogenes*, the bacterium becomes resistant to mesentericin Y105 (Robichon *et al.*, 1997). However, the mechanism remains unknown.

Even though bacteria may become resistant to cationic peptides, the future of peptide antibiotics is still positive and bright since most of attempts to find mutants, including those discussed above, have only shown an increased susceptibility to the cationic peptides examined.

CLINICAL EXPERIENCE AND TRIALS

There are as yet no published clinical trials and the following information is substantively derived from the claims and press releases of the biotechnology companies which have initiated these trials (Hancock, 1998; Hancock and Lehrer, 1998). In the early part of this decade, Magainin Pharmaceuticals Inc. (http://www.magainin.com) initiated phase I studies of its 22 amino acid magainin peptide MSI-78 (Jacob and Zasloff, 1994). After an abortive phase III study of efficacy against impetigo (a largely self-resolving disease given good hygiene), two phase III trials were initiated to compare the use, against polymicrobic diabetic foot ulcers, of topical MSI-78 (Locilex) cream. In late 1997, Magainin announced that this peptide showed efficacy equivalent to the conventional oral ofloxacin therapy in the 926 patients enrolled in the two phase III studies. Unfortunately, apparently due to manufacturing concerns, the new drug approval on Locilex was not given by the FDA.

In 1997, Intrabiotics Pharmaceuticals Inc (http://wwws.intrabiotics.com) initiated a phase I clinical trial of safety of IB-367 with the objective of testing efficacy against polymicrobic oral mucositis, a side effect of anti-cancer therapies, and in 1998 entered phase II trials. They announced that both phases had demonstrated safety and evidence for efficacy. They recently completed phase I clinical trials for aerosol delivery of IB-367 for use against *Pseudomonas aeruginosa* infections of cystic fibrosis patients. In 1998–9, Micrologix Biotech Inc (http://www.mbiotech.com) performed a phase I clinical trial of the efficacy of MBI-226 in prevention of central venous catheter-associated infections, through the sterilization of catheter insertion sites. The phase I results were quite exciting, with 99.9% decrease in bacterial numbers at skin sites for up to 3 days after single dose administration, and complete prevention of colonization of catheters in all 6 patients tested (cf. controls in which catheters became colonized in 5 of 6 patients). As a result of these promising results with a life threatening infection, Micrologix received fast track status from the FDA. Xoma is expected to initiate trials of Mycoprex, one of their BPI-derived peptides, against fungal infections within the next 12 months.

Xoma Co. (http://www.xoma.com) has developed rBPI-21 (Neuprex), a recombinantly-produced, 21 kDa N-terminal derivative of the cationic neutrophil-derived protein BPI (Kohn *et al.*, 1993). Neuprex is a cationic protein but is included here because it can be cleaved to produce active cationic antimicrobial peptides and shares many properties with the cationic peptides (Battafarano *et al.*, 1995). Clinical trials of Neuprex, administered via bolus and/or continuous infusion intravenous administration, were initiated in 1996 against pediatric meningococcemia. Both safety and efficacy have been demonstrated in early trials and pivotal phase III trials are now ongoing. Other phase II trials are underway examining efficacy against hemorrhagic trauma (leading to a reduced incidence of infections, organ dysfunction and death in 400 enrolled patients), infectious complications in hepatectomy, and as an adjunct to antibiotics in severe intra-abdominal infections. It was recently announced that the phase III clinical trial for traum has been abandoned due to lack of evidence of efficacy. A phase I clinical trial for cystic fibrosis infections is also underway.

Nisin, a 32 amino acid, lantibiotic bacteriocin from *Lactobacillus lactis* (Delves-Broughton *et al.*, 1996), has been used as a food additive, especially in soft cheeses, for more than 40 years. This cationic peptide has excellent activity against many Gram-positive bacteria but little activity against Gram-negative bacteria. Intriguingly, the first clinical trials were actually initiated for efficacy against the Gram-negative bacterium *Helicobacter pylori*, a major cause of gastritis, gastric ulcers and gastric cancers. Phase I studies demonstrated safety in 96 patients but no apparently this trial was abandoned.

COMPARISON WITH CONVENTIAL ANTIBIOTICS

Antimicrobial peptides will eventually be competing in the market place with conventional antibiotics, so a comparison seems warranted here. With the dramatic rise of antibiotic resistance, including the emergence of untreatable infections by multi-resistant tuberculosis and vancomycin-resistant *Enterococcus* (VRE) strains, and other almost untreatable infections such as methicillin resistant *Staphylococcus aureus* (MRSA) there is no doubt we need novel antimicrobials (Hancock and Knowles, 1999). Indeed the new streptogramin drug combination Synercid (quinpristin/dalfopristin) was given new drug approval in September, 1999, and thus became the first new structural class of antibiotics introduced into medical practice in the past 30 years. Resistance to this drug seems inevitable, and has already been observed in the clinic. Thus one can consider as a major asset the anti-resistance properties of antimicrobial peptides (i.e. relative difficulty of resistance development, activity against clinically important resistant organisms such as VRE and MRSA, etc, and synergy with conventional antibiotics against resistant organisms). The ability of cationic antimicrobial peptides to neutralize endotoxin (cf. conventional antibiotics that release endotoxin resulting in sepsis and endotoxaemia), can also be considered an asset. Further, the antimicrobial peptides offer an enormous wealth of diversity for the pharmaceutical chemist, there being 20^{20} possible variations of sequence of a 20-mer peptide alone (more if chemical modifications are pursued), and more than 500 template structures known from nature.

The deficits of antimicrobial peptides lie largely in the area of their unknown toxicities, pharmacokinetics, optimal formulations, routes of delivery, etc. (see Hancock, 1998 for a discussion of these issues). However, these are exactly the same issues that face any novel pharmaceutical product. At present, all of the clinical indications being investigated for peptide efficacy, are topical administrations. Another major issue is production, since these peptides are larger than all known conventional antibiotics. The only practical production methodology in the long run would appear to be recombinant production, and there are several methodologies that have been developed (Piers *et al.*, 1993; Hancock and Lehrer, 1998).

CONCLUDING REMARKS

Cationic peptides cover a far broader spectrum than most conventional antibiotics. Their ability to synergize with antibiotics and to neutralize endotoxin released by these antibiotics, and especially their activities against major antibiotic resistant pathogens of medical importance, offer an exciting future in human medicine. Cloning of genes encoding for cationic peptides under an inducible promoter and using gene therapy in organ specific sites will be another exciting aspect. However, prior to large-scale exploitation of cationic peptides in medical treatment, further basic research needs to be undertaken into their modes of action, their effectiveness and their safety and stability. Pharmacokienetic studies should be performed to assist with the design of compatible delivery formulations and to search for more subtle toxicities, since acute toxicity does not appear to be a problem (Hancock, 1997). The stability of cationic peptides to proteases in the body should be followed. With further improvement, peptide antibiotics will provide us with a novel tool in our ongoing battle against emerging pathogens and antibiotic resistance development.

ACKNOWLEDGEMENTS

We would like to acknowledge the funding of the Canadian Bacterial Diseases Network and of the Canadian Cystic fibrosis Foundation's SPARx program to the cationic peptide research in the laboratory of REWH. RH is the recipient of a Medical Research Council of Canada Distingushed Scientist Award.

REFERENCES

Agerberth, B., Lee, J.Y., Bergman, T., Carlquist, M., Boman, H.G., Mutt, V., *et al.* (1991) Amino acid sequence of PR-39. Isolation from pig intestine of a new member of the family of proline-arginine-rich antibacterial peptides. *Eur. J. Biochem.* **202**, 849–854.

Ahmad, I., Perkins, W.R., Lupan, D.M., Selsted, M.E. and Janoff, A.S. (1995) Liposomal entrapment of the neutrophil-derived peptide indolicidin endows it with *in vivo* antifungal activity. *Biochem. Biophys. Acta* **1237**, 109–114.

Alcouloumre, M.S., Ghannoum, M.A., Ibrahim, A.S., Selsted, M.E. and Edwards, J.R. JE. (1993) Fungicidal properties of defensin NP-1 and activity against *Cryptococcus neoformans in vitro. Antimicrob. Agents Chemother.* **37**, 2628–2632.

Andersons, D., Engstrom, A., Josephson, S., Hansson, L. and Steiner, H. (1991) biologically active and amidated cecropin produced in a baculovirus expression system from a fusion construct containing the antibody-binding part of protein A. *Biochem. J.* **280**, 219–224.

Andreu, D., Ubacin, J., Boman, I.A., Wåhlin, B., Wade, D., Merrifield, R.B., *et al.* (1992) Shortened cecropin A-melittin hybrids. Significant size reduction retaines potent antibacterial activity. *FEBS lett.* **296**, 190–194.

Appelmelk, B.J., An, Y.Q., Geerts, M., Thijs, B.G., De Boer, H.A., MacLaren, D.M., *et al.* (1994) Lactoferrin is a lipid A binding protein. *Infect. Immun.* **62**, 2628–2632.

Baker, S.J., Daniels, C. and Morona, R. (1997) PhoP/Q regulated genes in *Salmonella typi* identification of melittin sensitive mutants. *Microb. Pathog.* **22**, 165–179.

Battafarano, R.J., Dahlberg, P.S., Raetz, C.A., Johnston, J.W., Gray, B.H., Haseman, J.R., *et al.* (1995) Peptide derivatives of 3 distinct lipopolysaccharide binding proteins inhibit lipopolysaccharide-induced tumor necrosis factor-alpha secretion *in vitro. Surgery* **118**, 318–324.

Benachir, T. and Lafleur, M. (1995) Study of vesicle leakage induced by melittin. *Biochem. Biophy. Acta.* **1235**, 452–460.

Beutler, B. and Cerami, A. (1988) Turmor necrosis, cachexia, shock and inflammation: a common mediator. *Annu. Rev. Biochem.* **57**, 505–525.

Bevins, C.L. and Zasloff, M. (1990) Peptides from frog skin. *Annu, Rev. Biochem.* **59**, 395–410.

Bloch, C. and Richardson, M. (1991) A new family of small (5 kDa) protein inhibitors of insect α-amylases from seeds or sorghum (Sorghum bicolor (L) Moench) have sequence homologies with wheat γ-purothionins. *FEBS Lett.* **279**, 101–104.

Boman, H.G., Wade, D., Boman, I.A., Wåhlin, B. and Merrifield, R.B. (1989) Antibacterial and antimalarial properties of peptides that are cecropin-melittin hybrids. *FEBS Lett.* **259**, 103–106.

Boman, H.G. (1995) Peptide antibiotics and their role in innate immunity. *Annu, Rev. Immunol.* **13**, 61–92.

Brackett, D.J., Lerner M.R., Lacquement, M.A., He, R. and Pereira, H.A. (1997) A synthetic lipopolysaccharide-binding peptide based on the neutrophil-derived protein CAP37 prevents endotoxin-induced responses in conscious rats. *Infect. Immun.* **65**, 2803–2811.

Casteels, P., Ampe, C., Jacob, F., Vaeck, M. and Tempst, P. (1989) Apidaecins: antibacterial peptides from honeybee. *EMBO J.* **8**, 2387–2391.

Cho, Y., Turner, J.S., Dinh, N.N. and Lehrer, R.I. (1998) Activity of protegrins against Yeast-phase *Candida albicans. Infect. Immun.* **66**, 2486–2493.

Christensen, B., Fink, J., Merrifield, R.B. and Mauzerall, D. (1988) Channel-forming properties of cecropins and related model compounds incorporated into planar lipid membranes. *Proc. Natl. Acad. Sci. USA* **85**, 5072–5076.

Cintas, L.M., Casaus, P., Holo, H., Hernandez, P.E., Nes, I.F. and Havarstein, L.S. (1998) Enterocins L50A and L50B, two novel bacteriocins from *Enterococcus faecium* L50, are related to Staphylococcal hemolysin. *J. Bacteriol.* **180**, 1988–1994.

Cruciani, R.A., Barker, J.L., Zasloff, M., Chen, H.C. and Colamonici, O. (1991) Antibiotic magainins exert cytolytic activity against transformed cell lines through channel formation. *Proc. Natl. Acad. Sci. USA* **88**, 3792–3796.

Daher, K.A., Lehrer, R.I., Ganz, T. and Kronenberg, M. (1988) Isolation and characterization of human defensin cDNA clones. *Proc. Natl. Acad. Sci. USA* **85**, 7327–7331.

Daher, K.A., Selsted, M.E. and Lehrer, R.I. (1986) Direct inactivation of viruses by human granulocyte defensins. *J. Virol.* **60**, 1068–1074.

Darveau, R.P., Cunningham, M.D., Seachord, C.L., Cassiano-Clough, L., Cosand, W.L., Blake, J., *et al.* (1991) Beta-lactam antibiotics potentiate magainin 2 antimicrobial activity in vitro and *in vivo. Antimicrob. Agents Chemother.* **35**, 1153–1159.

Dathe, M., Wieprecht, T., Nikolenko, H., Handel., L., Maloy, W.L., MacDonald, D.L., *et al.* (1997) Hydrophobicity, hydrophobic moment and angle subtended by charged residues modulate antibacterial and hemolytic activity of amphipathic helical peptides. *FEBS Lett.* **403**, 208–212.

Delves-Broughton, J., Blackburn, P., Evans, R.J. and Hugenholtz, J. (1996) Applications of the bacteriocin nisin. *Antonie van Leeuwenhoek.* **69**, 193–202.

de Vos, W.M., Kuipers, O.P., van der Meer, J.R. and Siezen, R.J. (1995) Maturation pathway of nisin and other lantibiotics: post-translationally modified antimicrobial peptides exported by Gram-positive bacteria. *Mol Microbiol.* **17**, 427–437.

Diamond, G., Zasloff, M., Eck, H., Brasseur, M., Maloy, W.L. and Bevins, C.L. (1991) tracheal antimicrobial peptide, a cysteine-rich peptide from mammalian tracheal mucosa: peptide isolation and cloning of a cDNA. *Proc. Natl. Acad. Sci. USA.* **88**, 3952–3956.

Diamond, G., Russell, J.P. and Bevins, C.L. (1996) Inducible expression of antibiotic peptide gene in lipopolysaccharide-challenged tracheal epithelial cells. *Proc. Natl. Acad. Sci. USA.* **93**, 5156–5150.

Diaz-Achirica, P., Ubach, J., Guinea, A., Andreu, D. and Rivas, L. (1998) The plasma membrane of leishmania donovani promastigotes is the main target for CA(1– 8)M(1–18), a synmthetic cecropin-melittin hybrid peptide. *Biochem. J.* **330**, 453–460.

Diot, P., Gagnadoux, F., Martin, C., Ellataoui, H., Furet, Y., Breteau, M., Boissinot, E., *et al.* (1997) Nebulization and anti-*Pseudomonas aeruginosa* activity of colistin. *Eur. Respir. J.* **10**, 1995–1998.

Drabick, J.J., Bhattacharjee, A.K., Hoover, D.L., Siber, G.E., Morales, V.E., Young, L.D., *et al.* (1998) Covalent polymyxin B conjugate with human immunoglobulin G as an antiendotoxin reagent. *Antimicrob. Agents Chemother.* **42**, 583–588.

Dunn, R.D., Weston, K.M., Longhurst, T.J., Lilley, G.G., Rivett, D.E., Hudson, P.J., *et al.* (1996) Antigen binding and cytotoxic properties of a recombinant immunotoxin incorporating the lytic peptide, melittin. *Immunotechnology.* **2**, 229– 240.

Durvasula, R.V., Gumbs, A., Panackal, A., Kruglov, O., Aksoy, S., Merrifild, R.B., *et al.* (1997) Prevention of insect-borne disease: an approach using transgenic symbiotic bacteria. *Proc. Natl. Acad. Sci. USA* **94**, 3274–3278.

Falla, T., Karunaratne, D.N. and Hancock, R.E.W. (1996) Mode of action of the antimicrobial peptide indolicidin. *J. Biol. Chem.* **271**, 19298–19303.

Fath, M.J. and Kolter, R. (1993) ABC transporters: bacterial exporters. *Microbial Rev.* **57**, 995–1017.

Fehlbaum, P., Bulet, P., Michaut, L., Lagueux, M., Broekaer, W.F., Hetru, C., *et al.* (1994) Insect immunity. Septic injury of Drosophila induces the synthesis of a potent antifungal peptide with sequence homology to plant antifungal peptides. *J. Biol. Chem.* **269**, 33159–33163.

Frank, R.W., Gennaro, R., Schneider, K., Przybylski, M. and Romeo, D. (1990) Amino acid sequences of 2 proline-rich bactenecins-antimicrobial peptides of bovine neutrophils. *J. Biol. Chem.* **265**, 18871–18874.

Fremaux, A., Muriana, P. and Klaenhammer, T.R. (1993) Molecular analysis of the lactacin F operon. *Appl. Environ. Microbiol.* **59**, 3906–3915.

Ganz, T. and Lehrer, R.I. (1995) Defensins. *Pharmac. Ther.* **66**, 191–205.

Ganz, T., Lide, L., Valore, E.V. and Oren, A. (1993) Posttranslational processing and targeting of transgenic human of defensins in murine granulocyte, macrophage, fibroblast and pituitary adenoma cell lines. *Blood* **82**, 641–650.

Garcia-Olmedo, F., Rodriguez-Palenzuela, P., Hermandez-Lucas, C., Ponz, F., Marana, C., Carmona, MJ., *et al.* (1989) The thionins: a protein family that includes purothionins, viscotoxins and crambins. *Oxf. Surv. Plant Mol. Cell Biol.* **6**, 31–60.

Gazzano-Santoro, H., Parent, J.B., Conlon, P.J., Kasler, H.G., Tsai, C.M., Lill-Elghanian, D.A., *et al.* (1995) Characterization of the structural elements in lipid A required for binding of a recombinant fragment of bactericidal/permeability-increasing protein rBPI23. *Infect. Immun.* **63**, 2201–2205.

Goldman, M.J., Anderson, G.M., Stolzenberg, E.D., Kari, U.P., Zasloff, M. and Wilson, J. (1997) Human β-defensin-1 is a salt-sensitive antibiotic in lung that is inactivated in cystic fibrosis. *Cell* **88**, 553–560.

Goto, H. and Nakamura, S. (1980) Liberation of endotoxin from *Escherichia coli* by addition of antibiotics. *Japan. J. Med.* **50**, 35–43.

Gough, M., Hancock, R.E.W. and Kelly, N.M. (1996) Antiendotoxin activity of cationic peptide antimicrobial agents. *Infect. Immun.* **64**, 4922–4927.

Gray, W.R., Lugue, F., Galyean, R., Atherton, E., Sheppard, R.C., Stone, B.L., *et al.* (1984) Conotoxin GI: disulfide bridges, synthesis, and preparation of iodinized derivatives. *Biochem.* **23**, 2796–2802.

Groisman, E.A., Saier, M.H. and Ochman, H. (1992a) Molecular genetic analysis of the *Escherichia coli phoP* locus. *J. Bacteriol.* **174**, 486–491.

Groisman, E.A., Parra-Lopez, C., Salcedo, M., Lipps, C.J. and Heffron, F. (1992b) Resistance to host animicrobial peptides is necessay for *Salmonella* virulence. *Proc. Natl. Acad. Sci. USA* **89**, 11939–11943.

Groisman, E.A., Kayser, J. and Soncini, F.C. (1997) Regulation of polymyxin resistance and adaptation to low-Mg^{2+} Environments. *J. Bacteriol.* **179**, 7040–7045.

Gunn, J.S., Lim, K.B., Krueger, J., Kim, K., Guo, L., Hackett, M., *et al.* (1998) PmrA-PmrB-regulated genes necessary for 4-aminoarabinose lipid A modification and polymyxin resistance. *Mol. Microbiol.* **27**, 1171–1182.

Guo, L., Lim, K.B., Gunn, B., Bainbrdge, B., Darveau, R.P., Hackett, M., *et al.* (1997) Regulation of lipid A modifications by *Salmonella typhimurium* virulence genes *phoP-phoQ*. *Science* **276**, 250–253.

Gwadz, R.W., Kaslow, D., Lee, J.Y., Maloy, W.L., Zasloff, M. and Miller, L.H. (1989) Effects of magainins and cecropins on the sporogonic development of malaria parasites in mosquitoes. *Infect. Immun.* **57**, 2628–2633.

Habermann, E. and Jentsch, J. (1967) Sequenzanalyse des melittins aus den trytischen und peptische spaltstu''cken. *Hoppe Seyler's Z. Physio. Chem.* **348**, 37–50.

Hancock, R.E.W., Falla, T. and Brown, M.H. (1995) Cationic bactericidal peptides. *Adv. Microb. Physiol.* **37**, 135–175.

Hancock, R.E.W. (1997) Peptide antibiotics. *Lancet* **349**, 418–422.

Hancock, R.E.W. (1998) Therapeutic potential of cationic peptides. *Expert. Opinion Invest. Dis.* **7**, 167–174.

Hancock, R.E.W. and Knowles, D. (1998). Are we approaching the end of the antibiotic era? Editorial overview. *Curr. Opinion Microbiol.* **1**, 493–494.

Hancock, R.E.W. and Lehrer, R. (1998) Cationic peptides: A new source of antibiotics. *Trends in Biotechnol.* **16**, 82–88.

Harder, J., Bartels, J., Christphers, E. and Schroder, J.M. (1997) A peptide antibiotic from human skin. *Nature* **387**, 861.

Hauge, H.H., Nissen-Mwyer, J., Nes, I.F. and Eijsink, G.H. (1998) Amphiphilic α-helices are important structural motifs in the α and β peptides that constitute the bacteriocin lactococcin G. *Eur. J. Biochem.* **251**, 565–572.

Hellers, M., Gunne, H. and Steiner, H. (1991) Expression of post-translational processing or prececropin A using a baculovirus vector. *Eur. J. Biochem.* **199**, 435–439.

Helmerhorst, E.J., Hof, W.V.T., Veerman, E.C.I., Simoons-smit, I. and Amerongen, A.V.N. (1997) Synthetic histatin analogues with broad-spectrum antimicrobial activity. *Biochem. J.* **326**, 39–45.

Hernandez, C., Mor, A., Gagger, F., Nicolas, P., Hernandez, A., Benedetti, E.L., *et al.* (1992) Functional and structural damage in Leishmania mexicana exposed to the cationic peptide dermaseptin. *Eur. J. Cell Biol.* **59**, 414–424.

Hill, C.P., Yee, J., Selsted, M.E. and Eisenberg, D. (1991) Crystal structure of defensin HNP-3, an amphiphilic dimer. Mechanisms of membrane permeabilization. *Science* **251**, 1481–1485.

Hogg, C.M., Barr, J.G. and Webb, C.H. (1998) In-vitro activity of the combination of colistin and rifampicin against mutldrug-resistant strains of *Acenetobacter baumannii*. *J. Antimicrob. Chemother.* **41**, 494–495.

Holo, H., Nillsen, O. and Nes, I. (1991) Lactococcin A, a new bacteriocin from *Lactococcus lactis* subsp. cremoris: isolation and characterization of the protein and its gene. *J. Bacteriol.* **173**, 3879–3887.

Houston, M.E. Jr., Chao, H., Hodges, R.S., Sykes, B.D., Kay, C.M., Sonnichsen, F.D., *et al.* (1998) Binding of an oligopeptide to a specific plane of ice. *J. Biol. Chem.* **273**, 11714–11718.

Hucho, F., Tssetlin, V.I. and Machold, J. (1996) Theemerging three-dimensional structure of a receptor. The nicotinic acetylcholine receptor. *Eur. J. Biochem.* **239**, 539–557.

Jack, F.W., Tagg, J.R. and Ray, B. (1995) Bacteriocins of Gram-positive bacteria. *Microbiol. Rev.* **59**, 171–200.

Jacob, L. and Zasloff, M. (1994) Potential therapeutic application of magainins and other antimicrobial agents of animal origin. *Ciba Found. Symp.* **186**, 197–216.

Jaynes, J.M., Burton, C.A., Barr, S.B., Jeffers, G.W., Julian, R.G., White, KL., *et al.* (1988) *In vitro* cytocidal effect of novel lytic peptides on *Plasmodium falciparum* and *Trypanosoma cruzi*. *FASEB J.* **2**, 2878–2883.

Jaynes, J.M., Julian, G.R., Jeffers, G.W., White, K.J. and Enright, F.M. (1989) In vitro cytocodal effect of lytic peptides on several transformed mammalian cell lines. *Peptide Res.* **2**, 157–160.

Jensen, T., Pedersen, S.S., Garne, S., Heilmann, C., Hoiby, N. and Koch, C. (1987) Colistin inhalation therapy in cystic fibrosis patients with chronic *Pseudomonas aeruginosa* lung infection. *J. Antimicrob. Chemother.* **19**, 831–838.

Jones, D.E. and Bevins, C.L. (1992) Paneth cells of the human small intestine express an antimicrobial peptide gene. *J. Biol. Chem.* **267**, 23216–23225.

Jones, D.E. and Bevins, C.L. (1993) Defensin-6 mRNA in human Paneth cells. *FEBS Lett.* **315**, 187–192.

Jung, G. (1991) In *Nisin and Novel Lantibiotics: Proceedings of the first International Workshop on Lantibiotics*. Jung, G. and Sahl, H.G. (eds). Leiden: Escom Publishers, pp. 1–25.

Kagan, B.L., Ganz, T. and Lehrer, R.I. (1994) Defensins: a family of antimicrobial and cytotoxic peptides. *Toxicology* **87**, 131–149.

Kagan, B.L., Selsted, M.E., Ganz, T. and Lehrer, R.I. (1990) Antimicrobial defensin peptides form voltage-dependent ion-permeable channels in planar lipid bilayer membranes. *Proc. Natl. Acad. Sci. USA* **87**, 210–214.

Kirikae, T., Hirata, M., Yamasu, H., Kirikae, F., Tamura, H., Kayama, F., et al. (1998) Protective effects of a human 18-kilodalton cationic antimicrobial protein (CAP18)-derived peptide against murine endotoxemia. *Infect. Immun.* **66**, 1861–1868.

Kleinkauf, H. and von Dohren, H. (1988) Peptide antibiotics, β-lactams and related compounds. *CRC Critical Rev. Biotechnol.* **8**, 1–32.

Kohn, T.F., Ammons, W.S., Horowitz, A., Grinna, L., Theofan, G., Weickmann, J., et al. (1993) Protective effect of recombinant amino terminal fragment of bactericidal/permeability-increasing protein in experimental endotoxaemia. *J. Infect. Dis.* **168**, 1307–1310.

Kokryakov, V.N., Harwig, S.S., Panyutich, E.A., Shevchenko, A.A., Aleshina, G.M., Shamova, O.V., et al. (1993) Protegrins: leokocyte antimicrobial peptides that combine features of corticosatic defensins and techyplesins. *FEBS Lett.* **327**, 231–236.

Kondejewski, L.H., Farmer, S.W., Wishart, D.S., Hancock, R.E.W. and Hodges, R.S. (1996) Gramicidin S is active against both Gram-positive and Gram-negative bacteria. *Int. J. Pept. Prot. Res.* **47**, 460–466.

Kondejewski, L.H., Farmer, S.W., Wishart, D.S., Kay, C.M., Hancock, R.E.W. and Hodges, R.S. (1996) Modulation of structure and antimicrobial and hemolytic activity by ring size in cyclic gramicidin S analogs. *J. Biol. Chem.* **271**, 25261–25268.

Koo, J.C., Lee, S.Y., Chun, H.J., Cheong, Y.H., Choi, J.S., Kawabata, S.I., et al. (1998) Two hevein homologs isolated from the seed of *Pharbitis nil* L. exhibit potent antifungal activity. *Biochim. Biophys. Acta.* **1382**, 80–90.

Kordel, M., Benz, R. and Sahl, H.G. (1988) Mode of action of the staphylococcinlike peptide Pep5: voltage-dependent depolarization of bacterial and artificial membranes. *J. Bacteriol.* **170**, 84–88.

Lee, D.G., Park, J.H., Shin, S.Y., Lee, S.G., Lee, M.K., Lyong, K., et al. (1997) Design of novel analogue peptides with potent fungicidal but low hemolytic activity based on the cecropin A-melittin hybrid structure. *Biochem. & Mol. Biol. Int.* **43**, 489–498.

Lee, J.Y., Boman, A., Sun, C., Andersson, M., Jo''nvall, H., Mutt, V., et al. (1989) Antibacterial peptides from pig intestine: isolation of a mammalian cecropin. *Proc. Natl. Acad. Sci. USA.* **86**, 9159–9162.

Lehrer, R.I., Barton, A., Daher, K.A., Harwig, S.S.L. Ganz, T. and Selsted, M.E. (1989) Interaction of human defensins with *Escherichia coli*. Mechanism of bactericidal activity. *J. Clin. Invest.* **84**, 553–561.

Lehrer, R.I., Daher, K.A., Ganz, T. and Selsted, M.E. (1985) Direct inactivation of viruses by MCP-1 and MCP-2, natural peptide antibiotics from rabbit leukocytes. *J. Virol.* **54**, 467–672.

Lehrer, R.I., Szklarek, D., Ganz, T. and Selsted, M.E. (1986) Synergistic activity of rabbit granulocyte peptides against *Candida albicans*. *Infect. Immun.* **52**, 902–904.

Levitz, S. (1991) The ecology of *Cryptococus neoformans* and the spidemiology of cryptococcosis. *Rev. Infect. Dis.* **13**, 1163–1169.

Levitz, S.M., Selsted, M.E., Ganz, T., Lehrer, R.I. and Diamond, R.D. (1986) In vitro killing of spores and hyphae of

Aspergillus fumigatus and *Rhizopus oryzae* by rabbit neutrophil cationic peptides and bronchoalveolar macrophages. *J. Infect. Dis.* **154**, 483–489.

Lichtenstein, A., Ganz, T., Selsted, M.E. and Lehrer, R.I. (1986) In vitro tumor cell cytolysis mediated by peptide defensins of human and rabbit granulocytes. *Blood* **68**, 1407–1410.

Ludtke, S., He, K., Heller, W.T., Harroun, T.A., Yang, L. and Huang, H.W. (1996). Membrane pores induced by magainin. *Biochemistry* **35**, 13723–13728.

Lubke, L.L. and Garon, C.F. (1997) The antimicrobial agent melittin exhibits powerful in vitro inhibitory effects on the lyme disease spirochete. *Clin. Infect. Dis.* **25 (suppl 1)**, S48–51.

Mancheño, J.M., Oñaderra, M., del Pozo, A.M., Díaz-Achirica, P., Andreu, D., Rivas, L., et al. (1996) Release of lipid vesicle contents by an antibacterial cecropin A-melittin hybrid peptide. *Biochemistry* **35**, 9892–9899.

Mangavel, C., Maget-Dana, R., Tauc, P., Brochon, J.C., Sy, D. and Reynaud, J.A. (1998) Structural investigations of basic amphipathic model peptides in the presence of lipid vesicles studied by cicular dichroism, fluorescence, monolayers and modeling. *Biochem. Biophys. Acta* **1371**, 265–283.

Mangoni, M.E., Aumelas, A., Charnet, P., Roumestand, C., Chiche, L., Despaux, E., et al. (1996) Change in membrane permeability induced by protegrin 1: implication of disulphide bridges for pore formation. *FEBS lett.* **383**, 93–98.

Martinez, J.S., Olivera, B.M., Gray, W.R., Craig, A.G., Groebe, D.R., Abramson, S.N., et al. (1995) Alpha-conotoxin EI, a new nicotinic acetylcholine receptor antagonist with novel selectivity. *Biochem.* **34**, 14519–14526.

Marugg, J.D., Gonzalez, C.F., Kunka, B.S., Ledeboer, A.M., Pucci, M.J., Toonen, M.Y., et al. (1992) Cloning, expression, and nucleotide sequence of genes involved in production of pediocin PA-1, and bacteriocin from *Pediococcus acidilactici* PAC1.0. *Appl. Environ. Microbiol.* **58**, 2360–2367.

Masuda, M., Nakashima, H., Ueda, T., Naba, H., Ikoma, R., Otaka, A., et al. (1992) *Biochem. Biophys. Res. Commun.* **189**, 845–850.

Matsuzaki, K., Yoneyama, S. and Miyajima, K. (1997) Pore formation and translocation of melittin. *Biophys. J.* **73**, 831–838.

Mcleod, G.I. and Spector, M.P. (1996) Starvation- and stationary-phase induced resistance to the antimicrobial peptide polymyxin B in *Salmonella typhimurium* is RpoS ($δ^s$) independent and occurs through both *phoP*-dependent and -independent pathways. *J. Bacteriol.* **178**, 3683–3688.

Méndez, E., Rocher, A., Calero, M., Girbés, T., Cittores, L. and Soriano, F. (1996) Primary structure of ω-hordothionin, a member of a novel family of thionins from barley endosperm, and its inhibition of protein synthesis in eukaryotic and prokaryotic cell-free systems. *Eur. J. Biochem.* **239**, 67–73.

Merrifield, R.B. (1986) Solid phase synthesis. *Science* **232**, 341–347.

Miyakawa, Y., Ratnaker, P., Rao, R., Costello, M.L., Mathieu-Costello, O., Lehrer, R.I., et al. (1996) In vitro activity of the antimicrobial peptides human and rabbit defensins and porcine leukocyte protegrin against *Mycobacterium tuberculosis*. *Infect. Immun.* **64**, 926–932.

Miyasaki, K.T. and Lehrer, R.I. (1998) Beta-sheet antibiotic peptides as potential dental therapeutics. *Int. J. Antimicrob. Agents.* **9**, 269–280.

Miyata, T., Tokunaga, F., Yoneya, T., Yoshikawa, K., Iwanaga, S., Niwa, M., et al. (1989) Antimicrobial peptides, isolated from horseshoe crab hemocytes, tachyplesin II, and polyphemusins I and II: chemical structures and biological activity. *J. Biochem. (Tokyo)* **106**, 663–668.

Mor, A., Nguyen, V.H., Delfour, A., Migliore-Samour, D. and Nicolas, P (1991a) Isolation, amino acid sequence and synthesis of dermaseptin, a novel antimicrobial peptide of amphibian skin. *Biochemistry* **30**, 88824–8830.

Mor, A., Nguyen, V.H. and Nicolas, P. (1991b) Antifungal activity of dermaseptin, a novel vertebrate skin peptide. *J. Mycol. Med.* **1**, 5–10.

Mor, A. and Nicolas, P. (1994) The NH2-terminal helical domain 1–18 of dermaseptin is responsible for antimicrobial activity. *J. Biol. Chem.* **269**, 1934–1949.

Morikawa, N., Hagiwara, K. and Nakajima, T. (1992) Brevinins-1 and 2, unique antimicrobial peptides from the skin of the frog, *Rana brevidipora porsa. Biochem. Biophys. Res. Commun.* **189**, 184–190.

Morimoto, M., Mori, H., Otake, T., Ueba, N., Kunita, N., Niwa, M., et al. (1991) Inhibitory effect of tachyplesin I on the proliferation of human immunodeficient virus in vitro. *Chemotherapy* **37**, 206–211.

Morrison, D.C. and Jacob, D.M. (1976) Inhibition of lipopoly-saccharide–initiated activation of serum complement by polymyxin B. *Infect. Immun.* **13**, 298–301.

Murakami, T., Niwa, M., Tokunaga, F., Miyata, T. and Iwanaga, S. (1991) Direct inactivation of tachyplesin I and its isopeptides from horseshoe crab hemocytes. *Chemotherapy* **37**, 327–334.

Muta, T., Fujimoto, T., Nakajima, H. and Iwanaga, S. (1990) Tachyplesins isolated from hemocytes of Southeast Asian horseshoe crabs (Carcinoscorpius rotundicauda and Tachypleus gigas): identification of a new tachyplesin, tachyplesin III, and a processing intermediate of its precursor. *J. Biochem. (Tokyo)* **108**, 261–266.

Nakamura, T., Furunaka, H., Miyata, T., Tokunaga, F., Muta, T., Iwanaga, S., et al. (1988) Tachyplesin, a class of antimicrobial peptide from the hemocytes of the horseshoe crab (Tachypleus tridentatus). Isolation and chemical structure. *J. Biol. Chem.* **263**, 16708–16713.

Nicolas, P. and Mor, A. (1995) Peptides as weapons against microorganisms in the chemical defense system of vertebrates. *Annu. Rev. Microbiol.* **49**, 277–304.

Ohno, N. and Morrison, D.C. (1989) Effect of lipopolysaccharide chemotype structure on binding and inactivation of hen egg lysozyme. *Eur. J. Biochem.* **22**, 621–627.

Parrillo, J.E. (1993). Pathogenetic mechanisms of septic shock. *N. Engl. J. Med.* **328**, 1471–1477.

Patel, S.U., Osborn, R., Rees, S. and Thornton, J.M. (1998) Structural studies of *Impatiens balsamina* antimicrobial protein (Ib-AMP1). *Biochem.* **37**, 983–990.

Pawlak, M., Stankowski, S. and Schwarz, G. (1991) Melittin induced voltage-dependent conductance in DOPC lipid bilayers. *Biochim. Biophys. Acta.* **1062**, 94–102.

Perlman, D. and Bodansky, M. (1971) Biosynthesis of peptide antibiotics. *Annu. Rev. Biochem.* **40**, 449–464.

Piers, K.L., Brown, M.H. and Hancock, R.E.W. (1993) Recombinant DNA procedures for producing small antimicrobial cationic peptides in bacteria. *Gene* **134**, 7–13.

Piers, K.L., Brown, M.H. and Hancock, R.E.W. (1994) Improvement of outer-membrane and lipopolysaccharide-binding activities of an antimicrobial cationic peptide by C-terminal modification. *Antimicrob. Agent Chemother.* **38**, 2311–2316.

Pouny, Y., Rapaport, D., Mor, A., Nicolas, P. and Shai, Y. (1992) Interaction of antimicrobial dermaseptin and its fluorescently labeled analogues with phospholipid membranes. *Biochem.* **31**, 12416–12423.

Qu, X.D., Harwig, S.S.L., Oren, A., Shafer, W.M. and Lehrer, R.I. (1996) Susceptibility of Neisseria gonorrhoeae to protegrins. *Infect. Immun.* **64**, 1240–1245.

Radermacher, S.W., Schoop, V.M. and Schluesener, H.J. (1993) Bactenecin, a leukocyticantimicrobial peptide, is cytotoxic to neuronal and glial cells. *J. Neurosci Res.* **36**, 657–662.

Reichhart, J.M., Petit, I., Legrain, M., Dimarcq, J.L., Keppi, E., Lecocq, J.P., Hoffmann, J.A., et al. (1992) Expression and secretion in yeast of active insect defensin, an inducible antibacterial peptide from the fleshfly Phormia terranovae. *Invert. Reproduct. Develop.* **21**, 15–24.

Rex, S. and Schwarz, G. (1998) Quantitative studies on the melittin-induced leakage mechanism of lipid vesicles. *Biochemistry.* **37**, 2336–2345.

Robichon, D., Gouin, E., Debarbouille, M., Cossart, P., Cenatiempo, Y. and Hechard, Y. (1997) The *rpoN* (δ54) gene from *Listeria monocytogenes* is involved in resistance to mesentericin Y105, an antibacterial peptide from *Leuconostoc mesenteroides. J. Bacteriol.* **179**, 7591–7594.

Robinson, W.E., Jr., McDougall, B., Tran, D. and Selsted, M.E. (1998) Anti-HIV-1 activity of indolicidin, an antimicrobial peptide from neutrophils. *J. Leukoc. Biol.* **63**, 94–100.

Romeo, D., Skerlavaj, B., Bolognesi, M. and Gennaro, R. (1988) Structure and bactericidal activity of an antibiotic dodecapeptide purified from bovine neutrophils. *J. Biol. Chem.* **263**, 9573–9575.

Russell, J.P., Diamond, G., Tarver, A.P., Scanlin, T.F. and Bevins, C.L. (1996) Coordinate induction of two antibiotic genes in tracheal epithelial cells exposed to the inflammatory mediators lipopolysaccharide and tumor necrosis factor alpha. *Infect. Immun.* **64**, 1565–1568.

Saberwal, G. and Nagaraj, R. (1994) Cell-lytic and antibacterial peptides that act by perturbing the barrier function of membranes: facets of their conformational features, structure-function correlations and membrane-perturbing ability. *Biochem. Biophy. Acta* **1197**, 109–131.

Sahl, H.G., Jack, R.W. and Bierbaum, M. (1995) Biosynthesis and biological activities of lantibiotics with unique post-translational modifications. *Eur. J. Biochem.* **230**, 827–853.

Schneider, A., Stachelhaus, T. and Mahariel, M.A. (1998) Targetted alteration of the substrate specificity of peptide synthetases by rational module swapping. *Mol. Gen. Genet.* **257**, 308–318.

Schonwetter, B.S., Stolzenberg, E.D. and Zasloff, M.A. (1995) Epithelial antibiotics induced at sites of inflammation. *Science* **267**, 1645–1648.

Schumann, R.R., Leong, S.R., Flaggs, G.W., Gray, P.W., Wright, S.D., Mathison, J.C., et al. (1990) Structure and function of LPS-binding protein. *Science* **249**, 1429–1433.

Schuster, F.L. and Jacob, L.S. (1992) Effects of magainins on ameba and cyst stages of *Acanthamoeba polyphaga. Antimicrob. Agents Chemother.* **36**, 1263–1271.

Schwarz, G. and Arbuzova, A. (1995) Pore kinetics reflected in the dequencing of a lipid vesicle entrapped fluorescent dye. *Biochem. Biophy. Acta* **1239**, 51–57.

Segal, G.P., Lehrer, R.I. and Selsted, M.E. (1985) In vitro effect of phagocyte cationic peptides on *Coccidioides immitis. J. Infect. Dis.* **151**, 890–894.

Segrest, J.P., de-Loof, H., Dohlman, J.G., Brouillette, C.G. and Anantharamaiah, G.M. (1990) Amphiphilic α-helix motif: classes and properties. *Proteins* **8**, 103–117.

Selsted, M.E., Harwig, S.L., Ganz, T., Schilling, J.W. and Lehrer, R.I. (1985) Primary structures of three human neutrophil defensins. *J. Clin. Invest.* **76**, 1436–1439.

Selsted, M.E., Novotny, M.J., Morris, W.L., Tang, Y.Q., Smith, W. and Cullor, J.S. (1992) Indolicidin: a novel abcatericidal tridecapeptide amide from neutrophil. *J. Biol. Chem.* **267**, 4292–4295.

Selsted, M.E., Szklarek, D., Ganz, T. and Lehrer, R.I. (1985) Activity of rabbit leukocyte peptides against *Candida albicans. Infect. Immun.* **49**, 202–206.

Selsted, M.E., Tang, Y.Q., Morris, W.L., McGuire, P.A., Novotny, M.J., Smith, W., et al. (1993) Purification, primary structures and antibacterial activites of beta-defensins, a new family of antimicrobial peptides from bovine neutrophils. *J. Biol. Chem.* **268**, 6641–6648

Severina, E., Sevrin, A. and Tomasz, A. (1998) Antibacterial efficacy of nisin against multidrug-resistant Gram-positive pathogens. *J. Antimicrob. Chemother.* **41**, 341–347.

Shafer, W.M., Qu, X.-D., Waring, A.J. and Lehrer, R.I. (1998) Modulation of Neisseria gonorrhoeae susceptibility to vertebrate antibacterial peptides due to a member of the resistance/nodulation/division efflux pump family. *Proc. Natl. Acad. Sci. USA* **95**, 1829–1833.

Shai, Y. (1995) Molecular recognition between membrane-spanning polypeptides. *Trends Biochem. Sci.* **20**, 460–464.

Shimoda, M., Ohki, K., Shimamoto, Y. and Kohashi, O. (1995) Morphology of defensin-treated *Staphylococcus aureus.* *Infect. Immun.* **63**, 2886–2891.

Simmaco, M., Mignogna, G., Barra, D. and Bossa, F. (1993) Novel antimicrobial peptides from skin secretions of the European frog *Rana esculenta. FEBS lett.* **324**, 156–159.

Stein, T., Vater, J., Kruft, V., Otto, A., Wittmann-Liebold, B., Franke, P., *et al.* (1996) The multiple carrier model of nonribosomal peptide biosynthesis at modular multienzyme templates. *J. Biol. Chem.* **271**, 15428–15435.

Steinberg, D.A., Hurst, M.A., Fujii, C.A., Kung, A.C., Ho, J.F., Cheng, F.C., *et al.* (1997) Protegrin-1: a broad-spectrum, rapidly microbicidal peptide with *in vivo* activity. *Antimicrob. Agents. Chemother.* **41**, 1738–1742.

Steiner, H., Hultmark, D., Engström, A., Bennich, H. and Boman, H.G. (1981) Sequence and specificity of two antibacterial proteins involved in insect immunity. *Nature* **292**, 246–248.

Storici, P. and Zanetti, M. (1993) A novel cDNA sequence encoding a pig leukocyte antimicrobial peptide with a cathelin-like prosequence. *Biochem. Biophys. Res. Commun.* **196**, 1363–1368.

Tailor, R.H., Acland, D.P., Attenborough, S., Cammue, B.P.A., Evans, I.J., Osborn, R.W., *et al.* (1997) A novel family of small cysteine-rich antimicrobial peptides from seed of *Impatiens balsamina* is derived from a single precursor protein. *J. Biol. Chem.* **272**, 24480–24487.

Tamamura, H., Ishihara, T., Otaka, A., Murakami, T., Ibuka, T., Waki, M., *et al.* (1996) Analysis of the interaction of an anti-HIV peptide, T22([Tyr5,12, Lys7}-polyphemusin II), with gp120 and CD4 by surface plasmon resonance. *Biochim. Biophys. Acta* **1298**, 37–44.

Tamamura, H., Kuroda, M., Masuda, M., Otaka, A., Funakoshi, S., Nakashima, H., *et al.* (1993) A comparative study of the solution structures of tachyplesin I and a novel anti-HIV synthetic peptide, T22([Tyr5,12,Lys7]-polyphemusin II), determined by nuclear magnetic resonance. *Biochim. Biophys. Acta* **1163**, 209–216.

Terras, F.G.R., Schoofs, H.M.E., DeBolle, M.F.C., van Leuren, F., Rees, S.B., Vander Leyden, J., *et al.* (1992) Analysis of two novel classed of plant antifungal proteins from radish (Raphanus sativus L.) Seeds. *Proc. Natl. Acad. Sci. USA.* **267**, 15301–15309.

Tobias, P.S., Soldau, K. and Ulevitch, R.J. (1986) Isolation of a LPS binding acute phase reactant from rabbit serum. *J. Exp. Med.* **164**, 777–793.

Valore, E.V. and Ganz, T. (1992) Posttranslational processing of defensin in immature human myeloid cells. *Blood* **79**, 1538–1544.

Valore, E.V., Park, C.H., Quayle, A.J., Wiles, K.R., McCray, P.B Jr. and Ganz, T. (1998) Human beta-defensin-1: an antimicrobial peptide of urogenital tissues. *J. Clin. Invest.* **101**, 1633–1642.

Van der Hooven, H.W., Fogolari, F., Rollema, H.S., Konings, R.N., Hilbers, C.W. and van der Ven, F.J. (1993) NMR and circular dichroism studies of the lantibiotic nisin in non-aqueous environment. *FEBS lett.* **319**, 189–194.

Wachinger, M., Kleinschmidt, A., Winder, D., von Pechmann, N., Ludvigsen, A., Neumann, M., *et al.* (1998) Antimicrobial peptides melittin and cecropin inhibit replication of human immunodeficiency virus 1 by suppressing viral gene expression. *J. Gen. Virol.* **79**, 731–740.

Westerhoff, H.V., Zasloff, M., Rosner, J.L., Hendler, R.W., De Wall, A., Vaz Gomes, A., *et al.* (1995) Functional synergism of the magainins PGLa and magainin-2 in *Escherichia coli*, tumor cells and liposomes. *Eur. J. Biochem.* **228**, 257–264.

Wieprecht, T., Dathe, M., Beyermann, M., Krause, E., Maloy, W.L., MacDonald, D.L., *et al.* (1997) Peptide hydrophobicity controls the activity and selectivity of magainin 2 amide in interaction with membranes. *Biochemistry* **36**, 6124–6132.

Winder, D., Gunzburg, W.H., Erfle, V. and Salmons, B. (1998) Expression of antimicrobial peptides has an antitumour effect in human cells. *Biochem. Biophys. Res. Commun.* **242**, 608–612.

Wu, M. and Hancock, R.E.W. (1999a) Interaction of the cyclic antimicrobial cationic peptide bactenecin with the outer and cytoplasmic membrane. *J. Biochem.* **274**, 29–35.

Wu, M., Maier, E., Benz, R. and Hancock, R.E.W. (1999b) Mechanism of interaction of different classes of cationic antimicrobial peptides with palnar bilayers and with the cytoplasmic membrane of *Escherichia coli*. *Biochemistry* **38**, 7235–7242.

Yarus, S., Rosen, J.M., Cole, A.M. and Diamond, G. (1996) Production of active bovine tracheal antimicrobial peptide in milk of transgenic mice. *Proc. Natl. Acad. Sci. USA.* **93**, 14118–14121.

Yasin, B., Harwig, S.S.L., Lehrer, R.I. and Wagar, E.A. (1996) Susceptibility of *Chlamydia trachomatis* to protegrins and defensins. *Infect. Immun.* **64**, 709–713.

Zanetti, M., Del Sal, G., Storici, P., Schneider, C. and Romeo, D. (1993) The cDNA of the neutrophil antibiotic Bac5 predicts a pro-sequence homologous to a cystein proteinase inhibitor that is common to other neutrophil antibiotics. *J. Biol. Chem.* **268**, 522–526.

Zanetti, M., Storici, P., Tossi, A., Scocchi, M. and Gennaro, R. (1994) Molecular cloning and chemical synthesis of a novel antibacterial peptide derived from pig myeloid cells. *J. Biol. Chem.* **269**, 7855–7858.

Zasloff, M. (1987) Magainins, a class of antimicobial peptides from Xenopus skin: isolation, characterization of two active forms, and partial cDNA sequence of a precursor. *Proc. Natl. Acad. Sci.* **84**, 5449–5453.

Zasloff, M., Martin, B. and Chen, H.C. (1988) Antimicrobial activity of synthetic magainin peptides and several analogues. *Proc. Natl. Acad. Sci. USA* **85**, 910–913.

Zhang, L., Falla, T., Wu, M., Fidai, S., Burian, J., Kay, W. and Hancock, R.E.W. (1998) Determinants of recombinant production of antimicrobial cationic peptides and creation of peptide variants in bacteria. *Biochem. Biophys. Res. Commun.* **247**, 674–680.

Zhang, L., Benz, R. and Hancock, R.E.W. (1999) Influence of proline residues on the antibacterial and synergistic activies of α-helical peptides. *Biochemistry* **38**, 8102–8111.

Zhao, C., Liaw, L., Lee, I.H. and Lehrer, L.I. (1997) cDNA cloning of three cecropin-like antimicrobial peptides (stylelins) from the tunicate, Styela clava. *FEBS Lett.* **42**, 144–148.

Table 1. Amino acid sequence of selected cationic peptides

Name	Sequence	Origin
α-Helices		
Cecropin A	KWKLFKKIEKVGQNIRDGIIKAGPAVAVVGQATQIAK-NH2	Insect
Cecropin B	KWKVFKKIEKMGRNIRNGIVKAGPAIAVLGEAKAL-NH2	Insect
Melittin	GIGAVLKVLTTGLPALISWIKRKRQQ	Honey bee
Magainin 1	GIGKFLHSAGKFGKAPVGEIMKS	Frog
Magainin 2	GIGKFLHSAKKFGKAFVGEIMNS	Frog
Dermaseptin S1	ALWKTMLKKLGTMALHAGKAALGAAADTISQGTQ	Frog
Dermaseptin S2	ALWFTMLKKLGTMALHAGKAALGAAANTISQGTQ	Frog
Extended		
Indolicidin	ILKWPWPWPWRR	Bovine
Bac5	RFRPPIRRPPIRPPFYPPFRPPIRPPIFPPIRPPFRPPLRFP	Bovine
Bac7	RRIRPRPPRLPRPRPRPLPFPRPGPRPIPRPLPFPRPGPRPIP	Bovine
PR-39	RRRPRPPYLPRPRPPPFFPPRLPPRIPPGFPPRFPPRFP-NH2	Porcine
β-sheet structure		
α-defensins		
HNP-1	ACYCRIPACIAGERRYGTCIYQGRLWAFCC	Human
HNP-2	CYCRIPACIAGERRYGTCIYQGRLWAFCC	Human
HNP-3	DCYCRIPACIAGERRYGTCIYQGRLWAFCC	Human
HNP-4	VCSCRLVFCRRTELRVGNCLIGGVSFTYCCTRV	Human
HNP-5	QARATCYCRTGRCATRESLSGVCEISGRLYRLCCR	Human
HNP-6	TRAFTCHCRRS-CYSTEYSYGTCTVMGINHRFCCL	Human
β-defensins		
HBD-1	GLGKRSDHYNCVSSGGQCLYSACPIFTKIQGTCYRGKAKCCK	Human
HBD-2	GIGDPVTCLKSGAICHPVFCPRRYKQIGTCGLPGTKCCKKP	Human
TAP	NPVSCVRNKGICVPIRCPGSMKQIGTCVGRAVKCCRKK	Bovine
LAP	GVRNSQSCRRNKGICVPIRCPGSMRQIGTCLGAQVKCCRRK	Bovine
Two disulphides		
Protegrin-1	RGGRLCYCRRRFCVCVGR-NH2	Pig
Protegrin-2	RGGRLCYCRRRFCICV-NH2	Pig
Protegrin-3	RGGGLCYCRRRFCVCVGR-NH2	Pig
Protegrin-4	RGGRLCYCRGWICPCVGR-NH2	Pig
Protegrin-5	RGGRLCYCRPRFCVCVGR-NH2	Pig

Table 1. (*continued*)

Name	Sequence	Origin
Tachyplesin-1	KWCFRVCTRGTCYRRCR-NH2	Crab
Tachyplesin-2	RWCFRVCYRGICYRKCR-NH2	Crab
Tachyplesin-3	KWCFRVCYRGICYRKCR-NH2	Crab
Polyphemusin I	RRWCFRVCYRGFCYRKCR-NH2	Crab
Polyphemusin II	RRWCFRVCYKGFCYRKCR-NH2	Crab
Loop peptide		
Bactenecin	RLCRIVVIRVCR	Bovine
Brevinin-1	FLPVLAGIAAKVVPALFCKITKKC	Frog
Brevinin-1E	FLPLLAGLAANFLPKIFCKITRKC	Frog
Brevinin-1Ea	FLPAIFRMAAKVVPTIICSITKKC	Frog
Brevinin-1Eb	VIPFVASVAAEMMQHVYCAASRKC	Frog
Brevinin-1Ec	PFPLLAGLAANFFPKIFCKITRKC	Frog
Brevinin-2	GLLDSLKGFAATAGKGVLQSLLSTASCKLAKTC	Frog
Brevinin-2E	GIMDTLKNLAKTAGKGALQSLLNKASCKLSGQC	Frog
Brevinin-2Ea	GILDTLKLNAISAAKGAAQGLVNKASCKLSGQC	Frog
Esculentin-1	GIFSKLGRKKIKNLLISGLKNVGKEVGMDVVRTGIDIAGCKIKGEC	Frog
Esculentin-1a	GIFSKLAGKKIKNLLISGLKNVGKEVGMDVVRTGIDIAGCKIKGEC	Frog
Esculentin-1b	GIFSKLAGKKLKNLLISGLKNVGKEVGMDVVRTGIDIAGCKIKGEC	Frog
Esculentin-2a	GILSLVKGVAKLAGKGLAKEGGKFGLELIACKIAKQC	Frog
Esculentin-2b	GIFSLVKGAAKLAGKGLAKEGGKFGLELIACKIAKQC	Frog
Peptides of other structure		
1b-AMP1	QWGRRCCGWGPGRRYCVRWC	Plant
1b-AMP2	QYGRRCCNWGPGRRYCKRWC	Plant
1b-AMP3	QYRHRCCAWGPGRKYCKRWC	Plant
1b-AMP4	QWGRRCCGWGPGRRYCRRWC	Plant

15. Molecular Genetic and Combinatorial Biology Approaches to Produce Novel Antibiotics

Richard H. Baltz
CognoGen Biotechnology Consulting, Indianapolis, Indiana

INTRODUCTION

The Role of Secondary Metabolites as Antibiotics

The majority of antibiotics in clinical use or in clinical trials are secondary metabolites of microorganisms or semisynthetic derivatives of secondary metabolites (Strohl, 1997). They range in size from the relatively small β-lactams (penicillins and cephalosporins) which are composed primarily of cyclized tripeptides, to the more complex heptapeptide glycopeptides, the 14- and 16-membered macrolides, and the 13-amino acid cyclic lipopeptide daptomycin. Most of these molecules are too complex to synthesize chemically by economically feasible processes. Their complexity also precludes synthesizing them *de novo* by traditional or combinatorial chemistry. Indeed, a recent survey of natural product and synthetic compound databases indicated that 40% of natural products are not represented in synthetic compound databases (Henkel *et al.*, 1999). It is unlikely that combinatorial chemistry will ever match the complexity of microbial biosynthesis, which employs highly sophisticated enzymes, such as giant multi-domain polyketide synthases (PKSs), nonribosomal peptide synthetases (NRPSs), a multitude of glycosyltransferases that utilize diverse deoxysugars, and many other tailoring enzymes such as hydroxylases, haloperoxidases, acylases and methyltransferases. It is the goal of molecular genetic and combinatorial biology approaches to harness and further direct the evolution of complex biosynthetic pathways to expand the repertoire of secondary metabolites, such as antibiotics.

Historical Roots to Molecular Genetic and Combinatorial Biology Approaches

Molecular genetic and combinatorial biology approaches are based in part upon successful production of novel antibiotics by mutagenesis, mutasynthesis, mutational biosynthesis, hybrid biosynthesis, bioconversions, and *in vitro* enzymatic conversion of natural or unnatural substrates (Baltz, 1982; Baltz *et al.*, 1986; Delzer *et al.*, 1984; Dutten *et al.*, 1991; Huffman *et al.*, 1992; Nagaoka and Demain, 1975; Marshall and Wiley, 1986; Omura *et al.*, 1983; Queener *et al.*, 1978; Rinehart, 1979; Sadakane *et al.*, 1982; Toscano *et al.*, 1983; Zmijewski and Fayerman, 1995). These studies demonstrated that at least some enzymes in secondary metabolite biosynthesis have relaxed substrate specificities, thus facilitating the incorporation of related or modified substrates into novel products related to known natural products. Current molecular genetic approaches are also based upon recent observations indicating that the multidomain subunits of the giant type I polyketide synthases (Katz, 1997; Khosla, 1997) and peptide synthetases (Marahiel *et al.*, 1997; von Dohren *et al.*, 1997) can be modified by molecular genetic manipulation of the corresponding genes to produce functional hybrid enzymes that produce novel molecules. Also, individual subunits of type II polyketide synthases can be exchanged, and the hybrid multi-component enzymes are often functional (Hopwood, 1997; Khosla and Zawada, 1996).

Molecular Genetic and Combinatorial Biology Approaches to Generate Novel Antibiotics

A number of molecular genetic technologies are needed to carry out manipulations to modify antibiotic biosynthetic pathways. These include methods to identify and clone antibiotic biosynthetic genes, cloning vectors, methods to introduce cloned genes into actinomycetes, and methods to exchange gene sequences in precise locations in the chromosome. The term combinatorial biology, which has (unfortunately) been co-opted from combinatorial chemistry, suggests having the ability to do these manipulations on a grand scale and in a relatively random fashion. Combinatorial biology (or combinatorial biosynthesis) approaches that focus on modifying existing biological activities, however, are likely be fruitful at a modest scale relative to current combinatorial chemistry approaches.

In this chapter, I limit my discussions primarily to the actinomycetes, which produce the majority of naturally occurring antibiotics, and illustrate molecular genetic and combinatorial biosynthesis approaches by key examples, citing review articles when possible. I discuss the state-of-the-art and limitations of the current molecular genetic and combinatorial biosynthesis approaches, and assess the prospects for developing robust combinatorial biosynthesis methods to generate novel antibiotics.

MOLECULAR GENETIC TOOLS IN ACTINOMYCETES

Gene Transfer Methods

The standard method to introduce DNA into actinomycetes is protoplast transformation (Baltz, 1995 and References therein). Protoplast transformation works well in some actinomycetes, parti-

cularly those that do not express significant levels of restriction (Baltz, 1995; Baltz and Hosted, 1996; Baltz, 1997a). Since most streptomycetes (and presumably most other actinomycetes) express restriction, a more robust method is to introduce DNA into actinomycetes by conjugation from *Escherichia coli* (Mazodier *et al.*, 1989), a procedure that circumvents restriction (Baltz and Hosted, 1996; Baltz, 1998; Bierman *et al.*, 1992; Matsushima *et al.*, 1994). Bacteriophage FP43-mediated transduction of plasmid DNA (McHenney and Baltz, 1988; 1989; Matsushima *et al.*, 1989) can also circumvent restriction barriers by transducing cells that have been grown under conditions to minimize the expression of restriction, and by preparing transducing lysates on hosts that modify the plasmid for certain restriction systems. Transduction can be used to rapidly transfer cloned DNA into a large number of unrelated streptomycetes. This procedure might be useful for combinatorial biology to rapidly add cloned genes to many different streptomycete strains. Transduction is currently limited to pIJ702-derived plasmids lacking replication functions for *E. coli*.

DNA has been introduced into *Streptomyces* species and into *Saccharopolyspora erythraea* by electroporation (Mazy-Servais *et al.*, 1997; Fitzgerald *et al.*, 1998; English *et al.*, 1998; Pigac and Schrempf, 1995). It remains to be seen if electroporation has any advantages over protoplast transformation and conjugation, but it may have important applications in some actinomycetes.

Cloning Vectors

There are a variety of cloning vectors available for *Streptomyces* species. These include self-replicating vectors of high and low copy number, site-specific integrating vectors, conjugal vectors, vectors for insertion of large blocks of genes into the chromoosome, vectors for high level expression of cloned genes, and single-strand vectors (Baltz, 1998; Baltz and Hosted, 1996; Bierman *et al.*, 1992; Sosio *et al.*, 2000; Hillemann *et al.*, 1991; Hopwood *et al.*, 1987; Kieser and Hopwood, 1991; Meurer and Hutchinson, 1999; Rao *et al.*, 1987; Rowe *et al.*, 1998). The conjugal vectors that contain *oriT* from the broad host range gram-negative plasmid RK2 are useful to bypass restriction barriers in actinomycetes (Baltz, 1997a; Baltz and Hosted, 1996; Flett *et al.*, 1997). Plasmids that integrate site-specifically into plasmid or phage attachment sites, or into neutral genomic sites by homologous recombination, are useful to generate stable recombinants that exert little or no negative effects on secondary metabolite production (Baltz, 1997a; Baltz, 1998). Self-replicating plasmid cloning vectors often cause actinomycetes to produce reduced levels of secondary metabolites (Baltz, 1997a). Cosmids are useful to clone large blocks of secondary metabolite biosynthetic genes, up to ~50kb. In some cases, this is sufficient to clone whole pathways (see Table 1). Bacterial artificial chromosome (BAC) vectors are useful to clone much larger clusters of genes. BACs have been used to clone inserts averaging >50 kb, and as large as 140 kb (Sosio *et al.*, 2000). Since it is not uncommon for antibiotic biosynthetic pathways to span 50 to 100 kb (Table 1), BACs should be particularly useful to clone complete antibiotic biosynthetic pathways from both culturable and unculturable microorganisms. Combinations of these key features are particularly useful (e.g., cosmid or BAC vectors containing *oriT* and site-specific insertion sequences; Bierman *et al.*, 1992; Baltz, 1998).

Cloning Procedures for Antibiotic Biosynthetic Genes

The procedures to clone antibiotic biosynthetic genes are well documented (Jones, 1989; Meurer and Hutchinson, 1998; McHenney *et al.*, 1998). They are based on the observation that antibiotic biosynthetic and resistance genes are generally clustered in actinomycetes (Seno and Baltz, 1989). Therefore, if one gene in the pathway is cloned, it can be used as a hybridization probe to identify overlapping cosmids or BACs spanning the whole pathway. Cloning the initial gene in a pathway has been accomplished by: (i) purifying an enzyme in the biosynthetic pathway, obtaining partial amino

Table 1. Some Antibiotic and Other Secondary Metabolite Biosynthetic Pathways Cloned from Actinomycetes

Antibiotic	DNA (kb)	Producing Organism	Reference
Peptides			
cephamycin	~ 22	*S. clavuligerus*	Paradkar *et al.*, 1997
pristinamycin	~ 32	*S. pristinaespiralis*	de Crecy-Lagard *et al.*, 1997a, b
chloroeremomycin	~ 72	*A. orientalis*	van Wageningen *et al.*, 1998
daptomycin	~ 50	*S. roseosporus*	McHenney *et al.*, 1998
Polyketides (type I)			
erythromycin	~ 60	*S. erythraea*	Staunton and Wilkenson 1997
tylosin	~ 90	*S. fradiae*	Baltz and Seno 1988
rifamycin	~ 90	*A. mediterranei*	August *et al.*, 1998
avermectin	~ 90	*S. avermitilis*	Ikeda *et al.*, 1999
rapamycin	~107	*S. hygoscopicus*	Schwecke *et al.*, 1995
Polyketides (type II)			
tetracycline	~ 30	*S. rimosus*	Binnie *et al.*, 1989
doxorubicin	~ 45	*S. peucetius*	Hutchinson 1997
tetracenomycin	~ 13	*S. glaucescens*	Hutchinson 1997
actinorhodin	~ 25	*S. coelicolor*	Fernandez-Moreno *et al.*, 1992
granatacin	~ 38	*S. violaceoruber*	Ichinose *et al.*, 1998
Glycosides			
streptomycin	~ 36	*S. griseus*	Piepersberg 1997
lincomycin	~ 35	*S. lincolnensis*	Peschke *et al.*, 1995
Others			
mitomycin C	~ 55	*S. lavendualae*	Mao *et al.*, 1999

acid sequence, and preparing a degenerate probe to identify the gene in a library; (ii) cloning an antibiotic resistance gene; (iii) complementing a mutation in an antibiotic biosynthetic gene; (iv) identifying an antibiotic biosynthetic gene by transposon mutagenesis and cloning the sequences flanking the transposon insertion; and (v) using a gene or sequences designed from related genes as hybridization probes (Jones, 1989; McHenney et al., 1998). Further characterization of the pathway genes can be accomplished by DNA sequence analysis, insertional mutagenesis, complementation of mutants blocked in antibiotic biosynthesis, and, in some cases, heterologous expression of individual genes or the whole pathway. Table 1 summarizes examples of antibiotic biosynthetic gene pathways cloned and analyzed by these methods.

In vivo Recombination Procedures

Gene replacement

To modify the biosynthesis of antibiotics, appropriate DNA segments are often joined in vitro by standard recombinant DNA techniques, and segments of DNA are inserted into the chromosome by gene replacement, relying on the hosts recombination machinery. The problem of replacing a segment of DNA by double crossover is that the first crossover is easily selected, using the antibiotic resistance expressed on the incoming plasmid, but the second crossover often needs to be screened from a large number of recombinants. This problem has been alleviated for gene disruption by using single stranded DNA, which gives high frequencies of double crossovers (Hillemann et al., 1991; Oh and Chater, 1997; Onaka et al., 1998). This works if the desired recombinants can be selected for antibiotic resistance. The problem of inserting a silent functional allele at a precise location of the chromosome has been addressed by developing vectors for direct selection of recombinants containing double crossovers. One system employs the rpsL gene (Hosted and Baltz, 1997), the locus for mutation to streptomycin resistance (SmR). Since streptomycin sensitivity (SmS) is dominant over SmR, double crossovers can be selected by placing a SmR mutation in the chromosome and placing the

wild type (SmS) rpsL gene on the plasmid. Gene replacement can be accomplished by selecting for a single crossover that inserts a temperature sensitive plasmid expressing an antibiotic resistance marker into the chromosome by homologous recombination. Subsequent selection for SmR yields recombinants that have eliminated the plasmid (SmS) by a second crossover. A high percentage of the progeny will have undergone gene replacement. This procedure is particularly powerful if an antibiotic resistance gene is present in the gene targeted for replacement, since recombinants containing double crossovers will loose the antibiotic resistance associated with the target gene. An appealing aspect of the rpsL system is that certain SmR mutants cause enhanced antibiotic biosynthesis (Hesketh and Ochi, 1997; Hosoya et al., 1998; Shima et al., 1996), so the final recombinant will require no further genetic manipulation to maintain high level antibiotic production.

Another approach for direct selection for double crossovers is the use of 6-deoxy-D-glucose to select recombinants that have eliminated a functional glucose kinase (glkA) gene located on a phage or plasmid (Fisher et al., 1987; Buttner et al., 1990; van Wezel and Bibb, 1996). In this case, genetic manipulations are carried out in glkA defective strain. It is not known if glkA defective strains are proficient in producing high levels of secondary metabolites, so additional studies are needed to determine if glkA mutations influence antibiotic yields in highly productive strains.

Homeologous recombination

To carry out multiple recombinations between two or more segments of DNA containing multiple type I PKS or NRPS modules, a procedure that could generate combinatorial libraries in vivo, it will be necessary to carry out recombination between partially homologous sequences (homeologous recombination). Unfortunately, homeologous recombination occurs at very low frequencies in wild type Streptomyces species (Hosted and Baltz, 1996; Baltz, 1998). In genetic crosses between E. coli and Salmonella typyhmurium, homeologous recombination occurs at very low frequencies, and is controlled by mismatch repair. Disruption of genes encoding mismatch repair proteins leads to very high levels of homeologous recombination (Rayssiguier *et al.*, 1989). In *Streptomyces roseosporus*, mutants were selected that undergo high frequency homeologous recombination, but the nature of the mutations is unknown (Hosted and Baltz, 1996). Such mutants will be useful for random crossing over within the type I PKS and NRPS coding regions *in vivo*. The further development of methods to facilitate random homeologous recombination *in vivo* will be important to further advance combinatorial biology and directed evolution.

Heterologous expression from plasmid vectors

Another method to carry out combinatorial biology is to express antibiotic biosynthetic genes from plasmids introduced into a heterologous host. This procedure can be carried out in *S. lividans* on one, two or three compatible plasmids (McDaniel *et al.*, 1999a; Zierman and Betlach, 1999; Xue *et al.*, 1999), and will be discussed in more detail below.

In vitro Mutation and Recombination Procedures

Site-directed mutagenesis and DNA shuffling

There are a number of methods available for molecular evolution of enzymes and other proteins (Encell *et al.*, 1999; Matsuura *et al.*, 1999; Patten *et al.*, 1997; Skandalis *et al.*,1997; Zhao *et al.*, 1998a). A very powerful new method is DNA shuffling (Patten *et al.*, 1997; Stemmer, 1994a). This procedure employs error-prone PCR and random recombination of DNA fragments to rapidly evolve proteins with improved characteristics. DNA shuffling of single genes has been used to: improve the catalytic activity of an enzyme (Stemmer, 1994b); change the substrate specificity of an enzyme (Zhang *et al.*, 1997); improve the folding of a protein in a heterologous host (Crameri *et al.*, 1996); and improve the function of a multigene pathway (Crameri *et al.*, 1997). An even more powerful procedure is DNA family shuffling (Crameri *et al.*, 1998). Using this method, a family of related genes can be shuffled under conditions of error-prone PCR. DNA

family shuffling has been used to evolve a moxalactamase gene containing eight segments of DNA from three different bacterial genes and thirty-three new amino acid substitutions not present in any of the four starting genes. The new chimeric/mutant gene encoded an enzyme with 270- to 540-fold improved moxalactamase activity over the starting enzymes. DNA family shuffling has also been used to evolve a thymidine kinase gene from herpes simplex virus types 1 and 2 that is much more sensitive to AZT (Christians *et al.*, 1999), and to evolve a chimeric interferon from 20 human interferon-α genes that is more active against murine encephalomyocarditis virus than natural murine interferons (Chang *et al.*, 1999). This powerful new technology might be applied to secondary metabolite tailoring enzymes (e.g., acyltransferases, glycosyltransferases, haloperoxidases, hydroxylases and methyltransferases) to alter substrate specificity or, in some cases, to alter cofactor specificity (e.g., NDP-sugar specificity for glycosyltransferases and acyl-CoA specificity for acyltransferases). DNA family shuffling may also have applications in generating chimeric PKS and NRPS multienzymes, and in improving the catalytic properties of the chimeric enzymes.

Sources of Genes for Molecular Genetic and Combinatorial Biology Approaches

Genes cloned from actinomycetes

Many genes encoding antibiotic and other secondary metabolite biosynthetic pathways have been cloned and sequenced (see Table 1 for some examples). More extensive lists of cloned secondary metabolite pathway genes can be found in Meurer and Hutchinson (1999) and von Dohren and Kleinkauf (1997). Therefore, many secondary metabolite biosynthetic genes are available for molecular genetic and combinatorial biosynthesis studies. These include type I and type II PKS genes, NRPS genes, sugar biosynthetic and glyosyltransferase genes, and genes encoding tailoring functions such as acyltransferases, methyltransferases, hydroxylases, and haloperoxidases. Some examples of these will be discussed below.

Genes cloned from environmental samples

Another potential source for secondary metabolite biosynthetic genes is environmental samples (Handelsman et al., 1998; Seow et al., 1997; Wells, 1998). It has been estimated that >99% of microbes are unculturable using standard culturing conditions (Amann et al., 1995; Hugenholtz and Pace, 1996). It is understandable that certain microbes that grow in highly specialized niches will not grow readily on standard media. Indeed, several of the Archaea have been sequenced, and they contain small genomes (1.66 to 2.18 Mb) encoding restricted functions highly adapted to niches not readily occupied by other microbes (Bult et al., 1996; Klenk et al., 1997; Smith et al., 1997). Their streamlined genomes contain no obvious type I PKS or NRPS genes, consistent with the notion that if they grow in an extreme environment, they need not produce secondary metabolites to ward off nonexistent competitors. However, these microbes may be good sources for tailoring enzymes (Robertson et al., 1996).

The secondary metabolite producing actinomycetes, on the other hand, grow aerobically in nutritionally rich environments occupied by many other bacteria and fungi. They have very large genomes (~8 Mb) (Leblond et al., 1993, 1996; Lezhava et al., 1995, Redenbach et al., 1996; Reeves et al., 1998; Solenberg and Baltz, 1991), encoding extensive catabolic functions facilitating growth on a wide variety of simple and complex carbon, nitrogen, and phosphorus sources. This begs the question: what fraction of actinomycetes are unculturable on standard media? A recent survey of the bacterial diversity among cultivated and uncultivated bacteria indicated that among the ''cosmopolitan'' bacteria divisions, those that reside in a variety of habitats, there are four groups that have many cultivated members: the proteobacteria, actinobacteria (actinomycetes), low G+C gram positives and cytophagales. There are four groups that contain primarily uncultivated members: the acidobacterium, verrucomicrobia, green non-sulfur bacteria an OP11 (Hugenholtz et al., 1998). This suggests that the actinomycetes may comprise a group of bacteria that are cultivatable at frequencies substantially higher than the norm. It would be useful to determine what fraction of actinomycetes are in fact cultivatable. It will also be

instructive to obtain genomic sequence for some individual cultivatable strains that group phylogenetically with primarily uncultivated bacteria to determine the size of their genomes, and if their genomes encode secondary metabolites.

A potential technical issue associated with cloning DNA from environmental samples is that the most abundant microbes, regardless of whether they are culturable, will likely dominate the libraries. Microbes that are present in very low abundance in environmental samples will likely be represented in low abundance in environmental libraries, unless steps are taken to remove viable microbes, and dominant uncultivated microbes. Furthermore, the secondary metabolite biosynthetic or tailoring genes will represent only a fraction of the genes cloned. Therefore, it may be necessary to prescreen the DNA and/or the libraries generated from environmental samples, and to use a number of different hosts with different secondary metabolite biosynthetic capabilities to increase the likelihood of isolating novel antibiotics by cloning complete or partial pathways.

EXAMPLES OF MOLECULAR GENETIC MODIFICATIONS OF ANTIBIOTICS

Type I Polyketides

Modular organization of polyketide synthase genes

The genetics and mechanisms of biosynthesis of complex polyketides are now relatively well understood (Aparicio et al., 1996; August et al., 1998; Bohm et al., 1998; Cane et al., 1998; Gokhale et al., 1999a, b; Holzbaur et al., 1999; Hopwood, 1997; Hutchinson and Fugii, 1995; Ikeda and Omura, 1997; Ikeda et al., 1999; Katz, 1997; Khosla, 1997; Khosla et al., 1999; Konig et al., 1997; Staunton et al., 1996; Staunton and Wilkenson, 1997; Tang et al., 1998; Weissman et al., 1997, 1998). Type I PKSs are involved in the biosynthesis of many relatively reduced polyketide structures, including the macrolide antibiotics erythromycin, tylosin, spiramycin, josamycin and rifamycin, and other secondary metabolites such as avermectin, spinosyns, rapamycin, and FK506. The modular type I PKSs are composed of a small number of very large multifunctional enzymes, each composed of modules containing all of the enzymatic functions required to process the individual fatty acid building units. These functions include the specificity for the side-chain length, stereochemistry at chiral centers, reduction level at each condensation step, and determination of macrolide ring size (Hopwood, 1997; Hutchinson and Fujii, 1995; Katz, 1997; Khosla, 1997; Kao *et al.*, 1995). The important implication of the modular type I PKS mechanism of polyketide assembly (Figure 1) is that each enzymatic reaction is encoded separately in giant modular genes, and the code can be changed by genetic manipulations such as focused deletions, additions, exchanges, rearrangements and translocations. The most extensive molecular genetic studies have been carried out on the erythromycin PKS, so this system serves as a model for what might be done in many other PKSs.

Erythromycin and ketolides

Erythromycin (Figure 1) is a 14-membered macrolide antibiotic first discovered in the 1950's. Its structure has been known for over 40 years (Wiley *et al.*, 1957). Erythromycin is a good model for combinatorial biology research, since derivatives of erythromycin continue to be important agents to treat respiratory pathogens (Kirst, 1991). The late stages in the biosynthesis of erythromycin have been understood for some time (Seno and Hutchinson, 1986), and more recently the biosynthesis of erythronolide, the 14-member macrolide ring of erythromycin, has been studied extensively (Bohm *et al.*, 1998; Cane *et al.*, 1998; Gokhale *et al.*, 1999a, b; Holzbaur *et al.*, 1999; Hopwood, 1997; Hutchinson and Fugii, 1995; Katz, 1997; Khosla, 1997; Staunton *et al.*, 1996; Staunton and Wilkenson, 1997; Weissman *et al.*, 1997, 1998). A wide variety of modifications, or reprogramming, of the polyketide synthase have been made in the producing strain, *S. erythraea*, and in the gene cluster transplanted into *Streptomyces coelicolor* or *Streptomyces lividans* (Hopwood, 1997; Katz, 1997; Khosla, 1997; Leadlay, 1997; Staunton, 1998).

Working in *S. erythraea*, the Abbott group (Katz, 1997) demonstrated that deletions in the enoylreductase (ER) of module 4 or in the ketoreductase (KR) of module 5 of the 6-deoxyerythronolide B synthase (DEBS)-coding region result in the biosynthesis of novel 14-member macrolides, including an acid stable derivative. Acid stability is an important feature of second

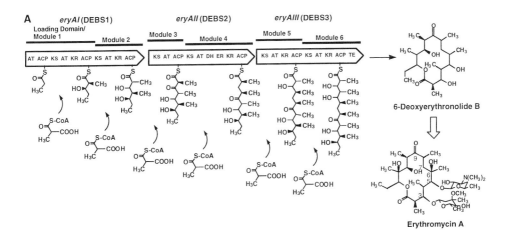

Figure 1. Biosynthesis of erythromycin by a type I polyketide synthase mechanism. The giant PKS genes *eryAI*, *eryAII*, and *eryAIII* encode three ~300kd proteins, DEBS1, DEBS2, and DEBS3, each of which has two modules containing the enzymatic functions required to process individual fatty acyl-CoA precursors. The fully extended chain is cyclised and released as 6-deoxyerythronolide B, which is further hydroxylated and glycosylated to give erythromycin A. AT, acyltransferase; ACP, acyl carrier protein; KS, ketosynthase; KR, ketoreductase; DH, dehydratase; and TE, thioesterase.

generation erythromycins (Kirst, 1991). They also made acyltransferase domain substitutions to generate desmethyl derivatives (Ruan *et al.*, 1997) and an ethyl-substituted derivative of erythromycin (Stassi *et al.*, 1998). The latter was particualrly interesting in that production of 6-ethylerythromycin A occurred only after a heterologous crotonyl-CoA reductase gene (*ccr*) from *Streptomyces collinus* was introduced into *S. erythraea* to provide a source for the butyryl-CoA needed to initiate biosynthesis of the ethyl-substituted polyketide. This points out an interesting dilemma: modification of the biosynthetic pathway is not always sufficient to assure production of the desired hybrid antibiotic. The cloning host also needs to provide all of the precursors and cofactors needed as building blocks.

Other modifications of erythronolide have been facilitated by the heterologous expression of the DEBS genes on a self-replicating plasmid in *S. coelicolor* (Kao *et al.*, 1994). Translocation of the thioesterase (TE) domain from the end of module 6 to the end of module 2 resulted in the production of novel 6-member triketides (Cortes *et al.*, 1995; Kao *et al.*, 1995), whereas translocation of the TE domain to the end of module 5 resulted in the production of a 12-member macrolide (Cortes *et al.*, 1995). The results of these and other studies (Bedford *et al.*, 1996; Oliynyk *et al.*, 1996) indicated that the side-chain length of building units, the reduction level of the polyketide, and the macrolide ring size can be modified by deletion mutation or by molecular surgery on the PKS coding region, by moving domains within a given pathway or by bringing in domains from heterologous pathways. This creates numerous possibilities for modifying the structures of existing molecules with important biological activities and for synthesizing totally new structures.

McDaniel *et al.* (1999a) have generated combinatorial libraries of erythronolide derivatives by modifying one, two or three carbon centers. They described the structures of 61 novel erythronolide derivatives, and estimated that >100 different compounds were generated, considering the minor components in the fermentation broths. Recently, this combinatorial approach has been streamlined by using two (Zierman and Betlach, 1999) or three plasmid systems (Xue *et al.*, 1999). The latter has provided a means to mix all combinations of

molecular genetic changes in the three individual giant PKS genes encoding the three subunits of the erythronolide PKS. However, the maximum yields for these processes with the unmodified erythronolide PKS were about 20 mg per liter, nearly 1000-fold lower than yields for typical macrolide fermentation production processes. In addition, yields from recombinants were often an additional 100-fold lower.

Another method to modify polyketide structures is to manipulate the loading module. The avermectin loading module has the capability to accept a large number of analogs related to normal branched-chain fatty acids (Denoya *et al.*, 1995; Dutten *et al.*, 1991). Marsden *et al.* (1998) exchanged the avermectin loading module for the erythronolide loading module in *S. erythraea*, and the recombinant produced six starter unit derivatives of natural erythromycin factors. This was an important advance in that it demonstrated that changing the starter unit did not interfere with the subsequent two glycosylations required for antibiotic activity. When the recombinant strain was fed short-chain fatty acids, an additional 12 erythromycin derivatives were generated (Pacey *et al.*, 1998). This approach could be coupled with the other modifications the PKS genes to generate a much larger combinatorial library of erythromycin derivatives. It might also be used to generate derivatives of other polyketides.

Yet another way to modify the starter unit in erythronolide biosynthesis is by mutational inactivation of the KS1 domain and feeding unnatural precursors (Jacobsen *et al.*, 1997). Feeding the mutant strain analogs of the diketide starter unit of erythromycin resulted in the production of novel polyketides in *S. coelicolor*. In one instance, the feeding of a triketide resulted in the production of a 16-member macrolide. They showed that some of the novel polyketides produced in *S. coelicolor* could be glycosylated by a *S. erythraea* mutant blocked in polyketide biosynthesis, producing active antibiotics.

A good example for a practical application of molecular genetic modification of erythronolide is the production of ketolides. Ketolides are semisynthetic 14-membered-ring derivatives of erythromycin that have a keto group at the three position rather than the sugar cladinose attached through a hydroxyl linkage (Agouridas *et al.*, 1998). The ketolide RU 64004 has potent antibacterial activity against many gram-positive pathogens, gram-negative upper respiratory pathogens, and mycoplasmas, including erythromycin-resistant and penicillin-resistant *S. pneumoniae*, many erythromycin-resistant and vancomycin-resistant enterococci, some erythromycin-resistant staphylococci, *Listeria monocytogenes, Haemophilus influenzae, Moraxella catarrhalis*, and *Streptococcus pyogenes* (Agouridas *et al.*, 1997, 1998; Ednie *et al.*, 1997; Schulin *et al.*, 1997). Ketolides are prepared from erythromycin A by acid cleavage of cladinose, oxidation of the 3-hydroxyl group to a keto group, followed by additional chemical modifications (Agouridas *et al.*, 1998). An alternative route to produce the 3-keto intermediate can be envisioned by modifying module 6 of the erythronolide gene cluster. Indeed, the Kosan group has accomplished the production of several 3-keto derivatives of erythronolide in *S. coelicolor* (McDaniel *et al.*, 1999). However, they have not shown if these molecules can be converted by *S. erythraea* to the glycosylated intermediate needed for chemical

modification to produce ketolides. Since cladinose is added to erythronolide before desosamine (Seno and Hutchinson, 1986), it is unlikely that *S. erythraea* is capable of glycosylating 3-keto-erythronolide. However, narbomycin and pikromycin are natural 14-membered macrolides containing a 3-keto group and desosamine attached at the 5 position. Furthermore, it was shown that the desosaminyltransferase from *Streptomyces venezuelae* can glycosylate 12-membered and 14-membered ketolides (Xue *et al.*, 1998a). Therefore, the desosaminyltransferase from a narbomycin- or pikromycin-producer might be more suitable to carry out the glycosylation (see below). Alternatively, the glycosyltransferase gene from *S. erythraea* might be modified by DNA family shuffling (Crameri *et al.*, 1998) with similar genes to generate a chimeric gene with the appropriate substrate specificity. This approach could potentially generate a large number of ketolide intermediates for chemical modification to further improve the properties of ketolides.

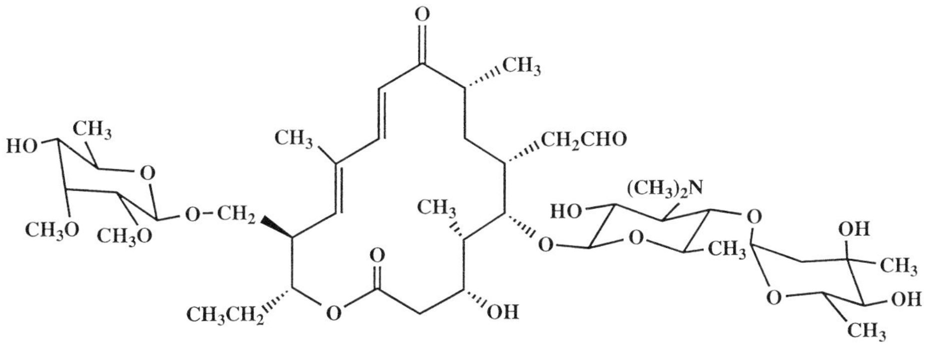

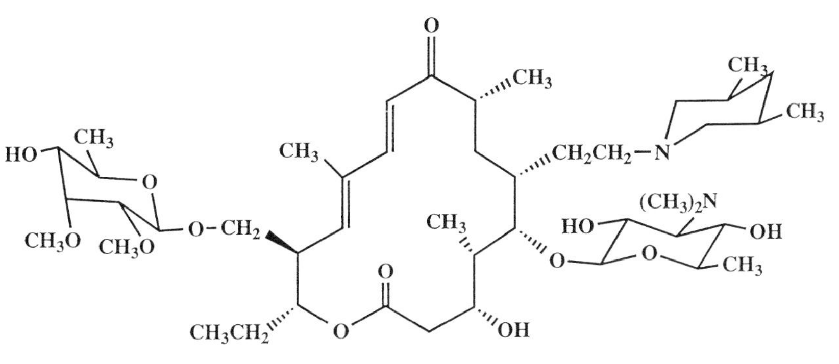

Figure 2.   The structures of tylosin and tilmicosin.

Tylosin and other 16-membered macrolides

Tylosin (Figure 2) is a macrolide antibiotic produced by *Streptomyces fradiae* (Baltz and Seno, 1988). Tylosin is composed of a 16-membered polyketide and three sugars, mycaminose, mycarose and mycinose. Whereas tylosin and the tylosin-derivatives 3-*O*-acetyl-4''-*O*-isovaleryltylosin and tilmicosin (Figure 2) are used in veterinary medicine (Kirst, 1994; Wilson, 1984), the related 16-membered macrolides, spiramycin and josamycin, are used in human medicine (Kirst, 1994; Nakayama, 1984; Olafsson *et al.*, 1999; Rubinstein and Keller, 1998). Interestingly, most clinical isolates of *S. pyogenes* and *S. pneumoniae* resistant to the 14-membered macrolides erythromycin, clarithromycin and roxithromycin, and to the 15-membered azithromycin, retained susceptibility to josamycin and spiramycin (Klugman *et al.*, 1998).

Tylosin is a good model for molecular genetic and combinatorial biosynthesis because: (i) tylosin can be bioconverted to highly active antibiotics by acylations at the 3-hydroxyl of the lactone and at the 4''-hydroxyl of mycarose (Okamoto *et al.*, 1980); (ii) mutants blocked in tylosin biosynthesis generated many novel antibiotics (Baltz and Seno, 1981, 1988), and many of these can be bioconverted to novel antibiotics by acylation (Kirst *et al.*, 1986); (iii) the genes involved in tylosin biosynthesis have been cloned, sequenced, and analysed in some detail (Baltz and Seno, 1988; Bate and Cundliffe, 1999; Bate *et al.*, 1999, 2000; Beckman *et al.*, 1989; Butler *et al.*, 1999; Cox *et al.*, 1986; Fish and Cundliffe, 1997; Fishman *et al.*, 1987; Gandecha *et al.*, 1997; Merson-Davies and Cundliffe, 1994; Wilson and Cundliffe, 1998); (iv) the biosynthetic pathway to tylosin is relatively well understood (Baltz *et al.*, 1983; Baltz and Seno, 1988); (v) the genes for self-resistance to tylosin in *S. fradiae* have been cloned, sequenced, and analyzed (Birmingham *et al.*, 1986; Fish and Cundliffe, 1996; Fouces *et al.*, 1999; Gandecha and Cundliffe, 1996; Keleman *et al.*, 1994; Kovalic *et al.*, 1994; Rosteck *et al.*, 1991; Zalacain and Cundliffe, 1989, 1991), (vi) the regulation of tylosin biosynthesis is highly sophisticated and is becoming better understood (Bate *et al.*, 1999; Fish and Cundliffe, 1997); and (vii) high tylosin producing mutants of *S. fradiae* are amenable to

facile molecular genetic manipulation (Baltz *et al.*, 1997; Baltz and Seno, 1988; Bierman *et al.*, 1992; Solenberg *et al.*, 1996), thus facilitating genetic modifications of the PKS in a genetic background proficient in macrolide oxidations and glycosylations, sugar methylations, and high-level production of glycosylated polyketide.

Spiramycin is also a good model for molecular genetic manipulation of 16-membered macrolides for many of the same reasons as tylosin. The producing organism, *S. ambofaciens*, is a particularly good host for gene cloning (Matsushima and Baltz, 1985), and macrolide biosynthetic genes have been cloned and analyzed in *S. ambofaciens* (Epp *et al.*, 1989; Kustoss *et al.*, 1996; Rao *et al.*, 1987). Kuhstoss *et al.* (1996) have exchanged the tylosin starter module, which specifies the incorporation of propionate, for the acetate starter module of platenolide in the spiramycin PKS in *S. ambofaciens*, and the recombinant produced 16-methyl platenolide I, a novel 16-membered polyketide. This validated the utility of *S. ambofaciens* as a suitable host for combinatorial biosynthesis.

Based upon the successful modifications in the erythromycin PKS, it seems reasonable to predict that numerous modifications of the PKS genes encoding 16-membered macrolides could yield functional enzymes. The translocation of the TE domain might convert the tylosin or spiramycin producing strains into producers of 14-membered macrolides with potentially beneficial attributes. Also, the tylosin pathway has a proof-reading thioesterase (Butler *et al.*, 1999) that might be manipulated to enhance the production of hybrid PKS proteins. The availability of cloned tylosin, spiramycin and erythromycin PKS genes, as well as other type I PKS genes and the rapamycin loading module, should provide the starting materials for a more extensive combinatorial biology approach to 14- and 16-membered macrolides.

Rifamycin

The complete ~90 kb cluster of genes involved in rifamycin biosynthesis in *Amycolatopsis mediterranei* has been cloned, sequenced and partially analysed (August *et al.*, 1998; Tang *et al.*, 1998). Polyketide biosynthesis initiates with a novel starter unit, 3-amino-5-hydroxy benzoic acid (AHBA). The rifamycin gene cluster also contains genes encoding the biosynthesis of AHBA. It is conceivable that the novel starter module and the AHBA genes could be used to generate additional chemical diversity in other polyketide pathways.

Type II Polyketides

Tetracycline, actinorhodin and related compounds

Unlike the more highly reduced macrolide antibiotics that are synthesized by modular type I PKS enzymes, the antibiotic tetracycline, and other polycyclic aromatic secondary metabolites such as actinorhodin, granaticin, tetracenomycin, jadomycin, and daunorubicin, are synthesized by type II PKSs (Hopwood, 1997; Hutchinson, 1997). The multienzyme type II PKSs are composed of several monofunctional proteins. The cyclic aromatic polyketides differ from the complex macrolides in that they do not require extensive reduction or reduction and dehydration cycles (Hutchinson and Fugii, 1995). Several type II PKS gene clusters have been cloned (Hopwood, 1997; Table 1), and many of the related enzymatic functions are interchangeable, allowing for the production of hybrid polyketides (Hopwood, 1997; Hutchinson and Fugii, 1995; Khosla and Zawada, 1996; Tsoi and Khosla, 1995). This combinatorial biosynthesis approach has generated a library of compounds from five PKS clusters (Tsoi and Khosla, 1995). Since some type II polyketides are glycosylated, additional diversity might be generated by coupling the hybrid PKS approach with glycosylations. Hybrid glycosylated anthracyclines have been generated by cloning PKS genes from *Streptomyces purpurascens* and *Streptomyces nogalater* into strains of *Streptomyces galilaeus* and *Streptomyces steffisburgensis* (Kunnari *et al.*, 1997; Niemi *et al.*, 1994; Ylihonko *et al.*, 1996). Because of the limited number of clinically useful antibiotics in this class, and the limited potential for generating large numbers of compounds by hybrid type II PKSs enzymes (Leadlay, 1997), manipulation of type II PKS genes may be less promising than manipulation of type I PKS genes and NRPS genes as means to generate novel antibiotics.

Peptide Antibiotics

Modular organization of peptide synthetases

Many antibiotics, including β-lactams (penicillins and cephalosporins), glycopeptides (vancomycin and teicoplanin), cyclic peptides (daptomycin and pristinamycin I), and other secondary metabolites contain a peptide backbone synthesized by NRPSs (Kleinkauf and von Dohren, 1990; Marahiel *et al.*, 1997; von Dohren, 1990; von Dohren and Kleinkauf, 1997; von Dohren *et al.*, 1997). Analogous to the modular type I PKS enzymes, the NRPS enzymes are composed of a small number of very large multifunctional enzymes composed of modules expressing all enzymatic functions required to process individual amino acids (Konz and Marahiel, 1999; Marahiel *et al.*, 1997; Stachelhaus and Marahiel, 1995; Stein and Vater, 1996; von Dohren *et al.*, 1997; Zuber and Marahiel, 1997). Also, like the type I PKS genes, the NRPS genes are organized as linear modules corresponding to the modular functions on the NRPS enzymes. Unlike ribosomal-mediated translation, the NRPS enzymes are not limited to the incorporation of L-amino acids. They can synthesize peptides containing D-amino acids, β-amino acids, hydroxy acids, and hydroxy-, *N*- and *C*-methylated amino acids, as well as L-amino acids, altogether accounting for about 300 different residues (Kleinkauf and von Dohren, 1996). The potential for combinatorial biosynthesis by this mechanism is enormous.

Module swaps

In *Bacillus subtilis*, the Leu module of a mono-modular subunit of the surfactin peptide synthetase coding region was substituted with modules of bacterial and fungal origin, and the hybrid genes encoded functional NRPSs with altered amino acid specificities that catalysed the biosynthesis of surfactin analogs with modified amino acid sequences (Stachelhaus *et al.*, 1995). Also, the minimal module, a core fragment of a module comprising the adenylation and thiolation domains, has been exchanged in a trimodular subunit of the surfactin NRPS coding region, yielding functional surfactin NRPS with altered substrate specificity (Schneider *et al.*, 1998).

Site-directed modifications of the adenylation domain

Recently, the crystal structure of the N-terminal adenylation subunit of the L-phe module of the *Bacillus brevis* gramicidin S synthetase complexed with AMP and L-phe has been determined (Conti *et al.*, 1997; Marahiel, 1997). Using this information, Stachelhaus *et al.* (1999) have determined the specificity-conferring code used by adenylation domains of NRPSs, based upon comparative studies of the amino acids lining the Phe binding site in the gramicidin PheA module and 160 other NRPS sequences in the databases. The 10 specificity-conferring amino acids are located in a 100 amino acid stretch in the adenylation domain. Using the deduced code for amino acid specificity, they introduced two mutations by site-directed mutagenesis in the PheA-encoding module and converted it to specify Leu. They also converted an Asp-specifying module to one specifying Asn by a single mutation. This methodology has the advantage that it results in minor changes in the overall protein structure, minimizing potential deleterious effects on protein-protein interactions in these giant enzymes.

Peptide antibiotics produced by actinomycetes

Daptomycin (Figure 3) is a cyclic lipopeptide produced by *S. roseosporus*. It is composed of a thirteen amino acid peptide cyclized to a ten amino

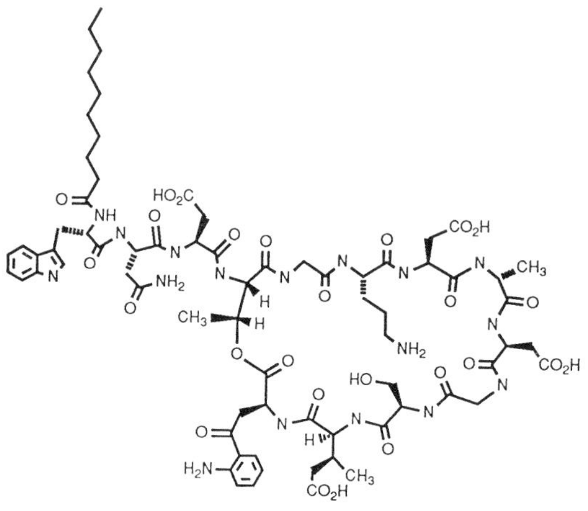

Figure 3. The structure of daptomycin.

acid ring and three amino acid tail. It has a ten carbon unit fatty acid (decanoic acid) attached to the *N*-terminal Trp. Daptomycin is a potent antibiotic with bacteriocidal activity against gram-positive pathogens, including methycillin-resistant *S. aureus*, vancomycin-resistant enterococci, and penicillin-resistant *S. pneumoniae* (Baltz, 1997b), and is currently in clinical trials sponsored by Cubist Pharmaceuticals. The genes for daptomycin biosynthesis have been cloned and partially sequenced (McHenney *et al.*, 1998). Since *S. roseosporus* can be manipulated genetically by a variety of techniques, including transposition mutagenesis, conjugation from *E. coli*, homeologous recombination, and gene replacement using the *rpsL* system (Hosted and Baltz, 1996, 1997; McHenney and Baltz, 1996; McHenney *et al.*, 1998), it represents a suitable model system to establish module swapping to generate novel derivatives of daptomycin.

The genes for the biosynthesis of the glycopeptide antibiotic chloroeremomycin (or A82846B; Figure 4) have been cloned and sequenced (Solenberg *et al.*, 1997a, b; van Wageningen *et al.*, 1997), and the NRPS genes have been identified. However, there are no reports of successful genetic manipulation of the producing strain, *A. orientalis*. *Amycolatopsis mediterranei* produces a related glycopeptide antibiotic, balhimycin. A transformation and gene replacement system has been developed for this strain (Pelzer *et al.*, 1997), and some of the balhimycin biosynthetic genes have been cloned and sequenced (Pelzer *et al.*, 1999). *A. mediterranei* may prove to be a suitable model system to generate hybrid NRPS genes to produce peptide analogs of vancomycin, chloroeremomycin, balhimycin and teicoplanin. Alternatively, the cloning and sequencing of the A47934 (Figure 4) NRPS genes from *Streptomyces toyocaensis* could provide the basis to carry out similar module swaps or site-directed mutagenesis in the NRPS genes in this strain, since *S. toyocaensis* is also readily manipulated by molecular genetic techniques and produces high levels of glycopeptides (Matsushima and Baltz, 1996; Solenberg *et al.*, 1997a).

Pristinamycins I (PI) and II (PII) are dissimilar bacteriostatic antibiotics produced by *Streptomyces pristinaespiralis* (de Crecy-Lagard *et al.*, 1997a; Sezonov *et al.*, 1997). In combination, PI and PII are bacteriocidal and highly active against staphylococci, streptococci, and *H. influenzae* (Sezonov

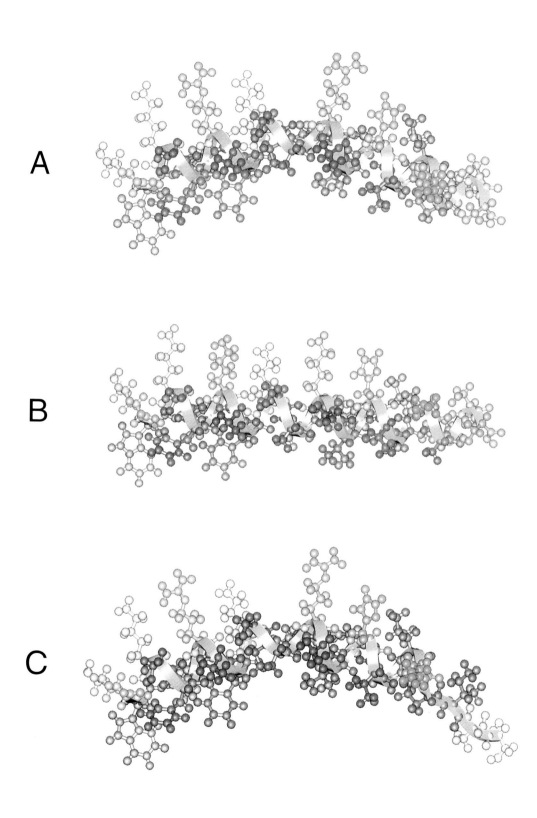

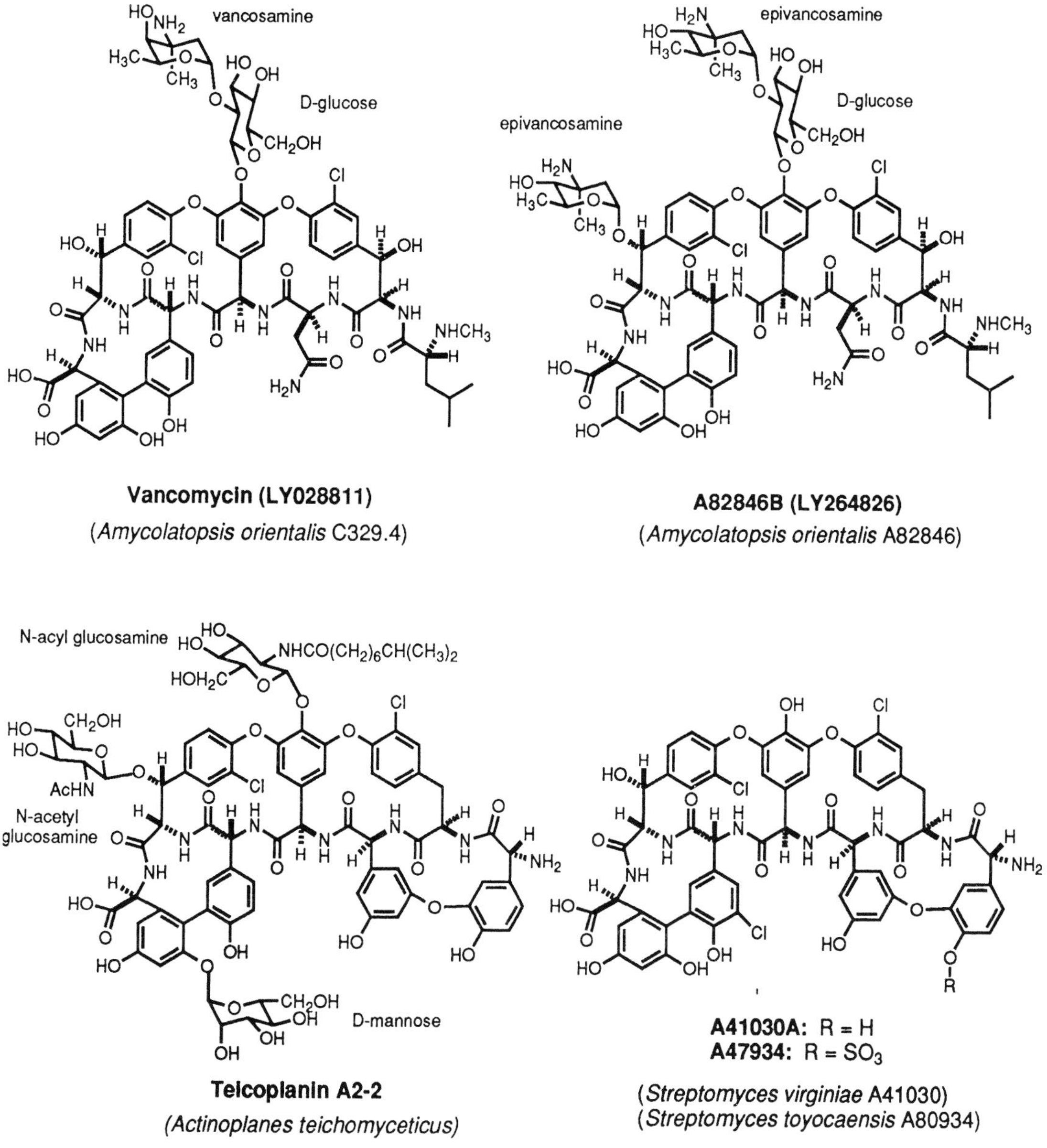

Figure 4. The structures of glycopeptide antibiotics.

et al., 1997). PI is composed of cyclohexadepsipeptides belonging to the B group of streptogramins, and PII of macrolactones of the A group of streptogramins. The genes for PI biosynthesis have been cloned and sequenced (Blanc et al., 1997; de Crecy-Lagard et al., 1997a, b), and could provide starting materials for molecular genetic modification of the peptide structure.

Acylations

Acylation of 16-membered macrolides

The activity of macrolide antibiotics can be modified by the acylation of hydroxyl groups on the polyketide or on a sugar attached to the polyketide. An example is the production of 3-O-acetyl 4″-O-isovaleryltylosin (AIV), a derivative of tylosin used in veterinary medicine (Kirst, 1994). AIV was initially produced by a two-step fermentation/bioconversion, using *S. fradiae* to produce tylosin and a mutant strain of *Streptomyces thermotolerans* blocked in polyketide biosynthesis to carry out the acylations of tylosin (Okamoto *et al.*, 1980). To streamline the process, Arisawa *et al.* (1994, 1996) cloned and sequenced genes from *S. thermotolerans* required for the acylations and introduced them into *S. fradiae*. When they inserted the macrolide 3-O-acyltransferase gene (*acyA*) into *S. fradiae* on the self-replicating plasmid, the recombinant produced 3-O-acetyltylosin (Arisawa *et al.*, 1994). Thus, the *acyA* gene expressed from its own promoter and produced enough enzyme to convert tylosin to 3-O-acetyltylosin.

The acylation of the 4″-hydroxyl position of the sugar mycarose was more difficult. The cloned *acyB1(carE)* gene that encodes the 4″-O-acyltransferase gene did not express enough enzyme to convert tylosin-related macrolides to the 4″-acylated products. Expression of *acyB1* requires the presence of *acyB2*, a regulatory gene adjacent to *acyB1* in *S. thermotolerans*, reading outward in the opposite direction, to express in *S. fradiae*. When *acyB1*, *acyB2* and *acyA* were introduced into *S. fradiae* on a replicating plasmid, the recombinant produced 3-O-acetyltylosin and 3-O-acetyl 4″-O-acetyltylosin, each at ~10% of the control (Arisawa *et al.*, 1996). When leucine was added to the medium to provide a precursor for the isovaleryl group, the predominant product was AIV, but the overall yield was <20% of control. It appears that the self-replicating plasmid containing the three *acy* genes caused a reduction in overall macrolide production. The solution to this problem may be to insert the *acy* genes into the chromosome in a neutral site.

Epp *et al.* (1989), who were the first to clone the *carE* (*acyB1*) gene encoding the 4″-O-acyltransferase, showed that the *carE* gene expressed from its own promoter was sufficient to catalyze the conversion of spiramycin to 4″-isovalerylspiramycin in *S. ambofaciens*. It appears that *S. ambofaciens* must express an *acyB2*-like function that is not expressed in *S. fradiae*. This example points out that different *Streptomyces* species can respond differently in their abilities to regulate the expression of heterologous genes.

Glycosylations

Many macrolide, glycopeptide, and aminoglycoside antibiotics have been identified as secondary metabolites produced by actinomycetes. It is evident that the antibiotic activity is often strongly influenced by sugar residues. For instance, tylactone, the macrolide precursor of tylosin devoid of the normal three sugars, lacks any detectable antibiotic activity (Kirst *et al.*, 1982). A47934, the sulfated aglycone of teicoplanin, has good *in vitro* antibiotic activity (Lancini and Cavalleri, 1990), but has not progressed to clinical studies. Piepersberg and Distler (1997) have compiled lists 45 D-6-deoxyhexoses and 62 L-6-deoxyhexoses found in naturally occuring secondary metabolites. Since many glycosyltransferase and sugar biosynthetic genes involved in secondary metabolism have been cloned and sequenced, and many sugar biosynthetic genes are clustered or partially clustered in actinomycetes (Bate and Cundliffe, 1999; Bate *et al.*, 1999; Gandecha *et al.*, 1997; Ichinose *et al.*, 1998; Kirsching *et al.*, 1997; Trefzer *et al.*, 1999; van Wageningen *et al.*, 1997; Xue *et al.*, 1998a), it is now feasible to construct novel combinations of macrolide or peptide backbones and sugar biosynthetic and attachment functions. The two main issues will be: expression of the cloned genes in heterologous hosts; and altering the substrate specificity of the glycosyltransferase enzymes to accept a broad array of substrates. DNA family shuffling could be a powerful tool to help generate an array of mutant/recombinant glycosyltransferases with the desired substrate specificities.

Glycosylation of macrolides

A number of different 6-deoxy sugars are present on 14- and 16-membered macrolides (Kirsching *et al.*, 1997; Nakagawa and Omura, 1984; Piepersberg and Distler, 1997; Trefzer *et al.*, 1999). It was shown many years ago that tylactone (protylonolide) could be taken up and glycosylated by *S. ambofaciens*, the producer of spiramycin, when the spiramycin polyketide biosynthesis was inhibited by cerulenin (Omura *et al.*, 1983). The resulting novel macrolide, chimeramycin, contained the lactone of tylosin, the forosamine of spiramycin, and the two sugars (mycaminose and mycarose) normally present on both tylosin and spiramycin. This demonstrated that the forosamyltransferase could use the tylosin lactone as a substrate. It has also been shown that one of the glycosyltransferases of *S. fradiae* can attach mycarose in place of mycaminose to tylactone in *tylB* mutants (Ōmura *et al.*, 1980; Jones *et al.*, 1982). Since all three glycosyltransferases involved in tylosin biosynthesis should be intact in *tylB* mutants (Merson-Davies and Cundliffe, 1994), it is not clear if the mycaminosyltransferase or mycarosyltransferase is carrying out the glycosylation. This enzyme, which has either relaxed NDP-sugar specificity or relaxed macrolide substrate specificity, might be a good starting point to initiate combinatorial glycosylations. DNA family shuffling technology should be applicable to modify both NDP-sugar cofactor specificities and macrolide substrate specificities. Since the the sugar biosynthetic and glycosyltransferase genes are well characterized in *S. fradiae* (Bate and Cundliffe, 1999; Bate *et al.*, 1999b; Fouces *et al.*, 1999; Gandecha *et al.*, 1997; Merson-Davies and Cundliffe, 1994; Wilson and Cundliffe, 1998), and since high tylosin-producing strains of *S. fradiae* can be manipulated genetically (Baltz and Seno, 1988; Baltz *et al.*, 1997; Bierman *et al.*, 1992; Solenberg *et al.*, 1996), *S. fradiae* should serve as a suitable host to explore combinatorial glycosylations.

Streptomyces venezuelae produces four macrolide compounds, the 14-membered narbomycin and pikromycin, and the 12-membered methymycin and neomethymycin. Narbomycin and pikromycin, and methymycin and neomethymycin differ from each other by hydroxylation patterns (Xue *et al.*, 1998a, 1998b). A single glycosyltransferase enzyme is capable of adding the amino sugar desosamine to both 12-membered and 14-membered lactones (Xue *et al.*, 1998a). Since this desosaminyltransferase enzyme has a relaxed substrate specificity and does not require a glycosylated 14-membered lactones as substrate, as seems to be the case for *S. erythraea* desosaminyltransferase (Seno and Hutchinson, 1986), and since narbomycin and pikromycin are natural ketolides, the *S. venezuelae* desosaminyltransferase gene *desVII* might be a good starting point for ketolide production (see below). It would also be a

candidate for DNA family shuffling with other glycosyltransferse genes for combinatorial glycosylations of macrolides and glycopeptides.

Glycosylation of glycopeptides

Glycopeptide antibiotics are potent bacteriocidal compounds active against gram-positive bacteria. Vancomycin is particularly important for the treatment of methycillin-resistant staphylococci. The glycopeptide antibiotics differ from each other in the core crosslinked heptapeptide and in glycosylation patterns (Lancini and Cavalleri, 1990; Nicas and Cooper, 1997). The simplest "glycopeptides" are comprised of chlorinated or chlorinated and sulfated heptapeptides devoid of sugars (Lancini and Cavaleri, 1990). These compounds are highly active as antibacterials *in vitro*, but lack sufficient *in vivo* activity to be pursued clinically. The clinically relevant glycopeptides, vancomycin and teicoplanin, contain two and three sugars, respectively (Nicas and Cooper, 1997). Since the heptapeptide core is sufficient to bind the terminal D-ala-D-ala of the stem peptide to block peptidoglycan biosynthesis, it appears that the glycosylations are needed to provide the appropriate pharmacokenetic/pharmacodynamic properties. Glycopeptides also form dimers, and it has been shown that the degree of dimer formation and membrane binding is correlated with *in vivo* activity (Allen *et al.*, 1997).

The natural product chloroeremomycin (A82846B), which is closely related to vancomycin (Nicas and Cooper, 1997), contains the vancomycin heptapeptide and chlorine residues, but contains an additional amino sugar (Figure 4). The two amino sugars present in chloroeremomycin are both epivancosamine rather than vancosamine, the amino sugar present on vancomycin. Chloroeremomycin has modest antibacterial activity against vancomycin-resistant enterococci (Cooper *et al.*, 1996), thus distinguishing it from other glycopeptides. This molecule has been modified chemically, and the p-chloro-biphenyl derivative LY333328 is highly active against vancomycin-resistant enterococci and other gram-positive pathogens (Balch *et al.*, 1998; Cooper *et al.*, 1996; Garcia-Garrot *et al.*, 1998; Kaatz *et al.*, 1998; Saleh-Mghir *et al.*, 1999). LY333328 has no greater binding affinity

for the terminal D-ala-D-ala in normal pentapeptide, or for D-ala-D-lac present in vancomycin-
resistant (VanR) enterococci, than vancomycin, but
is much more efficient at forming dimers and
binding to membranes (Allen *et al.*, 1997). Therefore, the glycosylations of the heptapeptide are
critical for *in vivo* activity, and to provide a scaffold
for further modifications to generate clinical
candidates active against vancomycin-resistant
pathogens.

Since there are three heptapeptide patterns and a
variety of glycosylation patterns for natural glycopeptides (Lancini and Caavalleri, 1990; Nicas and
Cooper, 1997), a random shuffling of these patterns
(Figure 5) could yield many different novel
compounds to use as substrates for further chemical
modification. Solenberg *et al.* (1997a, b) initiated
studies to address the concept of generating
recombinant strains to carry out unnatural glycosylations of A47934. A47934 has the same
heptapeptide as teicoplanin, but contains three
chlorine residues rather than two, and contains an
additional sulfate residue (Figure 4). A47934 posed
both strategic and tactical advantages for molecular
genetic modification. First, since it contains a
teicoplanin-like heptapeptide, it presents a different
peptide scaffold to further extend the important
observations made with chloroeremomycin modifications. If the three glycosylations of the

chloroeremomycin heptapeptide could be made on
the A47934 heptapeptide, this would provide a
novel starting material for chemical modification to
generate additional glycopeptides with activity
against VanR enterococci and other gram-positive
pathogens. It would also establish a basis for a
more extensive combinatorial biology approach to
generate novel glycopeptides. Secondly, A47934 is
produced by *S. toyocaensis*, an organism amenable
to facile molecular genetic manipulation by conjugation from *E. coli* and insertion of cloned DNA
into the ϕC31 *attB* site (Matsushima and Baltz,
1996).

Solenberg *et al.* (1997a, b) cloned the three
glycosyltransferase genes from the chloroeremo-
mycin-producer and two from the vancomycin-
producer. They showed by expression studies in
E. coli that the glucosyltransferase enzymes
encoded by the *gtfE* gene from the vancomycin-
producer, and the *gtfB* gene from the chloroer-
emomycin-producer, could attach glucose to the
vancomycin aglycone. However, only the enzyme
encoded by the *gtfE* gene could glycosylate
A47934. Insertion of the glucosyltransferase
genes into the chromosome of *S. toyocaensis*
under the control of the natural *A. orientalis*
promoters did not cause the production of
glucosylated A47934. However, insertion of a
strong promoter (*ermEp**) in front of the *gtfE*
caused the recombinant *S. toyocaensis* to produce
glucosyl-A47934. This demonstrated that the
heterologous glycosylation of a peptide substrate
related to the natural substrate is feasible *in vivo*.
There are two cautionary notes to consider
relative to random combinatorial biology: (i) only
one of the two highly related glucosyltransferase
enzymes was capable of using the unnatural
substrate; and (ii) expression *in vivo* required
fusing the glucosyltransferase gene to a strong
actinomycete promoter. Further advancements
will also require the cloning and expression of
genes involved in the biosynthesis of 6-deoxy
sugars such as vancosamine and epivancosamine.
The genes required for the conversion of 4-keto-
6-deoxy-D-glucose to epivancosamine have been
cloned and sequenced (van Wageningen *et al.*,
1998), and the open reading frames of four genes
are clustered and read in the same direction. A
fifth gene, which may be involved in sugar
methylation, is located outside of the four gene

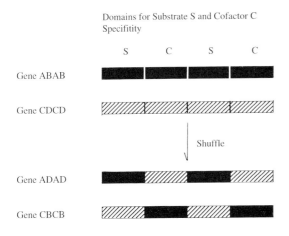

Figure 5. DNA shuffling of glycosyltransferase genes. The example shows how the hypothetical substrate and cofactor domains might be shuffled to give novel glycosyltransferase enzymes.

cluster. If the four genes are transcribed on a single mRNA, it might be feasible to engineer their coordinate expression by inserting a strong or regulatable actinomycete promoter upstream. If the fifth gene is needed, it could be added to generate a cassette for the biosynthesis of NDP-epivancosamine in *S. toyocaensis* or other actinomycete hosts.

Modification of NDP-sugar Biosynthesis

Another promising method to modify the sugars on antibiotics is to modify the NDP-sugar biosynthetic pathway. Madduri *et al.* (1998) demonstrated the feasibility of this approach by blocking the step in TDP-L-daunosamine biosynthesis encoded by the *dnmV* gene in *Streptomyces peucetius*, and complementing the mutation with the *avrE* gene from the NDP-oleandrose biosynthetic pathway from *S. avermitilis*, or with the *eryBIV* gene from the NDP-desosamine pathway in *S. erythraea*. Recombinants in both cases produced 4′-epidoxorubicin, containing the novel sugar L-4-epidaunosamine. In this example, the glycosyltransferase had broad enough cofactor specificity to use either of the 4′ epimers.

Borisova *et al.* (1999) and Zhao *et al.* (1998b) have made mutations in genes involved in NDP-desosamine biosynthesis, and some of the intermediates were used by the *S. venezualae* glycosyltransferase to generate novel derivatives of pikromycin and narbomycin. This implys that the glycosyltransferase (Xue *et al.*, 1998a) has broad macrolide substrate and broad NDP-sugar cofactor specificity. This suggests a potential route to synthesize ketolides biochemically.

Other sources of genes for glycosylations

There are many sources of sugar biosynthetic and glycosyltransferase genes for combinatorial glycosylations from gene clusters already cloned. Two other sugar biosynthetic pathways have been cloned and expressed in *S. lividans* (Kulowski *et al.*, 1999; Olano *et al.*, 1999). This should facilitate the generation of additional glycosylated polyketides, using glycosyltransferases with relaxed substrate/cofactor specifities.

Other Tailoring Enzymes

There are a number of other tailoring enzymes that modify peptides, polyketides, or sugar components of secondary metabolites. These include methyltransferases, hydroxylases, reductases, and haloperoxidases. These modifications are often critical for biological activity. Many genes encoding tailoring functions have been cloned and sequenced, and these could become starting materials for DNA family shuffling to alter substrate specificity. So far little attention has been paid to these enzymes important for combinatorial biosynthesis.

Mixed Peptide/Polyketides

There are a limited number of examples of antibiotics or other secondary metabolites that contain core structures composed of polyketide and peptide components. These include pristinamycin II (Thibaut *et al.*, 1995) and rapamycin (Aparicio *et al.*, 1996; Molnar *et al.*, 1996; Paiva and Demain, 1993; Schwecke *et al.*, 1995), for example. Mixed NRPS/PKS multidomain proteins have been reported in gram-negative bacteria (Gaisser and Hughes, 1997; Gehring *et al.*, 1998). The relatively small number of these compounds compared to core structures of pure polyketide or pure peptide may reflect the evolutionary history of secondary metabolite pathways. It is easy to imagine that current PKS or NRPS multimodular genes evolved by gene duplications/amplifications from more primative ancestral monomeric genes, and the resulting primative multimodules further evolved specificity by mutation and recombination. It is harder to imagine this process occurring with nonhomologous genes. Mixed PKS/NRPS multifunctional multimodular genes may have arisen by illegitimate recombination events. In the case of rapamycin, the addition of pipecolate to the completed polyketide chain is catalysed by a separate NRPS enzyme encoded by the *rapP* gene (Konig *et al.*, 1997). The constraints to natural evolution of mixed PKS/NRPS enzymes may not be a limitation for combinatorial biology, and the potential combinations unleashed by hybridizing the fundamental mechanisms of type I PKS and NRPS multienzymes is staggering. This coupled with glycosylations, acylations, and other tailoring

processes offers a wealth of potential for combinatorial biology approaches to novel antibiotics.

FUTURE PROSPECTS AND
CHALLENGES

The current state-of-the-art on molecular genetic
and combinatorial biology approaches to produce
novel antibiotics is very encouraging. A number
of molecular genetic tools to clone and express
antibiotic biosynthetic genes in heterologous
actinomycetes are in place, and many antibiotic
and other secondary metabolite gene clusters have
been cloned and sequenced. The mechanisms of
type I polyketide assembly and non-ribosomal
peptide assembly are now fairly well understood,
methods to genetically modify the biosynthetic
assembly code have been documented, and novel
compounds have been produced. Some progress is
being made on the heterologous expression of
complete NDP-sugar biosynthetic pathways and
glycosyltransferase genes, which will be necessary
to generate antibiotics with superior *in vivo*
activity, particularly macrolides, ketolides and
glycopeptides. DNA family shuffling offers a
powerful new technology to rapidly evolve
enzymes with superior characteristics, and this
methodology should play a critical role in
combinarorial biology. Many other important
advancements on other molecular genetic tools
and other secondary metabolites not discussed
here can be found in many of the review articles
cited.

While much progress has been made on the
molecular genetic tools needed for combinatorial
biosynthesis applications, there are still serious
limitations to the current technology. The model
systems (*S. coelicolor* and *S. lividans*) for expression of hybrid PKS genes are inadequate for large-
scale production of sufficient quantities of material
to support product development. Development of
more productive mutants of these strains is one
approach. Another approach is to develop industrial
antibiotic production strains for heterologous
expression of cloned genes. For type I PKS genes,
the obvious choices would be *S. erythraea*
(erythromycin producer), *S. fradiae* (tylosin producer) and *S. ambofaciens* (spiramycin producer).
These industrial strains have the potential to

produce ~1000-fold higher titers of macrolides
than *S. coelicolor* or *S. lividans* laboratory strains.
Likewise, *S. toyocaensis* may be an excellent host
for cloning glycopeptide genes, and other high-
producing pathway-specific hosts can be identified
and developed. Also, examples have been cited
indicating that different actinomycetes express
different levels of intermediates and regulatory
factors critical for expression of genes cloned from
heterologous hosts. Functionally identical enzymes
can display differences in substrate specificity that
can have a profound effect on the outcome of
generating hybrid glycopeptides. Also, little is
known about the global regulation, promoters,
and transcription factors needed to assure high
level expression of cloned genes in robust production strains. There is limited information on what
determines the efficiency of polyketide or peptide
assembly in chimeric pathways, and there is no
information yet available on the feasibility of
generating hybrid PKS/NRPS enzymes. Although
there is no published information on the improvement of enzyme activity in a chimeric PKS or
NRSP, DNA family shuffling offers an enticing
solution. These and a number of other issues
mentioned earlier in this chapter remain to be
addressed. Obviously, much more work is needed
at fundamental and applied levels to fully exploit
this potentially powerful approach to produce novel
antibiotics.

In the short term the most promising applications
of the current technology will be to modify known
antibiotic biosynthetic pathways (e.g., macrolides,
ketolides, glycopeptides, daptomycin, pristinamycins). In the longer term, greater efforts are needed
to develop combinatorial biology approaches that
truly mimic the natural evolutionary process of
mutation, recombination and selection for fitness,
but in a time frame consistent with drug development cycles, and with fitness defined by appropriate
antibiotic activities. This remains a daunting task,
and may require expanded funding by government
agencies, private endowments, venture capital, and
the biotechnology and pharmaceutical industries.
More actinomycete genomes need to be sequenced
to provide a better understanding of antibiotic
biosynthetic gene structure, organization, and
regulation, and to address the important linkages
between primary and secondary metabolism, particularly as it relates to precursor flux in highly

productive strains. Continued advancements will be facilitated by close collaborations between academic and industrial laboratories, which has been a strength of the actinomycete molecular genetics community in the past. More rapid exploitation of combinatorial biosynthesis might benefit from a multi-laboratory effort coordinated in a fashion similar to current international genomics projects.

ACKNOWLEGEMENT

I thank C.R. Hutchinson for providing the Figure on erythromycin biosynthesis.

REFERENCES

Agouridas, C., Bonnefoy, A. and Chantot, J.F. (1997) Antibacterial activity of RU 64004 (HMR 3004), a novel ketolide derivative active against respiratory pathogens. *Antimicrob. Agent Chemother.* **42**, 2149–2158.

Agouridas, C., Denis, A., Auger, J-M., Beneditti, Y., Bonnefoy, A., Bretin, F., *et al.* (1998) Synthesis and antimicrobial activity of ketolides (6-*O*-methyl-3-oxoerythromycin derivatives): a new class of antibacterials highly potent against macrolide-resistant and -susceptible respiratory pathogens. *J. Med. Chem.* **41**, 4080–4100.

Allen, N.E., LeTourneau, D.L. and Hobbs, J.N. (1997) The role of hydrophobic side chains as determinants of antibacterial activity of semisynthetic glycopeptide antibiotics. *J. Antibiot.* **50**, 677–684.

Amann, R.I., Ludwig, W. and Schleifer, K.-H. (1995) Phylogenetic identification and *in situ* detection of individual microbial cells without cultivation. *Microbiol. Rev.* **59**, 143–169.

Aparicio, J.F., Molnar, I., Schwecke, T., Konig, A., Haycock, S.F., Khaw, L.E., *et al.* (1996) Organization of the biosynthetic gene cluster for rapamycin in *Streptomyces hygroscopicus*: analysis of the enzymatic domains in the modular polyketide synthase. *Gene* **169**, 9–16.

Arisawa, A., Kawamura, N., Narita, T., Kojima, I., Okamura, K., Tsunekawa, H., *et al.* (1996) Direct fermentative production of acyltylosins by genetically-engineered strains of *Streptomyces fradiae*. *J. Antibiot.* **49**, 349–354.

Arisawa A., Kawamura, N., Takeda, K., Tsunekawa, H., Okamura, K. and Okamoto, R. (1994) Cloning of the macrolide antibiotic biosynthetic gene *acyA*, which encodes 3-*O*-acyltransferase, from *Streptomyces thermotolerans* and its use for direct fermentative production of a hybrid macrolide antibiotic. *Appl. Env. Microbiol.* **60**, 2657–2660.

August, P., Tang, L., Yoon, Y.J., Ning, S., Muller, R., Yu, T.-W., *et al.* (1998) Biosynthesis of the ansamycin antibiotic rifamycin: deductions from the molecular analysis of the *rif* biosynthetic gene cluster *of Amycolatopsis mediterranei* S699. *Chem. Biol.* **5**, 69–79.

Baltch, A.L., Smith, R.P., Ritz, W.J. and Bopp, L.H. (1998) Comparison of inhibitory and bacteriocidal activities and postantibiotic effects of LY333328 and ampicillin used singly and in combination against vancomycin-resistant

Enterococcus faecium. Antimicrob. Agent Chemother. **42**, 2564–2568.

Baltz, R.H. (1982) Genetics and biochemistry of tylosin production: a model for genetic engineering in antibiotic-producing *Streptomyces. Basic Life Sci.* **19**, 431–444.

Baltz, R.H. (1995) Gene expression in recombinant *Streptomyces. Bioprocess Technol.* **22**, 309–381.

Baltz, R.H. (1997a) Molecular genetic approaches to yield improvement in actinomycetes. In W.R. Strohl, (ed.), *Biotechnology of Antibiotics*, Marcel Dekker, New York, pp. 49–62.

Baltz, R.H. (1997b) Lipopeptide antibiotics produced by *Streptomyces roseosporus* and *Streptomyces fradiae*. In W.R. Strohl, (ed.), *Biotechnology of Antibiotics*, Marcel Dekker, New York, pp. 415–435.

Baltz, R.H. (1998) Genetic manipulation of antibiotic-producing *Streptomyces. Trends Microbiol.* **6**, 76–83.

Baltz, R.H. and Seno, E.T. (1981) Properties of *Streptomyces fradiae* mutants blocked in bisynthesis of the macrolide antibiotic tylosin. *Antimicrob. Agents Chemother.* **20**, 214–225.

Baltz, R.H. and Seno E.T. (1988) Genetics of *Streptomyces fradiae* and tylosin biosynthesis. *Ann. Rev. Microbiol.* **42**, 547–574.

Baltz, R.H. and Hosted, T.J. (1996) Molecular genetic methods for improving secondary-metabolite production in actinomycetes. *Trends Biotechnol.* **14**, 245–2449.

Baltz, R.H., Seno, E.T., Stonesifer, J. and Wild, G.M. (1983) Biosynthesis of the macrolide antibiotic tylosin: A preferred pathway from tylactone to tylosin. *J. Antibiot.* **36**, 131–141.

Baltz, R.H., Fayerman, J.T., Ingolia, T.D. and Rao, R.N. (1986) Production of novel antibiotics by gene cloning and protein engineering. In M. Inouye and R. Sarma, (eds.), *Protein Engineering: Applications in Science, Medicine, and Industry*, Academic Press, New York, pp. 365–381.

Baltz, R.H., McHenney, M.A., Cantwell, C.A., Queener, S.W. and Solenberg, P.J. (1997) Applications of transposition mutagenesis in antibiotic producing streptomycetes. *Antonie Leeuwenhoek* **71**, 179–187.

Bate, N. and Cundliffe, E. (1999) The mycinose-biosynthetic genes of *Streptomyces fradiae*, producer of tylosin. *J. Ind. Microbiol. Biotechnol.* **23**, 118–122.

Bate, N., Butler, A.R., Gandecha, A.R. and Cundliffe, E. (1999a) Multiple regulatory genes in the tylosin biosynthetic cluster of *Streptomyces fradiae*. *Chem. Biol.* **6**, 617–624.

Bate, N., Butler, A.R., Smith, I.P. and Cundliffe, E. (2000) The mycarose-biosynthetic genes of *Streptomyces fradiae*, producer of tylosin. *Microbiology.* **146**, 139–146.

Beckmann, R.J., Cox, K. and Seno, E.T. (1989) A cluster of tylosin biosynthetic genes is interrupted by a structurally unstable segment containing four repeated sequences. In C.L. Hershberger, S.W. Queener, and G. Hegeman, (eds.), *Genetics and Molecular Biology of Industrial Microorganisms*, American Society for Microbiology, Washington, D.C., pp. 176–186.

Bedford, D., Jacobson, J.R., Luo, G., Cane, D.E. and Khosla, Cc. (1996) A functional chimeric polyketide synthase generated via domain replacement. *Chem. Biol.* **3**, 827–831.

Belshaw, P.J., Walsh, C.T. and Stachelhaus, T. (1999) Aminoacyl-CoAs as probes of condensation domain selectivity in nonribosomal peptide synthesis. *Science* **284**, 486–489.

Bierman, M., Logan, R., O'Brien, K., Seno, E.T., Rao, R.N. and Schoner, B.E. (1992) Plasmid cloning vectors for the conjugal transfer of DNA from *Escherichia coli* to *Streptomyces* spp. *Gene* **116**, 43–49.

Binnie, C., Warren, M. and Butler, M.J. (1989) Cloning and heterologous expression in *Streptomyces lividans* of *Streptomyces rimosus* genes involved in oxytetracycline biosynthesis. *J. Bacteriol.* **171**, 887–895.

Birmingham, V.A., Cox, K.L., Larson, J.L., Fishman, S.E., Hershberger, C.L. and Seno, E.T. (1986) Cloning and expression of a tylosin resistance gene from a tylosin-producing strain of *Streptomyces fradiae*. *Mol. Gen. Genet.* **204**, 532–539.

Blanc, V., Gil, P., Bamas-Jacques, N., Lorenzon, M., Schleuniger, J., Bisch, D., *et al.* (1997) Identification and analysis of genes from *Streptomyces pristinaespiralis* encoding enzymes involved in the biosynthesis of the 4-dimethylamino-L-phenylalanine precursor of pristinamycin I. *Mol. Microbiol.* **23**, 191–202.

Bohm, I., Holzbaur, U., Hanefeld, U., Cortes, J., Staunton, J. and Leadlay, P.F. (1998) Engineering of a minimal polyketide synthase, and targeted alteration of the stereospecificity of polyketide chain extension. *Chem. Biol.* **5**, 407–412.

Borisova, S.A., Zhao, L. Sherman, D.H. and Liu, H.-w. (1999) Biosynthesis of desosamine: construction of a new macrolide carrying a genetically designed sugar moiety. *Org. Lett.* **1**, 133–136.

Bult, C.J., White, O., Olsen, G.J., Zhou, L., Fleischmann, R.D., Sutton, G.G., *et al.* (1996) Complete genome sequence of the methanogenic archaeon, *Methanococcus jannaschii. Science* **273**, 1058–1072.

Butler, A.R., Bate, N. and Cundliffe, E. (1999) Impact of thioesterase activity on tylosin biosynthesis in *Streptomyces fradiae. Chem. Biol.* **6**, 287–292.

Buttner, M.J., Chater, K.F. and Bibb, M.J. (1990) Cloning, disruption, and transcriptional analysis of three RNA polymerase sigma factor genes of *Streptomyces coelicolor* A3(2). *J. Bacteriol.* **172**, 3367–3378.

Cane, D.E., Walsh, C.T. and Khosla, C. (1998) Harnessing the biosynthetic code: combinations, permutations, and mutations. *Science* **282**, 63–68.

Chang, C.-C.J., Chen, T.T., Cox, B.W., Dawes, G.N., Stemmer, W.P.C., Punnonen, J., *et al.* (1999) Evolution of a cytokine using DNA family shuffling. *Nature Biotechnol.* **17**, 793–797.

Christians, F.C., Scapozza, A., Crameri, A., Folkers, G. and Stemmer, W.P.C. (1999) Directed evolution of thymidine kinase for AZT phosphorylation using DNA family shuffling. *Nature Biotechnol.* **17**, 259–264.

Conti, E., Stachelhaus, T., Marahiel, M.A. and Brick, P. (1997) Structural basis for the activation of phenylalanine in the non-ribosomal biosynthesis of gramacidin S. (1997) *EMBO J.* **16**, 4174–4183.

Cooper, R.D.G., Snyder, N.J., Zweifel, M.J., Staszak, M.A., Wilkie, S.C., Nicas, T.I., *et al.* (1996) Reductive alkylation of glycopeptide antibiotics: synthesis and antibacterial activity. *J. Antibiot.* **49**, 575–581.

Cortes, J., Wiesmann, K.E., Roberts, G.A., Brown, M.J., Staunton, J. and Leadlay, P.F. (1995) Repositioning of a domain in a modular polyketide synthase to promote specific chain cleavage. *Science* **268**, 1487–1489.

Cox, K.L., Fishman, S.E., Larson, J.L., Stanzak, R., Reynolds, P.A., Yeh, W.K., *et al.* (1986) The use of recombinant DNA techniques to study tylosin biosynthesis and resistance in *Streptomyces fradiae. J. Nat. Prod.* **49**, 971–980.

Crameri, A., Dawes, G., Rodriquez, E., Jr., Silver, S. and Stemmer, W.P.C. (1997) Molecular evolution of an arsenate detoxification pathway by DNA shuffling. *Nature Biotechnol.* **15**, 436–438.

Crameri, A., Raillard, S.-A., Bermudez, E. and Stemmer, W.P.C. (1998) DNA shuffling of a family of genes from diverse species accelerates directed evolution. *Nature* **391**, 288–291.

Crameri, A., Whitehorn, E.A., Tate, E. and Stemmer, W.P.C. (1996) Improved green fluorescent protein by molecular evolution using DNA shuffling. *Nature Biotechnol.* **14**, 315–319.

de Crecy-Lagard, V., Blanc, V., Gil, P., Naudin, L., Lorenzon, S., Famechon, A., *et al.* (1997a) Pristinamycin I biosynthesis in *Streptomyces pristinaespiralis*: molecular characterization of the first two structural peptide synthetase genes. *J. Bacteriol.* **179**, 705–713.

de Crecy-Lagard, V., Saurin, W., Thibaut, D., Gil, P., Naudin, L., Crouzet, J., *et al.* (1997b) Streptogramin B biosynthesis in *Streptomyces pristinaespiralis* and *Streptomyces virginiae*: molecular characterization of the last structural peptide synthetase gene. *Antimicrob. Agent Chemother.* **41**, 1904–1909.

Delzer, J., Fiedler, H.P., Muller, H., Zahner, H., Rathmann, R., Ernst, K., *et al.* (1984) New nikkomycins by mutasynthesis and directed fermentation. *J. Antibiot.* **37**, 80–82.

Denoya C.D., Fedechko R.W., Hafner E.W., McArthur H.A., Morgenstern M.R., Skinner D.D., *et al.* (1995) A second branched-chain alpha-keto acid dehydrogenase gene cluster (*bkdFGH*) from *Streptomyces avermitilis*: its relationship to avermectin biosynthesis and the construction of a *bkdF* mutant suitable for the production of novel antiparasitic avermectins. *J. Bacteriol.* **177**, 3504–3511

Dutten, C.J., Gibson, S.P., Goudie, A.C., Holdom, K.S., Pacey, M.S., Ruddock, J.C., *et al.* (1991) Novel avermectins produced by mutational biosynthesis. *J. Antibiotics* **44**, 357–365.

Ednie, L.M., Spangler, S.K., Jacobs, M.R. and Applebaum, P.C. (1997) Susceptibilities of 228 penicillin- and erythromycin-susceptible and -resistant pneumococci to RU 64004, a new ketolide, compared with susceptibilities to 16 other agents. *Antimicrob. Agent Chemother.* **41**, 1033–1036.

English, R.S., Lampel, J.S. and Vanden Boom, T.J. (1998) Transformation of *Saccharopolyspora erythraea* by electroporation of germling spores: construction of propionyl Co-A carboxylase mutants. *J. Ind. Microbiol. Biotechnol.* **21**, 219–224.

Encell, L.P., Landis, D.M. and Loeb, L.A. (1999) Improved enzymes for cancer gene therapy. *Nature Biotechnol.* **17**, 143–147.

Epp, J.K., Huber, M.L.B., Turner, J.R., Goodson, T and Schoner, B.E. (1989) Production of hybrid macrolide antibiotics in *Streptomyces ambofaciens* and *Streptomyces lividans* by introduction of a cloned carbomycin biosynthetic gene from *Streptomyces thermotolerans. Gene* **85**, 293–301.

Fernandez-Moreno, M.A., Martinez, E., Boto, L., Hopwood, D.A. and Malpartida, F. (1992) Nucleotide sequence and deduced functions of a set of cotranscribed genes of *Streptomyces coelicolor* A3(2) including the polyketide synthase for the antibiotic actinorhodin. *J. Biol. Cchem.* **267**, 19278–19290.

Fish, S.A. and Cundliffe, E. (1996) Structure-activity studies of tylosin-related macrolides. *J. Antibiot.* **49**, 1044–1048.

Fish, S.A. and Cundliffe, E. (1997) Stimulation of polyketide metabolism in *Streptomyces fradiae* by tylosin and its glycosylated precursors. *Microbiology* **143**, 3871–3876.

Fisher, S.H., Bruton, C.J. and Chater, K.F. (1987) The glucose kinase gene of *Streptomyces coelicolor* and its use in selecting spontaneous deletions for desired regions of the genome. *Mol. Gen. Genet.* **206**, 35–44.

Fishman, S.E., Cox, K., Larson, J.L., Reynolds, P.A., Seno, E.T., Yeh, W.-K., *et al.* (1987) Cloning genes for the biosynthesis of a macrolide antibiotic. *Proc. Nat. Acad. Sci. USA* **84**, 8248–8252.

Fitzgerald, N.B., English, R.S., Lampel, J.S. and Vanden Boom, T.J. (1998) Sonication-dependent electroporation of the erythromycin-producing bacterium *Saccharopolyspora erythraea*. *Appl. Environ. Microbiool.* **64**, 1580–1583.

Flett, F., Mersinias, V. and Smith, C.P. (1997) High efficiency intergeneric conjugal transfer of plasmid DNA from *Escherichia coli* to methyl DNA-restricting streptomycetes. *FEMS Microbiol. Lett.*, **155**, 223–229.

Fouces, R., Mellado, E., Diez, B. and Barredo, J.L. (1999) The tylosin biosynthetic cluster from *Streptomyces fradiae*: genetic organization of the left region. *Microbiology* **145**, 855–868.

Gaisser, S. and Hughes, C. (1997) A locus coding for putative nonribosomal peptide/polyketide synthase functions is mutated in a swarming-defective *Proteus mirabilis* strain. *Mol. Gen. Genet.* **253**, 415–427.

Gandecha, A.R. and Cundliffe, E. (1996) Molecular analysis of *tlrD*, an MLS resistance determinant from the tylosin producer, *Streptomyces fradiae*. *Gene* **180**, 173–176.

Gandecha, A.R., Large, S.L. and Cundliffe, E. (1997) Analysis of four tylosin biosynthetic genes from the *tylLM* region of the *Streptomyces fradiae* genome. *Gene* **184**, 197–203.

Garcia-Garrot, F., Cercenado, E., Alcala, L. and Bouza, E. (1998) *In vitro* activity of the new glycopeptide LY333328 against multiply resistant gram-positive clinical isolates. *Antimicrob. Agent. Chemother.* **42**, 2452–2455.

Gehring, A.M., DeMoll, E., Fetherston, J.D., Mori, I., Mayhew, G.F., Blattner, F.R., *et al.* (1998) Iron acquisition in plague: modular logic in enzymatic biogenesis of yersinia-bactin by *Yersinia pestis*. *Chem. Biol.* **5**, 573–586.

Gokhale, R.S., Hunziker, D., Cane, D.E. and Khosla, C. (1999a) Mechanism and specificity of the terminal thioesterase domain from the erythromycin polyketide synthase. *Chem. Biol.* **6**, 117–125.

Gokhale, R.S., Tsuji, S.Y., Cane, D.E. and Khosla, C. (1999b) Dissecting and exploiting intermodular communication in polyketide synthases. *Science* **284**, 482–485.

Handelsman, J., Rondon, M.R., Brady, S.F., Clardy, J. and Goodman, R.M. (1998) Molecular biological access to the chemistry of unknown soil microbes: a new frontier for natural products. *Chem. Biol.* **5**, R245–249.

Henkel, T., Brunne, R.M., Muller, H. and Reichel, F. (1999) Statistical investigation into the structural complementarity of natural products and synthetic compounds. *Angew. Chem. Int. Ed.* **38**, 643–647.

Hesketh, A. and Ochi, K. (1997) A novel method for improving *Streptomyces coelicolor* A3(2) for production of actinorhodin by introduction of *rpsL* (encoding ribosomal protein S12) mutations conferring resistance to streptomycin. *J. Antibiot.* **50**, 532–535.

Hillemann, D., Puhler, A. and Wohlleben, W. (1991) Gene disruption and gene replacement in *Streptomyces* via single stranded DNA transformation of integrative vectors. *Nucleic Acid Res.* **19**, 727–731.

Holzbaur, I.E., Harris, R.C., Bycroft, M., Cortes, J., Bisang, C., Staunton, J., *et al.* (1999) Molecular basis of Celmer's rules: the role of two ketoreductase domains in the control of chirality by the erythromycin modular polyketide synthase. *Chem. Biol.* **6**, 189–195.

Hopwood, D.A. (1997) Genetic contributions to understanding polyketide synthases. *Chem. Rev.* **97**, 2465–2479.

Hopwood, D.A., Bibb, M.J., Chater, K.F. and Kieser, T. (1987) Plasmid and phage vectors for gene cloning and analysis in Streptomyces. *Methods Enzymol.* **153**, 116–166.

Hosoya, Y., Muramatsu, H. and Ochi, K. (1998) Acquisition of certain streptomycin-resistant (str) mutations enhances antibiotic production in bacteria. *Antimicrob. Agent Chemother.* **42**, 2041–2047.

Hosted, T.J. and Baltz, R.H. (1996) Mutants of *Streptomyces roseosporus* that express enhanced recombination within partially homologous genes. *Microbiology* **142**, 2803–2813.

Hosted, T.J. and Baltz, R.H. (1997) Use of *rpsL* for dominance selection and gene replacement in *Streptomyces roseosporus*. *J. Bacteriol.* **179**, 180–186.

Huffman, G.W., Gesellchen, P.D., Turner, J.R., Rothenberger, R.B., Osborne, H.E., Miller, F.D., *et al.* (1992) Substrate specicity of isopenicillin synthase. *J. Med. Chem.* **35**, 1897–1914.

Hugenholtz, P. and Pace, N.R. (1996) Identifying microbial diversity in the natural environment: a molecular phylogenetic approach. *Trends Biotechnol.* **14**, 190–197.

Hugenholtz, P., Goebel, B.M. and Pace, N. (1998) Impact of culture-independent studies on the emegging phylogenetic view of bacterial diversity. *J. Bacteriol.* **180**, 4765–4774.

Hutchinson, C.R. (1997) Biosynthesis of daunorubison and tetracenomycin. *Chem. Rev.* **97**, 2525–2535.

Hutchinson, C.R. and Fujii, I. (1995) Polyketide synthase gene manipulation: a structure-function approach in engineering novel antibiotics. *Annu. Rev. Microbiol.* **49**, 201–238.

Ichinose, K., Bedford, D.J., Tornus, D., Bechthold, A., Bibb, M., Revill, W.P., *et al.* (1998) The granaticin gene cluster of *Streptomyces violaceoruber* Tu22: sequence analysis and expression in a heteologous host. *Chem. Biol.* **5**, 647–659.

Ikeda, H. and Omura, S. (1997) Avermectin biosynthesis. *Chem. Rev.* **97**, 2591–2609.

Ikeda, H., Nomoniya, T., Usami, M., Ohta, T. and Omura, S. (1999) Organization of the biosynthetic gene cluster of the polyketide anthelmintic macrolide avermectin in *Streptomyces avermitilis*. *Proc. Nat. Acad. Sci. USA* **96**, 9509–9514.

Jacobsen, J.R., Hutchinson, C.R., Cane, D.E. and Khosla, C. (1997) Precusror-directed biosynthesis of erythromycin analogs by an engineered polyketide synthase. *Science*, **277**, 367–369.

Jones, G. (1989) Cloning of *Streptomyces* genes involved in antibiotic synthesis and its regulation. In S. Shipiro (ed.), *Regulation of Secondary Metabolism in Actinomycetes*, CRC Press, Boca Raton, pp. 49–73.

Jones, N.D., Chaney, M.O., Kirst, H.A., Wild, G.M., Baltz, R.H., Hamill, R.L., *et al.* (1982) Novel fermentation products from *Streptomyces fradiae*: X-ray crystal structure of 5-O-mycarosyltylactone and proof of the absolute configuration of tylosin. *J. Antibiot.* **35**, 420–425.

Kaatz, G.W., Seo, S.M., Aeschlimann, J.R., Houlihan, H.H., Mercier, R.-C. and Ribak, M.J. (1998) Efficacy of LY333328 against experimental methicillin-resistant *Staphylococcus aureus* endocarditis. *Antimicrob. Agent Chemother.* **42**, 981–983.

Kao, C.M., Katz, L. and Khosla, C. (1994) Engineered biosynthesis of a complete macrolactone in a heterologous host. *Science* **265**, 509–512

Kao, C.M., Luo, G.L., Katz, L., Cane, D.E. and Khosla, C. (1995) Manipulation of macrolide ring size by directed mutagenesis of a modular polyketide synthase. *J. Am. Chem. Soc.* **117**, 9105–9106.

Katz, L. (1997) Manipulation of modular polyketide synthases. *Chem. Rev.* **97**, 2557–2575.

Keleman, G.H., Zalacain, M., Culebras, E., Seno, E.T. and Cundliffe, E. (1994) Transcriptional attenuation control of the tylosin-resistance gene *tlrA* in *Streptomyces fradiae*. *Mol. Microbiol.* **14**, 833–842.

Kempf, I., Reeve-Johnson, L., Gesbert, F. and Guittet, M. (1997) Efficacy of tilmicosin in the control of experimental

Mycoplasma gallisepticum infection in chickens. *Avian Dis.* **41**, 802–807.

Khosla, C. (1997) Harnessing the biosynthetic potential of modular polyketide synthases. *Chem. Rev.* **97**, 2577–2590.

Khosla, C. and Zawada, R.J.X. (1996) Generation of polyketide libraries via combinatorial biosynthesis. *Trends Biotechnol.* **14**, 335–341.

Khosla, C., Gokhale, R.S., Jacobsen, J.R. and Cane, D.E. (1999) Tolerance and specificity of polyketide synthases. *Annu. Rev. Biochem.* **68**, 219–253.

Kieser, T. and Hopwood, D.A. (1991) Genetic manipulation of *Streptomyces*: integrating vectors and gene replacement. *Methods Enzymol.* **204**, 430–458.

Kirsching, A., Bechtold, A.F-W. and Rohr, J. (1997) Chemical and biochemical aspects of deoxysugars and deoxysugar oligosaccharides. *Top. Curr. Chem.* **188**, 1–84.

Kirst, H.A. (1991) New macrolides: expanded horizons for an old class of antibiotics. *J. Antimicrob. Chemother.* **28**, 787–790.

Kirst, H.A. (1994) Semi-synthetic derivatives of 16-membered macrolide antibiotics. *Prog. Med. Chem.* **31**, 265–295.

Kirst, H.A., Wild, G.M., Baltz, R.H., Hamill, R.L., Ott, J.L., Counter, F.T., *et al.* (1982) Structure-activity studies among 16-membered macrolide antibiotics related to tylosin. *J. Antibiot.* **35**, 1675–1682.

Kirst, H.A., Debono, M., Willard, K.E., Trudell, B.A., Toth, T.E., Turner, J.R., *et al.* (1986) Preparation and evaluation of 3,4″-ester derivatives of 16-membered macrolide antibiotics related to tylosin. *J. Antibiot.* **39**, 1724–1735.

Kleinkauf, H. and von Dohren, H. (1990) Bioactive peptides – recent advances and trends. In H. Kleinkauf and H. von Dohren (eds.), *Biochemistry of Peptide Antibiotics*, Walter de Gruyter, Berlin, pp. 1–31.

Kleinkauf, H. and von Dohren, H. (1996) A nonribosomal system of peptide biosynthesis. *Eur. J. Biochem.* **236**, 135–151.

Klenk, H.P., Clayton, R.A., Tomb, J.F., White, O., Nelson, K.E., Ketchum, K.A., *et al.* (1997) The complete genome sequence of the hyperthermophilic, sulphate-reducing archaeon *Archaeoglobus fulgidus. Nature* **390**, 364–370.

Klugman, K.P., Capper, T., Widdowson, C.A., Koornhof, H.J. and Moser, W. (1998) Increased activity of 16-membered lactone ring macrolides against erythromycin-resistant *Streptoccoccus pyogenes* and *Streptococcus pneumoniae*: characterization of South African isolates. *J. Antimicrob. Chemother.* **42**, 729–734.

Konig., A., Schwecke, T., Molnar, I., A., Bohm, G.A., Lowden, P.A.S., Staunton, J., *et al.* (1997) The pipecolate-incorporating enzyme for the biosynthesis of the immunosuppressand rapamycin: nucleotide sequence analysis, disruption and heterologous expression of *rapP* from *Streptomyces hygroscopicus. Eur. J. Biochem.* **247**, 526–534.

Konz, D. and Marahiel, M.A. (1999) How do peptide synthetases generate structural diversity? *Chem. Biol.* **6**, 39–48. **PPS**

Kovalic, D., Giannattasio, R.B., Jin, H.-J. and Weisblum, B. (1994) 23s rRNA domain V, a fragment that can be specifically methylated *in vitro* by the *ermSF* (*tlrA*) methyltransferase. *J. Bacteriol.* **176**, 6992–6998.

Kuhstoss, S., Huber, M., Turner, J.R., Paschal, J.W. and Rao, R.N. (1996) Production of a novel polyketide through the construction of a hybrid polyketide synthase. *Gene* **183**, 231–236.

Kulowski, K., MacNeil, D. and Hutchinson, C.R. (1999) Minimal set for oleandrose biosynthesis and attachment to avermectin aglycones in heterologous host, *Streptomyces lividans*. Abstracts of the Annual Meeting of the Society for

Industrial Microbiology, Arlington, VA, August 1–5, p.113.

Kunnari. T., Tuikkanen, J., Hautala, A., Hakala, J., Ylihonko, K. and Mantsala, P. (1997) Isolation and characterization of 8-demethoxy steffimycins and generation o 2,8-demethoxy steffimycins in *Streptomyces steffisburgensis. J. Antibiot.* **50**, 496–501.

Lancini, C. and Cavalleri, B. (1990) Glycopeptide antibiotics of the vancomycin group. In H. Kleinkauf and H. von Dohren (eds.), *Biochemistry of Peptide Antibiotics*, Walter de Gruyter, Berlin, pp. 159–178.

Leadlay, P.F. (1997) Combinatorial approaches to polyketide biosynthesis. *Curr. Opin. Chem. Biol.* **1**, 162–168.

Leblond, P., Redenbach, M. and Cullum, J. (1993) Physical map of the *Streptomyces lividans* 66 genome and comparison with that of the related strain *Streptomyces coelicolor* A3(2). *J. Bacteriol.* **175**, 3422–3429.

Leblond, P., Fischer, G., Francou, F.-X.., Berger, F., Guerineau, M. and Decaris, B. (1996) The unstable region of *Streptomyces ambofaciens* includes 210 kb terminal inverted repeats flanking the extremities of the linear chromosomal DNA. *Mol. Microbiol.* **19**, 261–271.

Lezhava, A., Mizukami, T., Kajitani, T., Kameoka, D., Redenbach, M., Shinkawa, H., *et al.* (1995) Physical map of the linear chromosome of *Streptomyces griseus. J. Bacteriol.* **177**, 6492–6498.

Liu, H.-W. and Thorson, J.S. (1994) Pathways and mechanisms in the biogenesis of novel deoxysugars by bacteria. *Annu. Rev. Microbiol.* **48**, 223–256.

Madduri, K., Kennedy, J., Rivola, G., Inventi-Solari, A., Fillippini, S., Zanuso, G., *et al.* (1998) Production of the antutumor drug epirubicin (4′-epidoxorubicin) and its precursor by a genetically engineered strain of *Streptomyces peucetius. Nature Biotechnol.* **16**, 69–74.

Mao, Y., Varaglu, M. and Sherman, D.H. (1999) Molecular characterization and analysis of the biosynthetic gene cluster for the antitumor antibiotic mitomycin C from *Streptomyces lavendulae* NRRL 2564. *Chem. Biol.* **6**, 251–263.

Marahiel M. (1997) Protein templates for the biosynthesis of peptide antibiotics. *Chem. Biol.* **4**, 561–567.

Marahiel, M.A., Stachelhaus, T. and Mootz, H.D. (1997) Modular peptide synthases involved in nonribosomal peptide synthesis. *Chem. Rev.* **97**, 2651–2673.

Marsden, F.A., Wilkenson, B., Cortes, J., Dunster, N.J., Staunton, J. and Leadlay, P.F. (1998) Engineering broader specificity into an antibiotic-producing polyketide synthase. *Science* **279**, 199–202.

Marshall, V.P. and Wiley, P.F. (1986) Biomodification of antibiotics by *Streptomyces*. In S.W. Queener and L.E. Day (eds.), *The Bacteria, vol.IX, Antibiotic-Producing* Streptomyces, Academic Press, New York, pp. 323–353.

Matsushima, P. and Baltz, R.H. (1985) Efficient plasmid transformation of *Streptomyces ambofaciens* and *Streptomyces fradiae* protoplasts. *J. Bacteriol.* **163**, 180–185.

Matsushima, P. and Baltz, R.H. (1996) A gene cloning system for ''*Streptomyces toyocaensis*''. *Microbiology* **142**, 261–267.

Matsushima, P., Broughton, M.C., Turner, J.R. and Baltz, R.H. (1994) Conjugal transfer of cosmid DNA from *Escherichia coli to Saccharopolyspora spinosa*: effects of chromosomal insertions on macrolide A83543 production. *Gene* **146**, 39–45.

Matsushima, P., McHenney, M.A. and Baltz, R.H. (1989) Transduction and transformation of plasmid DNA in *Streptomyces fradiae* strains that express different levels of restriction. *J. Bacteriol.* **171**, 3080–3084.

Matsuura, T., Mijai, K., Trakulnaleamsai, S., Yomo, T., Shima, Y., Miki, S., *et al.* (1999) Evolutionary molecular engineering by random elongation mutagenesis. *Nature Biotechnol.* **17**, 58–61.

Mazodier, P., Petter, R. and Thompson, C. (1989) Intergeneric conjugation between *Escherichia coli* and *Streptomyces* species. *J. Bacteriol.* **171**, 3583–3585.

Mazy-Servais, C., Baczkowski, D. and Dusart, J. (1997) Electroporation of intact cells of *Streptomyces parvulus* and *Streptomyces vinaceus*. *FEMS Microbiol. Lett.* **15**, 135–138.

McDaniel, R., Thamchaipenet, A., Gufstafsson, C., Fu, H., Betlach, M., Betlach, M., *et al.* (1999a) *Proc. Nat. Acad. Sci. USA* **96**, 1846–1851.

McHenney, M.A. and Baltz, R.H. (1988) Transduction of plasmid DNA in *Streptomyces* and related genera by bacteriophage FP43. *J. Bacteriol.* **170**, 2276–2282.

McHenney, M.A. and Baltz, R.H. (1989) Transduction of plasmid DNA in macrolide producing streptomycetes. *J. Antibiotics* **42**, 1725–1727.

McHenney, M.A. and Baltz, R.H. (1996) Gene transfer and transposition mutagenesis in *Streptomyces roseosporus*: mapping of insertions that influence daptomycin or pigment production. *Microbiology* **142**, 2363–2373.

McHenney, M.A., Hosted, T.J., DeHoff, B.S., Rosteck, P.R., Jr. and Baltz, R.H. (1998) Molecular cloning and physical mapping of the daptomycin gene cluster from *Streptomyces roseosporus J. Bacteriol.* **180**, 143–151.

Merson-Davies, L. and Cundliffe, E. (1994) Analysis of five tylosin biosynthetic genes from the *tylIBA* region of the *Streptomyces fradiae* genome. *Mol. Microbiol.* **13**, 349–355.

Meurer, G. and Hutchinson, C.R. (1999) Genes for the biosynthesis of microbial secondary metabolites. In A.L. Demain and J.E. Davies (eds.), *Manual of Industrial Microbiology and Biotechnology, Second Edition*, ASM Press, Washington, D.C., pp. 740–758.

Molnar, I., Aparicio, J.F., Haycock, S.F., Khaw, L.E., Schwecke, T., Konig, A., *et al.* (1996) Organization of the biosynthetic gene cluster for rapamycin in *Streptomyces hygroscopicus*: analysis of genes flanking the polyketide synthase. *Gene* **169**, 1–7.

Nagaoka, K. and Demain, A.L. (1975) Mutational biosynthesis of a new antibiotic, streptomutin A, by an idiotroph of *Streptomyces griseus*. *J. Antibiot.* **28**, 627–635.

Nakagawa, A. and Omura, S. (1984) Structure and stereochemistry of macrolides. In S. Omura (ed.), *Macrolide Antibiotics: Chemistry, Biology, and Practice*, Academic Press, Tokyo, pp. 37–84.

Nakayama, I. (1984) Macrolides in clinical practice. In S. Omura (ed.), *Macrolide Antibiotics: Chemistry, Biology, and Practice*, Academic Press, Tokyo, pp. 261–300.

Nicas, T.I. and Cooper, R.D.G. (1997) Vancomycin and other glycopeptides. In W.R. Strohl (ed.), *Biotechnology of Antibiotics*, Marcel Dekker, New York, pp. 363–392.

Nicas, T.I., Flockowitsch, J.E., Preston, D.A., Snyder, N.S., Zweifel, M.J., Wilkie, S.C., *et al.* (1996) Semisynthetic glycopeptide antibiotics derived from LY264826 active against vancomycin-resistant enterococci. *Antimicrob. Agent Chemother.* **40**, 2194–2199.

Niemi, J., Ylihonko, K., Hakala, J., Parssinen, R., Kopio, A. and Mantsala. P. (1994) Hybrid anthracycline antibiotics: production of new anthracyllines by cloned genes from *Streptomyces purpurascens* in *Streptomyces galilaeus*. *Microbiology* **140**, 1351–1358.

Oh, S.-H. and Chater, K.F. (1997) Denaturation of circular or linear DNA facilitates targeted integration of *Streptomyces coelicolor* A3(2): possible relevance to other organisms. *J. Bacteriol.* **179**, 122–127.

Okamoto, R., Fukumoto, T., Nomura, H., Kiyoshima, K., Nakamura, K. and Takamatsu, A. (1980) Physicochemical properties of new acyl derivatives of tylosin produced by microbial transformation. *J. Antibiot.* **33**, 1300–1308.

Olafsson, S., Berstat, A., Bang, C.J., Nysaeter, G., Coll, P., Tefera, S., *et al.* (1999) Spiramycin is comparable to oxytetracycline in eradicating *H. pylori* when given with ranitidine bismuth citrate and metronidizole. *Aliment. Pharmacol. Ther.* **13**, 651–659.

Olano, C., Lomovskaya, N., Fonstein, L., Roll, J.T. and Hutchinson, C.R. (1999) A two-plasmid system for the glyosylation of polyketide antibiotics: bioconversion of epsilon-rhodomycinone to rhodomycin D. *Chem. Biol.* **6**, 845–855.

Oliynyk, M., Brown, M.J.B., Cortes, J., Staunton, J. and Leadlay, P.F. (1996) A hybrid modular polyketide synthase obtained by domain swapping. *Chem.. Biol.* **3**, 833–839.

Omura, S., Sadakane, N., Tanaka, Y. and Matsubara, H. (1983) Chimeramycins: new macrolide antibiotics produced by hybrid biosynthesis. *J. Antibiot.* **36**, 927–930.

Omura, S., Sadakane, N., Kitao, C., Matsubara, H. and Nakagawa, A. (1980) Production of mycarosyl protolonolide by a mycaminose idiotroph from the tylosin-producing strain *Streptomyces fradiae* KA-427. *J. Antibiot.* **33**, 913–914.

Onaka, H., Nagagawa, T. and Horinouchi, S. (1998) Involvement of two A-factor receptor homologues in *Streptomyces coelicolor* A3(2) in the regulation of secondary metabolism and morphogenesis. *Mol. Microbiol.* **28**, 743–753.

Pacey, M.S., Dirlam, J.P., Geldart, R.W., Leadlay, P.F., McArthur, H.A.I., McCormick, E.L., *et al.* (1998) Novel erythromycins from a recombinant *Saccharopolyspora erythraea* strain NRRL 2338 pIG1 I. Fermentation, isolation and biological activity. *J. Antibiot.* **51**, 1029–1034.

Paiva, N.L., Demain, A.L. and Roberts, M.F. (1993) The immediate precursor of the nitrogen-containing ring of rapamycin is free pipecolic acid. *Enzyme Microb. Technol.* **15**, 581–585.

Paradkar, A.S., Jensen, S.E. and Mosher, R.H. (1997) Comparative genetics and molecular biology of β-lactam biosynthesis. In W.R. Strohl (ed.), *Biotechnology of Antibiotics*, Marcel Dekker, New York, pp. 241–277.

Patten, P.A., Howard, R.J. and Stemmer, W.P.C. (1997) Applications of DNA shuffling to pharmaceuticals and vaccines. *Curr. Opin. Biotechnol.* **8**, 724–733.

Pelzer, S., Reichert, W., Heckmann, D. and Wohlleben, W. (1997) Cloning and analysis of a peptide synthetase gene of the balhimycin producer *Amycolatopsis mediterranei* DSM5908 and development of a gene disruption/replacement system. *J. Biotechnol.* **56**, 115–128.

Pelzer, S., Sussmuth, R., Heckmann, D., Recktenwald, J., Jung, G. and Wohlleben, W. (1999) Identification and analysis of the balhimycin biosynthetic gene cluster and its use for manipulating glycopeptide biosynthesis in *Amycolatopsis mediterranei* DSM5908. *Antimicrob. Agents Chemother.* **43**, 1565–1573.

Peschke, U., Schmidt, H., Zhang, H.Z. and Piepersberg, W. (1995) Molecular characterization of the lincomycin-production gene cluster of *Streptomyces lincolnensis. Mol. Microbiol.* **16**, 1137–1156.

Piepersberg, W. (1997) Molecular biology, biochemistry, and fermentation of aminoglycoside antibiotics. In W.R. Strohl (ed.) *Biotechnology of Antibiotics*, Marcel Dekker, New York, pp. 81–163.

Piepersberg, W. and Distler, J. (1997) Aminoglycosides and sugar components in other secondary metabolites. In H.-J. Rehm, G. Reed, A. Puhler and P. Stadler (eds.), *Biotechnology, 2nd edn., Vol. 7, Products of Secondary Metabolism*, VCH, Weinheim, pp. 399–488.

Pigac, J. and Schrempf, H. (1995) A simple and rapid method of transformation of *Streptomyces rimosus* R6 and other streptomycetes by electroporation. *Appl. Environ. Microbiol.* **61**, 352–356.

Queener, S.W., Sebek, O.K. and Vezina, C. (1978) Mutants blocked in antibiotic biosynthesis. *Annu. Rev. Microbiol.* **32**, 593–636.

Rao, R.N., Richardson, M.A. and Kuhstoss, S. (1987) Cosmid shuttle vectors for cloning and analysis of *Streptomyces* DNA. *Method Enzymol.* **153**, 166–198.

Rayssiguier, C., Thaler, D.S. and Radman, M. (1989) The barrier to recombination between *Escherichia coli* and *Salmonella typhimurium* is disrupted in mismatch-repair mutants. *Nature* **342**, 396–342.

Redenbach, M., Kieser, H.M., Denepaite, D., Eichner, A., Cullum, J., Kinashi, H., *et al.* (1996) A set of ordered cosmids and a detailed genetic and physical map for the 8 Mb *Streptomyces coelicolor* A3(2) chromosome. *Mol. Microbiol.* **21**, 77–96.

Reeves, A.R., Post, D.A. and Vanden Boom, T.J. (1998) Physical-genetic map of the erythromycin-producing organism *Saccharopolyspora erythraea*. *Microbiology* **144**, 2151–2159.

Reinhart, K.L., Jr. (1979) Biosynthesis and mutasynthesis of aminocyclitol antibiotics. *Jpn. J. Antibiot.* **32**, S32–46.

Robertson, D.E., Mathur, E.J., Swanson, R.V., Marrs, B.L. and Short, J.M. (1996) The discovery on new biocatalysts from microbial diversity. *SIM News* **46**, 3–8.

Rosteck, P.R., Jr., Reynolds, P.A. and Hershberger, C.L. (1991) Homology between proteins controlling *Streptomyces fradiae* tylosin resistance and ATP-binding transport. *Gene* **102**, 27–32.

Rowe, C.J., Cortes, J., Gaisser, S., Staunton, J. and Leadlay, P.F. (1998) Construction of new vectors for high-level expression in actinomycetes. *Gene* **216**, 215–223.

Ruan, X., Pereda, A., Stassi, D.L., Zeidner, D., Summers, R.G., Jackson, M., *et al.* (1997) Acyltransferase domain substitutions in erythromycin polyketide synthase yield novel erythromycin derivatives. *J. Bacteriol.* **179**, 6416–6425.

Rubinstein, E. and Keller, N. (1998) Spiramycin renaissance. *J. Antimicrob. Chemother.* **42**, 572–576.

Sadakane, N., Tanaka, Y. and Ōmura, S. (1982) Hybrid biosynthesis of derivatives of protylonolide and M-4365 by macrolide-producing microorganisms. *J. Antibiot.* **35**, 680–687.

Saleh-Mghir, A., Lefort, A., Petegnief, Y., Dautrey, S., Vallois, J-M., Le Guludec, D., *et al.* (1999) Activity and diffusion of LY333328 in experimental endocarditis due to vancomycin-resistant *Enterococcus faecalis. Antimicrob. Agent Chemother.* **43**, 115–120

Schneider, A., Stachelhaus, T. and Marahiel, M.A. (1998) Targeted alteration of the substrate specificity of peptide synthetases by rational module swapping. *Mol. Gen. Genet.* **257**, 308–318.

Schulin, T., Wennersten, R.C., Moellering, R.C., Jr. and Eliopoulos, G.M. (1997) *In vitro* activity of RU 64004, a new ketolide antibiotic, against gram-positive bacteria. *Antimicrob. Agent Chemother.* **41**, 1196–1202.

Schwecke, T., Aparicio, J.F., Molnar, I., Konig., A., Khaw, L.E., Haycock, S.F., *et al.* (1995) The biosynthetic gene cluster for the polyketide immunosuppressant rapamycin. *Proc. Nat. Acad. Sci. USA* **92**, 7839–7843.

Seno, E.T. and Baltz, R.H. (1989) Structural organization and regulation of antibiotic biosynthesis and resistance genes in actinomycetes. In S. Shipiro (ed.), *Regulation of Secondary Metabolism in Actinomycetes*, CRC Press, Boca Raton, pp. 1–48.

Seno, E.T. and Hutchinson, C.R. (1986) The biosynthesis of tylosin and erythromycin: model systems for studies of the genetics and biochemistry of antibiotic formation. In S.W. Queener and L.E. Day (eds.), *The Bacteria, vol.IX, Antibiotic-Producing* Streptomyces, Academic Press, New York, pp. 231–279.

Seow, K.T., Meurer, G., Gerwitz M., Wendt-Pienkowski, E., Hutchinson, C.R. and Davies, J. (1997) A study of iterative type II polyketide synthases, using bacterial genes cloned from soil DNA: a means to access and use genes from uncultured microorganisms. *J. Bacteriol.* **179**, 7360–7368.

Sezenov, G., Blanc, V., Bamas-Jacques, N., Friedmann, A., Pernodet, J.-L. and Guerineau, M. (1997) Complete conversion of antibiotic precursor to pristinamycin IIA by overexpression on *Streptomyces pristinaespiralis* biosynthetic genes. *Nat. Biotechnol.* **15**, 349–353.

Shima, J., Hesketh, A., Okamoto, S., Kawamoto, S. and Ochi, K. (1996). Induction of actinorhodin production by *rpsL* (encoding ribosomal protein S12) mutations that confer streptomycin resistance in *Streptomyces lividans* and *Streptomyces coelicolor* A3(2). *J. Bacteriol.* **178**, 7276–7284.

Skandalis, A., Encell, L.P. and Loeb, L.A. (1997) Creating novel enzymes by applied molecular evolution. *Chem. Biol.* **4**, 889–898.

Smith, D.R., Doucette-Stamm, L.A., Deloughery, C., Lee, H., Dubois, J., Aldredge, T., *et al.* (1997) Complete genome sequence of *Methanobacterium thermoautotrophicum* ΔH: functional analysis and comparative genomics. *J. Bacteriol.* **179**, 7135–7155.

Solenberg, P.J. and Baltz, R.H. (1991) Transposition of Tn5096 and other IS493 derivatives in *Streptomyces griseofuscus. J. Bacteriol.* **173**, 1096–1104.

Solenberg, P.J., Cantwell, C.A., Tietz, A.J., Mc Gilvray, D., Queener, S.W. and Baltz, R.H. (1996) Transposition mutagenesis in *Streptomyces fradiae*: identification of a neutral site for the stable insertion of DNA by transposon exchange. *Gene* **168**, 67–72.

Solenberg, P.J., Matsushima, P., Stack, D.R., Wilkie, S.C., Thompson, R.C. and Baltz, R.H. (1997a) Production of hybrid glycopeptide antibiotics *in vitro* and in *Streptomyces toyocaensis. Chem. Biol.* **4**, 195–202.

Solenberg, P.J., Matsushima, P., Stack, D.R., Wilkie, S.C., Thompson, R.C. and Baltz, R.H. (1997b) Glycosyltransferase genes from *Amycolatopsis orientalis* and their use to produce novel glycopeptide antibiotics. *Dev. Indust. Microbiol.* **34**, 115–121.

Sosio, M., Giusino, F., Cappelano, C., Bossi, E., Puglia, M.A. and Donadio, S. (2000) Artificial chromosomes for antibiotic-producing actinomycetes. *Nature Biotechnol.* **18**, 343–345.

Stachelhaus, T. and Marahiel, M.A. (1995) Modular structure of genes encoding multifunctional peptide synthetases required for non-ribosomal peptide synthesis. *FEMS Microbiol. Lett.* **125**, 3–14.

Stachelhaus, T., Schneider, A. and Marahiel, M.A. (1995) Rational design of peptide antibiotics by targeted replacement of bacterial and fungal domains. *Science* **269**, 69–72.

Stachelhaus, T., Mootz, H.D. and Marahiel, M.A. (1999) The specificity-conferring code of adenylation domains in non-ribosomal peptide synthetases. *Chem. Biol.* **6**, 493–505.

Stassi, D.L., Kakavas, S.J., Reynolds, K.A., Gunawardana, G., Swanson, S., Zeidner, D., *et al.* (1998) Ethyl-substituted erythromycin derivatives produced by directed metabolic engineering. *Proc. Nat. Acad. Sci. USA* **95**, 7305–7309.

Staunton, J., Caffrey, P., Aparicio, J.F., Roberts, G.A., Bethell, S.S. and Leadlay, P.F. (1996) Evidence for a double-helical structure for modular polyketide synthases. *Nature Struct. Biol.* **3**, 188–192.

Staunton, J. and Wilkinson, B. (1997) Biosynthesis of erythromycin and rapamycin. *Chem. Rev.* **97**, 2611–2629.

Staunton, J. (1998) Combinatorial biosynthesis of erythromycin and complex polyketides. *Curr. Opin. Chem. Biol.* **2**, 339–345.

Stein, T. and Vater, J. (1996) Amino acid activation and polymerization at modular multienzymes in nonribosomal peptide biosynthesis. *Amino Acids* **10**, 201–227.

Stemmer, W.P.C. (1994a) DNA shuffling by random fragmentation and reassembly: *In vitro* recombination for molecular evolution. *Proc. Nat. Acad. Sci. USA* **91**, 10747–10751.

Stemmer, W.P.C. (1994b) Rapid evolution of a protein *in vitro* by DNA shuffling. *Nature* **370**, 389–391.

Strohl, W.R. (1997) Industrial antibiotics: today and the future. In W.R. Strohl, (ed.), *Biotechnology of Antibiotics*, Marcel Dekker, New York, pp. 1–47.

Tang, L., Fu, . and McDaniel, R. (2000) Formation of functional heterologous complexes using subunits from the picromycin, erythromycin and oleandomycin polyketide synthases. *Chem. Biol.* **7**, 77–84.

Tang, L., Yoon, Y.J., Choi, C.-Y. and Hutchinson, C.R. (1998) Characterization of the enzymatic domains in the modular polyketide synthase involved in rifamycin B biosynthesis by *Amycolatopsis mediterranei*. *Gene* **216**, 255–265.

Thibaut, D., Bisch, D., Ratet, N., Maton, L., Couder, M., Debussche, L., *et al.* (1995) Purification of the two-enzyme system catalyzing the oxidation of the D-proline residue of pristinamyicn II$_B$ during the last step of pristinamycin II$_A$ biosynthesis. *J. Bacteriol.* **177**, 5199–5205.

Toscano, L., Fioriello, G., Spagnoli, R., Cappelletti, L. and Zanuso, G. (1983) New fluorinated erythromycins obtained by mutasynthesis. *J. Antibiot.* **36**, 1439–1450.

Trefzer, A., Salas, J.A. and Bechthold, A. (1999) Genes and enzymes involved in deoxysugar biosynthesis in bacteria. *Nat. Prod. Rep.* **16**, 283–299.

Tsoi, C.J. and Khosla, C. (1995) Combinatorial biosynthesis of unnatural natural products – the polyketide example. *Chem. Biol.* **2**, 355–362.

van Wageningen, A.M.A., Kirkpatrick, P.N., Williams, D.H., Harris, B.R., Kershaw, J.K., Lennard, N.L., *et al.* (1997) Sequencing and analysis of genes involved in the biosynthesis of a vancomycin group antibiotic. *Chem. Biol.* **5**, 155–162.

van Wezel, G.P. and Bibb, M.J. (1996) A novel plasmid vector that uses the glucose kinase gene (*glkA*) for the positive selection of stable gene disruptions in *Streptomyces*. *Gene* **182**, 229–230.

von Dohren, H. (1990) Compilation of peptide structures – a biogenetic approach. In H. Kleinkauf and H. von Dohren (eds.), *Biochemistry of Peptide Antibiotics*, Walter de Gruyter, Berlin, pp. 411–507.

von Dohren, H. and Kleinkauf, H. (1997) Enzymology of peptide synthetases. In W.R. Strohl (ed.), *Biotechnology of Antibiotics*, Marcel Dekker, New York, pp. 217–240.

von Dohren, H., Keller, U., Vater, J. and Zocher, R. (1997) Multifunctional peptide synthases. *Chem. Rev.* **97**, 2675–2705.

Wells, W.A. (1998) Digging in the dirt. *Chem. Biol.* **5**, R15–R16.

Weissman, K.J., Bycroft, M., Cutter, A.L., Hanefeld, U., Frost, E.J., Timoney, M.C., *et al.* (1998) Evaluating precursor-directed biosynthesis towards novel erythromycins through *in vitro* studies on a bimodular polyketide synthase. *Chem. Biol.* **5**, 743–754.

Weissman, K.J., Timoney, M.C., Bycroft, M., Grice, P., Hanefeld, U., Staunton, J., *et al.* (1997) The molecular basis of Celmer's rules: the stereochemistry of the condensation step in chain elongation on the erythromycin polyketide synthase. *Biochemistry* **36**, 13849–13855.

Wiley, P.F., Gerzon, K., Flynn, E.H., Sigal, M.V., Jr., Weaver, O., Quarck, C., *et al.* (1957) Erythromycin X. Structure of erythromycin. *J. Am. Chem. Soc.* **79**, 6062–6070.

Wilson, R.C. (1984) Macrolides in veterinary practice. In S. Ōmura (ed.), *Macrolide Antibiotics: Chemistry, Biology, and Practice*, Academic Press, Tokyo, pp. 301–347.

Wilson, V. and Cundliffe, E. (1998) Characterization and targeted disruption of a glycosyltransferase gene in the tylosin producer, *Streptomyces fradiae*. *Gene* **214**, 95–100.

Xue, Q., Hutchinson, C.R. and Santi, D.V. (1999) A multi-plasmid approach to preparing large libraries of polyketides. *Proc. Nat. Acad.Sci. USA* **96**, 11740–117745.

Xue, Y., Zhao, L., Liu, H.W. and Sherman, D.H. (1998a) A gene cluster for macrolide antibiotic biosynthesis in *Streptomyces venezuelae*: architecture of metabolic diversity. *Proc. Nat. Acad. Sci. USA* **95**, 12111–12116.

Xue, Y., Wilson, D., Zhao, L., Liu, H.W. and Sherman, D.H. (1998b) Hydroxylation of macrolactones YC-17 and narbomycin is mediated by the pikC-encoded cytochrome P450 in *Streptomyces venezuelae*. *Chem. Biol.* **5**, 661–667.

Ylihonko, K, Hakala, J., Kunnari, T. and Mantsala, P. (1996) Production of hybrid anthracycline antibiotics by heterologous expression of *Streptomyces nogalater* nogalamycin biosynthesis genes. *Microbiology* **142**, 1965–1972.

Zalacain, M. and Cundliffe, E. (1989) Methylation of 23s rRNA caused by *tlrA* (*ermSF*), a tylosin resistance determinant from *Streptomyces fradiae*. *J. Bacteriol.* **171**, 4254–4260.

Zalacain, M. and Cundliffe, E. (1991) Cloning of *tlrD*, a fourth resistance gene, from the tylosin producer, *Streptomyces fradiae*. *Gene* **97**, 137–142.

Zhang, J.-H., Dawes, G. and Stemmer, W.P.C. (1997) Directed evolution of fucosidase from a galactosidase by DNA shuffling and screening. *Proc. Nat. Acad. Sci. USA* **94**, 4504–4509.

Zhao, H., Giver, L., Shao, Z., Affholter, J.A. and Arnold, F.H. (1998a) Molecular evolution by staggered extension process (StEP) *in vitro* recombination. *Nature Biotechnol.* **16**, 258–261.

Zhao, L., Sherman, D.H. and Liu, H. (1998b) Biosynthesis of demosamine: construction of a new metymycin/neomethymycin analogue by deletion of a desosamine biosynthetic gene. *J. Am. Chem. Soc.* **120**, 10256–10257.

Ziermann, R. and Betlach, M.C. (2000) A two vector system for the production of recombinant polyketides in *Streptomyces*. *J. Ind. Microbiol. Biotechnol.* **24**, 46–50.

Zmijewski, M.J., Jr. and Fayerman, J.T. (1995) Glycopeptide antibiotics. In L.C. Vining (ed), *Genetics and Biochemistry of Antibiotic Production*, Butterworth Heineman, Boston, pp. 269–281.

Zuber, P. and Marahiel, M.A. (1997) Structure, function, and regulation of genes encoding multidomain peptide synthetases. In W.R. Strohl (ed.), *Biotechnology of Antibiotics*, Marcel Dekker, New York, pp. 187–216.

Index